AF449361

PRACTICAL
PETROLEUM
RESERVOIR ENGINEERING
METHODS

PRACTICAL

PETROLEUM RESERVOIR ENGINEERING METHODS

An Energy Conservation Science

H. C. Slider

Reservoir Engineering Consultant
Professor, Chemical Engineering Dept.
The Ohio State University

PennWell Books
PennWell Publishing Company
Tulsa, Oklahoma

Dedicated to Jennie,
an understanding wife
—most of the time.
Slip

Preface

Petroleum reservoir engineering deals with the problems of maximizing producing rate and ultimate recovery of oil and gas reservoirs. We face a constant threat of an energy crisis through gasoline shortage, curtailment of natural gas deliveries, shortage of home heating oil, manufacturing plant shut downs due to fuel shortage, and periodic electrical service interruptions. But few realize that all of these problems could be pushed many years into the future by some major breakthrough in reservoir engineering.

It is amazing, with the ever-increasing energy demand, that so few persons, even scientists, are aware of the inefficiency in oil and gas production. Most people believe that oil and gas is produced from underground caverns or holes and that once a well quits producing all of the oil and gas has been removed from that particular hole.

Not so. Practically all oil and gas is produced from the almost microscopic-size pores that exist in underground rocks. These rocks have a texture almost like concrete.

Imagine a block of concrete $\frac{1}{2}$ mile on each side and 30 ft thick that is saturated with oil. Now imagine drilling a 6-in. hole in the center of this oil-saturated block from which you remove the oil as it flows into the hole. It does not require much imagination to recognize that very little of the oil would flow into the hole; that flow would be at a very low rate; and that when the flow stopped we would have produced a very small fraction of the oil from the cement block. This is the problem of reservoir engineering.

Most commercial oil reservoirs contain oil that has gas in solution in the oil when production is initiated. As the pressure in the oil declines the gas will come out of solution just like the carbon dioxide bubbles out of soda pop when the bottle is opened. This expanding gas helps push the oil toward the well. Nevertheless when the oil quits flowing only about 15% of the oil in that reservoir may have been produced. It is seldom that as much as 25% of the oil can be produced from such a reservoir.

Some oil and gas reservoirs may have water moving into them from one or more sides. This water helps to move the oil toward the well but the displacement is very inefficient. A reservoir such as this may produce as much as 50% of its oil before the produced fluid is 100% water.

The job of the reservoir engineer is to produce oil and gas reservoirs in such a way that the recovery is maximized and the rate at which the petroleum is produced is maximized. Control of the amount of gas or water produced with oil, proper placement of wells, use of the proper distance between wells, injection of water or other fluids into the reservoir, and many other means are used to help maximize the recovery.

Nevertheless, with the present state of the art today, it has been estimated that when all of the reservoirs that have produced oil in the U.S. are abandoned only about one-third of the oil that we know existed in the reservoirs initially will have been produced. In other words, twice as much oil as has been produced will remain unproduceable in reservoirs. We know where this oil is; in most cases wells already exist but we do not know how to produce it.

This book provides the basic fundamentals and many of the useful practical reservoir engineering methods that can be used today. Some engineer or scientist who learns his reservoir engineering from this book may be the one to discover the key to economically unlocking the capillary, interfacial tension, and other forces that hold the petroleum in our reservoirs today. Such a breakthrough would fulfill the energy demands in this country for many years to come. In the meantime, practicing reservoir engineers throughout the world will carefully control the production of the existing reservoirs to coax from them as much oil and gas as possible.

This book is written for engineers and students who wish to learn and understand the methods of reservoir engineering. It is also aimed at the practicing petroleum engineer who unknowingly misuses reservoir-engineering techniques such as the Horner plot, Stiles prediction of flood behavior, gas-well-testing procedures set forth by state control agencies, productivity index tests, and others because he does not understand them. When misapplied, these techniques yield more wrong answers than right.

Most practicing engineers would prefer to do their engineering work on a "cookbook" basis that simply requires filling out forms. These engineers may not find this book to their liking because the author believes such an approach to reservoir engineering is impossible and often dangerous due to the wrong answers obtained.

On the other hand the engineers who are intrigued by theory, mathematical methods, and computer analyses but have very little knowledge of what is required of the practicing engineer may find that this book is too practical and contains too many rules of thumb for their taste in engineering. However, it is believed that they too will profit from this book by gaining a better understanding of the practical and physical limitations on the reservoir engineering methods.

The author's objective is to provide the simplest possible methods for obtaining accurate answers to practical problems. Unfortunately these methods cannot generally be applied by simply following a "cookbook," filling out a form, or supplying data to a canned computer program. On the other hand the author has tried to avoid the presentation of mathematically intriguing or scientifically interesting ideas that have little possibility for practical application.

Several practical but unpolished computer programs dealing with reservoir engineering techniques will be presented or discussed in this book. The author feels that the digital computer is a Godsend for reservoir engineering since so many of the practical problems must be solved by trial and error. Nevertheless, a computer program is pretty much of a "cookbook" solution to a problem and consequently only those computer programs which are completely understood by the engineer should be used by him. If the engineer cannot make the same calculations by hand, given enough time, he should avoid use of the program. Any engineer who answers "a computer method" to the question, "What technique did you use to make this calculation?" is grossly overestimating the ability of the computer. A computer printout has a tremendous influence in convincing managers that your analysis is correct. Except for this political advantage the digital computer should be looked upon simply as a very fast calculator.

A considerable amount of time and effort will be spent in this book in "unlearning" many of the well-known reservoir engineering concepts. For example, most engineers feel that a Horner plot should be used in virtually all pressure buildup analysis, when actually there are very few cases where the Horner plot is necessary or even theoretically correct. Everyone "knows" that if a pressure builds up to an unchanging value during a shut in period, that this unchanging value is the average reservoir pressure for the drainage area of this well, but it will be shown that this simply is not true. There are many such misconceptions that are well known and widely misused in the practice of reservoir engineering and it thus becomes necessary to discredit such ideas in order to avoid wrong answers.

Little effort has been expended in providing the book with a formal, logical organization. Rather than having each subject fit beautifully under a particular concise heading, it is the author's opinion that it is much more important to have the sequence of subjects introduced in such a way that the opportunity for understanding is maximized. This it appears, is seldom the same presentation sequence that would result from a formal concise, logical, outline-type organization. The only thing that appears to be gained by a formal organization is the ability of the reader to find the subject material he desires. This objective will be attained by use of an extensive subject index. Thus, the sequence of subjects covered in this book conforms to the sequence the author has found most effective in teaching the material presented and the outline organization based on the most popular criteria leaves much to be desired.

There is no effort in this book to cover everything that is known on a particular subject or even to reference all of the useful information on a particular subject. The author has simply tried to present the methods that he has found most effective in nearly 25 years of reservoir engineering practice and industry teaching. This does not mean that only the methods presented herein or referenced are good reservoir engineering methods. But it does mean that the methods presented herein are good, useful, practical reservoir engineering methods.

The author has attempted to keep the working equations presented in this text to a minimum since development of a multitude of special equations for all possible special situations adds considerable confusion to any scientific field. For example, the general material balance equation presented herein will provide the answers for more than 95% of the problems dealing with material balance without a multitude of special equations for conditions above the bubble point, below the bubble point, with water drive only, with gas cap drive only, etc.

The reader will note that problems are interspersed throughout the text. It is suggested that these problems be solved as they are encountered and the solution compared with the complete solution presented in the Appendix. If the text is conscientiously used in this fashion the reader will find that a minimum of instructor guidance will be required. The text was prepared in this way so that it could be used by the many engineers who work on reservoir engineering problems with little or no previous formal instruction in reservoir engineering or even petroleum engineering in general. The extensive nomenclature list, complete with specific units, will be found helpful for self-teaching.

Readers who desire a much easier (but much less effective) learning experience, can use the problems and Appendix C problem solutions as example problems. However, in taking the easy mental route, he will end up with a lesser practical working knowledge of the material than his more ambitious colleague who attempts to work all of the problems before turning to the prepared solutions.

One additional note should be made concerning this book. The author has inserted many personal opinions, evaluations, conclusions, etc. He is often criticized for doing this without specifically stating that these are personal thoughts. If the reader does not find a reference or logical proof of a particular statement he can safely assume that it is a personal opinion based on the author's experience and knowledge of the subject.

H. C. Slider
Columbus, Ohio
1975

Contents

Preface   **vii**

1 Reservoir Fluid Flow Fundamentals   **3**

Characteristics of Darcy's Equation 3
  Darcy's Law 3
  Absolute Permeability 5
  Multiple Phase Flow 7
Characteristics of Various Flow Regimes 14
  Steady State Flow Characteristics 14
  Unsteady State Characteristics 16
  Pseudo Steady State Characteristics 18
Steady State Flow 20
  Flow Equations for Specific Geometries 20
  Gas Flow Equations 25
General Problems in Fluid Flow Calculations 29
  Approximating Complex Geometries 29
  Determining "Average" Permeabilities 36
  Correcting for Static Pressure Differences 40
  Determining Effective Permeability 43
  Determining Viscosities and Flow Rates at Reservoir
    Conditions 45

2 Unsteady State and Pseudo Steady State Flow   **51**

Physical Description 51
Radial Diffusivity Equation 54
  General Solutions 59
Constant Terminal Rate Solution 60
  The Hurst-Van Everdingen p_{tD} Solution 61
  The Ei Function Solution 64
  Extension Equations for the Ei and p_{tD} Functions 67
  Choosing the Best Pressure Function 70
Constant Terminal Pressure Solution 72
Effective Compressibility 83
Superposition 85
  Accounting for the Effects of More Than One Well 85

Accounting for Rate Change Effects 87
Accounting for Pressure Change Effects 92
Simulating Boundary Effects 94
Pseudo Steady State Flow 97
Practical Flow Equations 99
Theoretical Basis for Pseudo Steady State Flow 101
Time Limits on Pseudo Steady State Flow 102

3 Well Pressure Behavior Analysis   **108**

Productivity Index Tests 108
Recommended Procedure for Interpreting P.I. Data 112
Pressure Buildup Analysis 116
"Unchanging" Pressure at Shut In 117
Finite Acting at Shut In 124
Infinite Acting at Shut In 129
Determining the Average Well Drainage Area Pressure 135
Two Rate Buildup Tests 144
Pressure Falloff Tests 149
Pressure Buildup Anomalies 153
Reservoir Limit Tests 160
Interpreting Drill Stem Test Data 169

4 Gas Reservoir Engineering   **179**

Natural Gas Properties 180
Evaluating z Factors 181
Gas Formation Volume Factor 188
Gas Density 188
Gas Compressibility 189
Gas Viscosity 191
Material Balance 191
The Gas Material Balance Equation 192
Material Balance Gas Production 193
Determining the Reservoir Pressure 196
Graphical Material Balance 205
Determining Water Encroachment 207
Fluid Flow in Gas Reservoirs 211
Darcy's Equation with Turbulence 211
Steady State Gas Flow 212
Pseudo Steady State Gas Flow 214
Conventional Back Pressure Tests 217
Well Spacing Effects 219
Radial Diffusivity Equation for Gas 221
Unsteady State Gas Flow—Constant Pressure Solution 223
Unsteady State Flow—Constant Rate Solution 224
Isochronal Testing 226
Determining Isochronal Data from Continuous Flow Data 232
Turbulence and Skin Evaluations from Continuous Flow Data 235
Flow Line Capacity 247
Pressure Buildup in Gas Wells 241

Equipment Capacity Limitations on Deliverability 244
  Flow Line Capacity 247
  Tubing or Casing Capacity 247
Predicting Reservoir Performance 252
Determining the Gas Reservoir Size 260

5 Fluid Distribution and Frontal Displacement   **264**

Initial Saturation Distribution in a Reservoir 265
  Determining the Original WOC and GOC in a Reservoir 266
  Capillary Pressure Data 268
  Calculation of the initial Saturation Distribution from
    Capillary Pressure Data 272
  Calculating Reservoir Capillary Pressure Data from
    Lab Data 275
  Averaging Capillary Pressure Data 277
Reservoir Saturation Distribution During Displacement 283
  General Characteristics of Fluid Displacement 284
  General Procedure for Calculating the Saturation
    Distribution During a Displacement 288
  Determining the Fractional Flow Curve 290
  The Buckley-Leverett Equations 294
  Determination of the Frontal Saturation by Material
    Balance 296
  Graphical Analysis of Fluid Displacement with Uniform
    Initial Saturations (Welge Method) 299
Interface Tilt During Displacement 306
  Initial Interface Tilt (Hydrodynamic Tilt) 307
  Interface Tilt During Linear Displacement 309
  Interface Tilt in a Radial Flow System: Coning 312

6 Material Balance   **318**

A General Material Balance Equation 319
  Material Balance in Gas Reservoirs 319
  Material Balance for Liquid Expansion 322
  Gas Liberation in the Reservoir 324
  Pressure-Volume-Temperature (PVT) Relationships 325
  Material Balance with Gas Liberation 328
  Including Pore Volume Changes in Material Balance 330
  Modifications of the General Material Balance Equation 333
General Difficulties in Applying Material Balance
  Equations 335
  Accuracy of Production Data 337
  Fluid Flow Fundamentals 337
  Accuracy of Reservoir Pressure Data 337
Predicting Gas Drive Behavior 338
  Behavior of the Produced Gas-Oil Ratio 339
  The Schilthius Method 342
  The Tarner Method 344
  A Recommended Procedure 345

The Muskat Method 348
Material Balance in Partially Saturated Reservoirs 349
Gas Cap Complications 352
Prediction of Water Drive Reservoir Behavior 353
Calculating Water Encroachment 353
The Prediction Procedure 355
Modeling the Reservoir Prediction Problem 359
Increasing Primary Recovery 360
Well Control 360
Total Reservoir Control 362

7 Decline Curve Analysis

365

Decline Rate Definition 366
Constant Percentage Decline 368
The Constant Percentage Decline Rate Equation 368
Determining Reserves During Constant Percentage
Decline 370
Hyperbolic Decline 373
The Hyperbolic Decline Equations 373
The Curve Fitting Method 375
Other Methods 381
Mobile Oil By Hyperbolic Decline 383
Harmonic Decline 384

8 Waterflooding and Its Variations

386

The Water Flood Displacement Mechanism 387
Predicting Total Flood Recovery 392
Determining Oil Saturation at the Start 393
Determining Gross Swept Volume, 394
Determining the Residual Oil Saturation 399
Determining the Ultimate Horizontal Sweep Efficiency 400
Determining the Ultimate Vertical Sweep Efficiency 403
Predicting Rate Versus Time Performance 413
The Recommended Method 414
Converting Reduced Time Data to Production Vs.
Injection 419
Recovery Curves for Individual Strata 421
Injectivity Curves for Individual Strata 423
Predicting Water Flood Performance by Analogy
(or from Pilot Flood Results) 428
Predicting Total Flood Recovery 429
Predicting Effective Injection Rates 430
Predicting Oil Production Rate Versus Time 433
Waterflooding Variations 436
Adjusting Mobility Ratios 436
Determining the Slug Distribution in the Various Zones 440
Determining Injectivity Curves for a Slug Type
Displacement 441
Use of Micellar Solutions 443

Appendix A. Nomenclature List 445

Appendix B. Empirical Reservoir Data 450

   1. Wylie Empirical Relative Permeability Equations 450
   2. Relative Permeability Ratio Data for Typical Reservoirs 451
   3. Compressibility and Gas in Solution for Water 457
   4. Empirical Determination of Oil Compressibility 458
   5. Compressibility of Natural Gases 460
   6. Formation Compressibility 460
   7. Gas Deviation Factors 461
   8. Pseudo Critical Pressures and Temperatures 462
   9. Gas Deviation Factor, z, vs. (p_r/z) 463
  10. Gas Turbulence Factors 464
  11. Physical Properties of Petroleum Components 465
  12. Effect of Condensate Volume on the Ratio of Surface Gas Gravity to Well Fluid Gravity 466
  13. Hyperbolic Decline Curve Functions 467
  14. Gas Viscosities 485
  15. Reservoir Oil Viscosities 486
  16. Reservoir Water Viscosities 487
  17. Gas in Solution or Bubble Point Pressure 488
  18. Gas Equivalent of Stock Tank Condensate 488
  19. Water Content of Natural Gas in the Reservoir 489
  20. Oil Formation Volume Factors 490

Appendix C. Problem solutions for problems in text 491

Subject Index 554

PRACTICAL
PETROLEUM
RESERVOIR ENGINEERING
METHODS

Reservoir Fluid Flow Fundamentals

Reservoir fluid flow is probably the one area of reservoir engineering most in need of a comprehensive treatment today. In addition to attempting to teach the fundamental concepts of fluid flow and their application, one of the principle objectives of this text will be to integrate the treatment of reservoir fluid flow so that the engineer can readily comprehend the relationship between steady state, pseudosteady state, and unsteady-state fluid flow and recognize the characteristics of each. In so doing, the limitations of each flow regime will become apparent.

All practical fluid-flow equations are based on two concepts—Darcy's equation and material balance. The simpler concepts of reservoir engineering are based on simply one of these concepts. However, the more complex concepts—and quite often the most useful ones—are based on both material balance and the Darcy equation.

Fluid flow can be characterized or categorized so many different ways it is virtually impossible to consider all of the possibilities. Fluid flow can be classified according to the geometric configuration involved, the compressibility of the fluids, the constancy of flow rates and pressure with respect to time, whether single or multiphase flow, and others. A particular working equation would probably apply to only one combination of each of these classifications, which makes it virtually impossible to consider all of the possible combinations. Consequently, this text will simply strive to demonstrate the means of accounting for the various classifications and will leave it to the engineer to devise his own derivations for peculiar cases that may be encountered.

CHARACTERISTICS OF DARCY'S EQUATION

Darcy's Law. This is simply an empirical relationship that was derived for the vertical flow of fluid through packed sand. It can be imagined that Darcy would be shocked to note how we have stretched his empirical relationship to fit our particular needs in reservoir engineering. The continued use of this empirical equation to solve our very complex reservoir-engineering problems seems analogous to using a horse to go to the moon. However, there appears to be no easy way to back up and start over with a more theoretical treatment of fluid flow, so, we will use Darcy's empirical relationship to the best of our ability.

Darcy simply showed that the apparent velocity of a fluid flowing through a porous media was proportional to the pressure gradient, dp/dx, and assumed it would be proportional to the reciprocal of the viscosity, μ. He then wrote an equation for the velocity of fluid traveling through a particular porous media and added a proportionality constant characteristic of that particular porous media, a permeability, k.

$$v' = -(k)\,\frac{1}{\mu}\left(\frac{\Delta p'}{\Delta x'}\right) \tag{1.1}$$

The negative sign is added because if x' is measured in the direction of flow, the pressure p' will decline as x' increases which results in a negative value for $(\Delta p'/\Delta x')$. Thus, the minus sign must be added to make the velocity v' positive. The primes, ('), on the v, p, and x in Equation 1.1 denote "Darcy units." Darcy's units are odd as compared with our modern oil-field units. If we substitute for the apparent velocity, v', the expression q'/A', we obtain the Darcy equation in the volumetric rate form in which q' is in cc per second, A' is the gross cross-sectional area in square centimeters, and the pressure gradient, dp/dx, is in atmospheres per centimeter. The viscosity μ is in centipoises.

$$q' = -\frac{kA'}{\mu}\frac{\Delta p'}{\Delta x'} \tag{1.2}$$

The units of the resulting constant, k, using this combination of units was termed by Mr. Darcy, the Darcy. This system of units, although close to the recommended international units being adopted, would be very difficult to use in the oil fields. Consequently, no attempt will be made in this text to state equations in the international units. Thus, the strange (to everyone except the oil producer) system of oil-field units will be employed. If we state the rate q as a function of reservoir barrels per day, A as a function of square feet and the pressure gradient, dp/dx, as a function of pounds per square inch per foot, Equation 1.2 can be changed to the more useful form

$$q = -\frac{1.127kA}{\mu}\frac{\Delta p}{\Delta x}. \tag{1.3}$$

This form will be used as the Darcy equation.

Two difficulties may arise in using this equation for those who are familiar with other fluid-flow work. First the rate q is stated in reservoir barrels per day whereas many other equations state the rate in stock-tank barrels per day. Secondly, the permeability is stated in Darcies rather than millidarcies (thousandths of a Darcy). This system of units will be employed throughout this text unless it is specifically stated that some other units are used for a special case.

Darcy's equation has some serious limitations which we must recognize. Darcy's experiments were run at relatively low flow rates and low pressure drops using small heads of water as a driving pressure. Consequently, as would be expected since the flow rate is proportional to the pressure drop, this expression does not apply to turbulent flow. This should be kept in mind throughout the use of Darcy's equation to derive the various working equations.

Each time the Darcy equation is used in the form of Equation 1.3 to derive some working equation we must realize that that particular equation will not apply to turbulent flow. This does not appear to be a serious practical limitation when we apply Darcy's equation to the flow of liquid. Generally speaking, liquid flow rates are too small for the additional pressure drop due to turbulence to have any great significance. However, just the opposite is true when we consider the flow of natural gas. In this case the flow rate is possibly 100 times as great as the flow of liquid under similar conditions due to the viscosity differences. Thus, in the flow of gas, an additional pressure drop generally does exist due to turbulence.

Various efforts have been made to adapt the Darcy equation to turbulent flow. One of these will be discussed in some detail in the chapter on gas reservoir engineering.

Absolute permeability. Undoubtedly, the Darcy work was never meant to apply to multiple-phase flow. It was simply meant to describe the flow of one fluid saturating 100% of the porous media. Under these conditions, the permeability to a particular fluid is independent of the nature (viscosity) of the fluid. In other words, the permeability to a 100% saturating fluid is a constant and characteristic of the porous media which we know as the absolute permeability. Reported differences of the permeability to different phases is believed to be due to the reaction of the porous media to one or more of the phases. Consequently, we can think of the absolute permeability as being a fixed characteristic of a particular porous media similar to the porosity or pore-size distribution.

Although the absolute permeability of a porous media is a physical property of the porous media, it appears to have some abstract qualities that are not exhibited by other properties such as porosity and pore-size distribution. There is an indication that the permeability of a naturally occurring porous media is a function of the size of the sample. It could be that such observations are simply a matter of undetected plugging of the pores caused by the cutting of the samples. However, the wide variations in permeability of the sandstones and limestones that appear to be homogeneous in nature is a well established fact.

A very uniform sandstone sample, for example a 6-in.-long, $3\frac{1}{2}$-in.-diameter core, can be cut into as many permeability plugs (normally 1 in. in length and $\frac{3}{4}$ in. in diameter) as possible and the measured permeabilities of this "homogeneous" sandstone will probably vary by as much as 300%. It is not clear how the effective or average permeability is related to the permeability distribution obtained in such a situation. Until some conclusive evidence is found to the contrary it is recommended that the arithmetic average permeability be used as the characteristic permeability of a formation.

Permeability distributions are often characterized as indicated in Fig. 1.1 which is a plot of permeability versus the cumulative fraction of permeability samples with a permeability greater than the subject permeability. In Fig. 1.1, note that 15 ft of the formation has a permeability greater than 110 md.

It is sometimes helpful to prepare a histogram of permeability distributions. If a geometric progression is used to define the limits of each permeability group considered, a distribution approaching a normal distribution may be obtained. Actually the shape of the histogram in Fig. 1.2 is typical of permeability distribution. Some engineers believe that the peak of the permeability histogram will best characterize the actual permeability behavior of a formation.

The permeability of pores of a particular size, for example the permeability of a fracture, can be calculated by mathematical means. The Craft and Hawkins text has an example of such calculations for a fracture and the fluid displacement chapter of this text derives a permeability for a bundle of capillary tubes of a particular size which also gives some

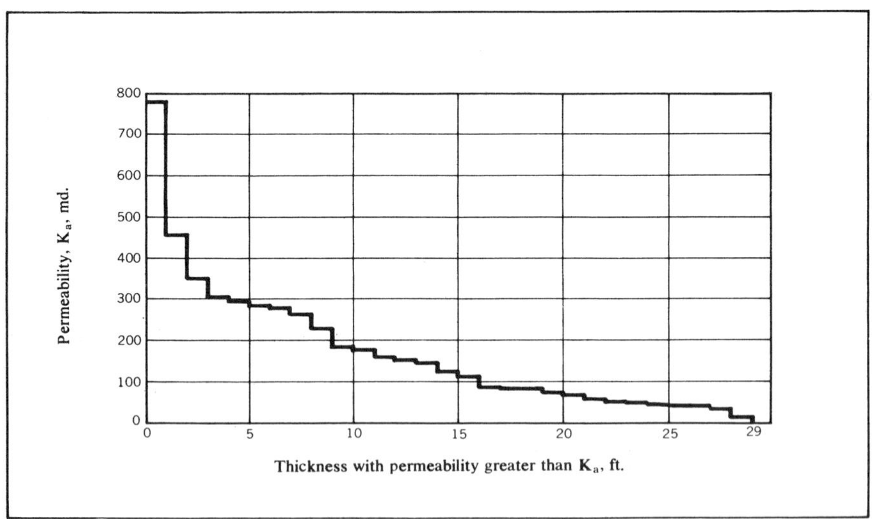

FIG. 1.1   Permeability distribution.

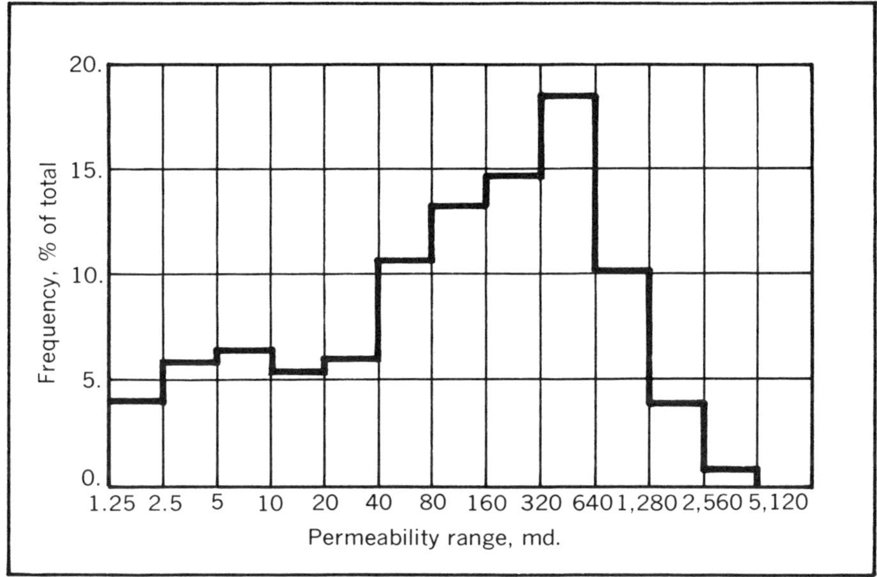

FIG. 1.2 Permeability histogram.

insight into the relationship between permeability and porosity for an idealized case.

Multiple-phase flow. Since all hydrocarbon-bearing reservoirs contain water which may or may not be mobile, it is necessary to extend the permeability concepts and use permeability as a function of the flowing-phase saturation. For example, the permeability of a particular porous media to oil, when the oil saturation is 50% and water saturation is 50%, may only be 45% of the permeability exhibited when the formation is 100% saturated with oil. At 60% oil saturation and 40% water, the permeability based on the flow rate of the oil may be 70% of the permeability of the formation when it is 100% saturated with oil. This ratio of a permeability at a particular saturation to the permeability at 100% saturation is termed the relative permeability, k_r. Consequently, the effective permeability to a particular phase will be the absolute permeability, k, multiplied by the relative permeability k_r. Thus

$$k_o = kk_{ro}; \quad k_g = kk_{rg}; \quad \text{and} \quad k_w = kk_{rw} \qquad (1.4)$$

In these relationships the subscript, r, refers to relative and the subscripts o, g, and w refer to oil, gas, and water. To check your understanding of the permeability and relative permeability concepts work the following problem and then check your methods against the solution in Appendix C.

Problem 1.1: Calculation of Permeability Data from Lab Tests

Given the following laboratory data from relative permeability tests:

Cross-sectional area of core, 5 sq cm
Core length, 3 cm
Viscosity of H_2O, 1.0 cp; Viscosity of oil, 1.25 cp
Pressure at outlet face of core, 1 atm
Pressure at inlet face of core, 2 atm
Saturation and rate data as follows:

Saturation, %		Flow rates, cc/sec	
Water	*Oil*	*Water*	*Oil*
100	0	0.50	0.00
90	10	0.30	0.00 critical
80	20	0.15	0.01
60	40	0.03	0.10
40	60	0.01	0.25
30	70	0.00 critical	0.38

What is the absolute permeability? What is the permeability to oil at a water saturation of 30%? Calculate the relative permeability values.

Answers: 300 md, 285 md, and the relative permeabilities plotted in Fig. 1.3.

To obtain effective permeabilities to use in flow equations it is necessary to multiply the absolute permeability by the relative permeability to the flow phase of interest according to Equations 1.4. The relative permeability data resulting from the calculations of Problem 1.1 are shown in Fig. 1.3. They exhibit typical relative permeability characteristics.

Note that at any particular saturation the relative permeabilities do not total 1.0. You might say that they mutually interfere with the flow of each other. Also note that both phases have an irreducible saturation at which the relative permeability to that phase is 0.0. As could be anticipated from the previous discussion, the relative permeabilities at 100% saturation of a particular phase are 1.0. This is generally an assumed value for k_{ro} because in running relative permeability data, the first point investigated is normally the 100% water saturation point and it is impossible to drive the water saturation below the irreducible water, in this case 0.3, without artificially cleaning the core and starting again with 100% oil saturation.

The shapes of the relative permeability curves are also characteristic of the wetting qualities of the two fluids. When water and oil are considered together, water is almost always the wetting phase. This means that the water, or wetting phase, would occupy the smallest pores while the nonwetting, or oil phase, would occupy the largest pores. This causes the shape of the relative permeability curves for the wetting and nonwetting phase to be different.

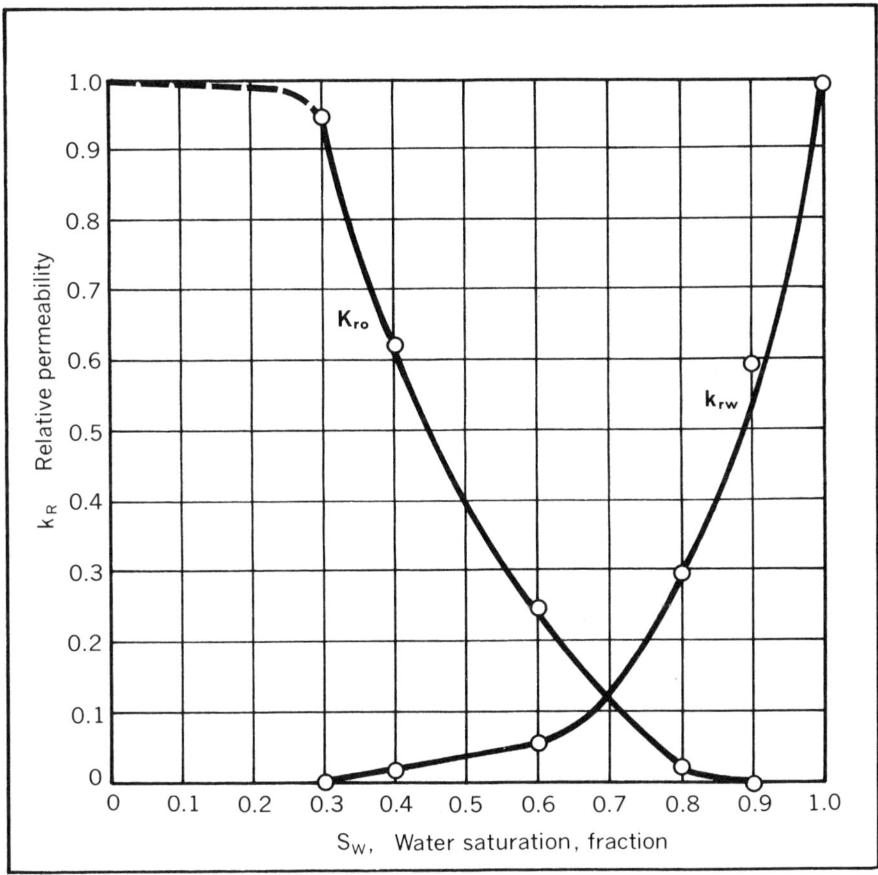

FIG. 1.3 Relative permeability vs water saturation.

Possibly this is best illustrated by looking at the relative permeability to one phase at the irreducible saturation of the other phase. The relative permeability to water at an irreducible oil saturation of 10% (90% water) is about 0.6 whereas the relative permeability to the nonwetting phase, oil, at the irreducible water saturation of 0.3 approaches 1.0. In this case it is 0.95. One practical effect of this observation is that we normally assume that the effective permeability of the nonwetting phase in the presence of an irreducible saturation of the wetting phase is equal to the absolute permeability. Consequently, oil flowing in the presence of connate water or an irreducible water saturation is assumed to have a permeability equal to the absolute permeability. Similarly, gas flowing in a reservoir in the presence of irreducible water saturation is assumed to have a permeability equal to the absolute permeability.

Also note that since the relative permeability to the wetting phase at the nonwetting-phase irreducible saturation is much less than the relative permeability to the nonwetting phase at the irreducible wetting-phase saturation, decidedly different shaped curves are obtained for the two phases. The wetting-phase relative permeability is concave upward while the nonwetting-phase relative permeability takes on an "S" shape curve.

Note that the same general observations apply to gas-oil relative-permeability data, as can be seen for a typical set of data in Fig. 1.4. Note that this might be termed gas-liquid relative permeability since it is plotted versus the liquid saturation. This is typical of gas-oil relative-permeability data in the presence of connate water. Since the connate or irreducible water normally occupies the smallest pores in the presence of oil and gas it appears to make little difference whether these pores are

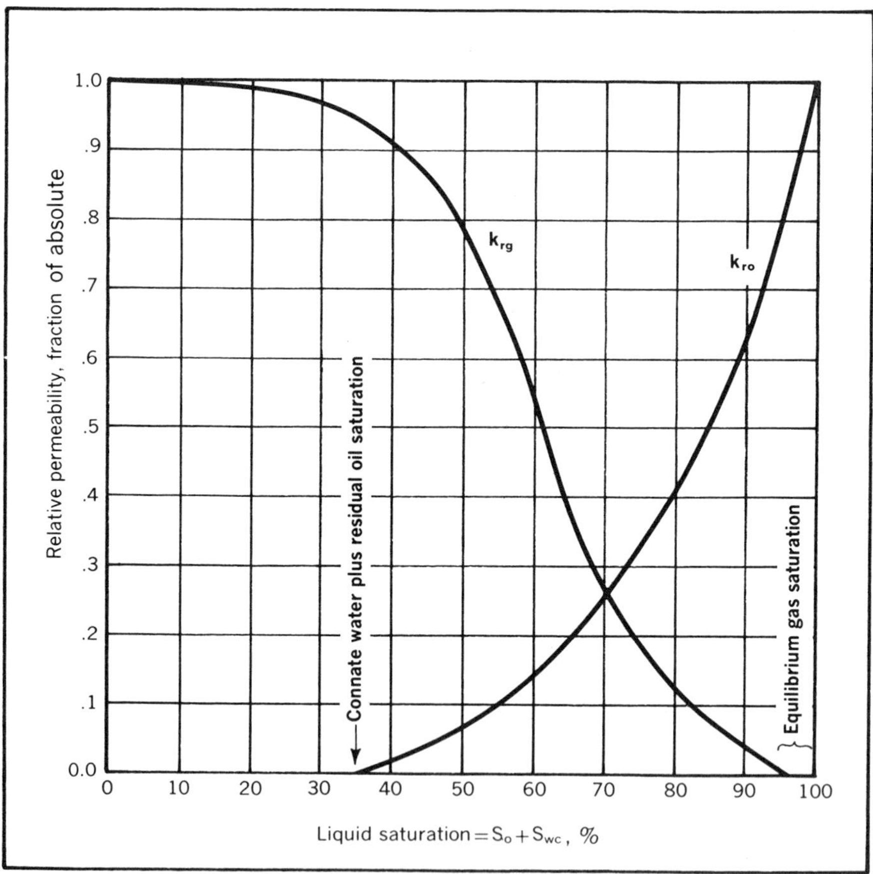

FIG. 1.4 Gas-oil relative permeability data.

occupied by water or oil which would also be essentially immobile in these small pores. Consequently, in applying the gas-oil relative-permeability data to a reservoir, the total liquid saturation is normally used as a basis for evaluating the relative permeability to the gas and oil.

Note that the relative-permeability curve representing oil changes completely from the shape of the relative-permeability curve for oil in the water-oil system. In the water-oil system, as noted previously, oil is normally the nonwetting phase whereas in the presence of gas the oil is the wetting phase. Consequently, in the presence of water only, the oil relative-permeability curve takes on an "S" shape whereas in the presence of gas the oil relative-permeability curve takes on the shape of the wetting phase, or is concave upward. Note further that the irreducible gas saturation is generally very small. As a matter of fact it is generally smaller than the irreducible gas saturation predicted by lab analysis. This is also called the equilibrium gas saturation.

It might be concluded that residual saturations in general are smaller in the reservoir than those predicted in the laboratory. That is, the irreducible water saturation, residual oil saturation, and equilibrium gas saturation are less in the reservoir than those predicted from laboratory relative-permeability measurements. This is probably the result of the time factors involved. In the lab we are faced with using finite displacement times whereas in the reservoir times actually approach infinite values for all practical purposes.

Three-phase relative permeability data has been run by several investigators (see Fig. 1.5). However, these data generally have no practical significance because they have such limited use. It is very difficult to find a situation where three phases flow concurrently in the reservoir.

For example, when water displaces oil and gas the oil soon forms a bank between the displaced gas and the advancing water. Consequently, the oil displaces gas with only gas and oil flowing at a particular point in the reservoir and water displaces oil with only oil and water flowing at other points of the reservoir. Only for very short periods of time is it possible for water to displace both oil and gas so that all three are flowing at the same point in the reservoir at the same time.

Experienced engineers may be confused by the fact that many wells produce at high rates of oil and water production with high gas-oil ratios greatly in excess of the solution gas that would be expected if there was no free gas flowing in the reservoir. These wells are obviously experiencing concurrent flow of free gas, oil, and water in the reservoir. However, this is generally the result of stratification. With the most permeable strata producing gas and oil and the strata with lesser permeabilities, or possibly with a lower structural position, producing water and oil. But there would still be few places in this well-drainage area where all three phases are flowing concurrently.

The one notable exception is when an oil reservoir exists in the transition zone someplace between the 100% water level in a reservoir

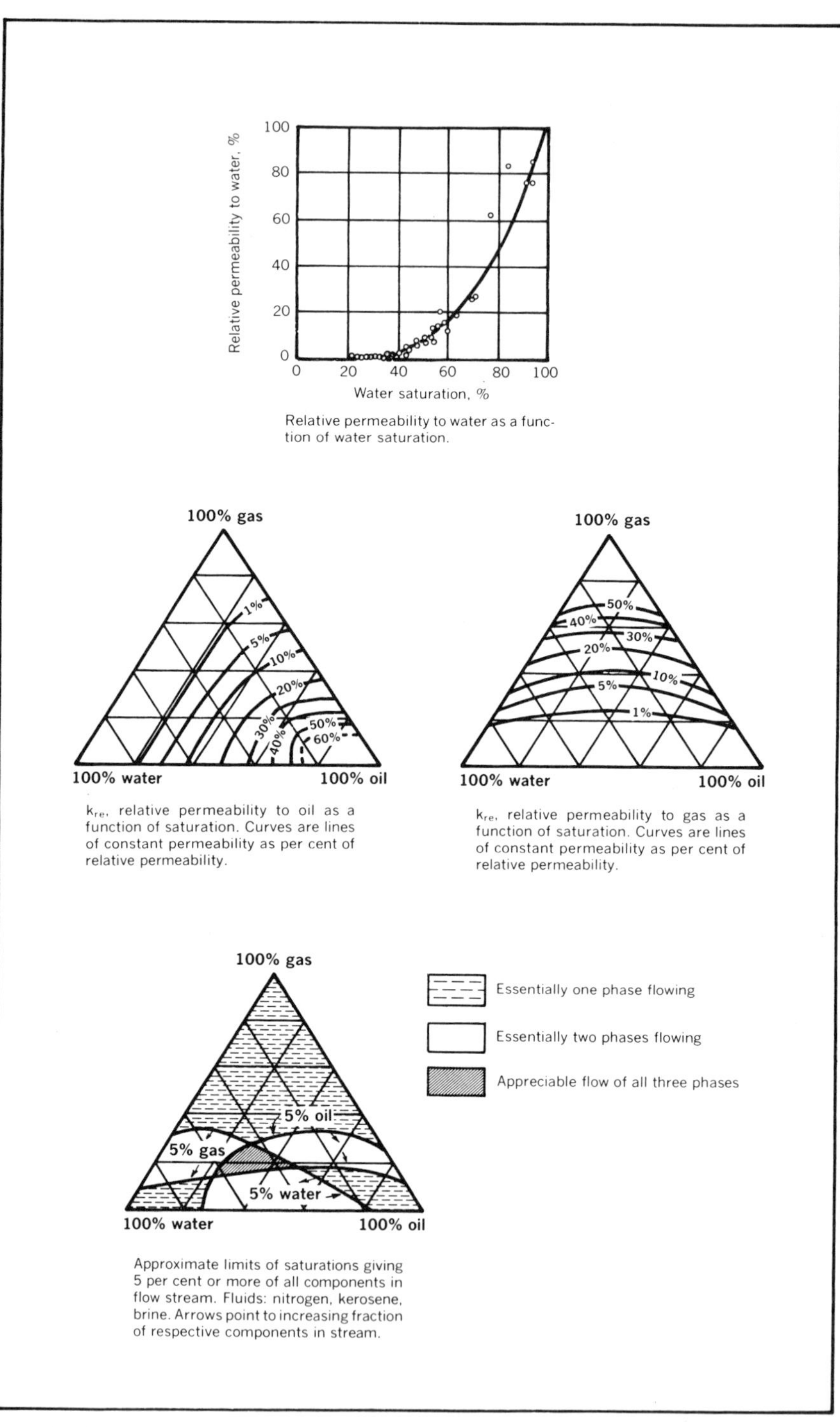

Relative permeability to water as a function of water saturation.

k_{re}, relative permeability to oil as a function of saturation. Curves are lines of constant permeability as per cent of relative permeability.

k_{re}, relative permeability to gas as a function of saturation. Curves are lines of constant permeability as per cent of relative permeability.

Approximate limits of saturations giving 5 per cent or more of all components in flow stream. Fluids: nitrogen, kerosene, brine. Arrows point to increasing fraction of respective components in stream.

FIG. 1.5 Example three phase relative permeability data (after Amyx, Bass, and Whiting[1], McGraw-Hill Book Company).

and the up-structure position characterized by an irreducible water saturation. Yet the down-structure reservoir is limited by a permeability barrier that does not permit active water encroachment. In such cases water and oil will flow concurrently and once the bubble-point pressure is reached free gas will form in the reservoir and will flow concurrently with the water and oil.

Such a formation is generally characterized by a declining produced water-oil ratio together with a decline in the rate of oil production and an increase in the produced gas-oil ratio. This is in contrast to a normal active-water-drive reservoir where the rate of oil production tends to stabilize after some initial period of decline followed by a period when total liquid production stays constant but the water cut continues to increase.

Generally speaking, obtaining relative-permeability data for a par-

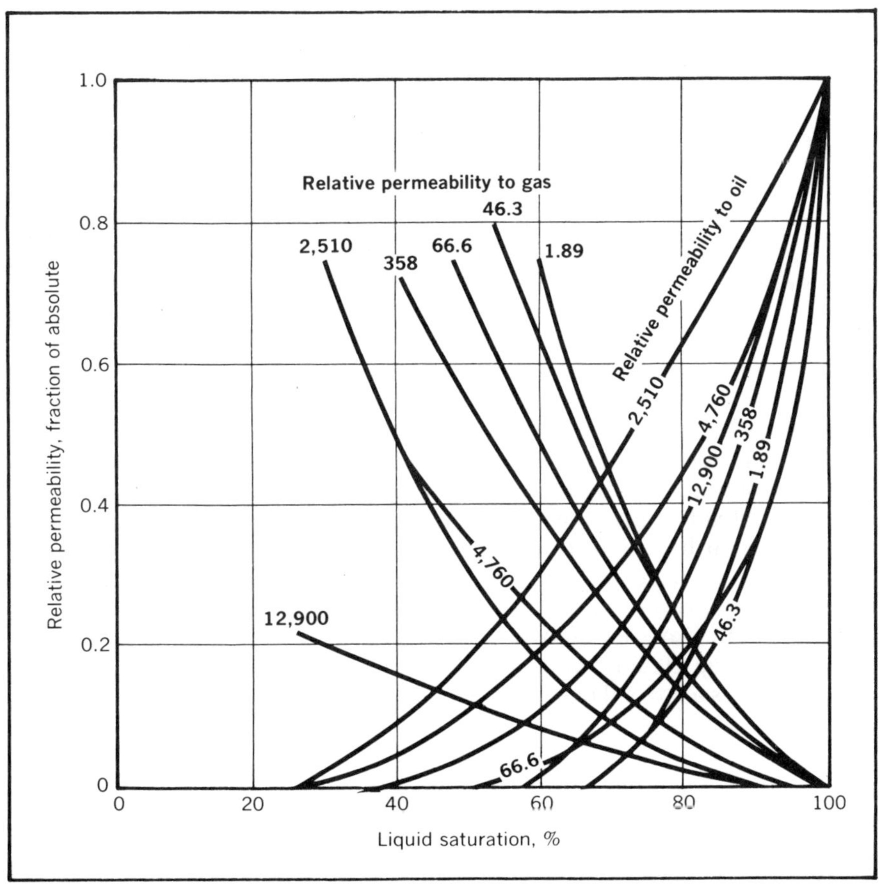

FIG. 1.6 Example family of relative permeability.

ticular formation presents a sizeable difficulty. Few companies have relative permeabilities run as a matter of course. This is understandable since relative-permeability data as obtained in the laboratory tend to be erratic, unreliable, and expensive.

Figure 1.6 on page 13 illustrates a portion of a suite of relative-permeability curves run for one reservoir by a major oil company. The interpretation of such data to obtain one relative permeability curve to apply to a particular formation presents a considerable task. If you must perform such a normalizing task it is suggested that it be done on the basis of permeability distribution. However, there is little reason to believe that the single relative-permeability curve so obtained will be representative of the behavior of the reservoir.

The writer has never seen a situation in which the laboratory relative-permeability data matched the indicated relative-permeability data of a solution-gas-drive reservoir. As is discussed in the material balance chapter, in the prediction of solution-gas-drive behavior the best results are obtained by calculating the relative-permeability characteristics from past performance of the reservoir and then extrapolating these relative-permeability characteristics to lower liquid saturations to predict the future behavior of the solution-gas-drive reservoir. Many engineers believe that relative-permeability data, as good as or better than lab-measured relative-permeability data, can be obtained by using empirical equations devised by Wylie and others. These equations along with other relative permeability data that might be used by analogy can be found in Appendix B.

CHARACTERISTICS OF VARIOUS FLOW REGIMES

It is convenient to group practical flow equations according to the flow regime which they represent—steady state, pseudosteady state, or unsteady state. Actually we will see that pseudosteady state is a special case of unsteady-state flow. Care should be exercised in defining exactly what is meant by any of these terms.

For the purposes of this text steady state will refer to the situation in which the pressure and the rate distribution in the reservoir remain constant with time. By contrast, unsteady state is the situation in which the pressure or the flow rate vary with time. As stated previously, pseudosteady state is a special case of unsteady state that resembles steady-state flow.

Steady-state flow characteristics.   Fig. 1.7 represents the pressure distribution and rate distribution during radial flow into a well that exhibits the characteristics of steady-state flow. This pressure and rate distribution will remain the same as long as the drainage area remains

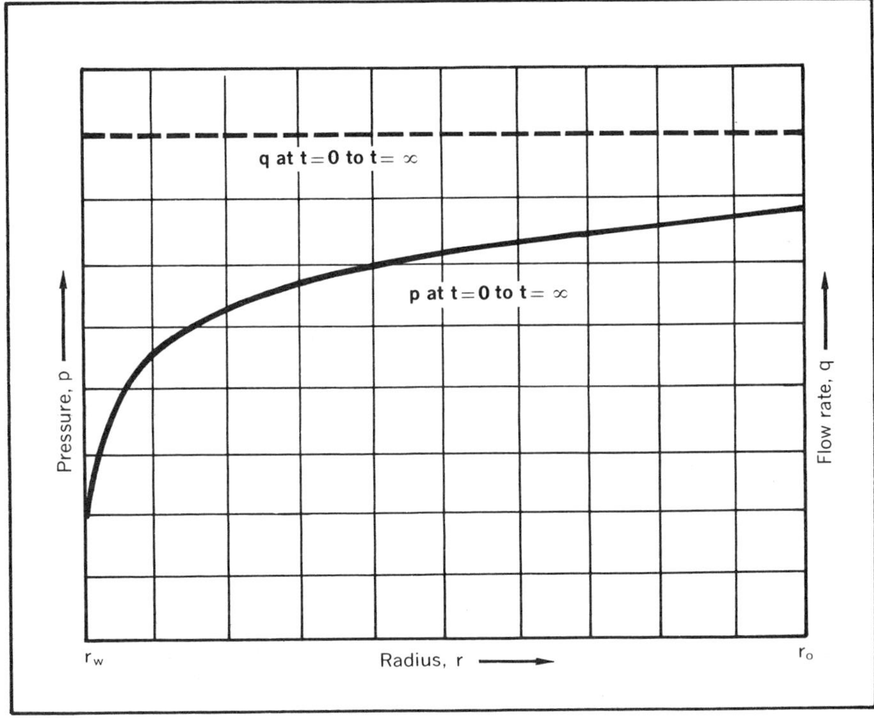

FIG. 1.7 Radial steady-state flow.

in steady-state flow. Note that Equation 1.3 can be solved for the pressure gradient $\Delta p/\Delta x$ at any radius.

$$\left(\frac{\Delta p}{\Delta r}\right)_r = \frac{q\mu}{1.127kA_r} \tag{1.5}$$

Note that the minus sign of Equation 1.3 has been dropped because the distance, r, is now measured against the direction of flow so that the pressure decreases with a decrease in radius and $\Delta p/\Delta r$ is then positive.

The cross-sectional area is subscripted with an r to indicate that it is a function of the radius; thus, the pressure gradient is also a function of the radius and is similarly subscripted. For a particular radius and a particular rate of flow, q, note that the slope of the plot of pressure versus radius, $(\Delta p/\Delta r)$, will remain constant as long as there is no change in the saturation which would change the effective permeability, k. Consequently, as long as the flow rate remains constant the pressure distribution will also remain constant.

This idea could be expanded to show that it would apply equally well for compressible fluids such as gas if the flow rate q is stated in mass units such as standard cubic feet. Thus, well pressure and flow-rate histories could be used to determine whether or not a well is in steady state. If the flow rate is constant and the bottom-hole pressure remains constant there would be little doubt that the drainage area of this well is in steady-state flow.

Note that for such a situation to be strictly true it is necessary that the flow across the external drainage radius r_e, be equal to the flow across the well radius at r_w and that the same fluid be crossing both radii. This is never strictly met in a reservoir but a strong water drive, whereby the water-influx rate equals the producing rate, will give a pressure and rate history almost identical to the one described in Fig. 1.7. Pressure maintainance by water injection down-dip or by gas injection up-dip would also approximate steady-state conditions as would most pattern water-floods after the initial stages of injection has passed.

Steady-state equations are also useful in analyzing the conditions near the well bore because even in an unsteady-state system the flow rate near the well bore is almost constant so that the conditions around the well bore are almost constant. Thus, steady-state flow equations can be applied to this portion of the reservoir without any significant error.

Unsteady-state characteristics.   Fig. 1.8 shows the pressure and rate distributions for a system similar to the radial steady-state system of Fig. 1.7 except that in this case all of the production is due to the expansion of the fluid in the reservoir. This causes the rate at r_e to be zero and the rate increases to a maximum at the well radius, r_w. In the steady-state case the flow across the outer boundary, r_e, was equal to the flow across r_w, the well radius. With flow across r_e zero the only energy causing the flow of fluid is the expansion of the fluids themselves. Initially the pressure is uniform throughout the reservoir at p_i. This represents the zero producing time.

Examine Fig. 1.8 which shows the pressure and rate history for an unsteady-state system. The production rate is controlled so that the pressure at the well is constant. After some short period of time of producing the well, at such a rate that the well pressure remains constant, we obtain a pressure distribution shown as p at t_1. Notice that at this small time only a small portion of the reservoir has been affected or has had a significant pressure drop.

Now remember that flow is taking place due to the expansion of the fluid. Consequently, if no pressure drop exists in the reservoir at a particular point, or outside of that point, no flow could be taking place at that particular radius. The fluid could not expand without a drop in pressure. Thus, as shown in the plot of q at t_1 the rate at r_e is zero and it increases with a reduction in radius until the maximum rate in the reservoir is obtained at r_w. Fig. 1.8 is schematic and not meant to be quantitative. The pressure and rate distributions at time t_1 represent simply an instant in time and the pressure and rate distributions move on through these

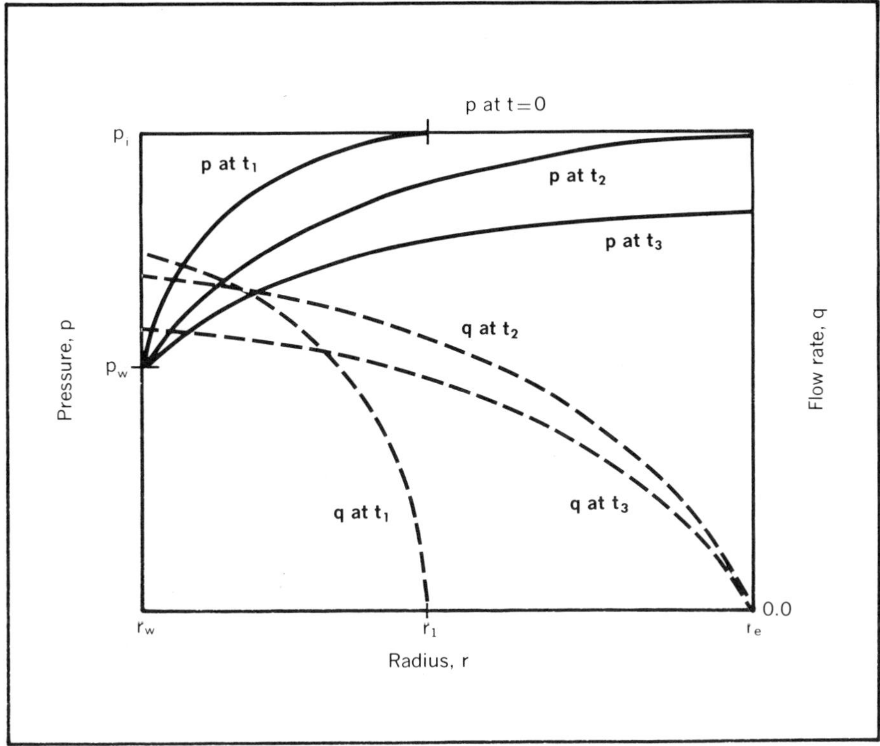

FIG. 1.8 Unsteady state radial flow with constant well pressure.

positions immediately as the production continues to affect more and more of the reservoir. That is, more and more of the reservoir continues to experience a significant pressure drop and is subjected to flow, until the entire reservoir is affected as shown by the pressure at t_2. The rate, q, at t_2 indicates that the flow rate at this time extends throughout the reservoir since all of the reservoir has been affected and has a significant pressure drop.

Notice that the rate at the well has declined somewhat from time t_1 to t_2 since the same pressure drop $(p_i - p_w)$, is effective over a much larger volume of the reservoir. Once the pressure in the entire reservoir has been affected the pressure will drop throughout the reservoir as production continues so that the pressure distribution might be as shown for p at t_3 in Fig. 1.8. The rate will have declined somewhat during time t_1 to t_2 due to the increase in the radius over which flow is taking place and it will continue to decline from t_2 to t_3 due to the fact that the total pressure drop from r_e to r_w, $(p_e - p_w)$, is declining. Fig. 1.8 is an example of unsteady-state flow by our definition since the pressure and the rate are both changing with time except for the one pressure that we have maintained artificially, the pressure at the well, p_w.

Unsteady-state flow can be said to cover all of the reservoir conditions not listed under steady state above. The particular system described here would be the sort of system that would represent a well flowing at its full capacity or against a constant choke size so that the well pressure in the bottom of the hole tends to remain constant. Then, if the reservoir experiences no water encroachment the pressure and rate distribution out to the wells drainage radius, r_e, would be similar to that shown in Fig. 1.8.

Note that from time $t = 0$ to time t_2, when a pressure drop is finally affected throughout the entire reservoir, the pressure and rate distributions would not be affected by the size of the reservoir or the position of the external drainage radius r_e. During this time we say that the reservoir is infinite acting because during this period the outer drainage radius, r_e, could mathematically be infinite. We will find that even in reservoir systems that are dominated by steady-state flow the affect of changes in well rates or well pressures at the well will be governed by unsteady-state flow equations until the changes have been in effect for a sufficient length of time to affect the entire reservoir and have the reservoir again reach a steady-state condition.

As might be expected, the mathematics governing unsteady-state flow with the pressure and rate varying both with time and with radius gives a complex expression that will be shown later as the radial diffusivity equation, a second-degree partial differential equation. However, general solutions for this radial diffusivity equation make it possible for us to simply insert constants into various expressions to obtain the rate behavior if constant-pressure conditions prevail, as in Fig. 1.8. If a constant well rate prevails, as in Fig. 1.9, another solution will let us analyze the pressure behavior. We will also learn to apply both the constant rate and constant-pressure solutions to variable rates and variable pressures by a technique known as superposition.

Pseudosteady state characteristics.   Fig. 1.9 illustrates the pressure and rate distribution for the same unsteady-state system discussed in Fig. 1.8 except that in this particular case the rate at the well, q_w, is held constant. This might be comparable to a prorated well or one that is pumping at a constant rate. Again at time $t = 0$ the pressure throughout the reservoir is uniform at p_i. Then after some short producing time t_1 at a constant rate only a small portion of the reservoir will have experienced a significant pressure drop and consequently the reservoir will be flowing only out to a radius r_1. As production continues at the constant rate, the entire reservoir will eventually experience a significant pressure drop as shown as p at t_2 in Fig. 1.9. Shortly after the entire reservoir pressure has been affected, a rather unexpected situation arises. The change in the pressure with time at all radii in the reservoir becomes uniform so that the pressure distributions at subsequent times are parallel as il-

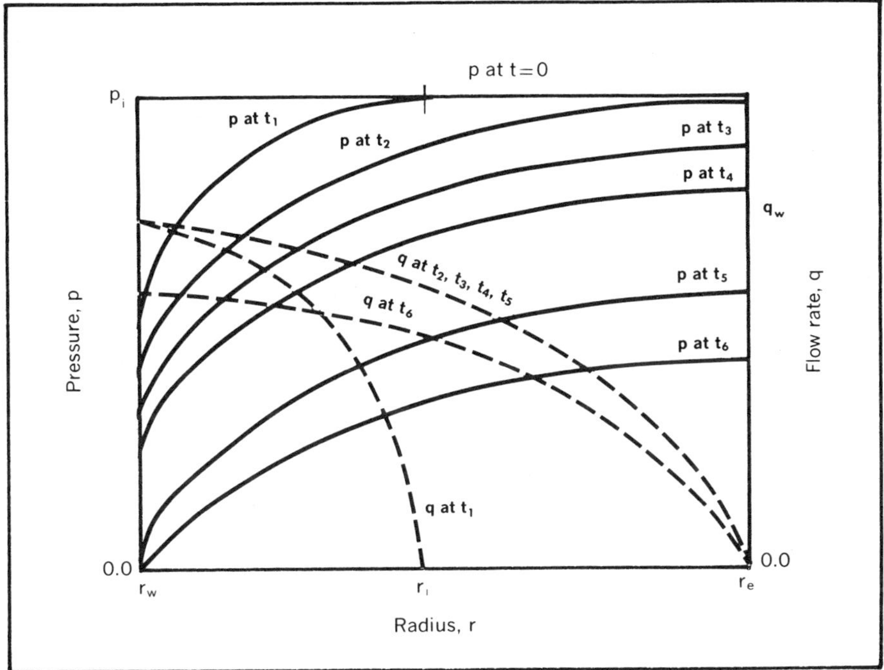

FIG. 1.9 Unsteady state radial flow with constant producing rate—pseudo steady state t_2 to t_5.

lustrated by the pressure distributions at times t_3, t_4, and t_5. This situation will continue with constant changes in pressure with time at all radii and with subsequent parallel pressure distributions until the reservoir is no longer able to sustain a constant flow rate at the well bore. This will occur when the pressure at the well, r_w, has reached its physical lower limit. Note that during the time when the change in pressure with time throughout the reservoir is constant, the rate distribution will remain constant. This can be seen by again examining Equation 1.3 written for the rate of flow at a particular radius q_r.

$$q_r = \frac{1.127kA_r}{\mu}\left(\frac{\Delta p}{\Delta r}\right)_r \qquad (1.6)$$

As noted previously for a particular radius, A_r will be a constant. Also unless some saturation change occurs in the reservoir the permeability, k, will remain constant. Now note that $(\Delta p/\Delta r)$ at any particular radius represents the slope of the pressure versus radius plot and that as long as the pressure distributions remain parallel, the slope of the plot at a particular radius will remain constant, and the rate at that radius will be constant.

This situation in the reservoir that exists after the reservoir has been produced at a constant rate for a long enough period of time to affect the entire reservoir, causing a constant change in pressure with time at all radii, and resulting in parallel pressure distributions and corresponding constant rate distributions, is termed pseudosteady-state flow. It is easy to see how the name was obtained. Since all of the terms in the Darcy equation appear to remain constant, or do remain constant, it is normal to assume that steady-state flow does exist. In fact, Craft and Hawkins[3] refer to this phenomena as steady-state flow from a bounded reservoir. However, by the definition of steady state previously cited, it will be seen that the absolute pressure is changing throughout the reservoir with time and thus the system would, by our definition, be unsteady state. Hence the term pseudo-, or false, steady state, a system that appears to represent steady state but is actually unsteady state.

Pseudosteady-state flow is recognized as a special case of unsteady-state flow because simplified equations similar in form to those for steady-state flow can be derived and used to describe the behavior of a reservoir during this period of time. Also, it will be seen that a period of the reservoir production covered by pseudosteady-state flow represents a large portion of the reservoir life. Some engineers prefer to call this stabilized flow but care should be used in assuming that stabilized flow is synonymous with pseudosteady state since the term stabilized is used in a broad sense to mean many different things.

STEADY STATE FLOW

As noted previously steady-state flow represents the situation that exists when the pressure and rate distributions throughout the reservoir do not change with time. In order for this to occur the mass flow rate into the reservoir must equal the mass flow rate out of the reservoir. These conditions may be closely approximated when a reservoir has a strong water drive, a large gas-cap drive, or is experiencing secondary recovery on a pattern basis. In many other instances the actual deviations from steady state are sufficiently minor that the use of steady-state equations is warranted. Rates that vary over long distances in the reservoir may be practically constant over short distances. Similarly other reservoir conditions that may change over long periods of time may be substantially unchanged over short periods of time. From Figs. 1.8 and 1.9 it can be seen that in unsteady-state flow at relatively large times, the flow rate near the well bore is virtually constant and thus steady-state flow equations can be used to represent short periods of time for flow around the well bore. By limiting the application to conditions near the well bore for short periods of time we make the flow rate practically constant.

Flow equations for specific geometries. Darcy's equation can be applied to specific geometries and compressibilities to obtain equations that are more readily applied than the basic Darcy equation. Some of these equations can be found in Table 1.1, a listing of steady state flow equations. The linear flow equation for liquid flow is probably the simplest of these equations since the compressibility of a liquid is so small that for the purposes of steady-state flow we can consider the liquid flow rate q as being constant. As noted previously dp/dx in Equation 1.3 is the slope of a plot of p versus x. This is a straight line for linear flow. Thus Δx is the length of the linear system, L, and Δp is the difference in the pressures, $p_2 - p_1$, (the pressure decreases as x increases). Then Darcy's equation becomes the linear flow equation for incompressible flow.

$$q = \frac{1.127kA}{\mu} \frac{(p_2 - p_1)}{L} \tag{1.7}$$

The radial incompressible flow equation is obtained by substituting for the cross-sectional area, A, the surface of a cylinder, $2\pi rh$, in the Darcy equation.

$$q = \frac{1.127k(2\pi rh)}{\mu} \left(\frac{\Delta p}{\Delta r}\right) \tag{1.8}$$

The finite difference form $\Delta p/\Delta r$ is used for convenience here. $\Delta p/\Delta r$ is the slope of a plot of pressure versus radius as previously noted, which is the slope of a tangent to a curve at a particular radius. Thus, $\Delta p/\Delta r$ is constant only for a very small change in the radius Δr. From Equation 1.8, the change in pressure, Δp, over a small change in radius, Δr, is

$$\Delta p = \frac{q\mu}{7.08kh} \left(\frac{\Delta r}{r}\right) \tag{1.9}$$

Note that $q\mu/7.08kh$ is the same for all Δr's when we are dealing with an incompressible fluid. Thus we can write Equation 1.9 for all the Δr cylinders in the reservoir from the well radius, r_w, to the external radius, r_e (see Fig. 1.10). If we sum all of the pressure drops from the drainage radius, r_e, to the well radius, r_w, we would get the total change in pressure, $p_e - p_w$. Since these sums are equivalent to the corresponding integrals, we can show that the sum of all the $\Delta r/r$ values is $\ln (r_e/r_w)$. Thus we obtain

$$\sum_{p_w}^{p_e} \Delta p = \frac{q\mu}{7.08kh} \sum_{r_w}^{r_e} (\Delta r/r) \tag{1.10}$$

$$\int_{p_w}^{p_e} \Delta p = \frac{q\mu}{7.08kh} \int_{r_w}^{r_e} (\Delta r/r) \tag{1.11}$$

$$p_e - p_w = \frac{q\mu}{7.08kh} \ln(r_e/r_w) \tag{1.12}$$

$$q = \frac{7.08kh}{\mu} \frac{(p_e - p_w)}{\ln(r_e/r_w)} \tag{1.13}$$

This equation is one of the best known and most used of the reservoir flow equations. As with practically all flow equations it should be remembered that as written here, the permeability, k, would be an average value of the undamaged permeability in the drainage area and the damaged, or improved, permeability around the well bore. This is true of all of the equations in Table 1.1. They could also be written as a function of the undamaged permeability by accounting for the additional pressure drop due to damage or improvement (negative pressure drop) separately.

$$q = \frac{7.08 k_{und} h (p_e - p_w - \Delta p_{skin})}{\mu \ln(r_e/r_w)} \qquad (1.14)$$

In this equation Δp_{skin} represents the additional pressure drop due to damage or improvement around the well bore. To test your understanding of radial flow and the Δp_{skin} concept, work the following problem and check your solution against the one found in Appendix C.

TABLE 1.1
FLOW EQUATIONS

Geometry	Steady state	
	Gas	Liquid
Linear	$q_g = \dfrac{0.112 A k (p_1^2 - p_2^2)}{T_f z \mu L}$	$q = \dfrac{1.127 kA}{\mu}\dfrac{(p_1 - p_2)}{L}$
Radial	$q_g = \dfrac{0.703 kh (p_2^2 - p_1^2)^n}{\mu T_f z \ln(r_2/r_1)}$	$q = \dfrac{7.08 kh (p_1 - p_2)}{\mu \ln(r_1/r_2)}$
Hemispherical	$q_g = \dfrac{0.703 k (p_2^2 - p_1^2)^n}{\mu T_f z \left(\dfrac{1}{r_1} - \dfrac{1}{r_2}\right)}$	$q = \dfrac{7.08 k (p_2 - p_1)}{\mu \left(\dfrac{1}{r_1} - \dfrac{1}{r_2}\right)}$
5-Spot		$q = \dfrac{3.541 kh (p_{wi} - p_{wp})}{\mu \left(\ln \dfrac{d}{r_w} - 0.619\right)}$
7-Spot		$q = \dfrac{4.721 kh (p_{wi} - p_{wp})}{\mu \left(\ln \dfrac{d}{r_w} - 0.569\right)}$

q　= reservoir b/d
q_g　= Mscfd
A　= ft²
k　= Darcys.

p　= psia.
p_{wi}　= psia at injection well
p_{wp}　= psia at producer.

μ　= cp
z　= compressibility factor.
L　= ft of length
T_f　= formation temperature in °R. = (°F. + 460)
r_e　= ft, external radius
r_w　= ft, well radius
d　= ft between input and producer

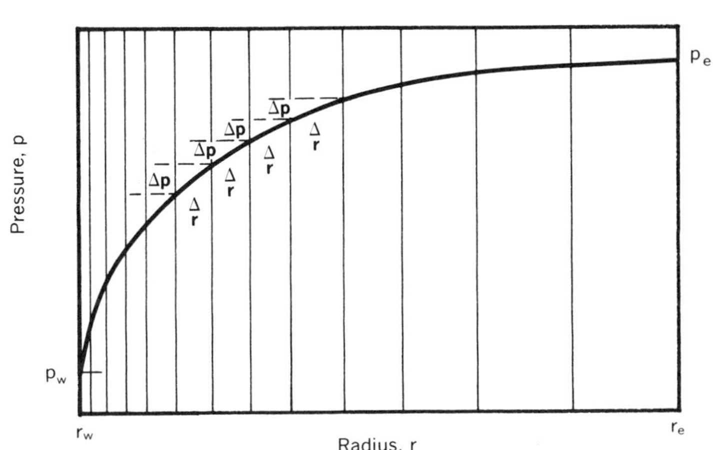

FIG. 1.10 Radial segments in radial flow.

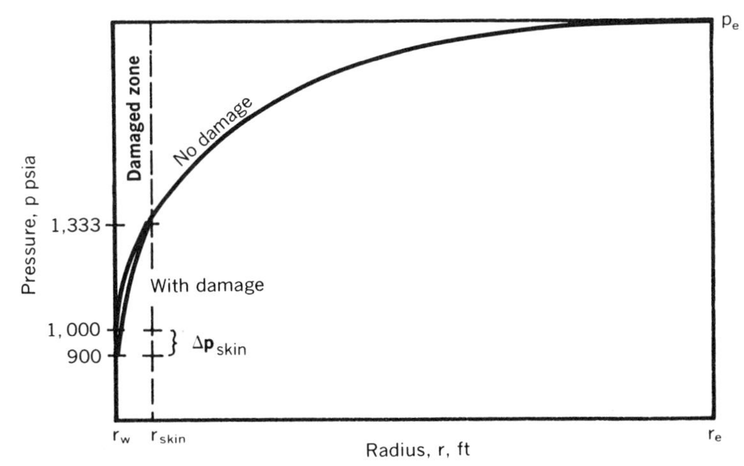

FIG. 1.11 Illustration of Δp_{skin}.

Remember that the rate in the equation is in reservoir barrels. Thus it is necessary to correct the rate of flow stated in stock-tank barrels per day to reservoir barrels by multiplying by the oil formation volume factor. This oil formation volume factor is simply the ratio between the amount of oil in the reservoir at the reservoir pressure and temperature and the amount of stock-tank oil that results from this mass of reservoir oil when the pressure and temperature are reduced to stock-tank conditions. This reduction in pressure and temperature causes liberation of gas and shrinkage of the liquid volume.

Ideally, oil formation volume factors are determined in the laboratory for a specific hydrocarbon system. However, generally lab data for your specific reservoir are not available and it is necessary to obtain estimates of these formation volume factors from empirical data. Such relationships are included in the Appendix.

Problem 1.2: Radial Steady-state Flow from a Damaged Well

Figure 1.11 shows the physical significance of this problem.

A. An undersaturated water-drive reservoir has produced for several years without any substantial change in the reservoir pressure. A pressure buildup curve in Well No. 3 indicates an undamaged reservoir permeability to oil of 10 md and a shut-in well indicated a pressure at the No. 3 well's drainage boundary of 2,000 psia. The flowing pressure in Well No. 3 is 900 psia while it is flowing at a rate of 80 b/d stock-tank oil (STO). What is Δp_{skin} if the following reservoir data are known?

Effective drainage radius = 700 ft
Well radius   = 0.7 ft
Average net thickness   = 6.9 ft
Reservoir oil viscosity   = 0.708 cp
Oil formation volume factor = 1.25 reservoir bbl/S T bbl

B. What would be the producing rate of this well if the pressure drop in the skin is reduced to zero by well treatments and the same well pressure (900 psia) is maintained?

C. What would be the total pressure drop in the damaged zone if the damage exists to a radius of 7.0 ft? The pressure drop from 700 ft to 7 ft can be calculated using Equation 1.13 and a "well radius" of 7 ft.

When thick reservoirs contain a bottom-water drive or a gas cap, the wells are often completed with perforations open in a very thin portion of the total thickness of the formation as far removed from the water or gas as possible. In these cases flow may be hemispherical in geometry in the vicinity of the well bore. The hemispherical flow equation for incompressible fluids can be derived in much the same way that the radial flow equation was derived. In this case the cross-sectional area A is $2\pi r^2$. When this expression is substituted into the Darcy equation, the variables separated, and the integration performed, the hemispherical steady-state incompressible-flow equation is obtained.

$$q = \frac{7.08k(p_2 - p_1)}{\mu \left(\dfrac{1}{r_1} - \dfrac{1}{r_2}\right)} \tag{1.15}$$

This equation is generally used in combination with a radial flow equation in the manner described in the section of this chapter called Approximating Complex Geometries. Care must be exercised in approximating the radii used in hemispherical flow. The internal radius, r_1, is not the well radius. It is more closely associated with the thickness of the formation that is open to production. The external radius, r_2, is a function of the total formation thickness.

If only a small section of the formation near the center is open to production the use of a spherical flow equation might be advisable. In this case we could simply multiply the right-hand side of Equation 1.15 by 2.0 to obtain a spherical flow equation.

When the cross-sectional area can be simply stated, as in the cases described above, it is a relatively simple matter to account for any particular flow geometry. However, many flow geometries do not lend themselves to such a simple treatment. For example the flow between the wells experienced during many waterfloods or other secondary recovery operations presents a much more complex system for analysis. Fig. 1.12 represents the streamline and pressure distribution for a quadrant of a five-spot flood pattern. It will be seen that there is no simple way of describing the cross-sectional area at any particular point in the reservoir. Nevertheless, the use of more complex mathematical methods can be used to analytically obtain the equation for the flow rate between the injection well and the producing well. Muskat[5] derives several of these pattern-flow equations in his text. Two of the most often used pattern-flow equations are included in Table 1.1.

Gas-flow equations. Gas-flow equations differ from liquid equations because the volumetric flow rate, q, varies with pressure due to the compressibility of the gas. Thus, 1.0 b/d of gas at 1,000 psi does not represent the same weight or mass of gas as 1.0 b/d at 100 psi. To make the gas rate of flow a constant it is stated in terms of the equivalent gas rate of flow when the gas is measured at standard conditions. This is equivalent to stating the gas rate of flow in mass units since the rate in standard cubic feet is directly proportional to the pounds of gas. By using the gas equation,

$$pV = ZnRT \tag{1.16}$$

we can determine the flow rate in barrels per day at the reservoir pressure and temperature as a function of the gas flow rate stated in thousands of standard cubic feet per day (Mscfd). In Equation 1.16, V is the volume, Z is the gas deviation factor, n the number of moles of gas, and R the gas constant for the particular system of units used. R is 10.73 if psi, cubic

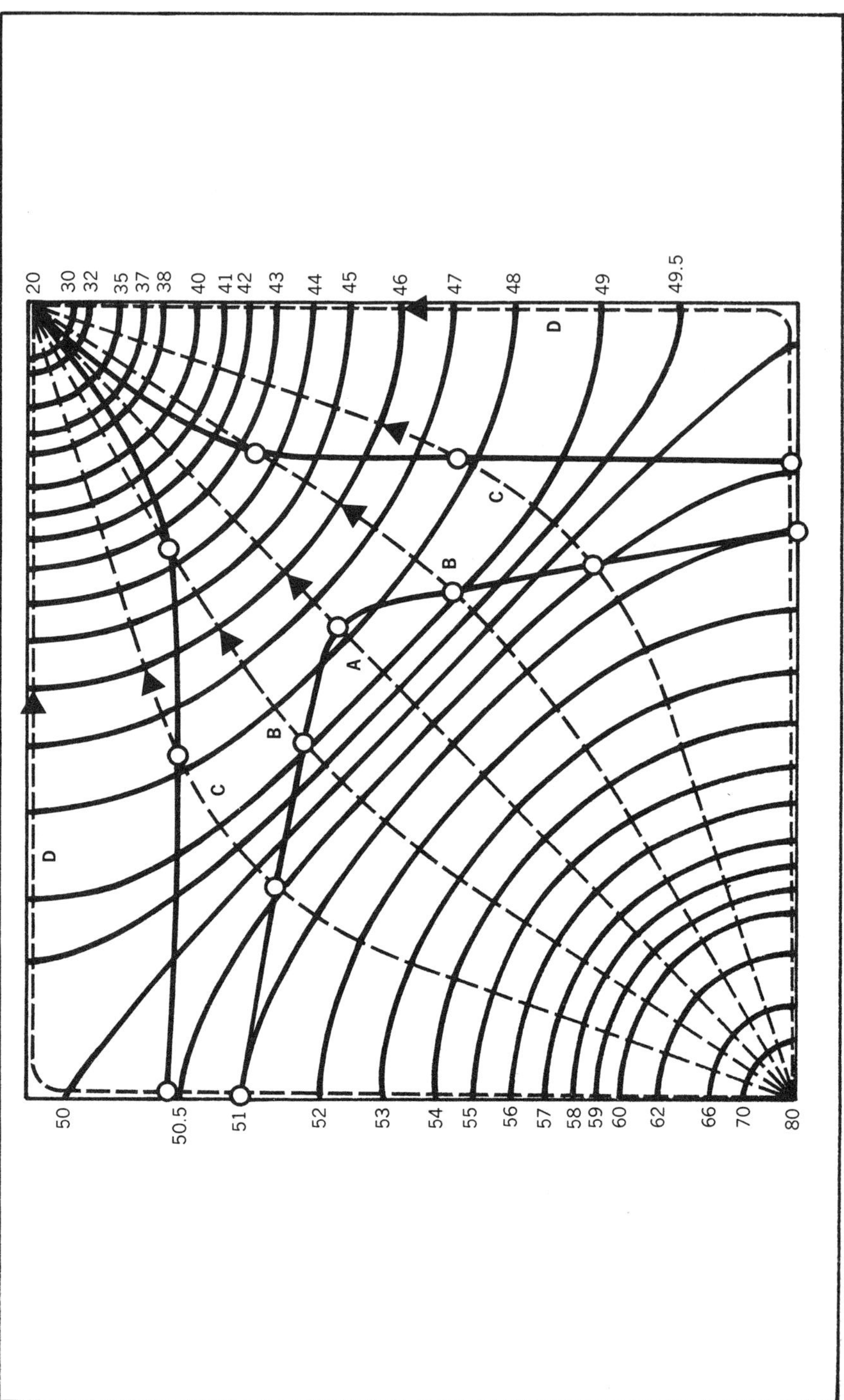

FIG. 1.12 Potentiometric model study of the five-spot network showing the isopotential lines, and two flood fronts. (After B. C. Craft and M. F. Hawkins, *Applied Petroleum Reservoir Engineering,* © 1959. By permission of Prentice-Hall, Inc., Englewood Cliffs, N.J.)

feet, and degrees Rankine are employed. By recognizing the fact that one mole of any gas at standard conditions (14.7 psia and 60° F), occupies 379 ft³ we can write an expression for V, the volume of 1.0 scf at a reservoir pressure p and a reservoir temperature T.

$$(V/scf) = Z(1/379)(10.73)T/p \qquad (1.17)$$

If we multiply this expression by the flow rate in thousands of standard cubic feet per day, q_g, and convert to barrels by dividing the expression by 5.615 ft³/bbl, we obtain the expression,

$$q = \frac{5.04q_gTz}{p}. \qquad (1.18)$$

Here T and p are the reservoir temperature and pressure respectively, q_g is the rate in Mscfd, and z is the gas deviation factor evaluated at the reservoir pressure and temperature. A discussion of the evaluation of the gas deviation factor is covered in Chapter 4 and in the Appendix.

Equation 1.18 can be used to derive gas-flow equations for the various geometries. For example the radial gas-flow equation can be derived from Equation 1.6, by substituting for q_r according to Equation 1.18 and substituting $2\pi rh$ for A_r as was done previously in deriving the radial incompressible flow equation,

$$\frac{5.04q_gTz}{p} = \frac{1.127k(2\pi rh)}{\mu}\frac{\Delta p}{\Delta r} \qquad (1.19)$$

Rearranging,

$$p\Delta p = \frac{5.04q_gTz\mu}{7.08kh}\frac{\Delta r}{r} \qquad (1.20)$$

Equation 1.20 would then apply to any small Δr increment of the radial flow system (Fig. 1.10) and by summing all of the $p\Delta p$ values and $\Delta r/r$ values for all of the Δr cylinders from the external boundary, r_e, to the well radius, r_w, we obtain the radial steady-state flow equation for gas.

$$\sum_{p_w}^{p_e} p\Delta p = \frac{5.04q_gTz\mu}{7.08kh}\sum_{r_w}^{r_e}\frac{\Delta r}{r} \qquad (1.21)$$

Integrating to obtain the summation terms,

$$\int_{p_w}^{p_e} p\Delta p = \frac{5.04q_gTz\mu}{7.08kh}\int_{r_w}^{r_e}\frac{\Delta r}{r} \qquad (1.22)$$

$$\left(\frac{p_e^2}{2} - \frac{p_w^2}{2}\right) = \frac{5.04q_gTz\mu}{7.08kh}(\ln r_e - \ln r_w) \qquad (1.23)$$

$$q_g = \frac{0.703kh(p_e^2 - p_w^2)}{\mu z T \ln(r_e/r_w)} \qquad (1.24)$$

Gas-flow equations are often confusing in that they use different pressure and temperature bases for q and different temperature scales for the reservoir temperature, T. Equation 1.24 uses q_g in thousands of standard cubic feet per day where the base pressure is 14.7 psia and the base temperature is 520° R. (60° F). The engineer should remember that Equation 1.24 is based on Darcy's equation which is not accurate for turbulent flow. Consequently, Equation 1.24 as written does not represent a practical equation for most gas wells. However, in the chapter on gas reservoir engineering it will be shown that an empirical exponent to the pressure squared term can be used to account for turbulence and other nonideal flow conditions such as the normal variation of temperature with gas expansion and the variation of the gas deviation factor and gas viscosity with the pressure.

Equation 1.24 then becomes

$$q_g = \frac{0.703kh(p_e^2 - p_w^2)^n}{\mu z T \ln(r_e/r_w)} \qquad (1.25)$$

The exponent "n" must be evaluated by well testing. The engineer is again reminded that this equation as with the radial incompressible steady-state flow equation is written in terms of an average permeability unless a separate term is added to account for the additional pressure drop caused by damage around the well bore.

Steady-state flow equations are seldom applied to gas flow because steady state seldom applies to a gas reservoir. Even when a gas reservoir has a strong water drive the vast difference in the viscosities of the two phases generally means that the water cannot encroach at a rate even approaching the rate of gas withdrawal. In the past many gas wells did produce under steady-state conditions because the gas production was restricted due to a lack of market. However, such a situation is uncommon today. Nevertheless, we will see that the pseudosteady-state flow equation can be applied to gas reservoirs and the equation is practically as simple as Equation 1.25.

The linear gas flow equation can be similarly derived by substituting for q in Equation 1.13 according to Equation 1.18 and recognizing the fact that the cross-sectional area will remain constant. The resulting equation,

$$\frac{5.04 q_g T z}{p} = \frac{1.127kA}{\mu} \frac{\Delta p}{\Delta x}, \qquad (1.26)$$

can be rearranged so that the variables are separated into $p\Delta p$ and Δx terms. Then the summation or integration of these terms from p_1 and x_1 to p_2 and x_2 will result in the equation,

$$q_g = \frac{0.112\ Ak(p_1^2 - p_2^2)}{Tz\mu L}.$$ (1.27)

Here L is the difference between x_1 and x_2.

Since turbulence is not normally a problem in linear flow, this equation is not used with an exponent on the Δp^2 term as was the case with the radial-flow equation 1.25. To explain further, at the very small cross-sectional areas represented in flow into a small well bore, the velocities become very high and turbulent flow is nearly always reached. However, in linear flow the cross-sectional area is constant and the velocities are not nearly as critical as regards turbulence. An equation similar to Equations 1.24 and 1.27 can also be derived for the hemispherical flow of gas. This equation is listed in Table 1.1.

GENERAL PROBLEMS IN FLUID-FLOW CALCULATIONS

There are various difficulties associated with practically all types of reservoir flow calculations. Having established some knowledge of fluid-flow fundamentals and the working equations for the steady-state flow regime, we will now consider several of the difficulties that arise in the application of these steady-state flow equations as well as the flow equations for pseudosteady state and unsteady state.

Approximating complex geometries. We have considered simple geometries of linear, radial, and hemispherical flow but these simple geometries are never encountered in these ideal forms in practice. Linear-flow equations are practically never applied by themselves to a reservoir problem. Linear flow in a reservoir without accompanying radial or spherical flow would imply production from a trough or a ditch in a reservoir which of course is absurd. Consequently, in practically all cases when linear-flow equations are used they are combined with radial and/or spherical-flow equations to account for the very high pressure drops that accompany the radial or spherical flow into a well.

Radial-flow equations alone do not exactly fit existing drainage areas in the United States where leases are relatively small and operators must protect against drainage across the lease lines. The system lends itself to an overlying rectilinear spacing of wells. This means that the drainage area of a particular well is normally approximated by a square or a rectangle. Generally speaking such a drainage system can be readily analyzed using radial-flow equations by simply calculating an effective radius from the drainage area.

$$\pi r_e^2 = (\text{Acres/well})(43,560)$$ (1.28)

$$r_e = [(\text{Acres/well})(43,560)/\pi]^{1/2}$$ (1.29)

This approximation seldom introduces any significant error because the deviation from radial flow is out near the extremity of the system where the cross-sectional area is so large that the pressure drop is extremely small. It can be shown that using 10-acre spacing and an 8-in. diameter well the maximum error possible would be 5% and this would be comparing a well radius of 330 ft, the distance to the nearest boundary, to a well radius of 457 ft, the distance to the corner of the 10-acre rectangle. A 5% error may appear to be large to engineers new to reservoir engineering. However, the uncertainties that accompany reservoir description (see appendix) are such that an answer with a 5% error compares favorably in accuracy with an electrical engineering error in the third significant figure.

Another means of applying a radial-flow equation to an actual flow problem is to correct the actual flow rate, representing flow from a part of a circle, to the equivalent full circle flow rate and then apply the flow equation as required. For example, in Fig. 1.13 is shown a small reservoir with a well near the apex of two intersecting faults forming an angle of 120° or $\frac{1}{3}$ of a full circle. If we wish to apply a radial-flow equation to this situation we can simply divide the actual q from the well by $\frac{1}{3}$ and use this as the q in Equation 1.13. This is possible because in radial flow there is no flow across any of the radial spokes of the system. Consequently, the radial-flow equations can be applied to a portion of a radial-flow system just as accurately as it can be applied to a full 360° radial system. The pressure distribution will be exactly the same in both cases.

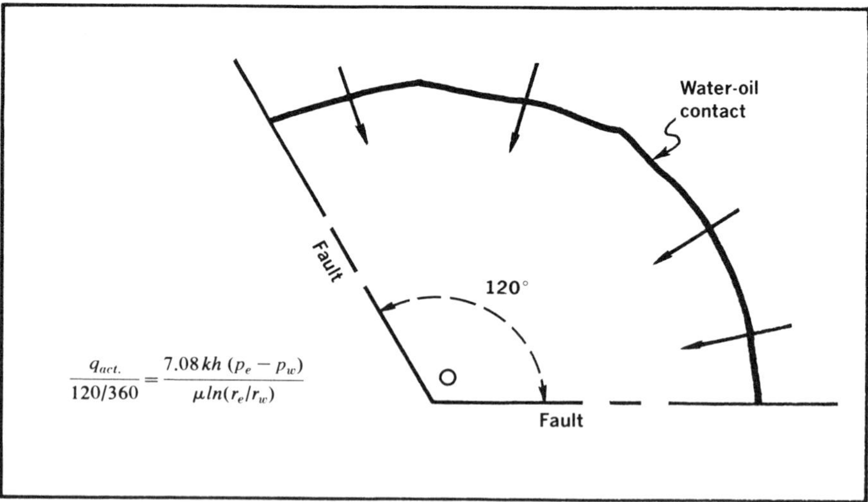

FIG. 1.13 Example application of a radial flow equation to flow from a portion of a circle.

When drainage areas are more complex the methods of obtaining flow equations become more complex. For example, suppose that a drainage area for a well is as shown in Fig. 1.14. We do not have a flow equation for this type of geometry but we can break this geometry into two simple geometries that encompass the same area. In this particular case the drainage area is modeled by using a combination of a linear and radial-flow system in series. The actual dimensions of the two systems are set by making certain that the total drainage area is equal to the total area of the model as indicated. Then with this model it can be assumed that the pressure drops in the two systems would be additive. Thus, the total pressure drop from the water-oil contact to the well radius would be equal to the pressure drop in the linear system represented by the first term plus the pressure drop in the radial system (half a circle) as indicated in the second term. If it was so desired, this resulting expression could then be solved for the flow rate q and would give a close approximation of the flow characteristics of this drainage area.

In general we observe that if a complex geometry can be broken into a combination of simple geometries, flow through which takes place in series, we can then write an expression for the pressure drop for each of the simple geometries and the total of these pressure drops will represent the total system pressure drop. In performing this type of approximation the engineer should be very careful to always include some radial or spherical flow in his model if the flow considered is terminated by a well. In other words, if you use the pressure in a well as one of the terminal flow pressures your modeling should include at least a portion of a radial or spherical flow system. There is no way to have flow into a

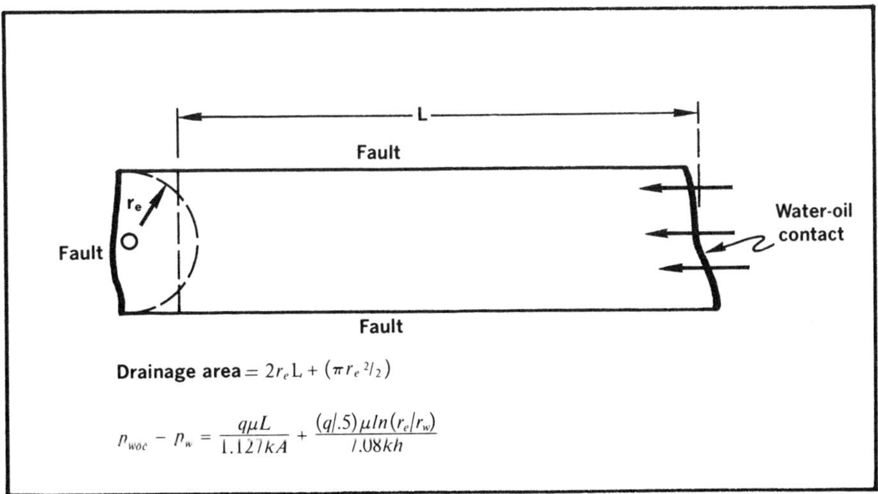

Drainage area $= 2r_e L + (\pi r_e^2/2)$

$$p_{woc} - p_w = \frac{q\mu L}{1.127kA} + \frac{(q/.5)\mu \ln(r_e/r_w)}{1.08kh}$$

FIG. 1.14 Approximating flow equations with two simple systems in series.

well under linear conditions. Such an assumption will introduce a gross and intolerable error in your analysis.

If the drainage area is broken up in such a way that the combination of simple geometries represents geometries with flow in parallel systems, then the rates are additive. Such a situation is illustrated in Fig. 1.15. This is similar to the geometry described in Fig. 1.13 but the distance to the water-oil contact is not uniform. Consequently, a closer approximation of the flow characteristics of this drainage system would be obtained by modeling the actual drainage geometry into a combination of two radial systems as indicated, with the total area of the two radial systems being equal to the drainage area of the actual system. When the flow geometry is modeled in this way, flow takes place through parallel systems and the flow rates in the two systems would be additive. This is the situation indicated in the flow equation on Fig. 1.15 with the first term representing $^{45}/_{360}$ of a radial flow system and the second term representing $^{60}/_{360}$ of a radial system.

Due to the nature of radial flow it is doubtful that there is a significant increase in accuracy gained by using two parallel radial sys-

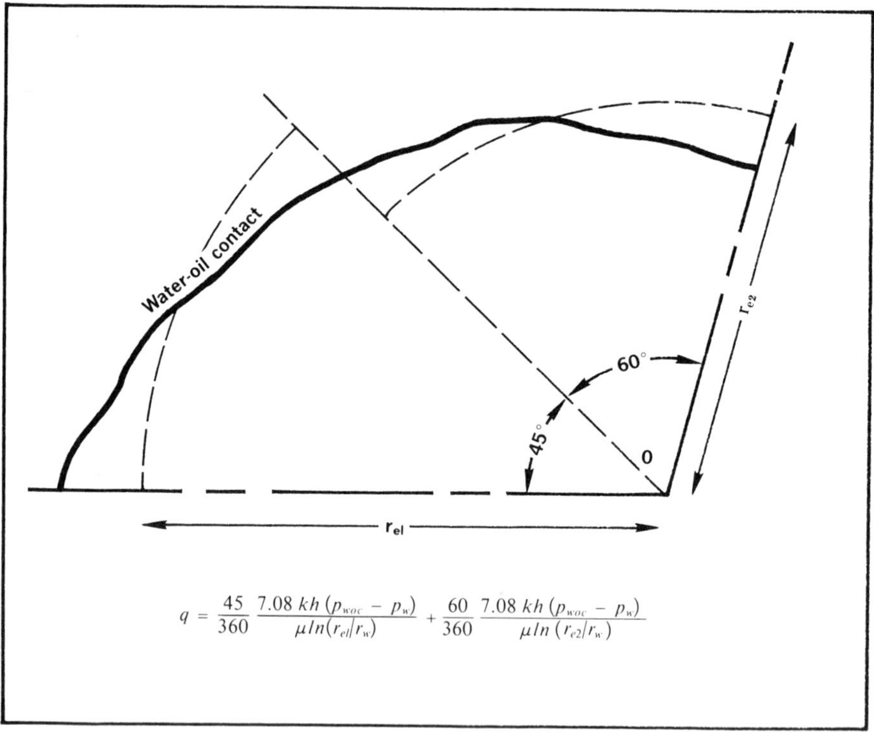

$$q = \frac{45}{360} \frac{7.08\,kh\,(p_{woc} - p_w)}{\mu\,ln(r_{e1}/r_w)} + \frac{60}{360} \frac{7.08\,kh\,(p_{woc} - p_w)}{\mu\,ln\,(r_{e2}/r_w)}$$

FIG. 1.15 Approximating flow equations with two simple systems in parallel.

tems as opposed to the use of one equivalent radial system with an angle of 105°. However, Fig. 1.15 does illustrate one more technique of approximating flow equations for a complex system.

To provide some indication of the accuracy or inaccuracy involved in approximating a flow system with a combination of simple geometries, we will derive an equation for the five-spot flow geometry and compare it with the exact analytical equation derived by Muskat.[5] Fig. 1.16 shows that a quadrant of a five spot can be approximated by two radial systems back to back whose total area is the same as the area of the quadrant, or each radial system should be equal to one-half the quadrant area.

$$\tfrac{1}{4}\,\pi(r_{ep})^2 = \tfrac{1}{4}\,\pi(r_{ei})^2 = d^2/4 \tag{1.30}$$

or
$$r_{ep} = r_{ei} = 0.57d \tag{1.31}$$

The total pressure drop for this quadrant, $p_{wi} - p_{wp}$, will then be equal to the pressure drop on the injection side plus the pressure drop on the

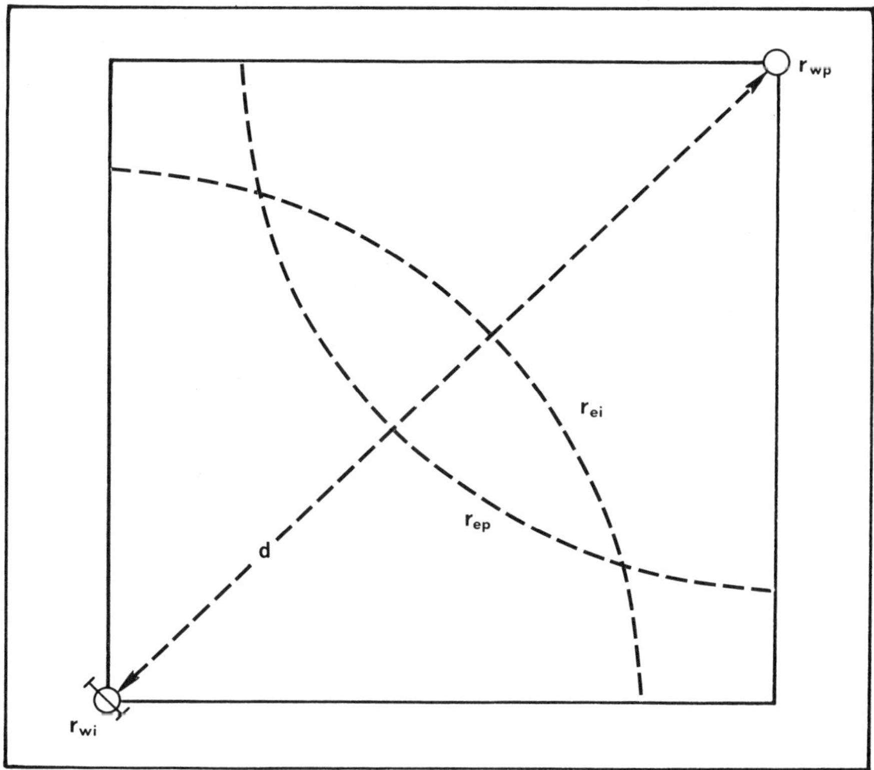

FIG. 1.16 Approximating five-spot flow with two radial systems.

producing side. Using the radial incompressible liquid-flow equation to denote these two pressure drops we would obtain

$$p_{wi} - p_{wp} = \frac{q\mu \ln(0.57d/r_{wi})}{7.08kh} + \frac{q\mu \ln(0.57d/r_{wp})}{7.08kh} \tag{1.32}$$

When this equation is solved for q, assuming that r_{wi} will equal r_{wp}, we obtain

$$q = \frac{3.54kh(p_{wi} - p_{wp})}{\mu[\ln(d/r_w) - 0.571]} \tag{1.33}$$

The exact analytical equation derived by Muskat is

$$q = \frac{3.54kh(p_{wi} - p_{wp})}{\mu[\ln(d/r_w) - 0.619]} \tag{1.34}$$

When the approximate Equation 1.33 is compared with the exact analytical Equation 1.34 we note that the only difference is in the constant in the denominator. When we consider that the log of the distance between wells divided by the well radius will normally be in the range of 7.0 to 10 we see that the error in modeling the five spot by using two radial systems introduces an error of less than 1%.

Again it should be emphasized that the approximation of this flow equation in this manner is performed to illustrate the accuracy that can be obtained with this and similar approximations of complex geometries by using a combination of simple geometries. The engineer should not use Equation 1.33 for his calculations since Equation 1.34 is more accurate.

When the geometry is very complex it may be necessary to break the geometry into so many segments that the segments are both parallel and in series. In such a case a digital computer is generally necessary to analyze the model. To illustrate the principals involved, suppose we desired to determine the pressure distribution in a five-spot flood pattern by this technique. We would find that all of the flood pattern is made up of units such as that illustrated in Fig. 1.17 which is one-half a quadrant of a five spot. In this figure there will be no flow across any of the outside figure boundaries except at the injection well, r_{wi}, and the production well, r_{wp}.

We can write an equation for flow into and out of each of the numbered segments shown in such a way that the equations are stated as functions of the pressures only. By assuming pressures at the injection and producing wells, we would have 15 equations, one for each numbered segment, and 15 unknowns, the pressures for the 15 points shown. These simultaneous equations can then be solved by various numerical procedures for the 15 pressures which define the pressure distribution. Once the pressure distribution is known the rate distribution can also be determined.

To illustrate this technique consider the steady-state flow equations

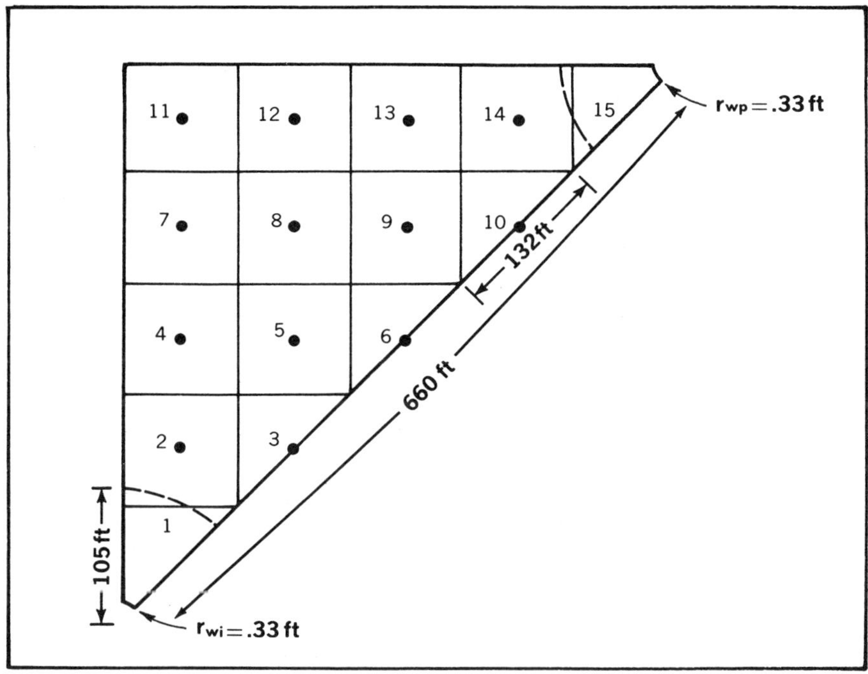

FIG. 1.17   Segmenting a five-spot for computer analysis.

governing flow into and out of segment No. 8 of Fig. 1.17. We will assume
that flow is upward and to the right. (The validity of this assumption
will not affect the results.) Based on this assumption we can say,

$$q_{7-8} + q_{5-8} = q_{8-12} + q_{8-9} \tag{1.35}$$

The subscript numbers are the segment numbers between which
flow is represented by that particular rate. Each rate could be found by
using the Darcy Equation.

$$q = \frac{1.127kA}{\mu} \frac{\Delta p}{\Delta x} \tag{1.3}$$

Now note that for the square grid, Δx and A will be the same for all
of the rates in Equation 1.35. If we assume only one fluid flowing and the
permeability the same in all directions, k/μ will be the same for all rates
in Equation 1.35. Under these conditions the rates would simply be
proportional to the respective pressure drops so we could write Equation
1.35 as,

$$(p_7 - p_8) + (p_5 - p_8) = (p_8 - p_{12}) + (p_8 - p_9). \tag{1.36}$$

This equation can be rearranged to

$$p_5 + p_7 - 4p_8 + p_9 + p_{12} = 0. \tag{1.37}$$

Note that such an equation can be written for each of the 15 segments. Many of the segment equations will only have two or three terms.

Segments 1 and 15 pose special problems. To approximate the pressure distribution into the very small well bore would require excessive linear grid segments. Consequently, these segments are treated as radial flow from the well radius to a radius of 93.2 ft or vice versa. Thus, flow across these segments is stated as flow from one radial boundary to the opposite radial boundary rather than from the center of one segment to the center of another as is the case with most of the segment flow equations.

Consequently, the pressure at grid "points" 1 and 15 actually are the pressure along the dotted lines near the numbers in Fig. 1.17. The dotted line represents the radial and the solid line the linear extremity of segment 2. A similar situation exists for segment 14.

When we assume the injection pressure at r_{wi} as 100 and the producing pressure at r_{wp} as 0.001 we can show that the equations for segments 1 and 15 are

$$110.2\ p_1 - 103.11\ p_2 = 708 \tag{1.38}$$

$$110.2\ p_{15} - 103.11\ p_{14} = 0.00708 \tag{1.39}$$

Note that in writing these two equations the Δx for the flow rate from point 1 to 2 and from 14 to 15 is one-half the Δx from one point to the other between the other segments. Once all the equations have been written the coefficients of the pressures and the constant terms can be fitted into an augmented matrix such that each column of the matrix represents the coefficient of a pressure point or the equation constant term and each matrix row represents an equation for a particular segment. For example -103.11 in Equation 1.38 might be equal to A (1,2) where A is the matrix value for row 1 and column 2 where row 1 represents the coefficients for the pressures representing the equation for segment 1 and column 2 means that the coefficient is for pressure point 2. Similarly, from Equation 1.37 the coefficient for p_5, which is 1.0, would be designated A (8, 5).

When a complete matrix has been generated, a digital computer subroutine such as the Gauss-Jordan or Gauss-Seidel methods[2] can be used to reduce the matrix or in effect solve for the pressures. Such subroutines can be obtained from IBM or probably any other digital computer company. The results for this example, (Fig. 1.17), are shown in Fig. 1.18. Comparison with the analytically obtained data of Fig. 1.12 will indicate close agreement. Accuracy would of course be greatly improved by using a finer grid.

Determining average permeabilities. Permeability varies widely from point to point in the reservoir. This may be a statistical variation

```
PRESSURE DISTRIBUTION IN A HALF-QUADRANT OF A 5-SPOT DURING STEADY STATE

FLOW, INPUT PRESSURE IS 100 PSI AND OUTLET PRESSURE IS 0.001 PSI.

                                        WELL PRESSURE IN IS:100.0 PSI
```

50.0	51.0	53.0	56.3	59.1
49.0	50.0	51.8	54.1	
47.0	48.2	50.0		
43.7	45.9			
40.9				

```
0.001PSI IS THE WELL PRESSURE OUT
```

FIG. 1.18 Results of computer calculation of Fig. 1.17 pressure grid.[6]

as observed previously in this chapter for a "homogeneous" formation or it may be a systematic variation as would occur due to stratification of a reservoir or a gradual change in permeability in a particular direction. Since sedimentary formations were laid down normally in ancient seas over extremely long periods of time by the deposition of sediments, it is normal that producing formations have different permeabilities at different vertical depths, or in the case of dipping formations, at different positions at right angles to the bedding plane. The drainage volume of a well may also have a systematic variation in permeability radially due to various degrees of damage that occurs during drilling or treating of the well. Since most flow equations require simply one permeability value, we need to determine the average permeability for such systematic variations.

Consider a vertical variation in permeability as shown in Fig. 1.19. This drawing shows three beds of different permeabilities and different thicknesses. Linear flow is depicted and it is obvious that the flow rate through these parallel beds would be equal to the total of the flow rates through the individual beds

$$q = q_1 + q_2 + q_3 \tag{1.40}$$

If we use the Darcy Equation to write an expression for each one of these flow rates including an expression for the total rate, q_{total}, based on an average permeability, k_{avg}, and the total thickness, h, we obtain

$$\frac{1.127k_{avg}hw\Delta p}{\mu L} = \frac{1.127k_1h_1w\Delta p}{\mu L} + \frac{1.127k_2h_2w\Delta p}{\mu L} + \frac{1.127k_3h_3w\Delta p}{\mu L} \tag{1.41}$$

Instead of writing the cross-sectional area A in each one of these terms the A was equated to hw to show that the width, w, would cancel

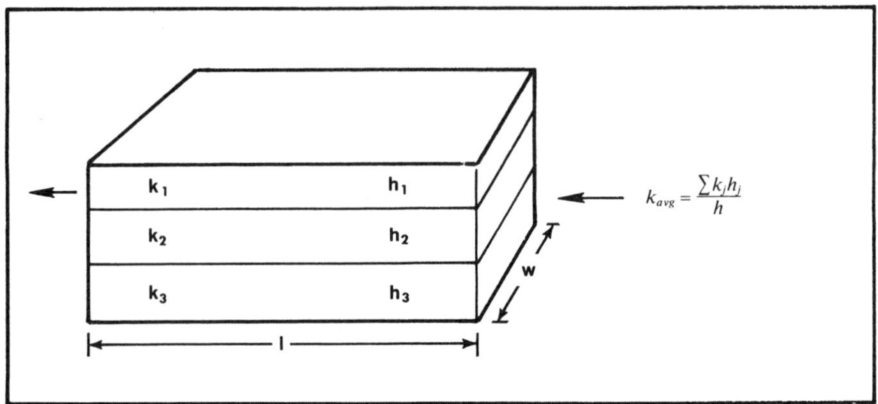

FIG. 1.19   Parallel flow.

along with all the other factors except the permeability and thickness. Thus Equation 1.41 simplifies to

$$k_{avg} = \frac{k_1 h_1 + k_2 h_2 + k_3 h_3}{h} \qquad (1.42)$$

When this approach is generalized we can simply write the sum of the kh products as one term

$$k_{avg} = \frac{\displaystyle\sum_{j=1}^{j=n} (k_j h_j)}{h} \qquad (1.43)$$

In this equation, n is the total number of zones considered.

If the same derivation is followed for radial flow, we find that exactly the same equation results. Consequently, Equation 1.43 applies to both parallel linear and radial flow.

If the variation in permeability is lateral rather than vertical, the pressure drops are added and not the flow rates as was the case for parallel flow. This situation is shown in Fig. 1.20. Using the radial incompressible flow equation to evaluate Δp for each term we find

$$\Delta p = \Delta p_1 + \Delta p_2 + \Delta p_3 \qquad (1.44)$$

$$\frac{q\mu\ln(r_e/r_w)}{7.08 k_{avg} h} = \frac{q\mu\ln(r_o/r_i)_1}{7.08 k_1 h} + \frac{q\mu\ln(r_o/r_i)_2}{7.08 k_2 h} + \frac{q\mu\ln(r_o/r_i)_3}{7.08 k_3 h} \qquad (1.45)$$

In this expression, r_o is the outer radius of the zone and r_i is the inner radius of the zone. By cancelling like terms in the equation it becomes

$$\frac{\ln(r_e/r_w)}{k_{avg}} = \frac{\ln(r_o/r_i)_1}{k_1} + \frac{\ln(r_o/r_i)_2}{k_2} + \frac{\ln(r_o/r_i)_3}{k_3} \qquad (1.46)$$

When solved for k_{avg} we obtain

$$k_{avg} = \frac{\ln(r_e/r_w)}{\displaystyle\sum_{j=1}^{j=n} [\ln(r_o/r_i)_j/k_j]} \tag{1.47}$$

There appears to be little likelihood that an equation would ever be needed to calculate k_{avg} for a lateral variation in permeability with linear flow, but for completeness we will consider this equation. It can be derived in exactly the same way that the series flow equation was derived for radial flow except that the Δp values would be determined from the linear flow equation rather than the radial flow equation. When this derivation is carried out, all of the factors except the lengths and per-

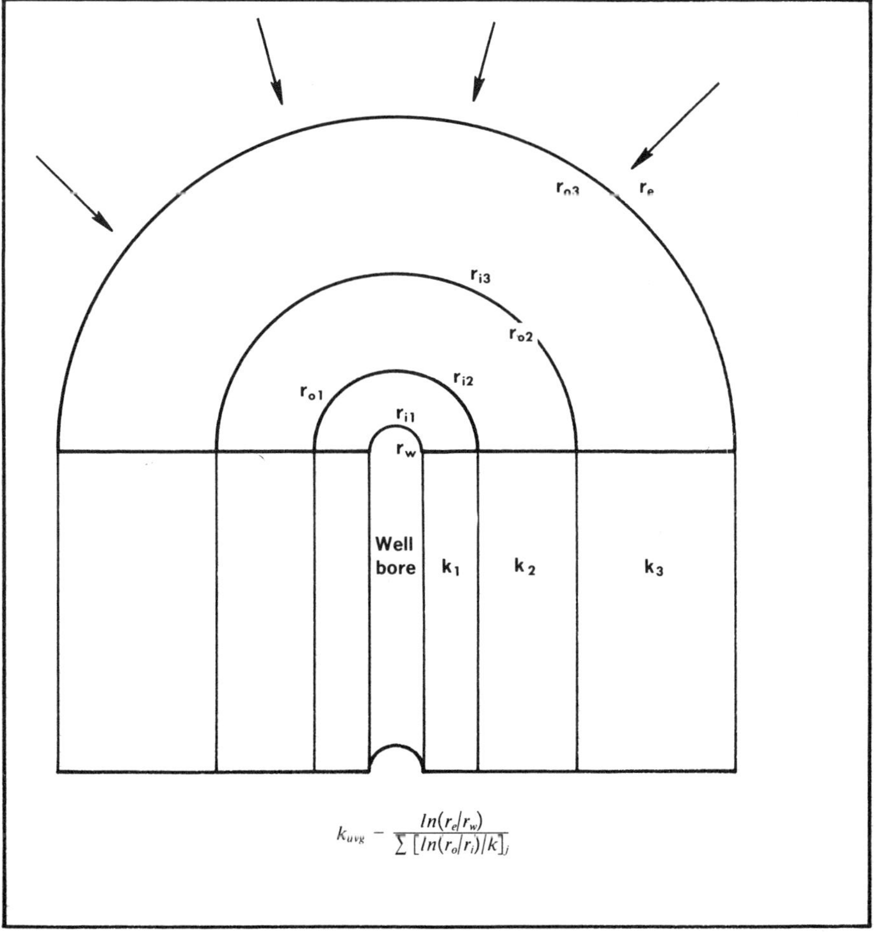

FIG. 1.20 Radial series flow.

meabilities of the individual zones cancel out and we are left with the general expression

$$k_{avg} = \frac{L}{\sum\limits_{j=1}^{j=n} \left(\frac{L}{K}\right)_j} \qquad (1.48)$$

To test your understanding of the permeability-averaging techniques work the following problem and check your work against the solution in Appendix C.

Problem 1.3: Determining Average Permeabilities

Part A

A stratified aquifer contains 2 ft of 200-md sand, 3 ft of 100-md sand and 5 ft of 10-md sand. What will be the steady-state water-injection rate into this 8-in. diameter well when the well pressure is 600 psia and the reservoir pressure at a distance of 100 ft is 500 psia? Assume $\mu = 1.0$ and $B_w = 1.0$ *Note:* steady-state rates can seldom be maintained in an aquifer.
Answer: 93 b/d

Part B

A pressure build-up analysis indicates an undamaged reservoir permeability of 75 md. If a well in this reservoir has well-bore damage to a radius of 10 ft that reduces the permeability to this radius to 10% of the undamaged permeability (7.5 md) what average permeability would be calculated from a flow test of this well (i.e., what is the average permeability of the well-drainage area)?

well radius = 1 ft
well drainage radius = 1,000 ft

Answer: 19 md

Correcting for static pressure differences. When Muskat published his book on fluid flow in the early 1940's he did not explain that the pressure drops he was discussing were the pressure drops due to flow alone and did not include the static pressure differences. As is the case with the Muskat equations, most of the equations used in this book deal specifically with the pressure drop due to fluid flow.

Darcy's equation can be modified to include the static pressure differences but the nature of the correction is such that the equation has very limited use except for linear flow. As we have already indicated linear flow has very limited use in reservoir engineering. Nevertheless the equation will be used in later portions of the book in discussing fluid displacement and the effects of gravity on fluid displacement so we will indicate the basis for the Darcy equation including the static pressure drop at this time.

The Darcy equation can be corrected to include the static pressure difference by simply stating the pressure drop due to fluid flow as a function of the total pressure drop and the static pressure difference. Fig. 1.21 indicates fluid flow up dip in a bed with a dip angle α. Note that if the flow

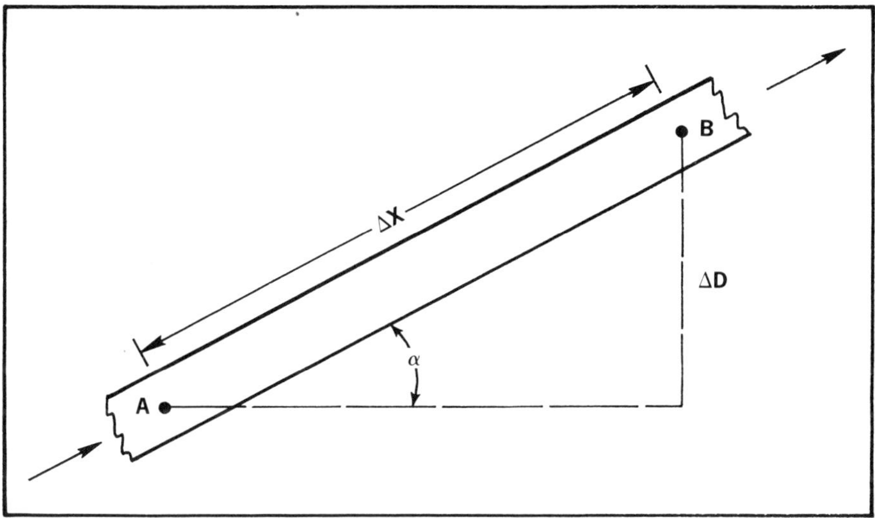

FIG. 1.21 Linear flow in a tilted reservoir.

rate was zero there would still be a pressure difference between A and B due to the difference in the static head of fluid, ΔD, between points A and B. Since the change in pressure with depth for fresh water is 0.433 psi/ft of depth the change in pressure with depth for a fluid whose specific gravity is γ would be 0.433 γ and the difference in static pressure between points A and B would then be 0.433 $\gamma\Delta D$. Consequently, we can determine the flowing pressure drop from the total pressure drop in a system from the equation

$$\Delta p_{flow} = \Delta p_{total} \pm 0.433 \, \gamma\Delta D \tag{1.49}$$

This equation is written with a positive sign for flow down dip since the pressure drop in the direction of flow for a negative rate would be negative which would give a positive absolute sign to the term. Or another way of saying this is that ΔD in the direction of flow would be negative which would give a positive sign to the second term. At any rate it seems simpler to write the equation with plus and minus signs rather than to attempt to define the terms more specifically.

Equation 1.49 can be used to correct total pressure differences to flowing pressure differences regardless of the prevailing flow geometry. Note also that the same result is obtained by correcting all of the pressures to some datum level in the reservoir and then using the differences in those pressures as the flowing pressure difference. This is the technique used by most companies for correcting data. For example, if the pressures at A and B in Fig. 1.21 are corrected to the datum level of A then the pressure at B would be corrected to the datum level at A by adding 0.433 $\gamma\Delta D$ to the pressure at B and the flowing pressure would be

$$\Delta p_{flow} = p_A - (p_B + 0.433 \, \gamma\Delta D) \tag{1.50}$$

Now note that the $p_A - p_B$ expression is simply the pressure drop and Equation 1.50 is equivalent to Equation 1.49.

As noted previously we can also modify Darcy's basic equation to include static pressure differences. When we substitute the right hand side of Equation 1.49 for Δp in the Darcy equation we obtain

$$q = \frac{1.127kA}{\mu} \frac{(\Delta p_{total} \pm 0.433\,\gamma\Delta D)}{\Delta x} \tag{1.51}$$

Now note that $\Delta p_{total}/\Delta x$ is the total pressure gradient and $\Delta D/\Delta x$ is the sine of the angle α (see Fig. 1.21). Consequently, Equation 1.51 can be written as

$$q = \frac{1.127kA}{\mu} \left[(\Delta p/\Delta x)_{total} \pm 0.433\,\gamma \sin \alpha\right] \tag{1.52}$$

Equation 1.52 — Darcy's equation as a function of the total pressure drop — has limited application. This is true because radial flow in a dipping formation has a different dip angle, α, along each radial ray. Consequently, for all flow geometries except linear flow, it is simpler to correct all reservoir pressures to some common datum level and use the difference in these pressures as the flowing pressure difference or use Equation 1.49 to obtain the pressure drop due to flow. However, Equation 1.52 will be used later to help identify the gravity forces that affect efficiency.

Try the following problem and check your solution against the one in Appendix C to make certain you understand static pressure corrections.

Problem 1.4: Correcting for Static Pressure Differences

Given: The following diagram of a producing well, A, and shut-in well, B, and reservoir data.

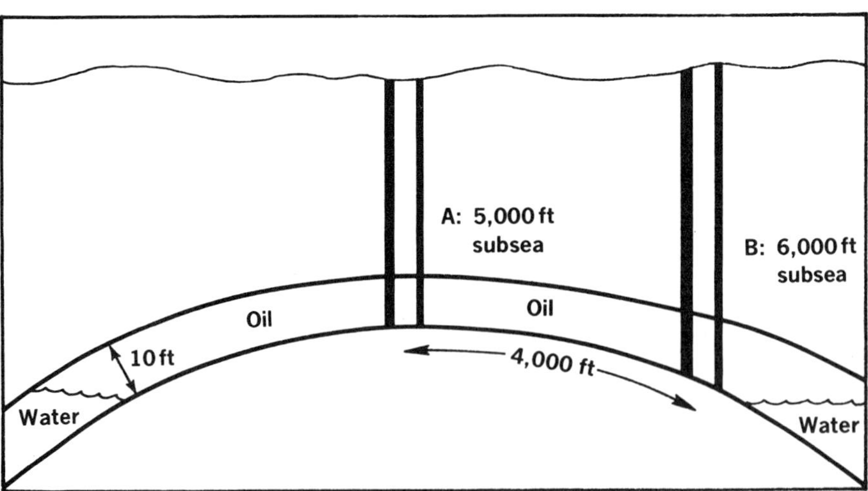

Reservoir Data:

Permeability to oil:	92 md
Oil reservoir viscosity:	0.708 cp
Well radius:	0.4 ft
Reservoir thickness	10 ft
Reservoir oil sp. gr.:	1.0
Saturation pressure:	2,000 psia
Flowing BHP at A:	2,250 psia
Shut-in BHP at B:	3,000 psia (shut-in well)

Water drive activity is such
　　that pressures do not vary
　　with time.
Assume radial flow and no
　　well damage.

Find:

A. The producing rate in reservoir b/d for the perfect dome indicated.
B. What would the producing rate be if the reservoir were flat ($r_e = 4,000$ ft.) and the same pressures prevailed?

Determining effective permeability. Probably the biggest single difficulty in applying flow equations is in determining the effective permeability. We have already seen that it is necessary to take into account the naturally occurring variation in permeability to determine the permeability of a "homogeneous" reservoir. We also discussed the difficulties of determining "average" permeabilities when the variation in permeability is systematic as with a vertical variation or lateral variation in permeability. We previously discussed the relative-permeability concept and the difficulties involved in obtaining representative relative-permeability data from the laboratory. Here again you are reminded that the Appendix contains empirical equations and data that can be used by analogy.

One other permeability effect or correction should be mentioned. It is seldom that the Klinkenberg[4] effect has any practical significance when compared with normal variations in permeability but to avoid confusion when the effect is discussed, the engineer should be familiar with the concept. Darcy's equation describes viscous fluid flow, and when conditions are such that viscous flow can not take place, Darcy's equation is inaccurate. Such a situation arises when gas flows at very low pressures. Remember that flow is through very small pores of the reservoir rock. When the pressure for a gas gets low enough the distance between gas molecules becomes comparable to the size of the pores. Under these circumstances viscous flow can not take place.

In viscous flow in a tube we have various cylinder surfaces in the pipe moving at different velocities or shearing the fluid at different rates. A velocity profile normally exists in a tube where viscous flow is taking place with the velocity at the surface of the tube being zero and the velocity at the center of the tube being a maximum. This is illustrated

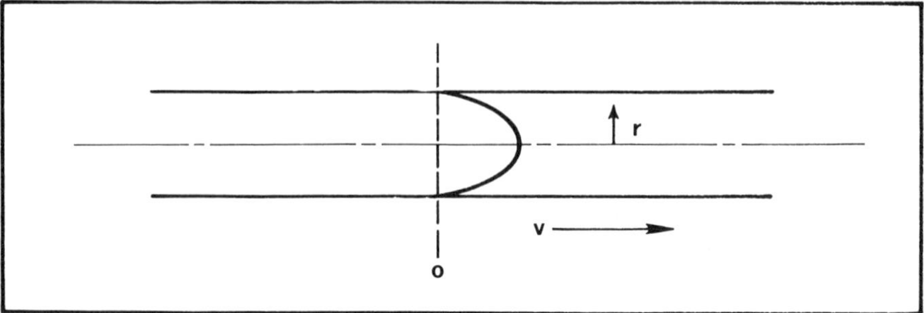

FIG. 1.22   Velocity profile for viscous flow in a tube.

in Fig. 1.22. However, if the distance between molecules is comparable to the pore radius r, only one molecule at a time can flow through the small tube, and there could be no velocity profile under these conditions. Since there is a wide variety of pore sizes in the normal petroleum reservoir, this critical condition is reached at various pressures for various pore sizes. Consequently, as the pressure gets lower the flow deviates more and more from the viscous flow of Darcy's equation.

This phenomena is presented graphically in Fig. 1.23 where the permeability to gas is plotted versus the reciprocal to the average of the pressure (p_m) used in measuring the permeability. It appears that theoretically this curve extrapolated to a reciprocal mean pressure of zero gives a permeability equivalent to the liquid permeability or the absolute permeability because this would represent an infinite mean pressure and gas acts like liquid at very high pressures. However, the Klinkenberg effect presented in this way is somewhat misleading since there is some minimum permeability to gas, equal to the absolute permeability, that is reached before an infinite mean pressure is reached. When the smallest pore in the porous media is in viscous flow the entire pore structure will

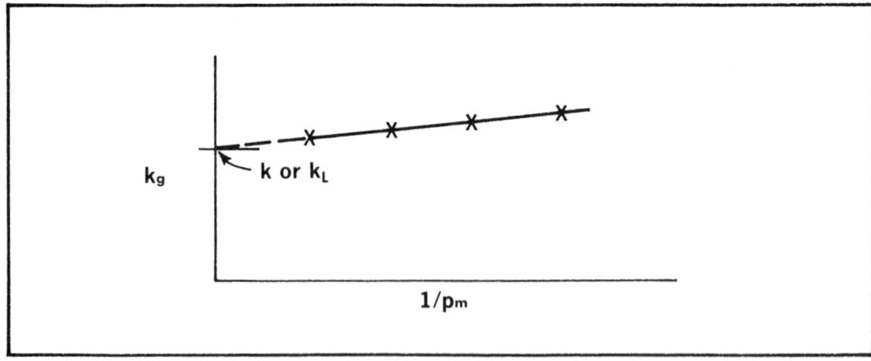

FIG. 1.23   Illustration of the Klinkenberg effect.

be experiencing viscous flow. Since there is a particular pressure for a particular pore size when this will occur, it is obvious that there is a mean pressure for a particular pressure above which no change in permeability is experienced. However, when a scale for the reciprocal pressure is chosen such that the Klinkenberg effect is obvious, the difference between this minimum gas permeability and the zero point on such a scale is insignificant.

The only real practical application of the Klinkenberg effect is one of interpreting laboratory data. In the laboratory it is most convenient to measure permeability with gas at a pressure as low as possible. Consequently, the engineer should be careful that the laboratory is running permeability data in a gas-pressure range where the Klinkenberg effect is negligible. Once this situation has been ascertained it would appear that for all practical purposes the engineer can forget the Klinkenberg effect until he changes core-analysis laboratories. For a more quantitative treatment of this subject the reader is referred to the Amyx, Bass, and Whiting text[1] which concerns itself almost entirely with the nature of the porous media and is thus an excellent reference on such subjects.

Determining viscosities and flow rates at reservoir conditions. The engineer is cautioned to carefully determine whether a flow equation requires a rate stated at reservoir conditions or in stock-tank barrels (STB). In this book all rates are stated in reservoir volumes and consequently stock-tank-barrel rates must be converted to reservoir volumes by multiplying the stock-tank-barrel rate by the oil formation volume factor. However, in other references and texts the rate q may refer to either stock-tank barrels or reservoir barrels. One way of determining the units of the rate, q, in a particular equation is to examine the flow equation to see if it contains the oil formation volume factor, B_o. If it does not contain B_o you can be certain that you must convert to reservoir volumes before substituting a number for the rate in the equation. On the other hand if the equation does contain the formation volume factor, B_o you can likewise be certain that the rate in the equation should be stated in stock-tank barrels.

If you do not have formation volume factors available for the particular reservoir you are studying, you are referred to the Appendix for means of obtaining these from empirical data. This is discussed in the chapter on material balance.

Many engineers introduce large errors into their calculations by failing to realize that the viscosity of liquids in the reservoir is much different than the viscosity of stock-tank liquids. This is true of both water and oil. In the case of oil we must remember that in the reservoir the stock-tank oil will have a large amount of gas in solution and this solution gas will greatly reduce the viscosity of the oil. Water may also have small amounts of gas in solution but this effect is generally very small. However, the effect of pressure and temperature on the viscosity of both oil and water is considerable. Here again the engineer is referred to the Appendix for empirical viscosity data that will permit him to evaluate the amount of gas in solution in the reservoir oil at various reservoir conditions, the effect of the gas on the viscosity, and the effect of the reservoir pressure and temperature on the viscosity.

Empirical gas-viscosity data are also included in the appendix. Here again the effect of pressure and temperature on the gas viscosity is pronounced and somewhat more complex than it is for liquids. At low pressures the gas viscosity decreases as the pressure decreases since the molecules become widely separated. However, at high pressures the gas reacts as does a liquid and a reduction in pressure increases the viscosity.

ADDITIONAL PROBLEMS

1.5 A well is producing at a rate of 748 STB/d of oil and no water with a BHP (bottom-hole pressure) of 900 psia from a reservoir with a very strong water drive (assume steady state). What is the formation thickness if the lab data of Problem 1.1 are applicable and a pressure drawdown analysis indicates that Δp_{skin} at this rate is 115 psia?

> Well bore diameter = 8 in.
> Distance between wells = 600 ft
> Estimated pressure midway between wells = 1,200 psia
> Reservoir oil viscosity = 2.5 cp
> Formation volume factor = 1.25 reservoir bbl/STB

1.6 Rework problem 1.5 assuming that this is a gas reservoir instead of an oil reservoir and that it is producing 100 million scfd with Δp_{skin} plus the additional pressure drop due to turbulence totaling 200 psia. No water is being produced.

> Reservoir gas deviation factor = 0.9
> Reservoir gas viscosity = 0.01 cp
> Reservoir temperature = 120° F.

1.7 The No. 1 well in Fig. 1.24 is located very near the fault and is offset in all other directions by similar wells at a distance of 600 ft. Find the average permeability-thickness product for this well's drainage area based on the absolute permeability. Material balance calculations indicate the current saturations to be $S_o = 0.565$, $S_{wc} = 0.300$, and $S_g = 0.135$.

> Well No. 1 producing rate = 750 STB/d
> Oil formation volume factor = 1.20
> Pressure midway between wells
> (from pressure build-up analysis) = 1,000 psia
> Reservoir oil viscosity = 2.5 cp
> Bottom-hole pressure = 800 psia
> Well radius = 0.3 ft
> Relative permeability data as in Fig. 1.25
> Assume no well damage and steady-state flow.

1.8 A gas-cap reservoir discovery well produces at an initial GOR of 2,000 scf/STB from a reservoir whose pressure is 2,000 psia and temperature is

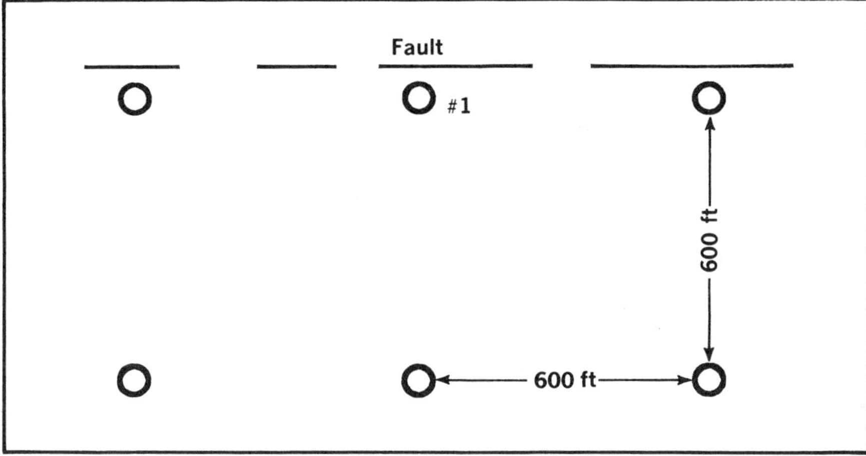

FIG. 1.24　Diagram for Problem 1.7.

120°F. Assuming that the oil zone contains no free gas and the gas zone contains no oil, estimate the gas-bearing thickness in the well given the following:

> Solution gas = 500 scf/STB
> B_o = 1.25
> k_g/k_o = 1.0 (assumes relative permeability to gas and oil
> 　　　　are both 1.0)
> μ_o = 1.2 cp
> μ_g = 0.02 cp
> Producing interval = 30 ft

1.9 Use the Wylie equations from the Appendix and calculate the relative-permeability data for the core sample analyzed in Problem 1.1. Assume the core is a well-cemented sandstone and that the irreducible water saturation is 30% as indicated in Problem 1.1.

1.10 Verify the constant 1.127 of Equation 1.3.

1.11 A bottom-hole heater is to be used in a well with a damaged well bore to decrease the oil viscosity. (A) What will be the producing rate if the effect of the heater is equivalent to reducing the viscosity in a 7-ft radius around the well from 7.08 to 0.708 cp. (B) Find Δp_{skin}. Assume steady-state flow and the following reservoir data.

> Pressure at the external boundary = 2,000 psia
> Pressure at the well = 950 psia
> Oil formation volume factor = 1.5
> K_o undamaged = 100 md
> K_o damaged = 10 md
> radius of damaged zone = 10 ft
> reservoir thickness = 10 ft
> drainage radius = 700 ft
> well radius = 0.7 ft

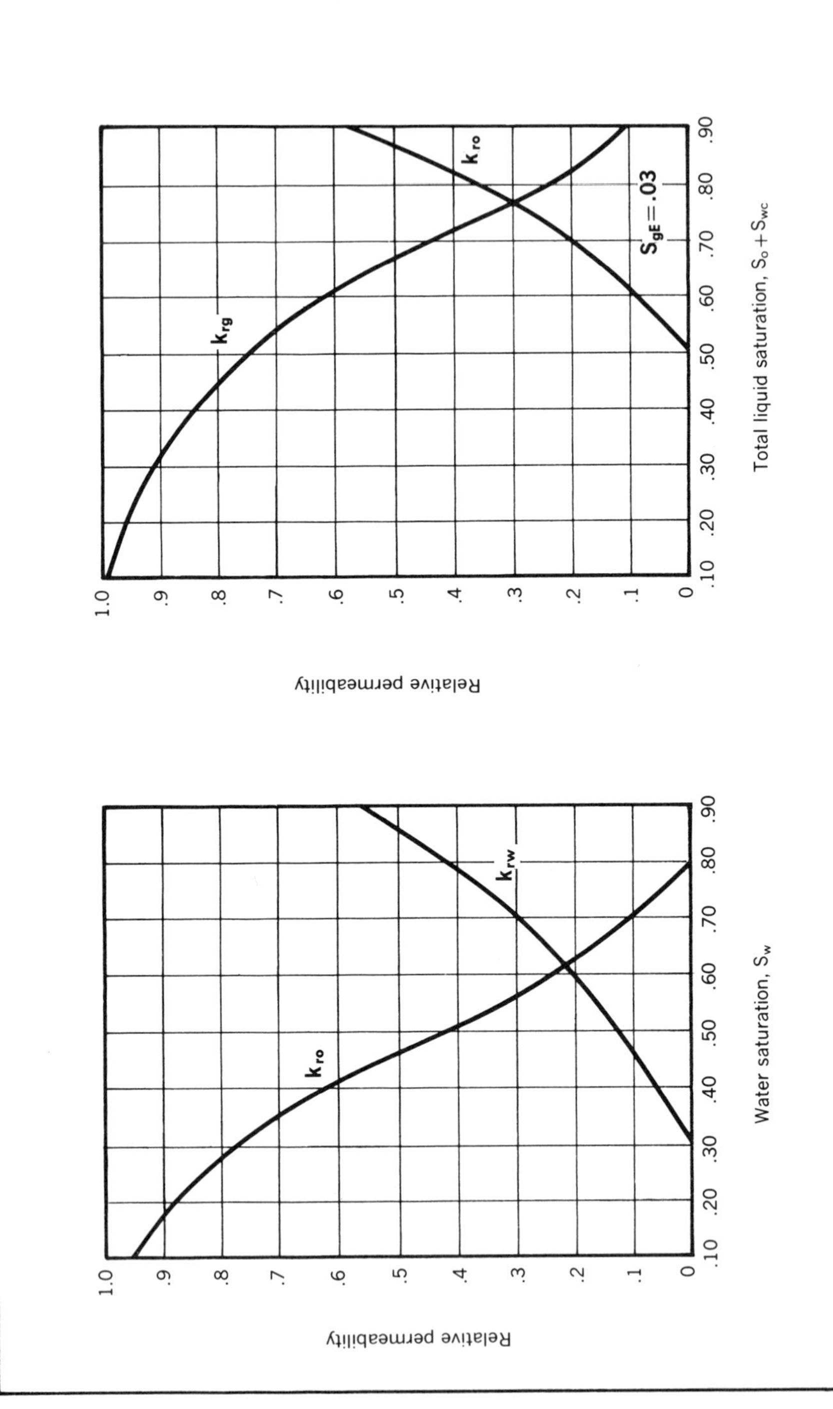

FIG. 1.25 Relative permeability data (synthetic) for Problem 1.7.

1.12 A well drains a radial reservoir consisting of two beds with undamaged permeabilities of 100 md and 75 md. Near the well the formation is damaged to a depth of 33 ft resulting in reduced permeabilities of 60 md and 30 md respectively. If the well radius is 0.33 ft and no vertical movement of fluid takes place what will be the steady-state flow rate of gas when the gas saturation is 30% and p_e and p_w are 1,000 and 800 psia, respectively. Assume there is no additional pressure drop due to turbulence. Bed thicknesses are 2 ft and 3 ft, respectively. The external drainage radius is 3,300 ft, and Fig. 1.25 is applicable.

$$\text{Gas viscosity} = 0.01 \text{ cp}$$
$$\text{Gas deviation factor} = 0.97$$
$$\text{Reservior temperature} = 140° \text{F.}$$

1.13 An undersaturated water-drive reservoir is shaped as shown in Fig. 1.26 with water encroachment as indicated. The pressure along A is 3,000 psia and the pressure at the well, B, is 2,500 psia. Other reservoir data are: permeability to oil, 92 md; oil sp. gr. 1.0; saturation pressure, 2,000 psia. Water-drive activity is such that pressures do not vary with time. Find the producing rate in reservoir b/d.

$$\text{Oil viscosity} = 2.0 \text{ cp}$$
$$\text{Well radius} = \tfrac{1}{3} \text{ ft}$$
$$\Delta p_{skin} = 105 \text{ psi}$$

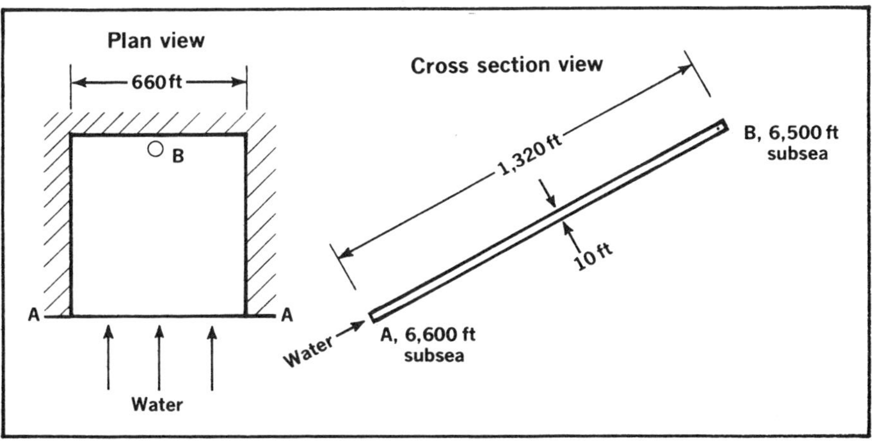

FIG. 1.26 Problem 1.13 diagrams (not to scale).

REFERENCES

1. J. W. Amyx, D. M. Bass, Jr., and R. L. Whiting, *Petroleum Reservoir Engineering*, McGraw-Hill Book Co., Inc., 1960, pp. 91–93.
2. Conte, S. D., *Elementary Numerical Analysis*, McGraw-Hill Book Co., New York, 1965, pp. 191–197.
3. B. C. Craft and M. F. Hawkins, *Applied Petroleum Reservoir Engineering*, Prentice-Hall, Inc., Englewood Cliffs, N. J., 1959, pp. 280–283.

4. L. J. Klinkenberg, "Permeability of Porous Media to Liquids and Gases," *Drilling and Production Practice*, API (1941), p. 200.

5. Muskat, M., *The Flow of Homogeneous Fluids Through Porous Media*, McGraw-Hill Book Co., Inc., New York, 1937.

6. T. J. Rusnak, Undergraduate, Ch. E. 442 Course Report, Columbus, Ohio, January 1971.

7. Sheridegger, Adrian E., *The Physics of Flow Through Porous Media*, Univ. of Toronto Press, Toronto, 1960.

Unsteady State and Pseudo Steady-State Flow

As defined in Chapter 1, unsteady-state flow is flow that occurs while the rates and/or the pressures change with time. Consequently it covers all reservoir flow except for the specific situation when the rates and pressures do not change with time.

Many engineers seem to be leary of unsteady-state flow applications, probably because the material is generally presented in a mathematical form not completely understood or even trusted by him. This chapter will endeavor to give physical meaning and mathematical understanding to the unsteady-state methods without going into the detailed math required for rigorous derivation of some of the expressions used. This chapter will also provide the reader with the guidelines and working equations of pseudosteady-state flow, the particular part of unsteady-state flow that embraces most of the productive life of a reservoir and lends itself to much simpler mathematical analysis. In fact the equations governing pseudosteady-state flow are about as simple to apply as steady-state equations.

A complete understanding of unsteady-state flow is a necessity for any competent reservoir engineer. It is the guts of such useful techniques as pressure buildup analysis, drawdown analysis, interference tests, reservoir boundary delineation, prediction of water encroachment, prediction of disposal-well behavior, gas-well testing, drill-stem-test (DST) analysis, etc.

PHYSICAL DESCRIPTION

Refer back to Figs. 1.8 and 1.9 which give typical pressure and rate plots versus radius at different times for a reservoir producing under unsteady-state conditions. The reservoir represented by these diagrams

51

is a circular drainage area with uniform permeability to the outer radius where the reservoir or well drainage ends abruptly with all production due to the expansion of the reservoir fluids. Lest the engineer be discouraged by the idealized nature of this reservoir, it should be explained that methods of applying the theoretical equations to nonideal (actual) reservoirs will be discussed and illustrated, but it is simpler to derive and describe the behavior of simple reservoirs initially.

Fig. 1.8 represents conditions that result from producing this idealized reservoir at a rate to keep the well pressure constant. This is comparable to flowing a well against a constant choke size or keeping a well pumped down to a particular level. Under these conditions note from Fig. 1.8 that at some small producing time the reservoir pressure is affected significantly only to a particular radius, r_1. Since the reservoir is producing only due to the expansion of the reservoir fluid, the flow rate in the reservoir at any radius greater than r_1 is zero. No pressure drop has occurred to affect an expansion of the fluid and the subsequent flow. However, as the flow from the well is continued, more and more of the reservoir is affected until eventually all of the reservoir has suffered a pressure drop.[1]

The time effect of the production and the fact that it takes some period of time before the entire reservoir is affected is one of the concepts with which engineers have the most trouble. Consequently, we will try to examine this concept in different ways to gain a physical understanding of the phenomena.

Consider the segmented reservoir represented in Fig. 2.1A. Initially the same pressure, p_i, will exist throughout the reservoir when production is initiated. Consider what happens at the well bore at time $t = 0$ when the pressure in the well or at the internal radius of ΔV_1 is dropped to p_w. This causes a pressure drop across the sand face and according to Darcy's equation flow will occur. As fluid flows from ΔV_1 to the well bore, the pressure drops in ΔV_1 which causes an expansion of the remaining fluid. This provides the fluid for flow into the well bore.

Once enough fluid has been removed from ΔV_1 to cause a significant pressure drop, there will exist a pressure difference between ΔV_1 and ΔV_2. According to Darcy's equation this will result in flow from ΔV_2 to ΔV_1. The flow of fluid from ΔV_2 will cause a pressure drop in ΔV_2 and a corresponding expansion of the remaining fluid in ΔV_2 which will provide the fluid for flow into ΔV_1. The flow of fluid from ΔV_2 to ΔV_1 will also tend to maintain the pressure in ΔV_1.

When enough flow has taken place from ΔV_2 to cause a significant pressure drop in ΔV_2 a pressure difference will exist between ΔV_3 and ΔV_2 and flow will take place from ΔV_3 to ΔV_2 which will tend to maintain the pressure in ΔV_2 and will eventually cause a pressure drop that is large enough to initiate flow from ΔV_4 to ΔV_3 due to the pressure drop that exists between the two segments.

Physically this is the process that takes place and it requires time for

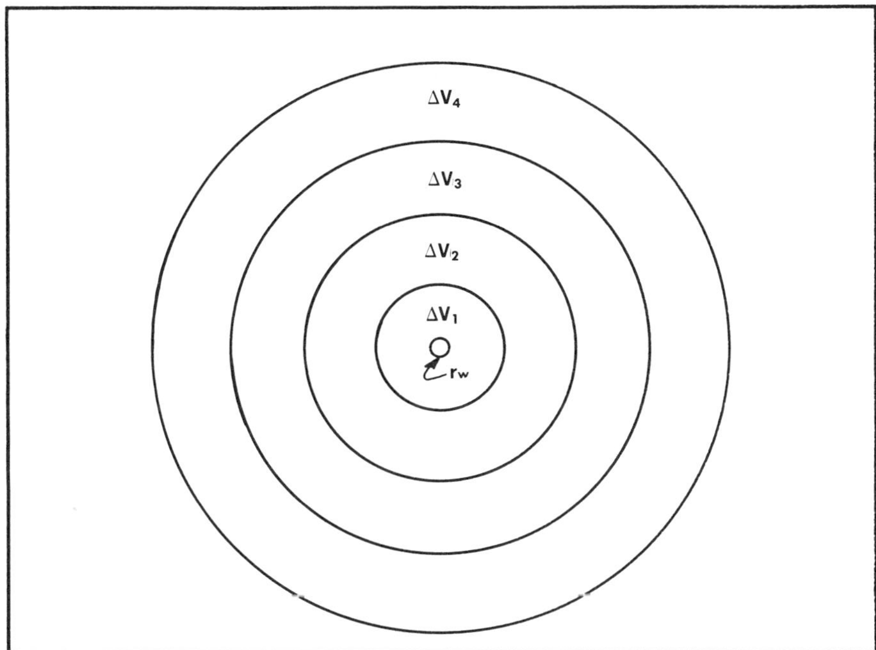

FIG. 2.1A Segmented circular reservoir—for discussion of the time effect in unsteady state flow.

the pressure effect to move across the reservoir. Note that as the pressure effect moves out into the reservoir it will continue to have a smaller effect on the pressure of each subsequent reservoir segment as the segment radii increase, because the increase in radius causes an increase in the segment size and thus a larger amount of fluid withdrawal is required to obtain the same pressure drop.

Hawkins explained the physical concepts of this time phenomenon by using a hydraulic analog.[2] He modeled a segmented reservoir such as that of Fig. 2.1A by representing the potential expansion capacity of each segment by a container with that volume. These containers are connected by pipes of such a size that they represent the relative resistance to flow between the various segments. Fig. 2.1B shows schematic representation of such a model of the reservoir of Fig. 2.1A. Note the relative sizes of the containers representing the different segments and the relative size of the pipes connecting the various segments. The only factor affecting the relative resistance to flow between the segments (pipe size) is the cross sectional area (A in Darcy's equation) that increases in proportion to the increase in radius.

To operate this model all of the containers are initially filled with water to the same level. Note that the level of the water represents the

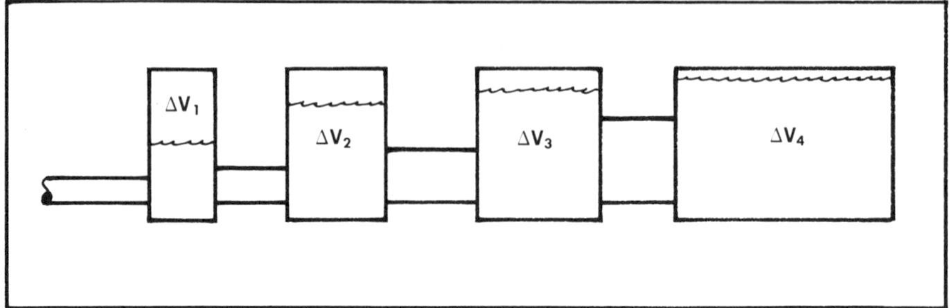

FIG. 2.1B   Hydraulic analog of slightly compressible unsteady state fluid flow.

pressure in each segment, and that as the water level is reduced the potential expansion of that particular segment is reduced. Thus, after filling each container to the same level, a valve out of ΔV_1 which represents flow into the well is opened and the model is permitted to discharge. It is then easy to see that flow will take place from ΔV_1 for a substantial period of time before the flow from ΔV_2 to ΔV_1 occurs and that flow from ΔV_3 to ΔV_2 and ΔV_4 to ΔV_3 are similarly delayed.

RADIAL DIFFUSIVITY EQUATION

Whereas steady-state flow is governed by very simple equations, unsteady-state flow is best described by a partial differential equation known as the radial diffusivity equation that may scare many practical engineers.

$$\frac{\partial p}{\partial t} = \eta \left[\frac{1}{r} \frac{\partial p}{\partial r} + \frac{\partial^2 p}{\partial r^2} \right] \tag{2.0}$$

The reservoir diffusivity constant, η, is $6.33k/\phi\mu c$. Written in a finite difference form the equation is much more meaningful to most engineers.

$$\frac{\Delta p}{\Delta t} = \eta \left[\frac{1}{r} \frac{\Delta p}{\Delta r} + \frac{\Delta(\Delta p/\Delta r)}{\Delta r} \right] \tag{2.1}$$

In this form it can be seen that the equation simply relates the slopes of three different plots at a particular time and a particular radius. These plots are pressure versus radius at different times, the slopes of the plot of pressure versus radius plotted versus the radius, and pressure versus time.

We are familiar with the nature of the plot of pressure versus radius at various times through the use of Figs. 1.7, 1.8, and 1.9. If we evaluated all the slopes representing different combinations of radius and time and then plotted the slopes versus the radius for various times we might obtain data similar to that of Fig. 2.2. The slope of these curves would then represent the term $\Delta(\Delta p/\Delta r)/\Delta r$, which is the change in the slope with respect to time. The term $\Delta p/\Delta t$ is the slope of a set of curves of p vs t where each curve represents a particular radius as illustrated in Fig. 2.3.

Note that only the pressure curve for the well radius declines from time zero. Every other curve initially has a period of time when the pressure remains at the initial pressure. This is due to the same time phenomenon described previously. A certain period of production must prevail before the pressure at a particular radius is affected. A pressure disturbance (increased or decreased) introduced at the well, diffuses throughout the reservoir; hence the name, diffusivity equation.

We should also note the significance of the zero value of $\Delta p/\Delta r$ in Fig. 2.2. This zero value occurs at the radius where the flow rate is zero and the initial pressure is as yet unaffected.

With the terms of Equations 2.0 and 2.1 having the physical significance described it should be obvious that the solution of the equation for a given reservoir is a series of curves relating the pressure, radius, and time that will satisfy the radial diffusivity equation.

The basis for the radial diffusivity equation can be easily shown by considering a plot of pressure versus the radius at two different times, t and $t + \Delta t$, as shown in Fig. 2.4. What we will first do is to determine the change in the number of barrels of fluid in a Δr increment of the reser-

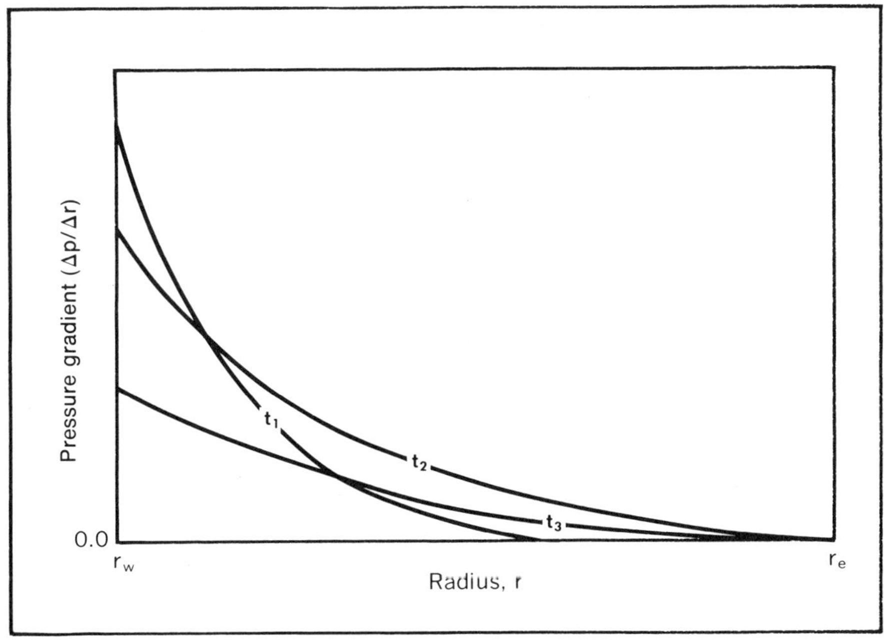

FIG. 2.2　Example plot of pressure gradient vs radius for unsteady-state flow.

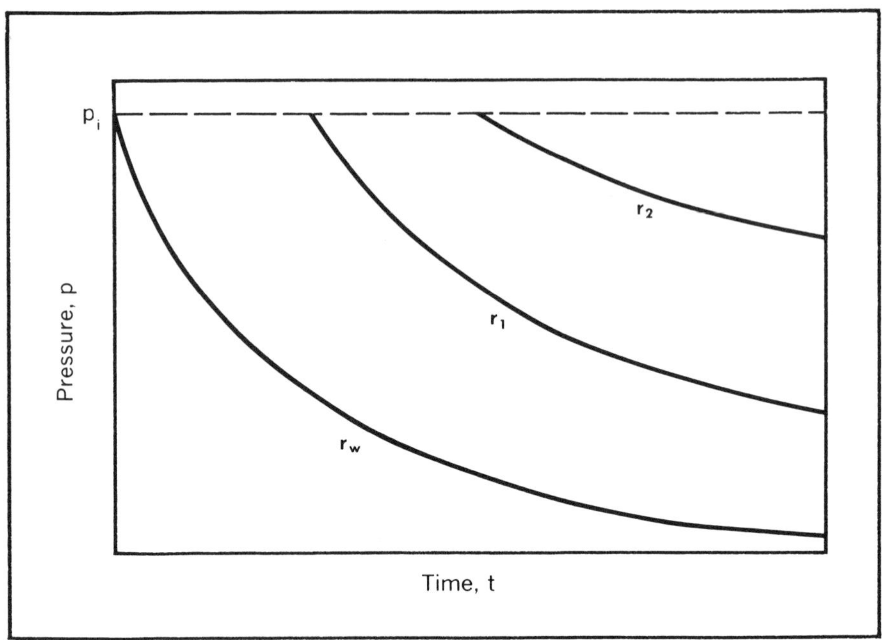

FIG. 2.3 Example plot of pressure vs time for unsteady-state flow.

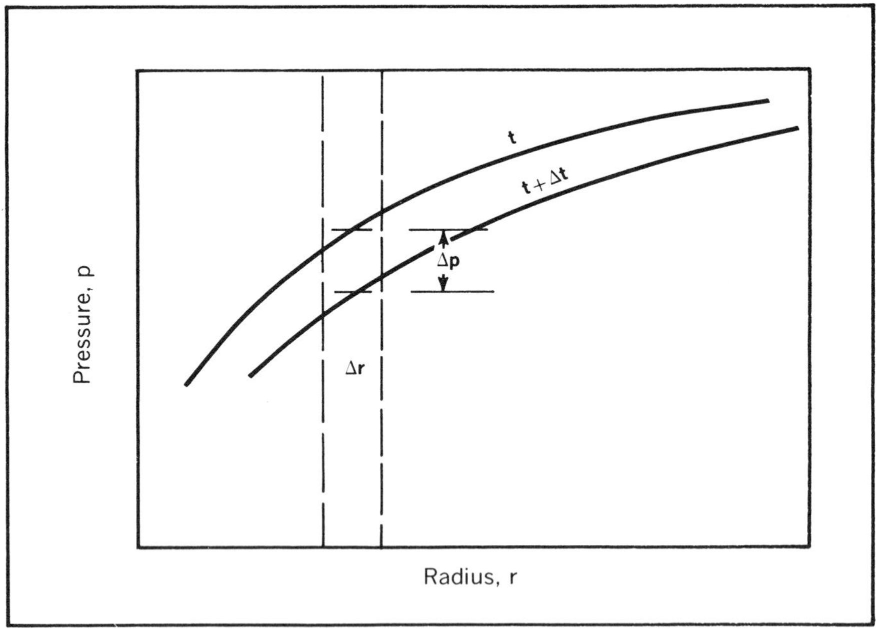

FIG. 2.4 Pressure vs. radius at successive times during unsteady state flow—for derivation of radial diffusivity equation.

voir during the time interval, Δt, when the average pressure declines by Δp. This will simply be equal to the expansion of the fluid in the increment caused by the drop in the average pressure, Δp. Thus, the volume of fluid, $2\pi r \Delta rh\phi$, when multiplied by the compressibility, c, will give the expansion of the fluid for 1.0 psi of pressure change. When this is multiplied by the average change in pressure per day, $\Delta p/\Delta t$, for this increment we obtain the contribution of this increment to the rate during this period of time. The constant 5.615 is used to correct the volume to barrels.

$$\Delta q = \frac{2\pi r \Delta rh\phi c}{5.615}\left(\frac{\Delta p}{\Delta t}\right) \tag{2.2}$$

In this expression the area of the Δr segment is calculated from the circumference of the segment, $2\pi r$, multiplied by its thickness, Δr. ϕ is the porosity, the fraction of the bulk volume that is pore space. The Δq of Equation 2.2 may be better understood by examining Fig. 2–5 which indicates a plot of the average rate versus radius during the time interval Δt and shows the physical significance of the Δq expression. Rearranging Equation 2.2 we obtain the slope of the q versus r plot.

$$\frac{\Delta q}{\Delta r} = \left(\frac{2\pi rh\phi c}{5.615}\right)\left(\frac{\Delta p}{\Delta t}\right) \tag{2.3}$$

We will obtain the radial diffusivity equation by first deriving an expression for the same slope of a plot of q versus r by differentiation of

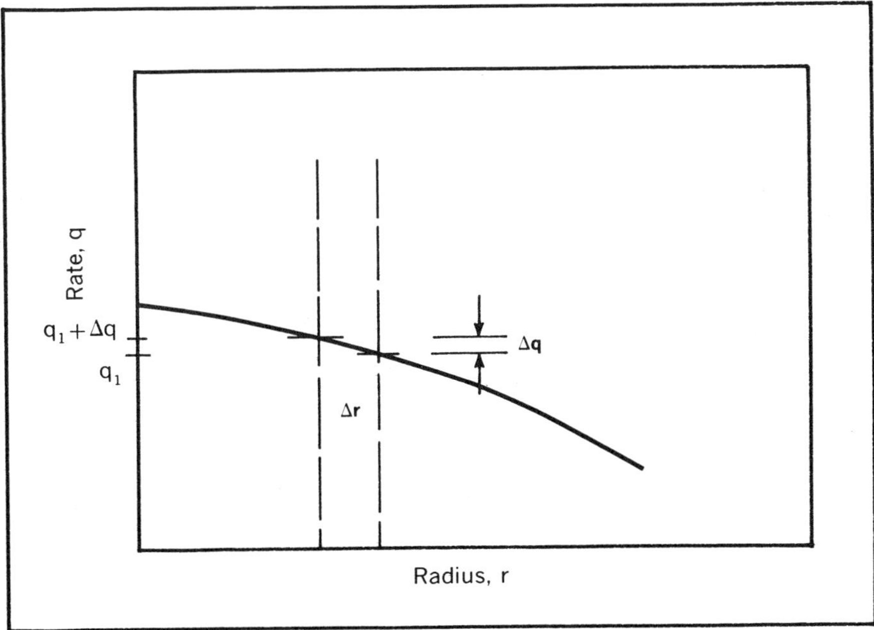

FIG. 2.5 Average rate vs. radius for Fig. 2.4—for derivation of radial diffusivity equation.

the Darcy equation with respect to the radius and then equating the resulting expression to the right-hand side of Equation 2.3. Darcy's equation written with the radial cross sectional area, $2\pi rh$, substituted for A is

$$q = \frac{1.127k(2\pi rh)}{\mu}\left(\frac{\Delta p}{\Delta r}\right) \tag{2.4}$$

Since the radius, r, and $\Delta p/\Delta r$ both vary with the rate, q, the equation for the differential of a product must be used.

$$\frac{\Delta(uv)}{\Delta x} = u\frac{\Delta v}{\Delta x} + v\frac{\Delta u}{\Delta x} \tag{2.5}$$

Letting $u = r$ and $v = (\Delta p/\Delta x)$ we find,

$$\frac{\Delta q}{\Delta r} = \frac{1.127k2\pi h}{\mu}\left[r\frac{\Delta(\Delta p/\Delta r)}{\Delta r} + \frac{\Delta p}{\Delta r}\right] \tag{2.6}$$

Now when the right-hand side of Equation 2.6 is equated to the right-hand side of Equation 2.3 the expression $2\pi h$ cancels. Rearranging the equation we obtain

$$\frac{\Delta p}{\Delta t} = \left(\frac{6.33k}{\phi\mu c}\right)\left[\frac{1}{r}\frac{\Delta p}{\Delta r} + \frac{\Delta(\Delta p/\Delta r)}{\Delta r}\right] \tag{2.7}$$

This equation is identical to Equation 2.1 if we recognize the diffusivity constant as being

$$\eta = \frac{6.33k}{\phi\mu c} \tag{2.8}$$

The constant 6.33 is the product of 1.127 and 5.615.

To most practical engineers the diffusivity equation seems to be something slightly ethereal dreamed up by reservoir engineers to solve their particular problems. However, it is doubtful if there is another equation in petroleum engineering that is used more widely by other types of engineers. The radial diffusivity equation describes almost all forms of unsteady-state mass and energy transfer. When applied to the flow of heat,[4] the pressure in Equation 2.1 becomes the temperature and the diffusivity constant describes the heat-transfer characteristics of the material. For molecular diffusion we substitute concentrations and use molecular activity characteristics in the constant. For electricity we use voltages and electrical characteristics, etc. Diffusivity equations are almost universal in their application.

Engineers who have difficulty understanding the time effect in unsteady state may find it easier to comprehend if they consider the application of the diffusivity equation to some other physical system such as the flow of heat. Consider a cylinder of lead with a small cylindrical hole in its center. Suppose this lead cylinder is heated to some elevated temperature and is then placed in an insulating atmosphere. Now consider the introduction of a coolant into the center hole of the cylinder in such a way that the temperature in the hole is maintained at a constant temperature. Physically it seems easy to believe that there will not be an instantaneous change in the temperature throughout the cylinder and that it will require some period of time for the temperature at the outer radius of the cylinder to be lowered. This is exactly analogous to the compressible flow of liquid in an enclosed circular reservoir with the pressure maintained at a constant value at the well and the pressure at the outer boundary requiring some time before it is affected by the lowering of the pressure at the well.

General solutions. The correct solution for applying the radial diffusivity equation to a particular reservoir is a combination of pressure, radius and time curves that will satisfy the radial diffusivity Equation 2.1. Any one who understands the physical significance of Equation 2.1 could eventually find a set of curves that would satisfy a particular reservoir situation. For many of us this would require a trial-and-error approach to the problem. Fortunately more scientific approaches are available.

It is possible to combine the reservoir constants with the time, pressure, and radius of Equation 2.1 so that the equation is stated in general terms that would apply to any reservoir of a particular size (r_e/r_w).

$$\frac{\Delta p_D}{\Delta t_{Dw}} = \frac{1}{r_D}\frac{\Delta p_D}{\Delta r_D} + \frac{\Delta(\Delta p_D/\Delta r_D)}{\Delta r_D} \tag{2.9}$$

In this equation the reservoir constants and variables are combined as follows:

$$t_{Dw} = \frac{\eta t}{r_w^2} \tag{2.10}$$

$$\Delta p_D = \frac{\Delta p}{(0.141 q\mu/kh)} \tag{2.11}$$

$$r_D = \frac{r}{r_w} \tag{2.12}$$

If the reader will substitute expressions for t, Δp, and r according to Equations 2.10, 2.11, and 2.12 into Equation 2.1, rearrangement will result in Equation 2.9. The grouping of the variables and constants according to Equations 2.10, 2.11, and 2.12 is somewhat arbitrary. However, the radial diffusivity constant and the well radius include all of the reservoir constants that would alter the time effect in a reservoir and Equation 2.11 contains all of the reservoir parameters that will affect the magnitude of the pressure drop at any particular radius and time. The latter point can be best illustrated by examining the radial steady-state flow equation written in terms of the pressure at any radius.

$$q = \frac{7.08kh}{\mu} \frac{(p_r - p_w)}{\ln(r/r_w)} \qquad (1.13A)$$

When solved for the pressure drop we obtain

$$p_r - p_w = \frac{0.141q\mu}{kh}\ln(r/r_w) \qquad (2.13)$$

This shows that the pressure drop in the reservoir is proportional to the group of terms, $0.141q\mu/kh$, which is used to normalize the pressure drop in Equation 2.11.

Based on Equation 2.9 a relationship between p_D, t_{Dw}, and r_D can be obtained that would apply to any reservoir of a particular size, r_e/r_w, that meets the characteristics of the particular solution. These solutions are dimensionless in character since Equations 2.10, 2.11, and 2.12 are dimensionless.

Many engineers are confused by the use of dimensionless groups. If this concept bothers you simply forget it and assume that the group of terms has units. If the general solutions are properly used you will never be aware of the fact that the groups are dimensionless except when you see the word used in describing the theory. Generally this text will prefer the term "reduced to dimensionless" since it seems to be less confusing.

General solutions can then be obtained by using the various groups of reduced terms of Equations 2.10 thru 2.12 rather than the corresponding time, pressure, and radius. Such a general solution can be applied to any reservoir by simply introducing the particular reservoir constants into the various groups of terms. These general solutions are of two types. The constant-terminal-rate solution provides the pressure change throughout a reservoir in which the rate at one extremity (terminal) of the radial reservoir is constant. The constant-terminal-pressure solution provides a knowledge of cumulative flow at any particular time for a reservoir in which the pressure at one of the reservoir extremities is constant.

CONSTANT-TERMINAL-RATE SOLUTION

The constant-terminal-rate solution is by far the most useful of the two general solutions to the radial diffusivity equation. It is an integral part of pressure buildup analysis, interference tests, drawdown analysis, isochronal testing of gas wells and many other important reservoir analysis techniques. This general type of solution can be obtained in two or three different forms by using slightly different assumptions and different methods of mathematical analysis. The various solutions are overlapping and all of them have their own particular use and limitations. It is then very important that we understand the limitations of each. Thus, each solution will be considered separately and then a general consideration of the entire problem will provide a practical means of determining when each solution should be used.

William Hurst is generally given credit for devising the first constant-rate solution in a general solution form.

The Hurst-Van Everdingen p_{tD} solution[3]. This constant-rate solution analyzes a reservoir model that is perfectly radial with the producing well in the exact center. Initially the pressure in the model is p_i and uniform throughout. Then at time zero a constant rate of production is begun from the well. It is then possible to calculate the pressure at the well at any time by using the p_{tD} functions of Fig. 2.6 or appropriate extension equations.

The examination of Fig. 2.6 will indicate that the p_{tD} values are plotted versus t_{Dw}, the reduced time based on the well radius, with a different curve representing each reservoir size. The reservoir size is represented in dimensionless form as the ratio of the external radius of the reservoir and the well radius. Do not be confused by the fact that the reservoir sizes shown in Fig. 2.6 are so small as to be absurd for a practical situation. Hurst was applying his data principally to the analysis of an aquifer that was supplying the water drive for an oil reservoir. Thus his "well" was the external boundary of the oil reservoir and dimensionless reservoir sizes of less than 10 were practical for his application.

Note that Fig. 2.6 contains a curve labeled with a dimensionless reservoir size of infinity. This has no practical physical meaning but it has considerable mathematical significance because it represents the reservoir prior to the time when the pressure at the outer boundary has been affected. Prior to the time when the pressure at the outer boundary is affected it makes no difference where the outer boundary is located. During this time, mathematically the radius is infinite. Reservoirs of different sizes will all behave exactly the same way (infinite acting) until the outer boundary of the smallest reservoir is affected. Consequently, note from Fig. 2.6 that the infinite reservoir curve forms the limiting curve for all of the other curves. This is as it should be since they all act the same until their outer boundaries are affected. Note further that the larger the r_e/r_w ratio of a particular curve, the longer it will follow the infinite curve.

The p_{tD} functions are used by first calculating the reduced time (based on the well radius) at which the pressure is desired using the appropriate real time in days in the expression and the well radius. The dimensionless reservoir size, r_e/r_w, is then calculated to determine which curve of Fig. 2.6 should be used. Using the calculated t_{Dw} and the dimensionless reservoir size, the corresponding pressure function, p_{tD}, can be obtained from Fig. 2.6. This value is put into equation 2.14, that

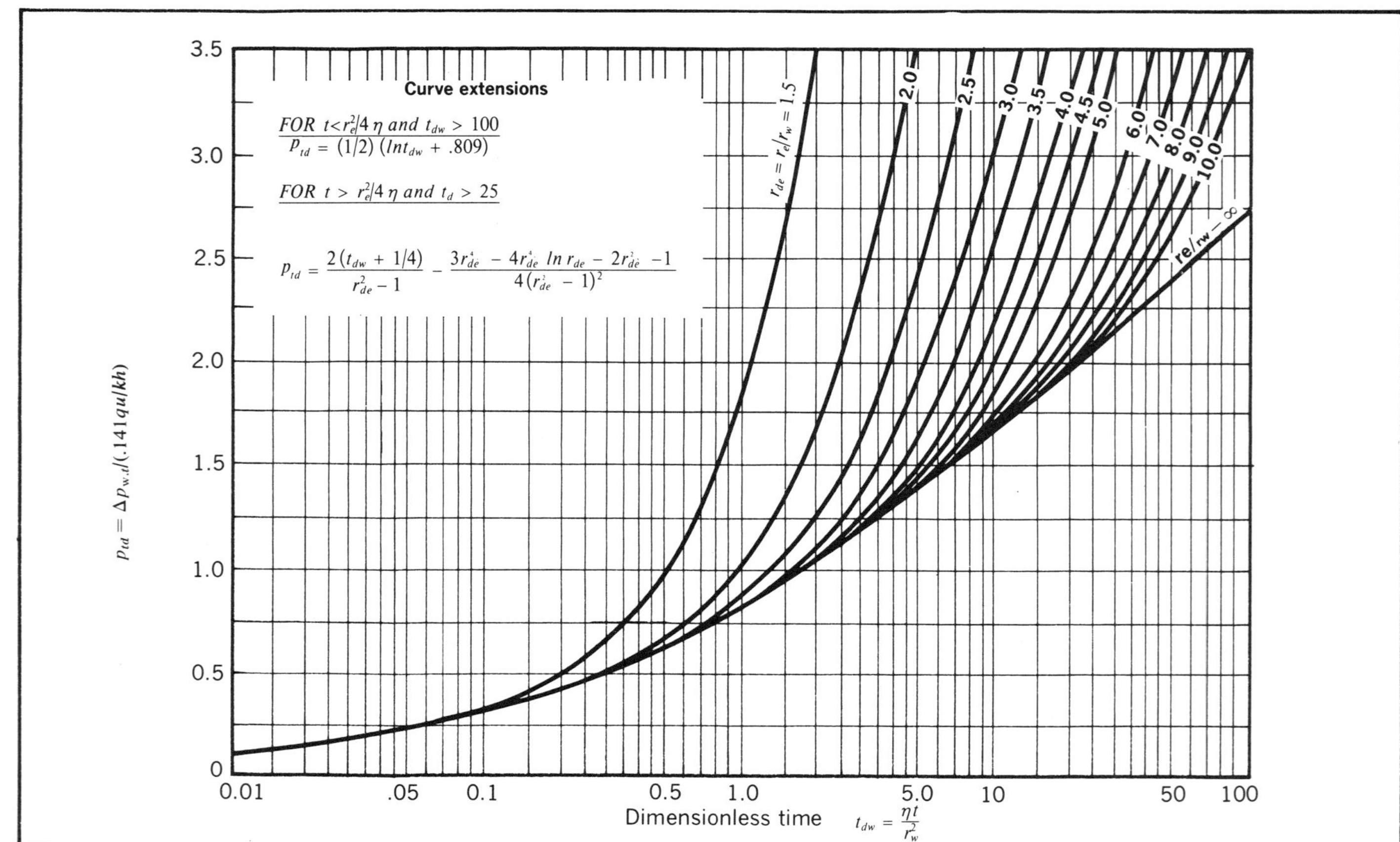

FIG. 2.6 Van Everdingen & Hurst PtD values.

defines the pressure function, to permit the calculation of the pressure drop and the corresponding pressure. The pressure function, PF, is defined similarly to that of Δp_D in Equation 2.11 except that the pressure drop, Δp, represents the drop in pressure from the initial pressure.

$$PF = \frac{p_i - p_{r,t}}{(0.141 q\mu/kh)} \qquad (2.14)$$

When this equation is applied specifically to the p_{tD} pressure function and is solved for the desired pressure, we obtain

$$p_{w,t} = p_i - \frac{0.141 q\mu}{kh} p_{tD} \qquad (2.15)$$

The practical significance of the Hurst-Van Everdingen solution can probably best be emphasized by a theoretical problem. In working the following problem, first calculate the reduced time based on the time and radius at which the pressure is desired. Remember the time unit of the equation is days. Then find p_{tD} from Fig. 2.6. In the problem you can assume the reservoir is infinite acting after only 6 min of production. The p_{tD} value obtained is then used as in Equation 2.15 to obtain $p_{w,t}$.

Problem 2.1A: Application of the p_{tD} Function

The discovery well in a reservoir is put on drill-stem test and flows at a rate of 18 STB/d.

What will be the well pressure after 6 min of production if the reservoir characteristics are as follows:

$k = 0.1$ md	$c = 6.33 \times 10^{-6}$/psi
$\phi = 0.0695$	$\mu_o = 0.4$ cp
$r_w = 0.5$ ft	$h = 141$ ft
$p_i = 3,000$ psia	
$r_e = 3,000$ ft	
$B_o = 1.39$	
$P_s = 1,000$ psia	
(saturation pressure)	
$\Delta p_{skin} = 0$	

Problem 2.1A is not a practical problem. It is simply designed to illustrate the physical significance of the p_{tD} function. Also the characteristics of the reservoir were chosen to demonstrate a variety of ideas in subsequent parts of the problem. However, the engineer should recognize that the problem could be turned around slightly to give a practical problem. Normally in a drill-stem test you will know the flow rate and the well pressure at a particular time. This would then permit the calculation of one or a combination of the reservoir properties such as the permeability or the permeability-thickness product.

In applying the Hurst-Van Everdingen p_{tD} functions the engineer should carefully notice that pressure can be calculated only at the radius where the flow rate is known. Normally this then restricts the application to the calculation of well pressures. You may occasionally be able to make an application to the calculation of pressure at the outer boundary of an oil reservoir by applying the solution to the surrounding aquifer but such applications are unusual. The important point is that mathematically the p_{tD} function can be applied to calculate the pressure only at the point in the reservoir where the rate is constant. We can show this by writing Equation 2.15 and subscripting the rate with the same r that is used to subscript the pressure.

$$p_{r,t} = p_i - \frac{0.141 q_r \mu}{kh} p_{tD} \qquad (2.16)$$

This is the same r used in the reduced time, t_D.

If we desire to calculate the pressure at some point in the reservoir where the rate is not known (and is not constant) we must use a different mathematical solution.

The Ei function solution.[5]   The Ei function solution can be used for the purpose of calculating pressure at any point in the reservoir using the flow rate at the well. This solution takes the same form as the p_{tD} solution but it is based on a different set of assumptions and a different mathematical approach.

It has been shown that for an infinite-acting reservoir and $\eta t / r_w^2$ greater than 100

$$PF = (1/2)\left(-Ei\frac{-1}{4t_D}\right) \qquad (2.17)$$

where t_D is calculated using the radius at which the pressure is desired. In working with the p_{tD} function we defined t_D as being based on the well radius. In that case, that was the only radius at which the pressure could be calculated because the p_{tD} application is restricted to the radius where the rate is known. However, since the Ei function solution lets us calculate the pressure at any radius, we must be careful to always base t_D on the radius at which the pressure is desired. The Ei pressure function can then be used with the rate at the well to determine the pressure at any point in the reservoir. Equation 2.14 can then be modified by substituting the right-hand side of Equation 2.17 for (PF) to obtain

$$p_{r,t} = p_i - \frac{0.141 q_w \mu}{kh}\left[(1/2)\left(-Ei\frac{-1}{4t_D}\right)\right] \qquad (2.18)$$

Note the two very important conditions that must be met before Equations 2.17 and 2.18 can be used. The reservoir must be infinite acting and $\eta t / r_w^2$ must be greater than 100. These limitations are the result of assumptions that must be made to mathematically obtain the Ei function solution. The first limit, an infinite-acting reservoir, is self explanatory. However, a surprising number of practicing engineers and researchers either are unaware of this limit or prefer to ignore it since no mention is made of the limitation when various applications of the Ei function solution are used or proposed.

The other limit on the Ei function solution is not nearly as often a practical limiting factor but it is also generally ignored. The limit that $\eta t/r_w^2$ must exceed 100 is an arbitrary limit that may be shortened or extended timewise depending upon the accuracy desired in a particular situation. The limit arises because it is necessary to assume production is from a point rather than a finite radius, r_w. Consequently, a certain amount of time must transpire before the effect of the inaccuracy becomes negligible. For all practical purposes this occurs when the rate at the well is equal to the theoretical rate at a radius of zero. In Chapter 1 we examined the pressure-versus-radius distribution at various times and Figs. 1.8 and 1.9 show that as producing time gets larger the slope of the rate-versus-radius curve at the well bore becomes almost flat.

Fig. 2.7 illustrates this point by comparing the rate at some theoretical zero radius with the rate at the well radius. In this particular example we might assume that $\eta t/r_w^2$ becomes greater than 100 at a time of about t_2. We will later obtain a better quantitative feel for this time limit on the Ei function solution.

The Ei function is a math function known as the exponential integral. Like other functions that occur quite often in mathematical

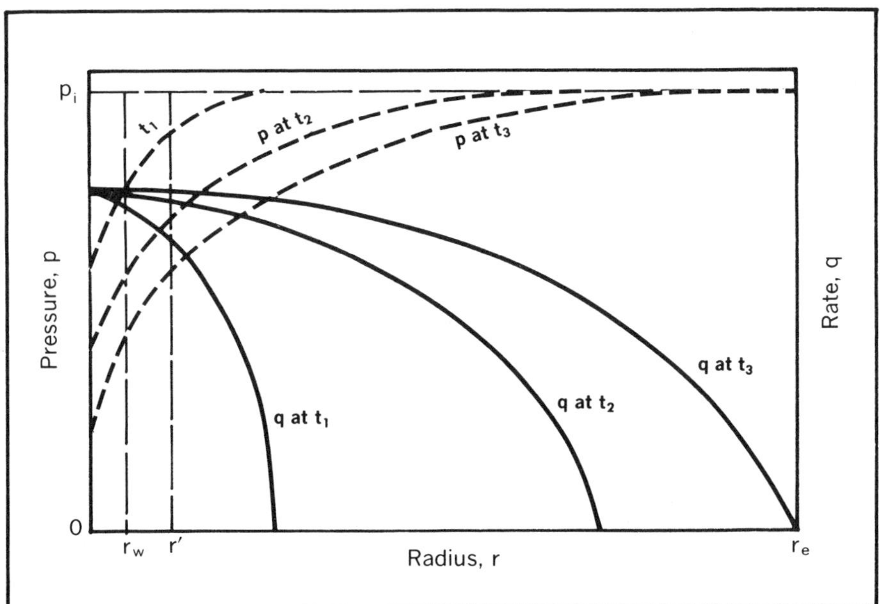

FIG. 2.7 Comparison of rates near the well bore during unsteady state flow with a constant rate at the well.

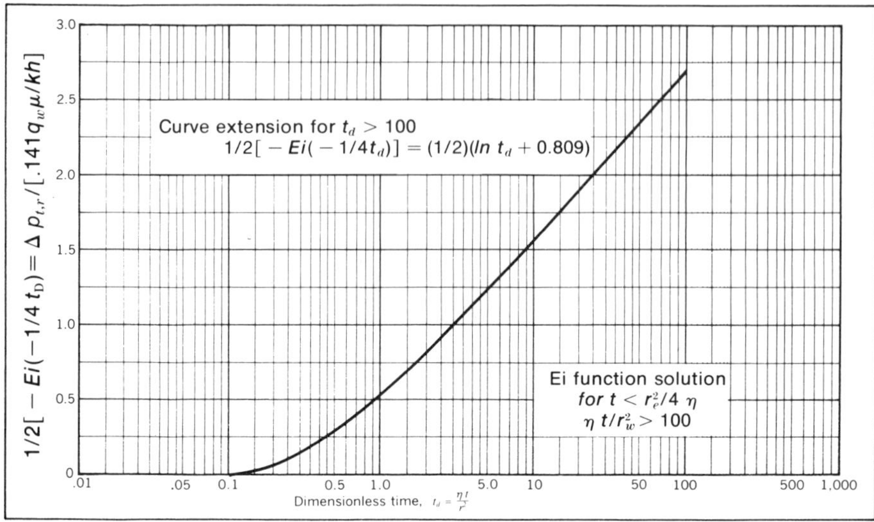

FIG. 2.8 Ei function solution.

analysis, values for different arguments (in this case $-1/4t_D$) are included in some of the more extensive math tables. Most of the better-known reservoir engineering texts contain a graph or table of the exponential integral. However, in this text, the entire Ei function, $(1/2)\left(-Ei\dfrac{-1}{4t_D}\right)$, is plotted versus the reduced time, t_D (Fig. 2.8). These data are often misinterpreted by the engineer. The Ei function includes the $1/2$ value; and the time scale is for t_D, *not* $4t_D$.

When applying the Ei function solution the first step in the calculation is to determine if the function limitations are met—infinite-acting reservoir and $\eta t/r_w^2$ greater than 100. Mistakes are often made in applying Ei functions when calculating the group of terms $\eta t/r_w^2$. Many engineers try to use the radius at which the pressure is being calculated rather than r_w in this group of terms. The fallacy of doing this is apparent when the reason for the test is considered. We are simply comparing conditions at the zero radius with conditions or times based on the well radius.

Because of the necessary assumption that production is from a zero radius or a point, the Ei function solution is often referred to as the point-source solution. Thus, the test to see if $\eta t/r_w^2$ is greater than 100 may be thought of or referred to as the test to see if the point-source assumption is valid. Be careful that you do not confuse this test with the calculation of t_D in which the radius is the radius at which the pressure is desired. Thus, $\eta t/r_w^2$ may or may not be the same as t_D, the basis for the Ei function. A problem may help to clarify the limitations and physical meaning of the Ei function solution. Try to calculate all of the answers before checking your solution in Appendix C.

Problem 2.1B: Application of the Ei Function

In example 2.1A a DST flow period at a rate of 18 STB/d was described and the pressure at the well after 6 min of production was calculated. What would be the pressure at a radius of 5 ft at this time? What will be the pressure at a radius of 5 ft after 1 hr of production? What will be the pressure at 50 ft after 1 hr of production?

In Problem 2.1B we saw that the application of the Ei function solution clearly shows that for all practical purposes the pressure is unaffected for a period of time at any radius removed from the well radius. However, it should be carefully noted by the engineer that there is some finite pressure drop almost immediately at all parts of the reservoir. This is easily seen when we examine the nature of the exponential integral.

We find that for any finite value of $(-1/4t_D)$ there is a finite value for $(1/2)\left[-\text{Ei}\dfrac{-1}{4t_D}\right]$. Consequently, we recognize there is a pressure drop at a particular radius but that there is no *significant* pressure drop at a particular radius at a particular time. The magnitude of "significant" seems to be a function of the measuring device in use. In a technique known as pulse testing the measuring device used will detect changes of about 0.001 psi. Consequently, a significant pressure drop is much smaller than it is for other purposes.

Since background changes in reservoir pressure caused by tides, noise, and other unexplained anomalies are greater than the disturbances we are trying to analyze, the use of pulse tests may be promising but it is still in the experimental or at best the development stage.

Extension equations for the Ei function and p_{td} function. Note that pressure functions provided in Figs. 2.6 and 2.8 only extend to a t_D of 100. Values for the functions for reduced times greater than 100 can be directly calculated by equation. These extension equations are listed on the illustrations. In the case of the p_{tD} functions note that two extension equations are necessary to cover the functions for the reservoirs when they are finite acting (the pressure at the outer boundary has been affected) and when they are infinite acting. In Fig. 2.6 it is indicated that a reservoir becomes finite acting when the producing time in days is greater than $r_e^2/4\eta$. We will show later that this is an acceptable approximation for the time when a reservoir can be considered finite acting.[6]

The limits set on this p_{tD} equation are again simply arbitrary limits found to give acceptable answers based on the average engineering opinion. Consequently, for $t > r_e^2/4\eta$ and $t_{Dw} > 25$ we can calculate the p_{tD} function from the equation

$$p_{tD} = \frac{2\left(t_D + \dfrac{1}{4}\right)}{r_{De}^2 - 1} - \frac{3r_{De}^4 - 4r_{De}^4 \ln r_{De} - 2r_{De}^2 - 1}{4(r_{De}^2 - 1)^2} \tag{2.19}$$

This equation is actually the first two terms of the general p_{tD} equation that defines all of the p_{tD} functions.[3] The third and last term is written to include some Bessel functions. Fortunately when t_D becomes large enough relative to the reservoir size r_{De}, which is r_e/r_w, the Bessel function term approaches zero. The accuracy of Equation 2.19 can be seen by examining Fig. 2.9 where the curves are based on the complete p_{tD} equation (not included in this text) and the large dots represent points calculated using Equation 2.19 for r_D of 10.

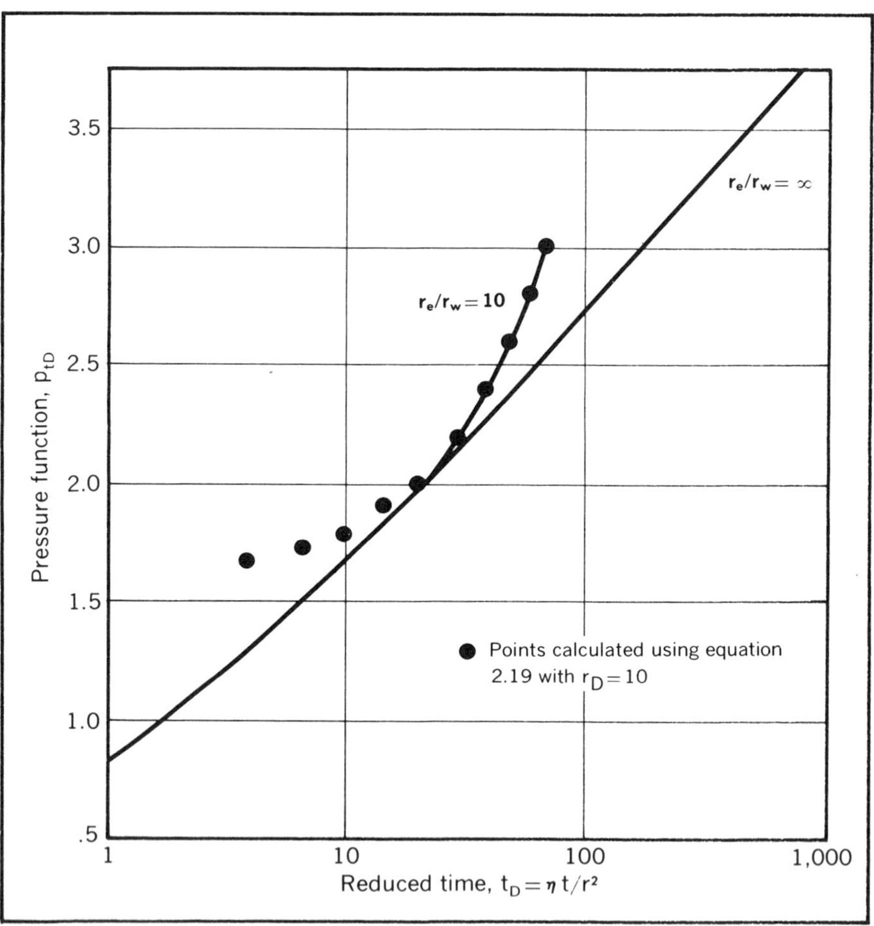

FIG. 2.9 Comparison of p_{tD} values with those calculated by Equation 2.19.

It is also useful to recognize that the significant portion of Equation 2.19 can be reduced to

$$p_{tD} = \frac{2t_D}{r_{De}^2} + \ln r_{De} - \frac{3}{4} \qquad (2.20)$$

This equation is determined by recognizing that many of the numbers in Equation 2.19 are insignificant compared with other numbers in the equation for a practical case. For example, when $t_D > 100$ and $r_{De} > 10$ note that the $\frac{1}{4}$ in the numerator of the first term of Equation 2.20 is insignificant when compared to t_D of 100. Also that the 1.0 in the denominator of the first term is insignificant when compared with the r_{De}^2 of 100, and r_{De}^2 values in the second term are insignificant when compared with the r_{De}^4 values of this term. When all such comparisons are taken into account Equation 2.20 results.

When the reservoir is infinite acting and the reduced time is greater than 100, Fig. 2.6 gives an extension equation of

$$p_{tD} = \frac{1}{2}(\ln t_D + 0.809) \qquad (2.21)$$

If we again use our as yet unproven critical time $t = r_e^2/4\eta$ as the time when a reservoir becomes finite acting, we can then use Equation 2.21 to calculate the p_{tD} function whenever the time is less than this critical time and the reduced time is greater than 100. This equation is used by practicing engineers and researchers almost completely without regard for the fact that it applies only to infinite-acting reservoirs — an excuse for disregarding the limit by those who come up with new means of misapplying the expression.

Now consider the extension equation listed on Fig. 2.8. This represents the pressure function for an infinite acting reservoir when the reduced time, t_D, is greater than 100. The extension equation is

$$\frac{1}{2}\left(-\text{Ei}\frac{-1}{4t_D}\right) = \frac{1}{2}(\ln t_D + 0.809) \qquad (2.22)$$

When we compare Equations 2.21 and 2.22 we find a rather unexpected situation. Both of these equations apply to an infinite-acting reservoir when the reduced time is greater than 100. However, Equation 2.21 can be used to obtain a pressure function that is usable only at the point in the reservoir where the rate is known, essentially the well bore, while Equation 2.22 evaluates a pressure function that can be used to calculate the reservoir pressure at any point in the reservoir. Although these expressions represent the same reservoir times and different purposes, the expressions are unexpectedly the same.

Possibly this unusual situation can be better appreciated by examining Fig. 2.10 which is a plot of the p_{tD}, Ei function, and the natural log expression represented by the right-hand side of Equations 2.22 and

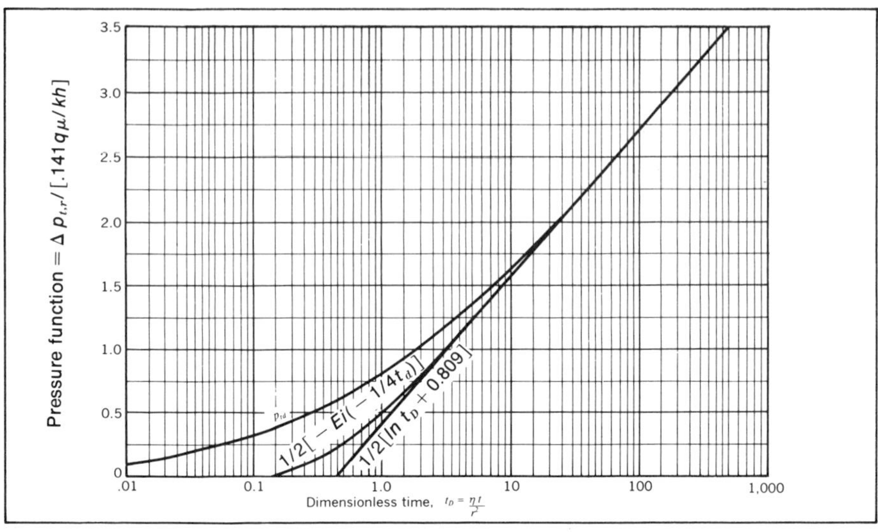

FIG. 2.10   Comparison of pressure functions for infinite reservoir.

2.23. This plot shows that all three of the functions become indistinguishable, one from the other, at some reduced time between 25 and 100. The reason that the p_{tD} function and Ei function become the same for large values of t_D, may be found by studying Fig. 2.7, which was used to explain the lower time limit on the point-source solution, and Equations 2.16 and 2.18 that are concerned with the application of the individual functions to calculate the pressure at some particular point in the reservoir. Note from Fig. 2.7 that when the producing time is large enough the rate at r_w and at any larger radius, r', will become essentially the same just as the theoretical rate at $r = 0$ and at r_w become the same to permit the use of the point-source solution.

Now again examine Equations 2.16 and 2.18 and note that the only difference in the equations, except for the difference in pressure functions, is that one of the equations uses the rate at the well and the other uses the rate at the radius where the pressure is being calculated. Consequently, if the rate at the well has been equal to the rate at the radius of interest for a long enough period of time, the two equations should give the same pressure. Furthermore, if they are to result in the same calculated pressure, the two pressure functions must be equal. Fig. 2.10 and Equations 2.21 and 2.22 show that the two functions will be equal when the reduced time, t_D, is greater than 100 or thereabouts.

Choosing the best pressure function. There are actually only two pressure-function solutions for the radial diffusivity equation. However, since both of the basic solutions can be approximated by the same log equation when the reservoir is infinite acting and the reduced time is greater than 100, it is common to think in terms of there being three solutions, the p_{tD}, Ei, and natural log solutions. The simplest of these to use is of course the log solution since it does not require any graphs and the expression can be mathematically manipulated to obtain other expressions.

Since all three of the pressure-function expressions have different limitations it is often confusing to determine which of the three should be used. The p_{tD} function can be used for infinite and finite acting reservoirs to find the pressure at the well or wherever the rate in the reservoir is known, the Ei solution can be used to find the pressure at any point in the reservoir but only if the reservoir is infinite acting and $\eta t/r_w^2$ is greater than 100, and the log function can be used if the reservoir is infinite acting and t_D is greater than 100. Since it is so easy to choose the wrong solution or one that is more difficult to evaluate, a "decision" flow sheet, Table 2.1, is presented to aid the engineer in obtaining the correct and easiest solution. "Is t_D greater than 100?", is a limit on the log function; "Is q_r known?", is the test for applicability of the p_{tD} function to determine if the rate is known at the radius where you desire the pressure; "Is t greater than $r_e^2/4\eta$?", is the test to determine if the reservoir is infinite or finite acting; and "is $\eta t/r_w^2$ greater than 100," is one of the limitations on the Ei solution. You will note that some reference is made to the pseudosteady-state equations. This matter will be discussed later in the chapter.

The use of Table 2.1 can be best illustrated by simple problems.

Problem 2.1C: Choice of A Pressure Function Solution

Using the reservoir data and production rate of Problems 2.1A and 2.1B find the well pressure after 100 minutes of production. What will be the well pressure after 1,000 days of production?

Occasionally an engineer may need to use pressure functions for some geometry other than radial. The original Hurst-Van Everdingen paper[3] contains a derivation of the linear diffusivity equation and pressure functions for the linear system. However, the linear system is so seldom of use in reservoir engineering that they have not been included here.

The engineer should remember that any of the radial flow equations or solutions can be used for pie-shaped radial systems by simply adjusting the rate to that of an equivalent full radial system. This is done by dividing the actual rate by the fraction of a circle through which flow is taking place. This technique was illustrated in Chapter 1 for a radial steady-state system.

The constant-terminal-rate flow of gas will be discussed in the Chapter on Gas Reservoir Engineering. The same pressure function solutions are used but they are used to predict the change in the pressure squared,

Table 2.1

CHOICE OF UNSTEADY-STATE PRESSURE FUNCTIONS

"Use p_{tD}" refers to functions of Fig. 2.6
"Use Ei" refers to functions of Fig. 2.8
"Use ln" refers to $\frac{1}{2}(\ln t_D + 0.809)$.

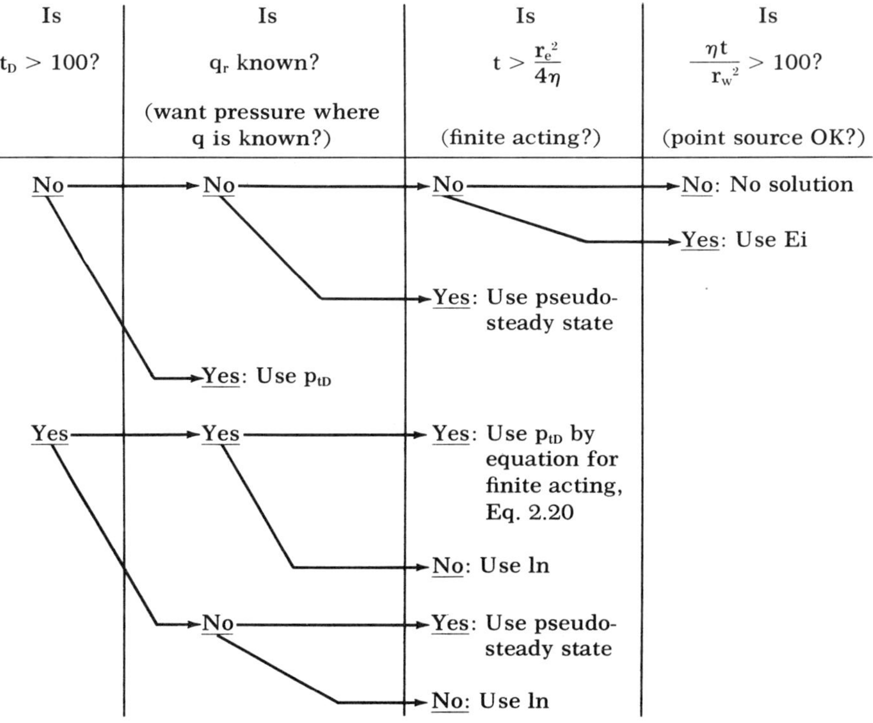

$\Delta(p^2)$, rather than the change in pressure as we have been doing with liquids.

CONSTANT-TERMINAL-PRESSURE SOLUTION[3]

In the constant-rate solution the rate was known to be constant at some part of the reservoir and the pressures were calculated throughout the reservoir. Conversely, in the constant-terminal-pressure solution the pressure is known to be constant at some point in the reservoir and the cumulative flow at any particular time across the subject radius can be evaluated. The constant-pressure solution is not as confusing as the constant-rate solution simply because we do not know as much about the constant-pressure solution. Only one constant-pressure solution is available so we are not faced with the decision as to which solution to use.

The behavior of a radial reservoir during constant-pressure operation was described in Chapter 1, Fig. 1.8. The constant-pressure solution is used most widely for calculating the water encroachment into the original oil zone or gas zone from the water drive of a reservoir.[7] The solution is also useful for predicting the disposal rate of water injected into an aquifer. The solution is not used as widely as is probably justified. It appears to be an excellent means of predicting the recovery rates from gas reservoirs prior to the time when the gas wells reach pseudosteady state.

The constant-pressure solution takes the form of a relationship between the reduced time and the reduced cumulative flow, Q.

$$Q_{tD} = \frac{Q}{1.12\phi hcr^2 \Delta p} \tag{2.23}$$

Equation 2.23 can be obtained by first solving Equation 2.11 for the rate and Equation 2.10 for the time t.

$$q = \frac{\Delta p}{0.141(\Delta p_D)\mu/kh} \tag{2.24}$$

$$t = t_D\, r^2/\eta \tag{2.25}$$

Now recognizing that for a constant Δp the rate, q, would vary, we could define the cumulative flow at any time as

$$Q = \sum_0^t q\Delta t \tag{2.26}$$

Substituting into this equation according to Equations 2.24 and 2.25 with η written as $6.33k/\phi\mu c$, and simplifying, we obtain

$$Q = 1.12\phi hcr^2\Delta p \sum_{t_D=0}^{t_D=t_D} \frac{1}{\Delta p_D}\Delta t_D \tag{2.27}$$

Thus, comparing Equations 2.23 and 2.27 we see that

$$Q_{tD} = \sum_{t_D=0}^{t_D=t_D} \frac{1}{\Delta p_D}\Delta t_D. \tag{2.28}$$

Remember that this equation must be applied by superposition which is discussed later. Consequently, the relationship is not as simple as it may appear to be at first glance.

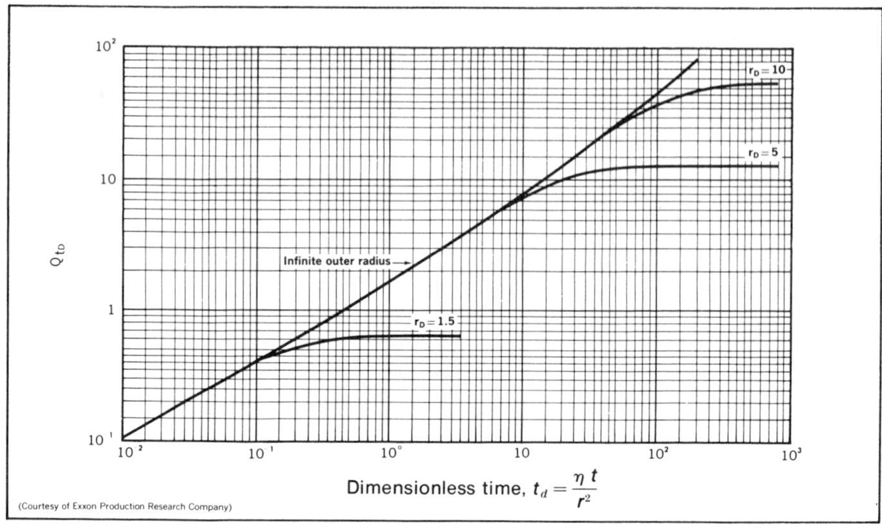

FIG. 2.11A   Constant pressure functions. (Courtesy Esso Production Research Co.)

Some resulting data are plotted in Figs. 211A, B, C, and D. Unfortunately the mathematical relationship between t_D and Q_{tD} is such that no simple relationship exists between the two over any significant range of reduced-time values as was the case with the pressure functions. This makes it necessary to either prepare a great number of graphs or put the data in tabular form as in Tables 2.2 and 2.3.

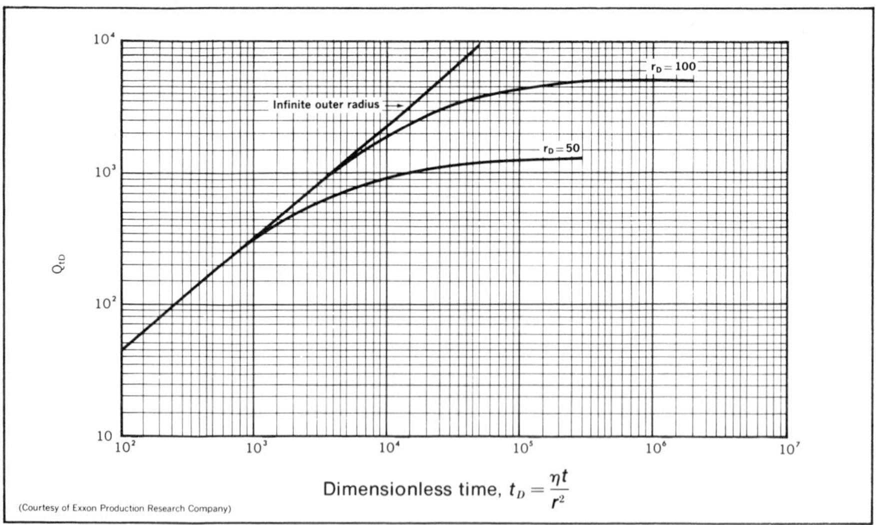

FIG. 2.11B   Constant pressure functions. (Courtesy Esso Production Research Co.)

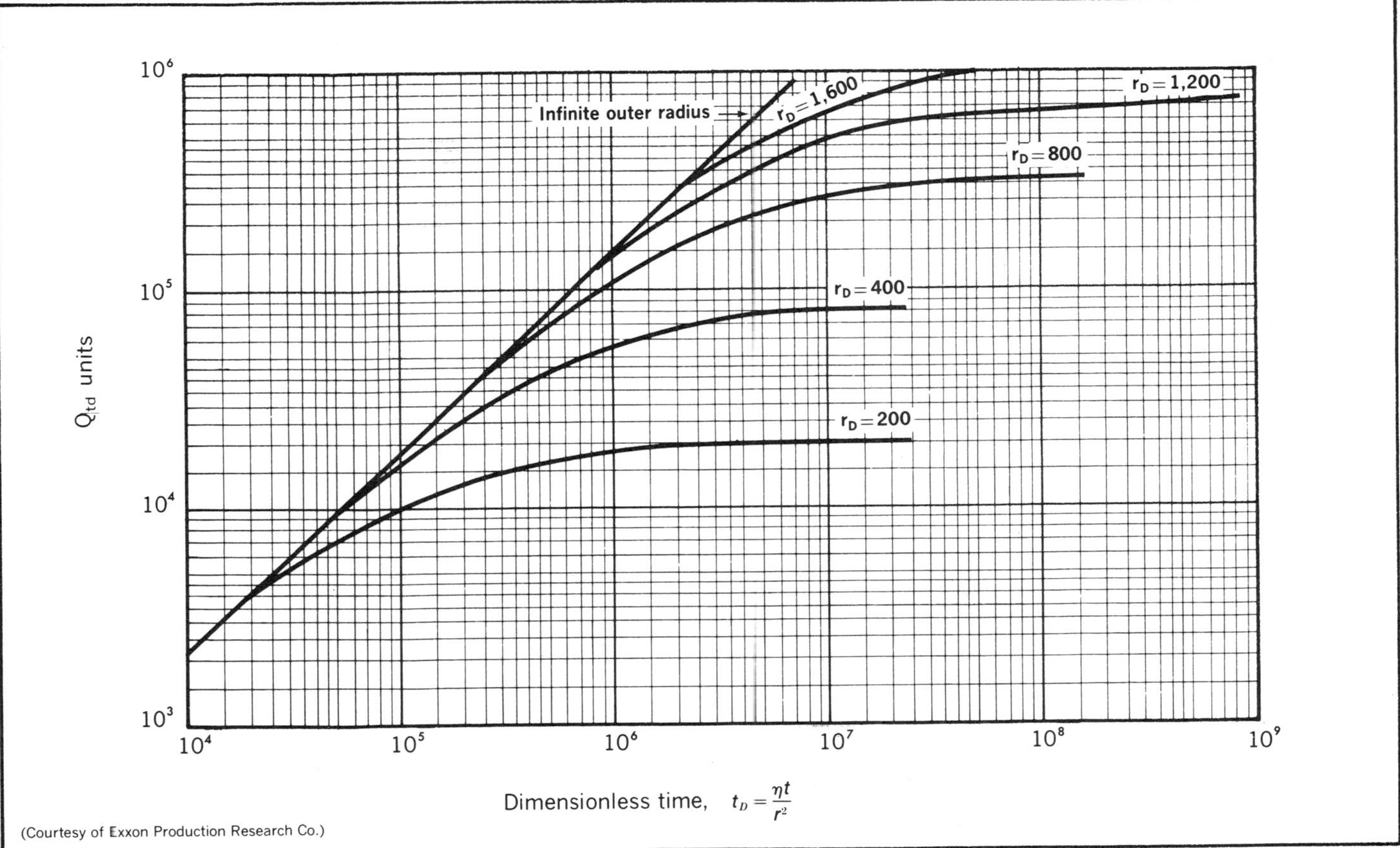

FIG. 2.11C Constant pressure functions. (Courtesy Esso Production Research Co.)

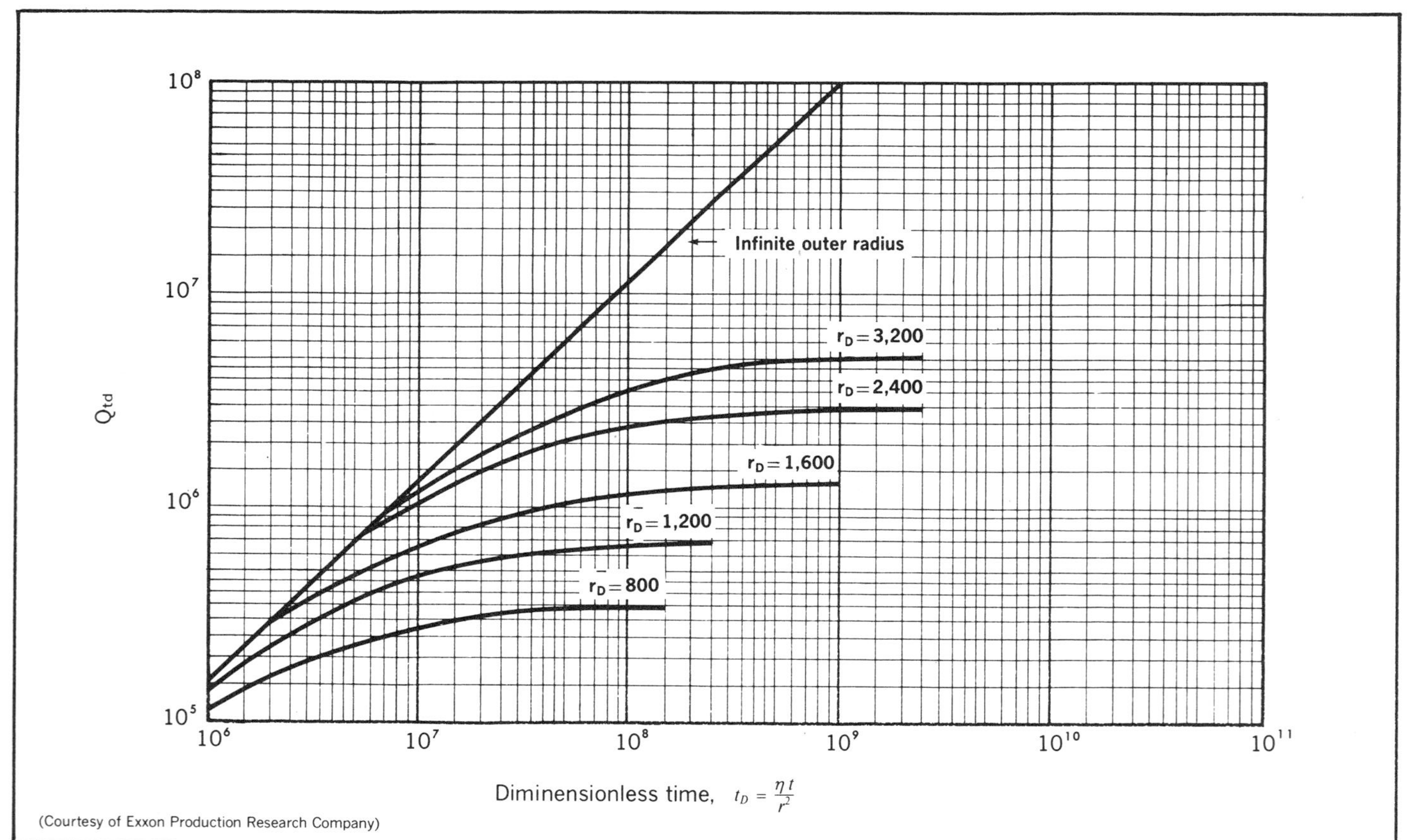

FIG. 2.11D Constant pressure functions. (Courtesy Esso Production Research Co.)

Table 2.2

HURST-VAN EVERDINGEN CONSTANT-PRESSURE Q_{tD} FUNCTIONS FOR
INFINITE-ACTING RADIAL RESERVOIRS

t_D	Q_{tD}	t_D	Q_{tD}	t_D	Q_{tD}	t_D	Q_{tD}
0.00	0.000	41	21.298	96	41.735	355	121.966
0.01	0.112	42	21.701	97	42.084	360	123.403
0.05	0.278	43	22.101	98	42.433	365	124.838
0.10	0.404	44	22.500	99	42.781	370	126.270
0.15	0.520	45	22.897	100	43.129	375	127.699
0.20	0.606	46	23.291	105	44.858	380	129.126
0.25	0.689	47	23.684	110	46.574	385	130.550
0.30	0.758	48	24.076	115	48.277	390	131.972
0.40	0.898	49	24.466	120	49.968	395	133.391
0.50	1.020	50	24.855	125	51.648	400	134.808
0.60	1.140	51	25.244	130	53.317	405	136.223
0.70	1.251	52	25.633	135	54.976	410	137.635
0.80	1.359	53	26.020	140	56.625	415	139.045
0.90	1.469	54	26.406	145	58.265	420	140.453
		55	26.791	150	59.895	425	141.859
1	1.569	56	27.174	155	61.517	430	143.262
2	2.447	57	27.555	160	63.131	435	144.664
3	3.202	58	27.935	165	64.737	440	146.064
4	3.893	59	28.314	170	66.336	445	147.461
5	4.539	60	28.691	175	67.928	450	148.856
6	5.153	61	29.068	180	69.512	455	150.249
7	5.743	62	29.443	185	71.090	460	151.640
8	6.314	63	29.818	190	72.661	465	153.029
9	6.869	64	30.192	195	74.226	470	154.416
10	7.411	65	30.565	200	75.785	475	155.801
11	7.940	66	30.937	205	77.338	480	157.184
12	8.457	67	31.308	210	78.886	485	158.565
13	8.964	68	31.679	215	80.428	490	159.945
14	9.461	69	32.048	220	81.965	495	161.322
15	9.949	70	32.417	225	83.497	500	162.698
16	10.434	71	32.785	230	85.023	510	165.444
17	10.913	72	33.151	235	86.545	520	168.183
18	11.386	73	33.517	240	88.062	525	169.549
19	11.855	74	33.883	245	89.575	530	170.914
20	12.319	75	34.247	250	91.084	540	173.639
21	12.778	76	34.611	255	92.589	550	176.357
22	13.233	77	34.974	260	94.090	560	179.069
23	13.684	78	35.336	265	95.588	570	181.774
24	14.131	79	35.697	270	97.081	575	183.124
25	14.573	80	36.058	275	98.571	580	184.473
26	15.013	81	36.418	280	100.057	590	187.166
27	15.450	82	36.777	285	101.540	600	189.852
28	15.883	83	37.136	290	103.019	610	192.533
29	16.313	84	37.494	295	104.495	620	195.208
30	16.742	85	37.851	300	105.968	625	196.544
31	17.167	86	38.207	305	107.437	630	197.878
32	17.590	87	38.563	310	108.904	640	200.542
33	18.011	88	38.919	315	110.367	650	203.201
34	18.429	89	39.272	320	111.827	660	205.854
35	18.845	90	39.626	325	113.284	670	208.502
36	19.259	91	39.979	330	114.738	675	209.825
37	19.671	92	40.331	335	116.189	680	211.145
38	20.080	93	40.684	340	117.638	690	213.784
39	20.488	94	41.034	345	119.083	700	216.417
40	20.894	95	41.385	350	120.526	710	219.046

t_D	Q_{tD}	t_D	Q_{tD}	t_D	Q_{tD}	t_D	Q_{tD}
720	221.670	1,175	337.142	1,900	510.861	4,050	990.108
725	222.980	1,180	338.376	1,925	516.695	4,100	1,000.858
730	224.289	1,190	340.843	1,950	522.520	4,150	1,011.595
740	226.904	1,200	343.308	1,975	528.337	4,200	1,022.318
750	229.514	1,210	345.770	2,000	534.145	4,250	1,033.028
760	232.120	1,220	348.230	2,025	539.945	4,300	1,043.724
770	234.721	1,225	349.460	2,050	545.737	4,350	1,054.409
775	236.020	1,230	350.688	2,075	551.522	4,400	1,065.082
780	237.318	1,240	353.144	2,100	557.299	4,450	1,075.743
790	239.912	1,250	355.597	2,125	563.068	4,500	1,086.390
800	242.501	1,260	358.048	2,150	568.830	4,550	1,097.024
810	245.086	1,270	360.496	2,175	574.585	4,600	1,107.646
820	247.668	1,275	361.720	2,200	580.332	4,650	1,118.257
825	248.957	1,280	362.942	2,225	586.072	4,700	1,128.854
830	250.245	1,290	365.386	2,250	591.806	4,750	1,139.439
840	252.819	1,300	367.828	2,275	597.532	4,800	1,150.012
850	255.388	1,310	370.267	2,300	603.252	4,850	1,160.574
860	257.953	1,320	372.704	2,325	608.965	4,900	1,171.125
870	260.515	1,325	373.922	2,350	614.672	4,950	1,181.666
875	261.795	1,330	375.139	2,375	620.372	5,000	1,192.198
880	263.073	1,340	377.572	2,400	626.066	5,100	1,213.222
890	265.629	1,350	380.003	2,425	631.755	5,200	1,234.203
900	268.181	1,360	382.432	2,450	637.437	5,300	1,255.141
910	270.729	1,370	384.859	2,475	643.113	5,400	1,276.037
920	273.274	1,375	386.070	2,500	648.781	5,500	1,296.893
925	274.545	1,380	387.283	2,550	660.093	5,600	1,317.709
930	275.815	1,390	389.705	2,600	671.379	5,700	1,338.486
940	278.353	1,400	392.125	2,650	682.640	5,800	1,359.225
950	280.888	1,410	394.543	2,700	693.877	5,900	1,379.927
960	283.420	1,420	396.959	2,750	705.090	6,000	1,400.593
970	285.948	1,425	398.167	2,800	716.280	6,100	1,421.224
975	287.211	1,430	399.373	2,850	727.449	6,200	1,441.820
980	288.473	1,440	401.786	2,900	738.598	6,300	1,462.383
990	290.995	1,450	404.197	2,950	749.725	6,400	1,482.912
1,000	293.514	1,460	406.606	3,000	760.833	6,500	1,503.408
1,010	296.030	1,470	409.013	3,050	771.922	6,600	1,523.872
1,020	298.543	1,475	410.214	3,100	782.992	6,700	1,544.305
1,025	299.799	1,480	411.418	3,150	794.042	6,800	1,564.706
1,030	301.053	1,490	413.820	3,200	805.075	6,900	1,585.077
1,040	303.560	1,500	416.220	3,250	816.090	7,000	1,605.418
1,050	306.065	1,525	422.214	3,300	827.088	7,100	1,625.729
1,060	308.567	1,550	428.196	3,350	838.067	7,200	1,646.011
1,070	311.066	1,575	434.168	3,400	849.028	7,300	1,666.265
1,075	312.314	1,600	440.128	3,450	859.974	7,400	1,686.490
1,080	313.562	1,625	446.077	3,500	870.903	7,500	1,706.688
1,090	316.055	1,650	452.016	3,550	881.816	7,600	1,726.859
1,100	318.545	1,675	457.945	3,600	892.712	7,700	1,747.002
1,110	321.032	1,700	463.863	3,650	903.594	7,800	1,767.120
1,120	323.517	1,725	469.771	3,700	914.459	7,900	1,787.212
1,125	324.760	1,750	475.669	3,750	925.309	8,000	1,807.278
1,130	326.000	1,775	481.558	3,800	936.144	8,100	1,827.319
1,140	328.480	1,800	487.437	3,850	946.966	8,200	1,847.336
1,150	330.958	1,825	493.307	3,900	957.773	8,300	1,867.329
1,160	333.433	1,850	499.167	3,950	968.566	8,400	1,887.298
1,170	335.906	1,875	505.019	4,000	979.344	8,500	1,907.243

Table 2.2 Continuation

t_D	Q_{tD}	t_D	Q_{tD}
8,600	1,927.166	2.5×10^7	2.961×10^6
8,700	1,947.065	3.0×10^7	3.517×10^6
8,800	1,966.942	4.0×10^7	4.610×10^6
8,900	1,986.796	5.0×10^7	5.689×10^6
9,000	2,006.628	6.0×10^7	6.758×10^6
9,100	2,026.438	7.0×10^7	7.816×10^6
9,200	2,046.227	8.0×10^7	8.866×10^6
9,300	2,065.996	9.0×10^7	9.911×10^6
9,400	2,085.744	1.0×10^8	1.095×10^7
9,500	2,105.473	1.5×10^8	1.604×10^7
9,600	2,125.184	2.0×10^8	2.108×10^7
9,700	2,144.878	2.5×10^8	2.607×10^7
9,800	2,164.555	3.0×10^8	3.100×10^7
9,900	2,184.216	4.0×10^8	4.071×10^7
10,000	2,203.861	5.0×10^8	5.032×10^7
12,500	2,688.967	6.0×10^8	5.984×10^7
15,000	3,164.780	7.0×10^8	6.928×10^7
17,500	3,633.368	8.0×10^8	7.865×10^7
20,000	4,095.800	9.0×10^8	8.797×10^7
25,000	5,005.726	1.0×10^9	9.725×10^7
30,000	5,899.508	1.5×10^9	1.429×10^8
35,000	6,780.247	2.0×10^9	1.880×10^8
40,000	7,650.096	2.5×10^9	2.328×10^8
50,000	9,363.099	3.0×10^9	2.771×10^8
60,000	11,047.299	4.0×10^9	3.645×10^8
70,000	12,708.358	5.0×10^9	4.510×10^8
75,000	13,531.457	6.0×10^9	5.368×10^8
80,000	14,350.121	7.0×10^9	6.220×10^8
90,000	15,975.389	8.0×10^9	7.066×10^8
100,000	17,586.284	9.0×10^9	7.909×10^8
125,000	21,560.732	1.0×10^{10}	8.747×10^8
1.5×10^5	2.538×10^4	1.5×10^{10}	1.288×10^9
2.0×10^5	3.308×10^4	2.0×10^{10}	1.697×10^9
2.5×10^5	4.066×10^4	2.5×10^{10}	2.103×10^9
3.0×10^5	4.817×10^4	3.0×10^{10}	2.505×10^9
4.0×10^5	6.267×10^4	4.0×10^{10}	3.299×10^9
5.0×10^5	7.699×10^4	5.0×10^{10}	4.087×10^9
6.0×10^5	9.113×10^4	6.0×10^{10}	4.868×10^9
7.0×10^5	1.051×10^5	7.0×10^{10}	5.643×10^9
8.0×10^5	1.189×10^5	8.0×10^{10}	6.414×10^9
9.0×10^5	1.326×10^5	9.0×10^{10}	7.183×10^9
1.0×10^6	1.462×10^5	1.0×10^{11}	7.948×10^9
1.5×10^6	2.126×10^5	1.5×10^{11}	1.17×10^{10}
2.0×10^6	2.781×10^5	2.0×10^{11}	1.55×10^{10}
2.5×10^6	3.427×10^5	2.5×10^{11}	1.92×10^{10}
3.0×10^6	4.064×10^5	3.0×10^{11}	2.29×10^{10}
4.0×10^6	5.313×10^5	4.0×10^{11}	3.02×10^{10}
5.0×10^6	6.544×10^5	5.0×10^{11}	3.75×10^{10}
6.0×10^6	7.761×10^5	6.0×10^{11}	4.47×10^{10}
7.0×10^6	8.965×10^5	7.0×10^{11}	5.19×10^{10}
8.0×10^6	1.016×10^6	8.0×10^{11}	5.89×10^{10}
9.0×10^6	1.134×10^6	9.0×10^{11}	6.58×10^{10}
1.0×10^7	1.252×10^6	1.0×10^{12}	7.28×10^{10}
1.5×10^7	1.828×10^6	1.5×10^{12}	1.08×10^{11}
2.0×10^7	2.398×10^6	2.0×10^{12}	1.42×10^{11}

Table 2.3

HURST-VAN EVERDINGEN CONSTANT-PRESSURE Q_{tD} FUNCTIONS FOR FINITE-ACTING RADIAL RESERVOIRS

$r_{De}=1.5$		$r_{De}=2.0$		$r_{De}=2.5$		$r_{De}=3.0$		$r_{De}=3.5$		$r_{De}=4.0$		$r_{De}=4.5$		$r_{De}=5$		$r_{De}=6$		$r_{De}=7$		$r_{De}=8$		$r_{De}=9$		$r_{De}=10$	
t_D	Q_{tD}	t_D	Q_{tD}	t_D	Q_{tD}	t_D	Q_{tD}	t_D	Q_{tD}	t_D	Q_{tD}	t_D	Q_{tD}	t_D	Q_{tD}	t_D	Q_{tD}	t_D	Q_{tD}	t_D	Q_{tD}	t_D	Q_{tD}	t_D	Q_{tD}
5.0×10^{-2}	0.276	5.00×10^{-2}	0.278	1.0×10^{-1}	0.408	3.0×10^{-1}	0.755	1.00	1.571	2.00	2.442	2.5	2.835	3.0	3.195	6.0	5.148	9.00	6.861	9	6.861	10	7.417	15	9.965
6.0×10^{-2}	0.304	7.50×10^{-2}	0.345	1.5×10^{-1}	0.509	4.0×10^{-1}	0.895	1.20	1.761	2.20	2.598	3.0	3.196	3.5	3.542	6.5	5.440	9.50	7.127	10	7.398	15	9.945	20	12.32
7.0×10^{-2}	0.330	1.00×10^{-1}	0.404	2.0×10^{-1}	0.599	5.0×10^{-1}	1.023	1.40	1.940	2.40	2.748	3.5	3.537	4.0	3.875	7.0	5.724	10	7.389	11	7.920	20	12.26	22	13.22
8.0×10^{-2}	0.354	1.25×10^{-1}	0.458	2.5×10^{-1}	0.681	6.0×10^{-1}	1.143	1.60	2.111	2.60	2.893	4.0	3.859	4.5	4.193	7.5	6.002	11	7.902	12	8.431	22	13.13	24	14.09
9.0×10^{-2}	0.375	1.50×10^{-1}	0.507	3.0×10^{-1}	0.758	7.0×10^{-1}	1.256	1.80	2.273	2.80	3.034	4.5	4.165	5.0	4.499	8.0	6.273	12	8.397	13	8.930	24	13.98	26	14.95
1.0×10^{-1}	0.395	1.75×10^{-1}	0.553	3.5×10^{-1}	0.829	8.0×10^{-1}	1.363	2.00	2.427	3.00	3.170	5.0	4.454	5.5	4.792	8.5	6.537	13	8.876	14	9.418	26	14.79	28	15.78
1.1×10^{-1}	0.414	2.00×10^{-1}	0.597	4.0×10^{-1}	0.897	9.0×10^{-1}	1.465	2.20	2.574	3.25	3.334	5.5	4.727	6.0	5.074	9.0	6.795	14	9.341	15	9.895	28	15.59	30	16.59
1.2×10^{-1}	0.431	2.25×10^{-1}	0.638	4.5×10^{-1}	0.962	1.00	1.563	2.40	2.715	3.50	3.493	6.0	4.986	6.5	5.345	9.5	7.047	15	9.791	16	10.361	30	16.35	32	17.38
1.3×10^{-1}	0.446	2.50×10^{-1}	0.678	5.0×10^{-1}	1.024	1.25	1.791	2.60	2.849	3.75	3.645	6.5	5.231	7.0	5.605	10.0	7.293	16	10.23	17	10.82	32	17.10	34	18.16
1.4×10^{-1}	0.461	2.75×10^{-1}	0.715	5.5×10^{-1}	1.083	1.50	1.997	2.80	2.976	4.00	3.792	7.0	5.464	7.5	5.854	10.5	7.533	17	10.65	18	11.26	34	17.82	36	18.91
1.5×10^{-1}	0.474	3.00×10^{-1}	0.751	6.0×10^{-1}	1.140	1.75	2.184	3.00	3.098	4.25	3.932	7.5	5.684	8.0	6.094	11	7.767	18	11.06	19	11.70	36	18.52	38	19.65
1.6×10^{-1}	0.486	3.25×10^{-1}	0.785	6.5×10^{-1}	1.195	2.00	2.353	3.25	3.242	4.50	4.068	8.0	5.892	8.5	6.325	12	8.220	19	11.46	20	12.13	38	19.19	40	20.37
1.7×10^{-1}	0.497	3.50×10^{-1}	0.817	7.0×10^{-1}	1.248	2.25	2.507	3.50	3.379	4.75	4.198	8.5	6.089	9.0	6.547	13	8.651	20	11.85	22	12.95	40	19.85	42	21.07
1.8×10^{-1}	0.507	3.75×10^{-1}	0.848	7.5×10^{-1}	1.299	2.50	2.646	3.75	3.507	5.00	4.323	9.0	6.276	9.5	6.760	14	9.063	22	12.58	24	13.74	42	20.48	44	21.76
1.9×10^{-1}	0.517	4.00×10^{-1}	0.877	8.0×10^{-1}	1.348	2.75	2.772	4.00	3.628	5.50	4.560	9.5	6.453	10	6.965	15	9.456	24	13.27	26	14.50	44	21.00	46	22.42
2.0×10^{-1}	0.525	4.25×10^{-1}	0.905	8.5×10^{-1}	1.395	3.00	2.886	4.25	3.742	6.00	4.779	10	6.621	11	7.350	16	9.829	26	13.92	28	15.23	46	21.69	48	23.07
2.1×10^{-1}	0.533	4.50×10^{-1}	0.932	9.0×10^{-1}	1.440	3.25	2.990	4.50	3.850	6.50	4.982	11	6.930	12	7.706	17	10.19	28	14.53	30	15.92	48	22.26	50	23.71
2.2×10^{-1}	0.541	4.75×10^{-1}	0.958	9.5×10^{-1}	1.484	3.50	3.084	4.75	3.951	7.00	5.169	12	7.208	13	8.035	18	10.53	30	15.11	34	17.22	50	22.82	52	24.33
2.3×10^{-1}	0.548	5.00×10^{-1}	0.983	1.0	1.526	3.75	3.170	5.00	4.047	7.50	5.343	13	7.457	14	8.339	19	10.85	35	16.39	38	18.41	52	23.36	54	24.94
2.4×10^{-1}	0.554	5.50×10^{-1}	1.028	1.1	1.605	4.00	3.247	5.50	4.222	8.00	5.504	14	7.680	15	8.620	20	11.16	40	17.49	40	18.97	54	23.89	56	25.53
2.5×10^{-1}	0.559	6.00×10^{-1}	1.070	1.2	1.679	4.25	3.317	6.00	4.378	8.50	5.653	15	7.880	16	8.879	22	11.74	45	18.43	45	20.26	56	24.39	58	26.11
2.6×10^{-1}	0.565	6.50×10^{-1}	1.108	1.3	1.747	4.50	3.381	6.50	4.516	9.00	5.790	16	8.060	18	9.338	24	12.26	50	19.24	50	21.42	58	24.88	60	26.67
2.8×10^{-1}	0.574	7.00×10^{-1}	1.143	1.4	1.811	4.75	3.439	7.00	4.639	9.50	5.917	18	8.365	20	9.731	25	12.50	60	20.51	55	22.46	60	25.36	65	28.02
3.0×10^{-1}	0.582	7.50×10^{-1}	1.174	1.5	1.870	5.00	3.491	7.50	4.749	10	6.035	20	8.611	22	10.07	31	13.74	70	21.43	60	23.40	65	26.48	70	29.29
3.2×10^{-1}	0.588	8.00×10^{-1}	1.203	1.6	1.924	5.50	3.581	8.00	4.846	11	6.246	22	8.809	24	10.35	35	14.40	80	22.13	70	24.98	70	27.52	75	30.49

3.4×10^{-1}	0.594	9.00×10^{-1}	1.253	1.7	1.975	6.00	3.656	8.50	4.932	12	6.425	24	8.968	26	10.59	39	14.93	90	22.63	80	26.26	75	28.48	80	31.61
3.6×10^{-1}	0.599	1.0	1.295	1.8	2.022	6.50	3.717	9.00	5.009	13	6.580	26	9.097	28	10.80	51	16.05	100	23.00	90	27.28	80	29.36	85	32.67
3.8×10^{-1}	0.603	1.1	1.330	2.0	2.106	7.00	3.767	9.50	5.078	14	6.712	28	9.200	30	10.98	60	16.56	120	23.47	100	28.11	85	30.18	90	33.66
4.0×10^{-1}	0.606	1.2	1.358	2.2	2.178	7.50	3.809	10.00	5.138	15	6.825	30	9.283	34	11.26	70	16.91	140	23.71	120	29.31	90	30.93	95	34.60
4.5×10^{-1}	0.613	1.3	1.382	2.4	2.241	8.00	3.843	11	5.241	16	6.922	34	9.404	38	11.46	80	17.14	160	23.85	140	30.08	95	31.63	100	35.48
5.0×10^{-1}	0.617	1.4	1.402	2.6	2.294	9.00	3.894	12	5.321	17	7.004	38	9.481	42	11.61	90	17.27	180	23.92	160	30.58	100	32.27	120	38.51
6.0×10^{-1}	0.621	1.6	1.432	2.8	2.340	10.00	3.928	13	5.385	18	7.076	42	9.532	46	11.71	100	17.36	200	23.96	180	30.91	120	34.39	140	40.89
7.0×10^{-1}	0.623	1.7	1.444	3.0	2.380	11.00	3.951	14	5.435	20	7.189	46	9.565	50	11.79	110	17.41	500	24.00	200	31.12	140	35.92	160	42.75
8.0×10^{-1}	0.624	1.8	1.453	3.4	2.444	12.00	3.967	15	5.476	22	7.272	50	9.586	60	11.91	120	17.45			240	31.34	160	37.04	180	44.21
		2.0	1.468	3.8	2.491	14.00	3.985	16	5.506	24	7.332	60	9.612	70	11.96	130	17.46			280	31.43	180	37.85	200	45.36
		2.5	1.487	4.2	2.525	16.00	3.993	17	5.531	26	7.377	70	9.621	80	11.98	140	17.48			320	31.47	200	38.44	240	46.95
		3.0	1.495	4.6	2.551	18.00	3.997	18	5.551	30	7.434	80	9.623	90	11.99	150	17.49			360	31.49	240	39.17	280	47.94
		4.0	1.499	5.0	2.570	20.00	3.999	20	5.579	34	7.464	90	9.624	100	12.00	160	17.49			400	31.50	280	39.56	320	48.51
		5.0	1.500	6.0	2.599	22.00	3.999	25	5.611	38	7.481	100	9.625	120	12.0	180	17.50			500	31.50	320	39.77	360	48.91
				7.0	2.613	24.00	4.000	30	5.621	42	7.490					200	17.50					360	39.88	400	49.11
				8.0	2.619			35	5.624	46	7.494					220	17.50					400	39.94	440	49.28
				9.0	2.622			40	5.625	50	7.497											440	39.97	480	49.36
				10.0	2.624																	480	39.98		

Table 2.2 is for an infinite-acting reservoir and Table 2.3 is for a finite-acting reservoir. However, note that the data of Table 2.3 are very limited as far as reservoir size is concerned. From a practical standpoint the data of Table 2.3 can be used only for predicting water encroachment from an aquifer into the hydrocarbon-bearing portion of the reservoir since the dimensionless reservoir sizes only range from 1.5 to 10. These small values of r_D would be completely unrealistic if we wanted to calculate cumulative flow into a well.

The data of Table 2.3 represent the original published data of the Hurst-Van Everdingen paper.[3] However, many of the major oil companies have extended these data for their own use using the equations from the subject paper. Any competent programmer with a knowledge of Bessel functions can generate data for any range of dimensionless reservoir sizes needed. Without these additional data the application of the Q_{tD} function is somewhat limited. Fortunately the Exxon Production Research Company permitted the use of their data in this book (Figs. 2.11A, B, C, and D) so the applications permitted by the original Hurst-Van Everdingen data can be greatly extended.

The application of the constant-pressure solution is similar to the application of the constant-rate data. The reduced time, t_D, is calculated based on the radius at which the cumulative flow is desired and the constant pressure is known. The t_D value also incorporates the time at which the cumulative flow is desired. The dimensionless reservoir size, r_D, is calculated so that the appropriate data can be used. It may also be helpful to calculate the critical time when the reservoir begins acting as finite, $r_e^2/4\eta$. Then the corresponding reduced cumulative flow value can be interpolated from Tables 2.2 or 2.3 or read from Figs. 2.11A, B, C, or D. The resulting Q_{tD} can be converted to barrels by using Equation 2.23 which defines the reduced cumulative flow expression. It is convenient to rearrange the equation so that the cumulative flow in barrels, Q, can be directly calculated.

$$Q = 1.12\phi hcr^2\Delta p Q_{tD} \qquad (2.29)$$

In applying this equation remember that the radius is the radius across which the cumulative flow is desired and Δp is the pressure drop at this radius. When the equation is applied to flow at the well remember that Δp should exclude Δp_{skin}. A problem will help to clarify the application of these data. c_e is the effective water compressibility. This concept will be discussed in the next section.

Problem 2.2: Cumulative Water Disposal in an Aquifer

An aquifer has the following characteristics:

p_i = 1,000 psia	μ_w = 0.5 cp
k_w = 10 md	c_e = 6.33 × 10⁻⁶/psi
h = 14.1 ft	r_e = 10,000 ft
ϕ = 0.2	Δp_{skin} = 200 psi
r_w = 0.5 ft	

A. Show that for the purposes of water injection this aquifer will still be acting as an infinite reservoir after 100 days of injection.
B. If the injection pump is capable of maintaining a BHP of 2,200 psia, how much water can be disposed of in the aquifer in 100 days?

This example would be more realistic if the cumulative disposal was calculated for several times and the cumulative curve was changed to a rate relationship so that the engineer could predict the disposal rate versus time.

Also note that a dimensionless rate function could be used to directly determine the rate without first determining cumulative values. The reduced pressure change, Δp_D of Equation 2.11 can be thought of as the pressure change per reduced flow rate. By this definition we could then show that the reduced rate is

$$q_{tD} = \frac{q}{(7.08kh\Delta p/\mu)} \tag{2.30}$$

or

$$q_{tD} = \frac{1}{\Delta p_D} \tag{2.31}$$

Equation 2.30 can also be obtained by dividing the reduced volume, Q_{tD} of Equation 2.23 by the reduced time, t_D of Equation 2.25.

To obtain the reduced rates versus the reduced time it would simply be necessary to differentiate the Q_{tD} versus t_D curves which directly give the q_{tD} values which could then be plotted versus the reduced time.[10] Some companies make use of reduced-rate data in this way.

We will see in the Chapter on Gas Reservoir Engineering that the constant-pressure solution can also be applied to the flow of gas. In fact, as was previously suggested, it appears that much more use should be made of the constant-pressure solutions in predicting the deliverability of gas reservoirs prior to the time when the well reaches pseudosteady state which may require an extended period of time for a tight gas reservoir.

EFFECTIVE COMPRESSIBILITY

In discussing compressibility previously we have referred to it very simply as water compressibility or oil compressibility or gas compressibility. However, the concept is a little more complex than this.

By referring to the derivation of the radial diffusivity equation you will see that the value required for the compressibility of a system is the amount of fluid forced from a unit of pore volume for 1.0 psi of pressure drop. If the pore volume was occupied 100% by one fluid and remained constant then we would indeed require an evaluation of only the compressibility of the one fluid. However, an oil or gas-producing reservoir will always contain interstitial water and the change in the pore volume that accompanies a decline in reservoir pressure may cause as much fluid to be forced from the pore volume as does the compressibility of the liquid. Consequently, we desire one value, an effective compressibility, c_e, that can be applied to the pore volume to give the amount of fluid forced from the pore volume with each 1.0 psi of pressure change. This change in volume, $c_e V_p$, would then be equal to the expansion of the oil, water, and gas in the pore volume plus the reduction in the pore volume resulting from the decline in the reservoir pressure.

$$c_e V_p = c_o S_o V_p + c_g S_g V_p + c_w S_w V_p + c_f V_p \qquad (2.32)$$

In this equation, c_f is the change in pore volume per unit pressure change. Since the pore volume, V_p, appears in each term of Equation 2.32 we can show that,

$$c_e = c_o S_o + c_g S_g + c_w S_w + c_f \qquad (2.33)$$

The change in pore volume that accompanies the change in the reservoir pressure is due primarily to two things. As the reservoir pressure declines the matrix or solid material in the porous media expands since the solid material is surrounded by the reservoir pressure. Secondly, the solid portion of the reservoir rock supports a large part of the weight of the rocks and fluids that exist above the subject reservoir. This overburden force tends to compress the reservoir formation and it is opposed by the mechanical strength of the rock and the reservoir pressure in the pores. As the reservoir pressure declines, resistance to the force of the overburden pressure is reduced and the thickness of the reservoir is slightly reduced. Thus, the volume of the solid material is increased, the bulk volume is decreased (due to the reduction of the thickness) and this results in a reduction in the pore volume. This concept is more thoroughly explained in Chapter 6.

The compressibility of oil, gas, water, and the formation or pore volume can all be determined from empirical data in the Appendix. However, notice that it is seldom that all terms in Equation 2.33 have significant values. When the reservoir pressure is above the saturation pressure, the gas saturation would be zero. On the other hand when the reservoir gas saturation is more than 2 or 3%, the gas-compressibility term dominates the value of the effective compressibility and the other terms are insignificant. This situation exists due to the great difference in the magnitude of the compressibility of gas as compared with the compressibilities of liquids and the formation.

The compressibility of an ideal gas is equal to the reciprocal of the gas pressure. This is shown in Chapter 4. Consequently, we can use the reciprocal of the pressure as an approximation of the gas compressibility. Thus, gas compressibilities will be in the range of 10^{-3} to 10^{-4}/psi while liquid and formation compressibilities are in the range of 10^{-5} to 10^{-6}. Therefore, even a gas saturation of 1.0% will generally mean that the gas contributes as much to the effective compressibility as all the other terms combined. As the gas saturation increases the contribution of the other terms becomes insignificant.

SUPERPOSITION

If we were limited in our practical applications of the constant rate and constant-pressure solutions to situations in which we simply had one well producing from a reservoir at a constant rate or constant pressure, the two solutions to the radial diffusivity equation would hardly be worthwhile. However, by using the concept of superposition we can account for the effects of producing from more than one well, the effects of rate and pressure changes, and we can even extend the use of infinite-acting solutions by artificially setting up boundary conditions. The concept of superposition is not the product of a reservoir engineer. The same technique is used in applying many of the equations governing the transfer of energy and mass.

Accounting for the effects of more than one well[8]. In applying the constant-rate solution we often wish to account for the effects of more than one well on the pressure at some point in the reservoir. We can do this by simply evaluating the pressure drop caused by each producing rate as though the others did not exist and then adding the pressure drops caused by the individual effects to obtain the total pressure drop resulting from all the rates. In other words we simply superimpose one effect upon the other; hence the name superposition. It is undoubtedly simpler to say that we add the effects.

To illustrate the meaning of this concept consider Fig. 2.12 which

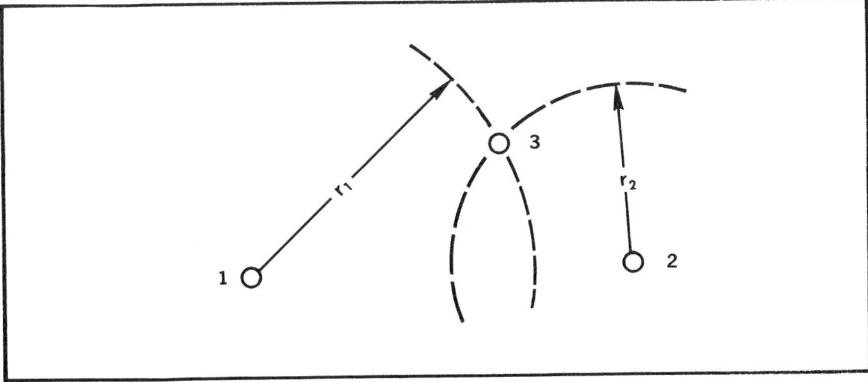

FIG. 2.12 Diagram for discussion of pressure drop caused by production from three wells.

shows the distance of two wells from the third well in a reservoir. If these three wells have been producing for a short enough period of time that they are all still infinite acting we could determine the pressure drop that would be caused by the production of Well 3 at its well radius by assuming that the other two wells were not there. We could also calculate the pressure drop that would be caused by the production from Well 2 at a radius r_2 by using the Ei function solution since we desire the pressure at a radius r_2 removed from the well bore where the rate is known. This calculation would be made by assuming that Wells 1 and 3 were not there. We could again calculate the pressure drop caused by production from Well 1 at a radius r_1 again using the Ei function solution as though Wells 2 and 3 did not exist. The actual pressure drop at Well 3 caused by the production of all three wells concurrently would be the sum of the individually calculated pressure drops described.

In making such an application it is necessary that all producing times terminate at the time of the desired pressure calculation and that all effects are still infinite acting. It is not necessary that all wells start producing at the same time. In fact it is unusual for such a coincidence to occur. There is no limit to the number of wells that could be accounted for in a calculation such as this. Also, by investigating the pressure at various points in the reservoir at the same time it is possible to describe the complete pressure distribution at any particular time or the pressure history at any particular point or both. A problem along these lines should clarify the physical significance of this discussion.

Problem 2.3A: Pressure Resulting from More Than One Well in a Reservoir

Wells 1 and 2 are drilled in an undeveloped reservoir. No. 1 is completed and produced at a constant rate of 540 STB/d for 52.5 days before No. 2 is completed with initial production of 1,080 STB/d. Given the listed reservoir data, what would you expect the pressure to be in No. 2 after it has produced for 10 days? At this time No. 1 has produced for 62.5 days.

$p_i = 3,000$ psia	$p_s = 2,235$ psia
$K = 100$ md	$u_o = 0.5$ cp
$h = 14.1$ ft	$c_e = 6.33 \times 10^{-5}$
$\phi = 0.20$	$B_o = 1.39$ (assume constant)
$r_w = 0.5$	$\Delta p_{skin} = 0.0$
$\eta = 10^5$	Distance between wells $= 2,500$ ft

The nearest reservoir boundary is 10,000 ft from each well.

Many reservoir engineers have had to determine whether or not two wells are producing from the same reservoir. The common approach is to put one well on production and observe the effect of this production on the other well. The engineer reasons that if he can see the effect of the production in the observation well in a reasonable length of time he can conclude that the wells are in the same reservoir and conversely, if he does not see any effect in the observation well in a reasonable length of time the wells are not in the same reservoir. Many such engineers have reached an erroneous conclusion that the wells were not in the same reservoir and were subsequently proven wrong because of geological information available from further exploration or very similar pressure histories of the two wells. In either case the engineer could have saved himself the error and embarrassment by making a simple calculation to check the magnitude of the interference that he was trying to read from a pressure gauge.

Such a calculation involves determining the magnitude of the pressure change that you have confidence you can note and then determining how long it will take to obtain such a pressure drop at the point in the reservoir that you are observing. In this case the normal calculating procedure is reversed. Using the assumed pressure drop the corresponding Ei function is calculated from Equation 2.18. The corresponding reduced time can then be read from Fig. 2.8 and the producing time necessary to obtain the desired pressure drop can be calculated from the definition of the reduced time, Equation 2.25. This procedure is illustrated in the following problem. Check your work against the solution in the Appendix.

Problem 2.3B: Calculation of the Time Necessary to Obtain a Particular Pressure Drop in the Reservoir

In planning an interference test in the reservoir of Problem 2.3A we need to know how long it will take the production from Well 1 to cause a pressure drop of 7.5 psi in Well 2. Assume this occurs before production is initiated in No. 2.

Accounting for rate change effects. In addition to accounting for the affects of withdrawing from more than one point in the reservoir in applying the constant-rate solution we also account for the effect of rate changes. Why this is possible may be best explained by referring to Fig. 2.13. In this figure two wells are shown a distance d apart. The number one well has produced for 20 days at a rate of 150 b/d and the number two well has produced for 10 days at a rate of 100 b/d. From the previous discussion we know that we could calculate the pressure in either well by accounting for the affect of each well for the specified rate and time and at the appropriate radius as though the other well was not there. Such a calculation could be made regardless of the magnitude of the distance d between wells. Consequently, we could perform such an analysis even if the two wells existed at the same point in the reservoir (d = 0).

On the surface it appears that such an assumption (d = 0) is ridiculous, but on further consideration you will see that this idea can be used to model a rate change or a series of rate changes in such a way that we can analyze the effect of these changes. Or, if we had two wells producing at different rates for different lengths of time at the same point in the reservoir, the effect on the reservoir pressure would be the same as one well producing at a rate history equal to the sum of the individual

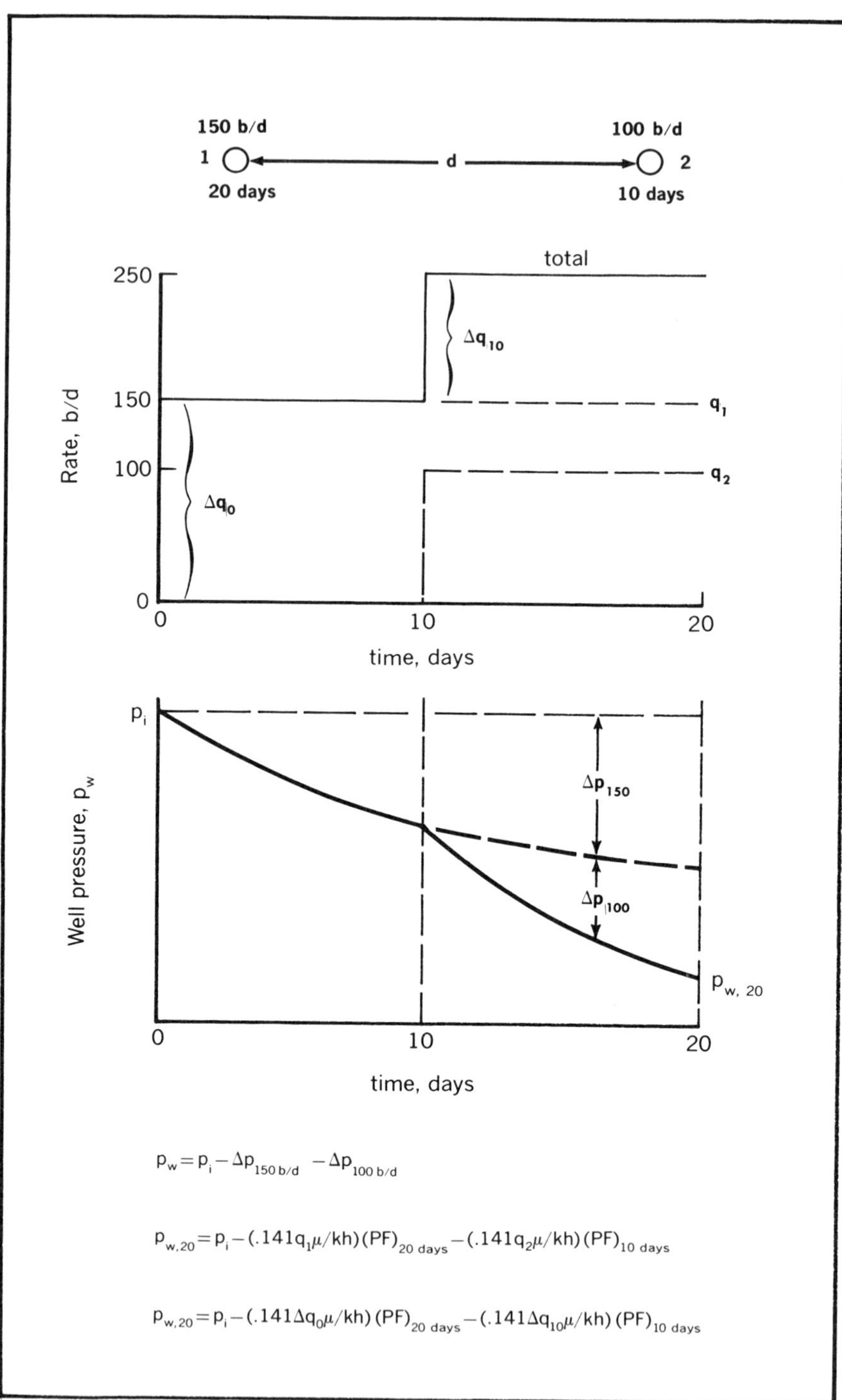

FIG. 2.13 Diagram for discussion of pressure drop due to variable rate.

well rates. In the particular case being considered, Fig. 2.13, the effect of the two wells producing as indicated with the distance between them being zero would give the same effect as one well producing with a history equivalent to the total production of the two wells, that is, producing at 150 b/d for 10 days and then the rate being increased for the next 10 days to 250 b/d.

This then provides a procedure for analyzing the effects of a rate change in a well. What must be done is to model this rate change by considering a number of wells at the same point in the reservoir that will give a total production equal to the one-well rate history. Mechanically this can be done by having each well represent the rate change producing for the time remaining to the time of interest. Referring again to Fig. 2.13, if we had one well producing at the rate history marked "total" we could analyze it by considering the affects of a well producing at the first rate change at time equal zero (150–0), for the remaining time of 20 days plus the effect of a second well producing at the second rate change (250–150), for the remaining time of 10 days.

Study of this technique will show that it can be extended to any number of rate changes either positive or negative. Note that regardless of how many terms or rate changes are involved for a single well each of the pressure-drop terms will contain the same group of constants, $0.141\mu/kh$, and thus the mathematical expression can be written as

$$p_{w,t} = p_i - \frac{0.141\mu}{kh} \sum_{j=1}^{j=m} [\Delta q_j (PF)_j] \qquad (2.34)$$

In using this expression the engineer must understand the meaning of the Δq_j and must recognize that PF_j must be based on the remaining length of time. Equation 2.34 can be written in a much more explicit mathematical form but experience shows that the more complex math form is more often misunderstood and misused than the cited form of 2.34 which requires an understanding of the equation before it can be used.

The rate change, Δq, of Equation 2.34 can be defined as $(q_{new} - q_{old})$ in order to keep the rate change sign correct. As stated previously some of the rate changes will be negative which will result in a pressure-increase affect.

To examine the meaning of a negative rate-change effect refer to Fig. 2.14, which indicates a rate of 250 b/d for 10 days with a reduction in rate for the next 10 days to 150 b/d. Note that if the 250 b/d rate was continued for the entire 20 days the resulting pressure would be the initial pressure less the pressure drop caused by 250 b/d of production for a period of 20 days. Obviously this pressure would be too low for the actual history shown because the well did not produce at 250 b/d for the full 20 days. Consequently, some means must be available for reducing this pressure drop. This is accomplished by considering the effect of the negative rate change from 250 to 150 as being the same as the effect of

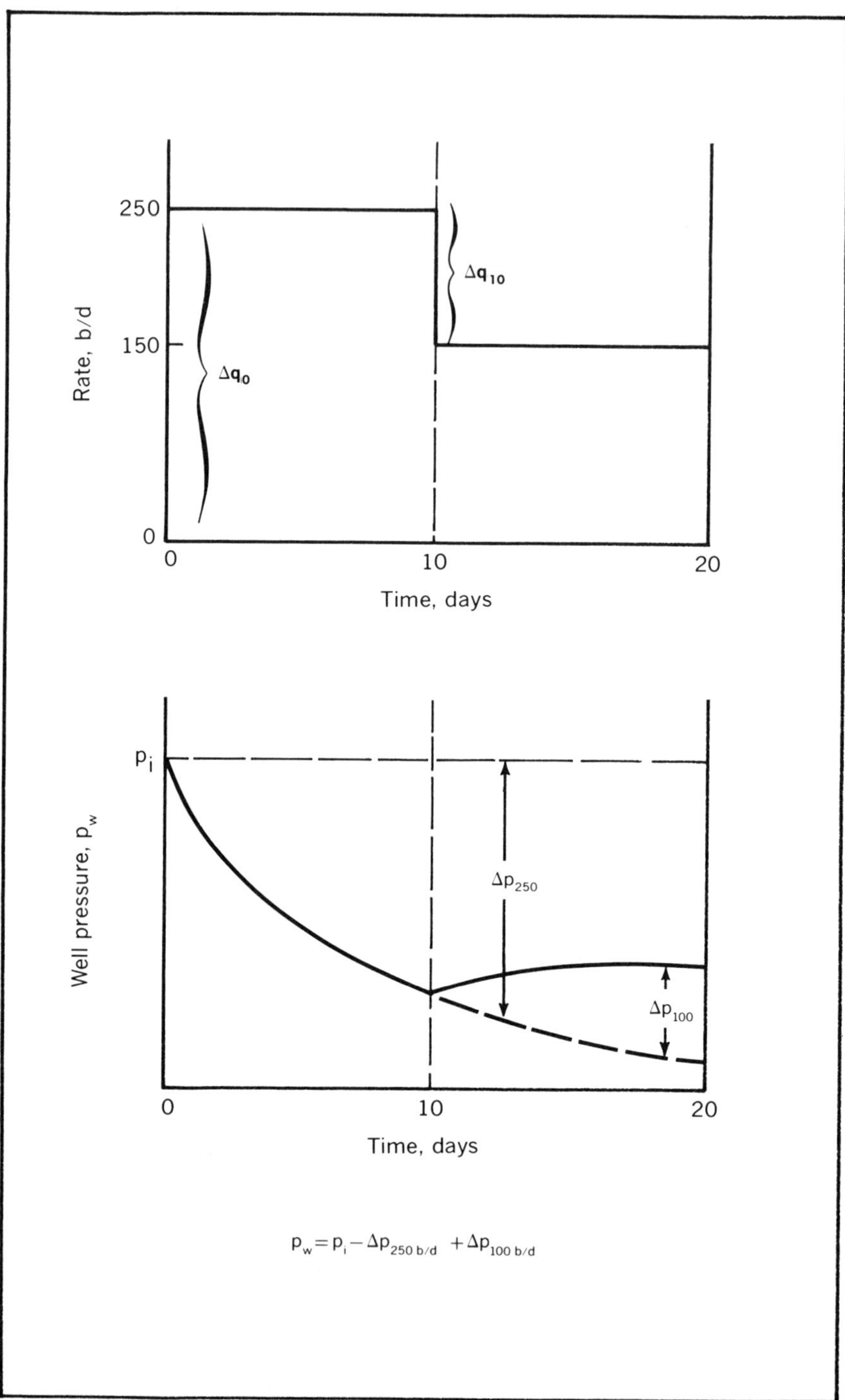

FIG. 2.14 The effect of a negative rate change on pressure.

an injection well injecting at a rate of 100 b/d. In other words a negative producing rate is the same as an injection rate and causes a corresponding pressure increase. Thus, the actual pressure resulting from the rate history of Fig. 2.14 will be the initial pressure less the reduction in pressure caused by a well producing at a rate of 250 b/d for 20 days plus the increase in pressure resulting from an injection well operating for a time of 10 days at 100 b/d. This is of course the same as the result that would be obtained by applying Equation 2.34.

Engineers evaluating the effect of rate changes or pressure changes often want to "simplify" the calculations. This simplification that is attempted is one of calculating the effect of a particular rate for the period of time it prevails and then adding to it the effect of the next rate for the period of time it prevails. When applied to Fig. 2.14, this simplification would lead to a well pressure equal to the initial pressure less the effect of a rate of 250 b/d for 10 days less the effect of a rate of 150 b/d for 10 days. This type analysis is incorrect! The production (or injection) times must be continuous to the time of interest, the time when you want to know the pressure. When the times are not continuous, as in the example cited, a negative rate change is introduced at the time when the consideration of the positive rate change is discontinued (at 10 days in the example). In other words to make the calculation correct we would have to add the effect of a negative rate change of 250 b/d for 10 days which would cause a pressure increase. But to make the total rate history correct we would have to add an additional 250 b/d for the last 10 days and then we would be right back where we started. Another way of explaining this problem is to recognize that there is no way of freezing or fixing a pressure distribution established in a reservoir that is other than a uniform pressure throughout. The pressure distribution will not remain fixed in the reservoir when you stop production. As soon as you discontinue production the effect of the negative rate change will immediately start a buildup back to a uniform static-pressure distribution throughout the reservoir.

Note that we can put together the effect of withdrawal from more than one point in the reservoir and the variation in rate to obtain the effect of varying the rate at more than one point in the reservoir. We could then apply the same Equation 2.34 to evaluate these effects by recognizing that each j counter has both a radius and time associated with it. A particular rate change, Δq_j, is then associated with a particular radius from the point where the rate is known as well as a particular time. Both the radius and time would then have to be reflected in the evaluation of the appropriate pressure function, PF_j. A problem will help to clarify this application.

Problem 2.3C: Accounting for Variations in Producing Rates

In Problem 2.3A assume that after Well 1 has produced for 62.5 days at 540 STB/d its production rate is reduced to 180 STB/d. What would be the pressure in Well 2 after a total of 72.5 days of production? At this time No. 2 will have produced at 1,080 STB/d for 20 days.

Accounting for pressure-change effects. Superposition is also used in applying the constant-pressure-rate case. Pressure changes are accounted for in applying this solution in much the same way that rate changes are accounted for in applying the constant rate case. Assume that a well for some reason experienced the pressure history shown in Fig. 2.15. In this case three distinct pressure changes occur so that it would be necessary to apply the constant-pressure solution to each individual pressure change. This would be the same as applying Equation 2.29 a total of three times.

The initial pressure drop from 5,000 to 3,000 psia would be effective for a total time of 30 days so that a t_D equivalent to 30 days and the well radius would be used to evaluate the reduced cumulative function, Q_{tD}. This would be used in Equation 2.29 with the pressure drop of 2,000 psi to calculate the production due to this pressure drop. This is the production that would accumulate if there was no further change in the pressure, if the 3,000 psia pressure prevailed for the entire 30 days.

However, this is not the case. At 10 days the pressure in the well was dropped further to a pressure of 2,000 psia or an additional pressure drop of 1,000 psi. Note that this could be considered effective for the remainder of the time, or 20 days. Therefore, the cumulative production due to this additional pressure drop is calculated by first calculating the reduced time, t_D, based on the well radius and a time of 20 days; finding Q_{tD} for this t_D from the appropriate table; and using this value in Equation 2.29 along with the 1,000-psi pressure drop to calculate the production due to this additional pressure drop of 1,000 psi acting for 20 days. Then if there is no further pressure change the sum of the two calculations is the production.

But we know that the production will not be this great because the pressure returns to 3,000 psia for the last 10 days. This increase in pressure will reduce the production that would have occurred if the pressure increase had not taken place. Therefore, we treat this as a negative pressure drop which will result in a negative production value when it is applied in Equation 2.24 with the pressure increase of -1000 psi and the Q_{tD} based on a t_D which in turn is based on a time of 10 days and the well radius. The sum of the production due to these three pressure changes will be the cumulative production at the end of 30 days.

If we generalize this procedure we could say that we are simply applying Equation 2.24 to each pressure drop or that

$$Q = \sum_{j=1}^{j=m} (1.12\phi hcr^2 \Delta p_j Q_{tDj}) \qquad (2.35)$$

or

$$Q = 1.12\phi hcr^2 \sum_{j=1}^{j=m} (\Delta p_j Q_{tDj}) \qquad (2.36)$$

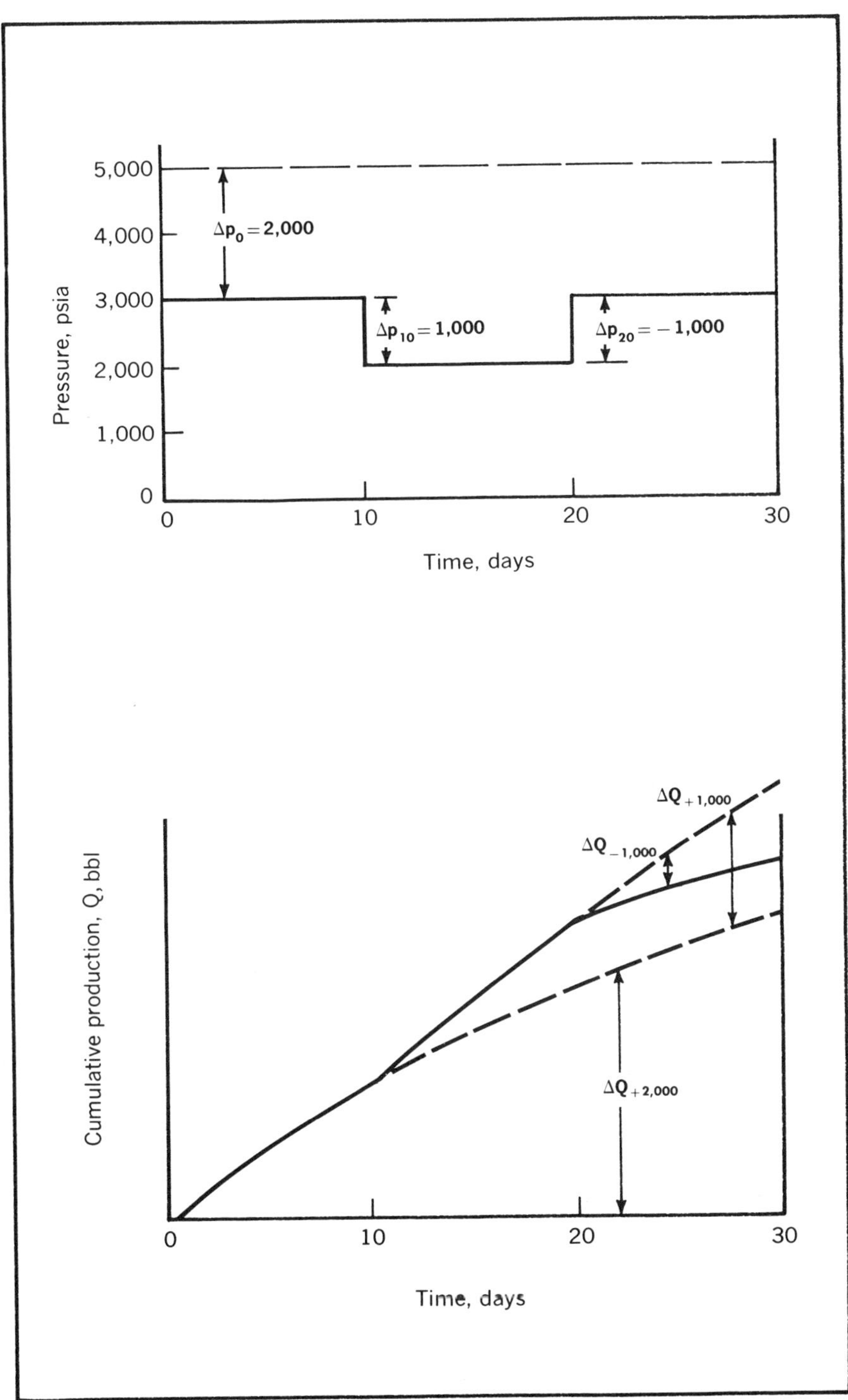

FIG. 2.15 Effect of pressure changes on cumulative production.

where

$$\Delta p_j = p_{old} - p_{new} \tag{2.37}$$

As was noted in applying superposition in the constant-rate case all times must be continuous to the time of interest.

Simulating boundary effects.   One of the most useful applications of the superposition concept is in applying infinite-acting solutions to reservoirs that are limited in one or more directions. This is done by finding some configuration of wells in an infinite-acting reservoir that will give the same drainage configuration as the one with the existing reservoir boundaries. Consider the simplest case, a plane fault boundary in an otherwise infinite-acting reservoir, as illustrated in Fig. 2.16. Assume that we desire to know the pressure behavior of this well. We could use the infinite-acting constant-rate solution to analyze this well pressure until the effects of the boundary are felt at the well but thereafter we do not have pressure functions for this flow geometry.

However, we can model this reservoir with an infinite-acting solution if we can discover some combination of wells in an infinite-acting system that will limit the drainage or flow along the boundary. The right-hand side of Fig. 2.16 indicates such a configuration of wells. If two wells with the same production capacity as the actual well were spaced so that the distance between them was twice the actual distance to the fault, then there would be no flow across the plane surface midway between the two wells in the infinite-acting system and the flow configuration in the drainage area of each well would be the same as the flow configuration for the actual well.

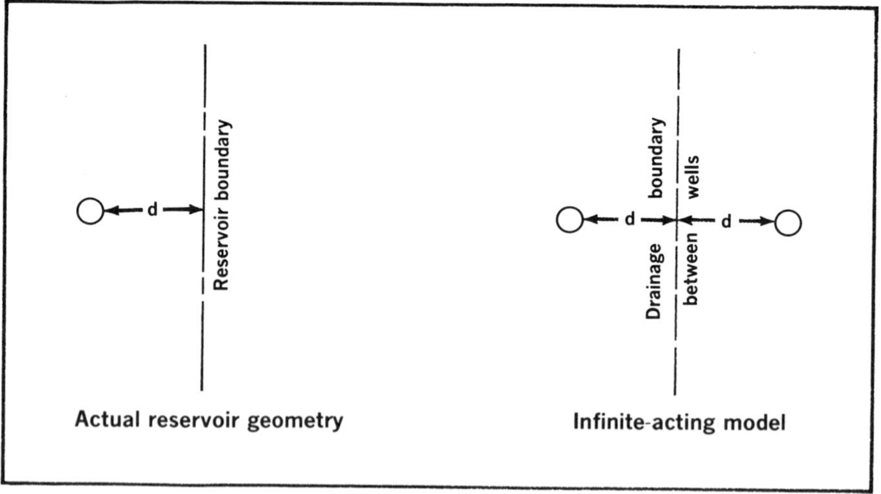

FIG. 2.16   Simulating a plane barrier in an otherwise infinite-acting reservoir.

Another means of physically understanding why the infinite model gives the same pressure behavior as the actual well is to assume that you actually have two wells in an infinite-acting reservoir each producing at the same rate for the same length of time. Under such conditions it is clear that any oil produced from the area to the right of the drainage boundary will be produced from the right-hand well and any fluid produced from the area to the left of the drainage boundary will be produced from the left-hand well. This then means that there is no flow across the drainage boundary. Now if there is no flow across this plane, a barrier could be put in the reservoir along this plane without interfering with the production of either well—the pressure behavior with a barrier in the reservoir would be exactly the same as the pressure behavior of the two wells in an infinite-acting system. Consequently, if we analyze the pressure behavior of the two wells in the infinite-acting system we will then have the pressure behavior of the one well with the reservoir boundary.

The pressure in the well will then be the initial reservoir pressure less the pressure drop caused by the well's production at the well radius less the pressure drop caused by the production from the second well at a radius of 2d.

$$p_w = p_i - \frac{0.141q\mu}{kh}(PF)_w - \frac{0.141q\mu}{kh}(PF)_{2d} \qquad (2.38)$$

In this expression the subscripts "w" and "2d" indicate that the pressure functions should be evaluated at reduced times based on radii of r_w and 2d, respectively. Note that setting up artificial boundaries such as this can be accomplished only as long as the reservoir is infinite acting since the analysis requires the calculation of pressure drops away from the well bore. This requires the Ei function solution which can only be used for an infinite-acting reservoir. Try the following problem and check your solution in Appendix C.

Problem 2.3D: Simulating Boundary Effects

Assume that the reservoir of Problem 2.3A contains a limiting fault 1,250 ft from the No. 1 well. What will be the well pressure in No. 1 after it has produced 52.5 days at a rate of 540 STB/d?

The additional wells that are considered in simulating a boundary are often referred to as "image" or "ghost" wells. Almost any boundary configuration can be simulated with the proposed technique but many of the configurations require computer facilities to be practical since a very large number of ghost wells must be included. Two boundaries at right angles to each other are simple to simulate.

For example, consider Fig. 2.17. On this diagram is indicated the real well and the two boundaries and the three image wells required to analyze the pressure behavior. Note that two wells are not enough to model

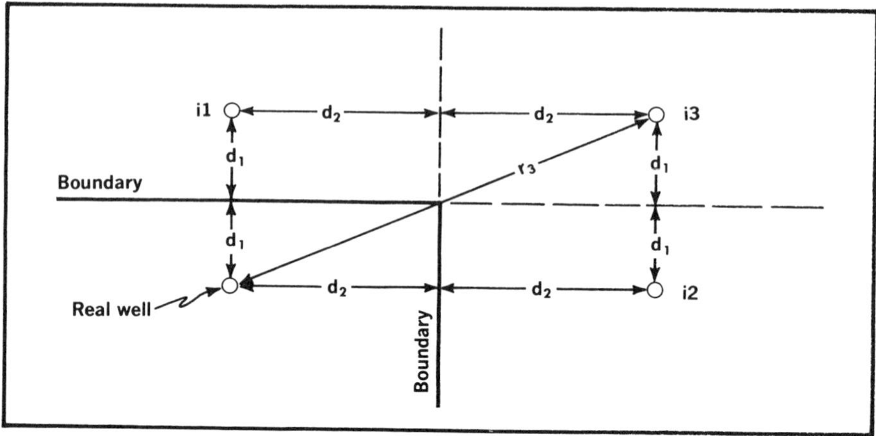

FIG. 2.17 Simulating two plane perpendicular barriers in an otherwise infinite-acting reservoir.

the two drainage boundaries. If you doubt this, consider just the three wells, the real well and the two image wells, i1 and i2, in an infinite system and delineate the drainage area of each well. You will find that the drainage areas do not conform to the required drainage boundaries. Consequently, four pressure-drop terms are required to determine the pressure in the real well. Note that the radius at which the effect of i3 is needed is the distance r_3 which is the hypotenuse of the right triangle indicated in the figure.

$$p_w = p_i - \frac{0.141q\mu}{kh}(PF)_w - \frac{0.141q\mu}{kh}(PF)_{2d1} - \frac{0.141q\mu}{kh}(PF)_{2d2}$$

$$- \frac{0.141q\mu}{kh}(PF)_{r_3} \tag{2.39}$$

Even a very simple geometry can cause considerable difficulty in simulating boundaries. Consider Fig. 2.18A. Here the real well is between two parallel and equidistant boundaries. If we add image well i1 we will stop drainage across the right boundary. Then if we add image wells i2 and i3 we will stop flow across the left-hand boundary. But with only image wells i1, i2, and i3, we would not have production balanced across the right boundary so that we would have to add wells i4 and i5. Then flow would not be balanced across the left boundary and we would have to add wells i6 and i7. This unbalances flow across the right boundary etc., etc. It may appear that no solution is possible but remember that when the image wells in the pattern get far enough away from the real well they will have no effect on the real well pressure during the time of interest. Consequently, we can determine what this limiting distance is for a particular time and know that we will not improve the accuracy

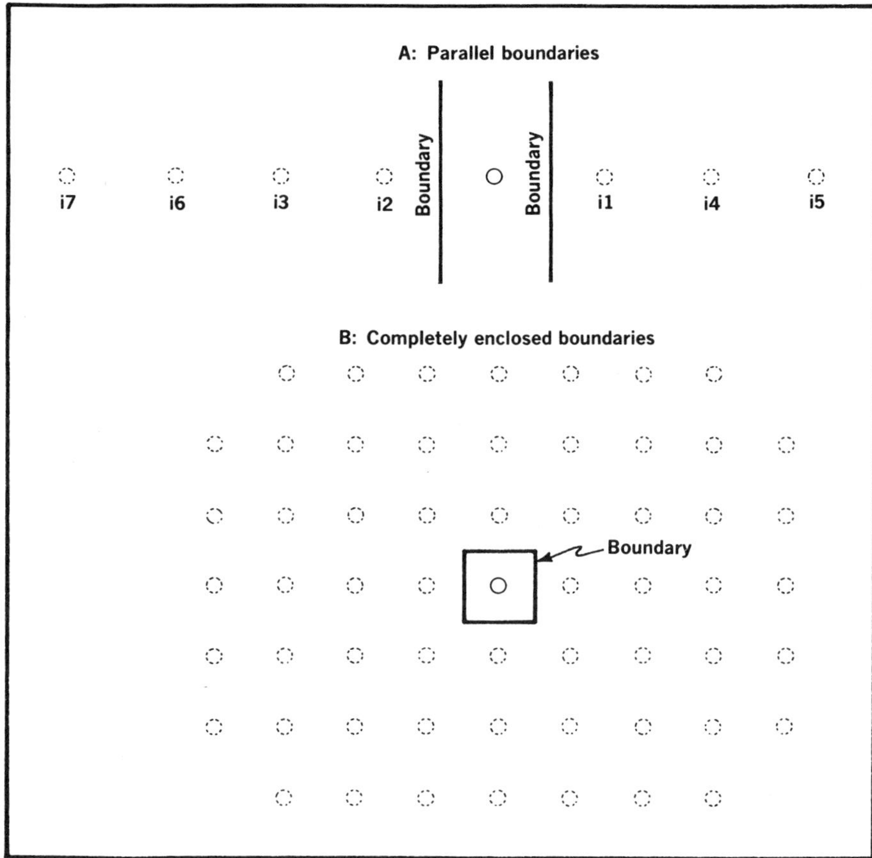

FIG. 2.18 Simulation requiring infinite ghost well arrays.

of the calculation for a particular time by adding additional image wells. This technique lends itself quite readily to a computer application.

Note from Fig. 2.18B that it is possible to analyze a completely enclosed drainage area using an infinite-acting solution if a large enough grid of image wells is considered.

The Q_{tD} functions that we have considered do not lend themselves to this type of boundary simulation.

PSEUDOSTEADY STATE FLOW

Pseudosteady-state flow is actually the finite-acting portion of the constant-rate solution to the radial diffusivity equation. Thus, it could be analyzed by using the same pressure-function solution previously described under the constant-rate solution section. However, this special case of the constant-rate solution lends itself to much simpler equations and methods so that it is important that we look at it as a special case. Although it is recognized as a special case it appears that more reservoirs spend more of their history in pseudosteady-state than in any other of the flow regimes.

If the reader will again refer to Fig. 1.9 in Chapter 1, he will find that after a well has produced at a constant rate for a long enough period of time to affect the pressure throughout the entire drainage area, the change in pressure with time becomes the same throughout the reservoir and the pressure distributions become parallel at successive periods. We will show later, mathematically, that this is true.

In this chapter we will talk principally about pseudosteady-state flow in a radial geometry. However, pseudosteady-state flow will occur regardless of the geometry. Very irregular geometries will also reach pseudosteady state whenever they have been produced long enough for the entire drainage area to be affected.

Since the pressure is declining uniformly with time at all points in the reservoir in pseudosteady-state flow we can equate this change of pressure with time to the material-balance equivalent since we know that all production and flow is due to the expansion of the fluid as the pressure declines. Consequently, the daily producing rate per unit of pressure drop will simply be the reservoir pore volume, $\pi r_e^2 h \phi$ for radial flow, multiplied by the compressibility, c. When this product is multiplied by the daily pressure change, $\Delta p / \Delta t$, we obtain the production rate in barrels per day.

$$q_w = \frac{(\pi r_e^2 h \phi)c(\Delta p / \Delta t)_{pseudo}}{5.615} \qquad (2.40)$$

This can be rearranged to obtain the change in pressure with time during pseudosteady-state radial flow.

$$(\Delta p / \Delta t)_{pseudo} = \frac{1.79q}{\phi h c r_e^2} \qquad (2.41)$$

Since Equations 2.40 and 2.41 are material-balance equations, c must represent the total effective compressibility and q the total rate of reservoir voidage including the oil, gas, and water produced. Also note that the change in pressure with time during pseudosteady-state flow will remain constant at all radii and for all times as long as the flow rate, q, and the compressibility, c, remain constant. Since both of these conditions are requirements for the constant-rate solution to the radial diffusivity equation and pseudosteady state is a part of this solution, the change in pressure with time during pseudosteady-state flow is constant.

Equation 2.41 is an overlooked but useful equation. It may be the only practical means of determining the effective compressibility of the reservoir when it contains substantial gas and oil saturation. Simply determining an average compressibility based on the saturations and the compressibility of gas, water, and the formation looks good on paper; however, it seldom appears to be correct. This is probably due to the difficulty of accurately determining the gas-production data which is generally the case when gas is not being metered for sale. This difficulty is discussed in later chapters.

Note that Equation 2.41 is basically a material-balance equation rather than a flow equation. The rate must then represent the total reservoir voidage rate and include gas if free gas is flowing in the reservoir. The total rate would then be the oil rate $q_o B_o$ plus the free gas rate $(R-R_s)B_g$. In this expression R is the producing gas-oil ratio and R_s is the gas in solution in the oil. Also note that an effective radius can be used in Equation 2.41 without sacrificing any accuracy regardless of the irregularity of the drainage area. Again this is true because the equation is basically a material-balance equation rather than a flow equation.

Equation 2.41 can also be written in terms of the total drainage volume in cubic feet, V_b. In this form it can be applied to any flow geometry.

$$(\Delta p/\Delta t)_{\text{pseudo}} = \frac{5.615q}{c\phi V_b} \tag{2.42}$$

Since the pressure gradient, $(\Delta p/\Delta r)$, at any particular radius in the reservoir remains constant, all of the factors in the radial form of the Darcy equation,

$$q_r = \frac{1.127kA_r}{\mu}\left(\frac{\Delta p}{\Delta r}\right)_r \tag{1.6}$$

are constant leading many engineers to define this flow regime as steady state. Since the pressures are actually declining, flow is unsteady state by the definitions of this text, but it appears to be steady state — hence the term, pseudo-, or false, steady state.

Practical flow equations.   Flow equations similar in form and simplicity to steady-state equations can be easily derived for pseudosteady-state flow. We use the same techniques employed in Chapter I to account for geometry and compressibility. The geometry is accounted for by substituting $2\pi rh$ for the cross-sectional area, A, and the compressibility of the fluid is accounted for by substituting a function of the well rate, q_w, for the general flow rate, q, in the Darcy equation. In Equation 2.40 it was shown that the flow rate, q_w is simply the pore volume multiplied by the compressibility and the change in pressure in time. The flow rate at any radius would be exactly the same except that the volume providing the flow rate at a particular radius is due only to the volume outside the particular radius. Thus, we could state the flow rate at any radius as

$$q_r = \frac{(\pi r_e^2 - \pi r^2)h\phi c(\Delta p/\Delta t)_{\text{pseudo}}}{5.615} \tag{2.43}$$

Now if we take a ratio of this equation and Equation 2.40 we obtain the relationship between q_w, the rate at the well, and q_r, the rate at any radius.

$$\frac{q_r}{q_w} = \frac{r_e^2 - r^2}{r_e^2} = 1 - \frac{r^2}{r_e^2} \tag{2.44}$$

$$q_r = q_w\left(1 - \frac{r^2}{r_e^2}\right) \tag{2.45}$$

Substituting this expression for q and $2\pi rh$ for A_r in Darcy's equation and rearranging we obtain

$$q_w\left(1 - \frac{r^2}{r_e^2}\right)\frac{\Delta r}{r} = \frac{7.08kh}{\mu}\Delta p \tag{2.46}$$

When the equation is integrated from well conditions to conditions at any radius, r

$$q_w\int_{r_w}^{r}(\Delta r/r) - q_w\int_{r_w}^{r}(r\Delta r/r_e^2) = \frac{7.08kh}{\mu}\int_{p_w}^{p_r}\Delta p \tag{2.47}$$

$$q_w\ln(r/r_w) - q_w\left(\frac{r^2}{2r_e^2} - \frac{r_w^2}{2r_e^2}\right) = \frac{7.08kh}{\mu}(p_r - p_w) \tag{2.48}$$

$$q_w = \frac{7.08kh}{\mu}\frac{(p_r - p_w)}{\left[\ln\left(\dfrac{r}{r_w}\right) - \dfrac{r^2}{2r_e^2} + \dfrac{r_w^2}{2r_e^2}\right]} \tag{2.49}$$

Equation 2.49 represents the most accurate form of the pseudo-steady-state flow equation. However, the term relating the squares of the well radius and external radius is normally insignificant when compared with the other terms in the denominator and it is thus assumed to be zero to obtain

$$q_w = \frac{7.08kh}{\mu}\frac{(p_r - p_w)}{\left(\ln\dfrac{r}{r_w} - \dfrac{r^2}{2r_e^2}\right)} \tag{2.50}$$

The form of the pseudosteady-state equation most often encountered in the literature is Equation 2.50 written in terms of the external radius and pressure at the external radius.

$$q_w = \frac{7.08kh}{\mu}\frac{(p_e - p_w)}{\left(\ln\dfrac{r_e}{r_w} - \dfrac{1}{2}\right)} \tag{2.51}$$

Care should be taken by the engineer to avoid misapplication of Equation 2.51. It can be applied only when using the pressure and radius at the external boundary of the reservoir. When other pressures and radii are involved Equation 2.50 must be used. Since the general steady-state radial flow equation similar to 2.51 can be applied between any two radii, the engineer often erroneously assumes that the same is true of this pseudo steady state equation.

In Table 2.1, which helps determine which constant rate solution is best suited for a specific case, one or two places will be found where the engineer is advised to use pseudosteady-state flow for the calculation. There are options here, ways depending on the data available. If the well pressure or external pressure is known, the pressure at the radius of interest can be determined directly by solving Equation 2.48. However, in the type of application covered by Table 2.1, we would only know the initial well pressure and when pseudosteady state is needed the reservoir is finite acting and the pressure is desired away from the well bore. Under these conditions the Ei function cannot be used because the reservoir is no longer infinite acting and the P_{tD} function cannot be used because we do not know the flow rate at the radius where the pressure is desired. In such a case the pressure at the well at the desired time can be calculated using the P_{tD} function and the pressure at the desired radius can be calculated by Equation 2.50.

Theoretical basis for pseudosteady-state flow. The pseudosteady-state flow equations were derived by assuming that pseudosteady-state flow does exist or more specifically that the change in pressure with time is constant throughout the reservoir. We will now show that based on the Hurst-Van Everdingen p_{tD} function solutions to the radial diffusivity equation the change in pressure with time is constant and equal to the change in pressure with time required for pseudosteady-state material balance as shown in Equation 2.41.

We have been assuming all along that pseudosteady-state flow begins about the time when the entire pressure has been affected by production. After this period of production at a constant rate we found that the p_{tD} function could be evaluated by using Equation 2.20. If we use this definition of the p_{tD} function, and determine the reservoir pressure at the well by the constant-rate solution when this p_{tD} equation applies, we would simply substitute the right-hand side of Equation 2.20 for p_{tD} in Equation 2.16.

$$p_{w,t} = p_i - \frac{0.141 q_w \mu}{kh} \left(\frac{2t_D}{r_{De}^2} + \ln r_D - \frac{3}{4} \right) \qquad (2.52)$$

Now note that only one term in the equation changes with time so that we could group all the other terms that are constant as far as time is concerned.

$$p_w = (\text{constant}) - \left(\frac{0.141 q_w \mu}{kh} \frac{2t_D}{r_{De}^2} \right) \qquad (2.53)$$

Now if we expand the expression for t_D, and r_D we can derive a relationship between p_w and real time.

$$p_w = (\text{constant}) - \frac{0.141 q_w \mu}{kh} \frac{(2)(6.33)kt}{\phi \mu c r_w^2 (r_e^2/r_w^2)} \qquad (2.54)$$

$$p_w = (\text{constant}) - \frac{1.79 q_w t}{\phi h c r_e^2} \qquad (2.55)$$

$$(\Delta p_w/\Delta t) = \frac{1.79 q_w}{\phi h c r_e^2} \qquad (2.56)$$

Equation 2.56 is obtained from Equation 2.55 by recognizing that a plot of p_w versus t would give a line whose slope would be the right-hand side of Equation 2.56. Thus, the change in well pressure with time when the p_{tD} function is governed by Equation 2.20 is exactly the same as the change in pressure with time required for pseudosteady-state flow, as shown by Equation 2.41. Consequently, whenever Equation 2.20 governs the p_{tD} function we are assured that we have pseudosteady-state flow.

Time limits on pseudosteady-state flow.[6]   If we can now determine when Equation 2.20 governs the p_{tD} function we can find the lower limit on pseudosteady-state flow. When the constant-rate solutions were studied previously in this chapter we simply used a time limit of $t = r_e^2/4\eta$. Consequently, our problem is one of deriving this time limit that we have previously assumed.

To do this, consider the problem of defining all of the p_{tD} values for the entire range of producing times for a particular sized reservoir with a particular r_D. If we exclude the period when the reduced time, t_D, is less than 100 we can conclude that during the early producing times when the reservoir is infinite acting the p_{tD} values would be governed by Equation 2.21 which we have referred to as the log solution. Also, we would know that at large times when the reservoir is finite acting the p_{tD} function would be governed by Equation 2.20 as previously described. This situation is illustrated by Fig. 2.19. The problem is: when should we switch from the use of the infinite-acting solution to the finite solution which is pseudosteady state? Fortunately the two solutions come very close together so the choice of a time to switch from one to the other is not critical as far as quantitative work is concerned. The real time representing the t_D at which we choose to switch is the critical time that is often referred to as the stabilization-time equation or simply the time equation. Many odd and sometimes seemingly ridiculous approaches have been used to derive this time equation. The approach presented here seems to be better defined than any of the others.

We will choose to switch from one solution to the other when the two solid line curves, Fig. 2.19, come closest together. This will occur when the difference between the two is a minimum. Fig. 2.20 shows what a plot of the difference versus the reduced time would look like. To find the

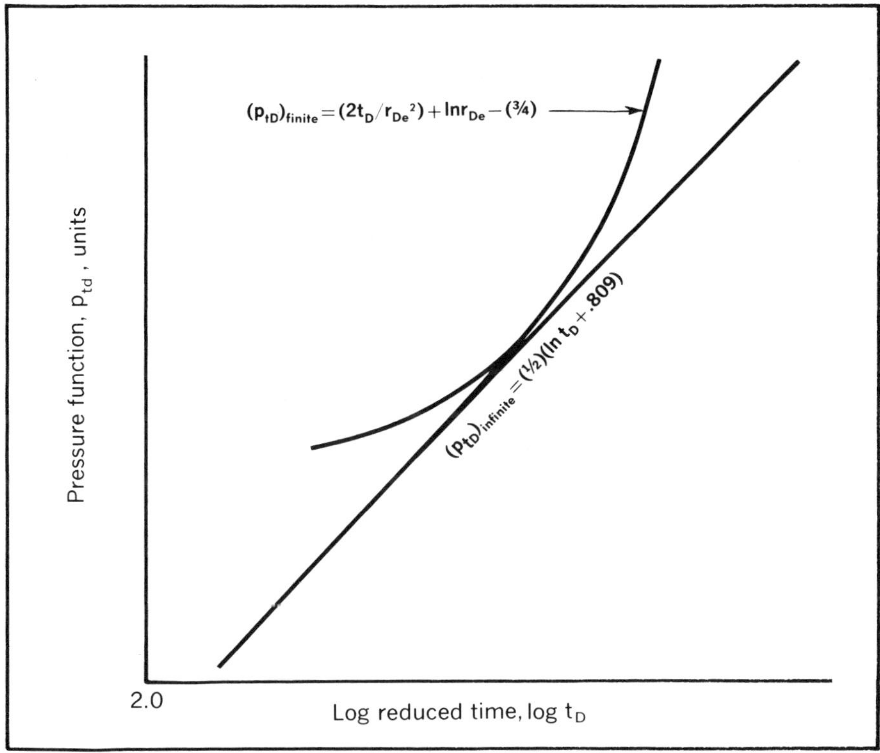

FIG. 2.19 Comparison of infinite and finite-acting pressure functions.

point where the difference is a minimum we can evaluate the point on the curve where the slope is zero.

$$\text{Difference} = \left(\frac{2t_{Ds}}{r_{De}^2} + \ln r_{De} - \frac{3}{4}\right) - \frac{1}{2}(\ln t_{Ds} + 0.809) \qquad (2.57)$$

$$(\Delta\text{Difference}/\Delta t_D) = \frac{2}{r_{De}^2} - \frac{1}{2t_{Ds}} = 0 \qquad (2.58)$$

$$t_{Ds} = \frac{r_{De}^2}{4} \qquad (2.59)$$

Now when t_{Ds} and r_{De} are expanded we obtain the previously used time limit on the infinite-acting behavior of the reservoir.

$$\frac{\eta t_s}{r_w^2} = \frac{r_e^2/r_w^2}{4} \qquad (2.60)$$

$$t_s = \frac{r_e^2}{4\eta} \qquad (2.61)$$

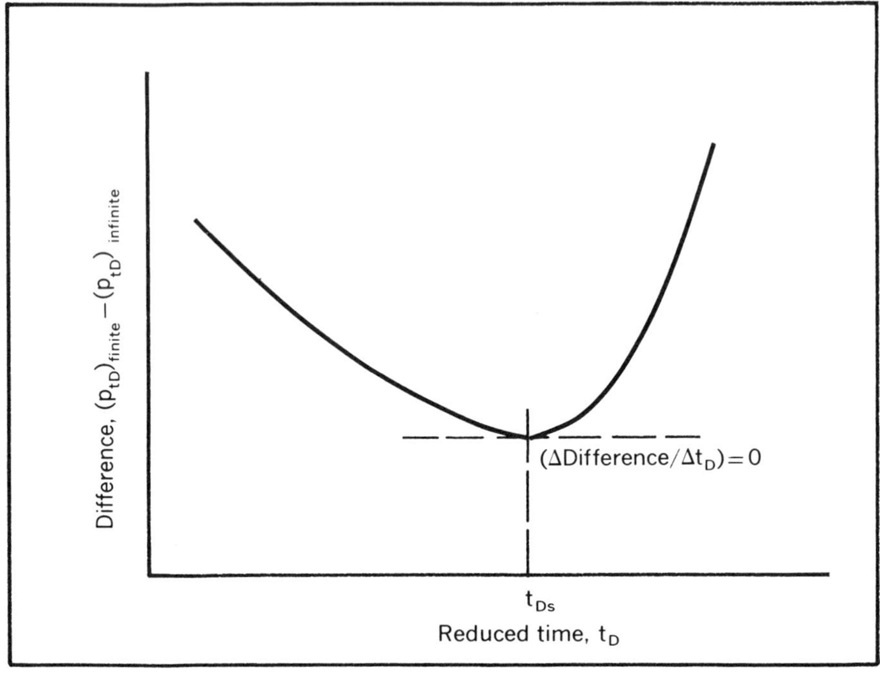

FIG. 2.20   Plot demonstrating the basis of the critical time equation.

The subscript s refers to stabilized which is a term used synonymously with pseudosteady state by many engineers. When we expand the diffusivity constant, η, we obtain the time equation that is most often seen in the literature.

$$t_s = r_e^2 \left/ \frac{(4)(6.33)k}{\phi\mu c}\right.$$ (2.62)

$$t_s = \frac{0.04\phi\mu c r_e^2}{k}$$ (2.63)

The constant, 0.04, in this equation does not exactly fit the reciprocal of (4)(6.33); however, the method used to derive this critical relationship is not exact. The time defined should be used simply as a guide. Since this constant 0.04 has been in common use for several years the author has chosen to continue its use rather than introduce another constant that would not greatly increase the accuracy of the critical time.

One confusing aspect of the time equation for many engineers is that the equation does not contain q. In other words, intuitively we expect the time of travel of a disturbance through the reservoir to be a function of the magnitude of the disturbance. When the mathematics of the situation are studied carefully we find that this is one more instance in which our intuition leads us astray. The magnitude of the pressure drop is affected by the rate which is a part of the conversion factor from the pressure function (see Equation 2.16). However, the speed of travel of the disturbance through the reservoir is independent of the magnitude of the disturbance. An analogous situation which we do take for granted is the time travel of sound through the air. We accept the fact for example that the sound from thunder travels about 1 mile in 5 sec regardless of the loudness of the thunder (magnitude of the disturbance). The loudness of the sound we hear is a function of the loudness of the thunder but the time it takes the noise to reach us is independent of the loudness.

Although the time equation was derived to obtain the lower time limit on pseudosteady-state flow it can also be used as an excellent means of determining how long is required for a well that is normally in steady state to reach steady state again after a rate change. This information is often very useful in analyzing well behavior in a reservoir with a very strong water drive. Note also that following a rate change in a well that is in pseudosteady state the same period of time denoted in Equation 2.63 is required to again establish pseudosteady-state flow. The normal well in a solution-gas-drive reservoir producing at capacity will continue to reach a series of pseudosteady states as the rate continues to decline due to pressure decline in the reservoir. The rate of production decline will be so small relative to the time to reach pseudosteady state that the well will, for all practical purposes, continue to produce in pseudosteady state.

To emphasize the physical significance of some of the pseudosteady-state equations, some example calculations will be helpful. Check your problem solution against the Appendix.

Problem 2.4: Analysis of a Pseudosteady-State Flow Test

A reservoir fluid with a viscosity of 0.5 cp and a saturation pressure of 2,500 psia is flowing from a radius, r_e, of 2,000 ft and a pressure of 3,500 psia (determined from a pressure buildup test) to a well radius of 0.5 ft and a pressure of 3,200 psia. The bottom-hole pressure has been declining at a constant rate of 2 psi/day, the producing rate is a constant 850 STB/d, the oil formation volume factor is 1.1 and the effective compressibility is 10^{-5}/psi.

A. What is the porosity-thickness?
B. What is the permeability-thickness?
C. If the bottom-hole pressure had been remaining constant and there was evidence of a water drive (steady-state flow), what would be the permeability-thickness?

Note that in the above example the error introduced in calculating the capacity, kh, for the formation using a steady-state equation rather than an unsteady-state equation amounts to only about 6% which is not overly important in reservoir calculations. This is probably typical of the error introduced by assuming steady-state flow when pseudosteady state actually exists in a reservoir. This is because flow around the well bore, where most of the pressure drop occurs, is practically steady state (the rates and pressures are constant) when the reservoir is in unsteady state or pseudosteady state. Consequently, the calculations are dominated by this pressure drop in the vicinity of the well bore.

Another point in connection with pseudosteady state often confusing to the engineer is that given the same pressure drop from the outer boundary to the well, a well in pseudosteady state will produce at a greater rate than will a well in steady state. Mathematically this is due to the constant, $\frac{1}{2}$, in the denominator of Equation 2.51 which is the only difference between the radial pseudosteady state and steady-state equations. Physically or mechanically the pseudo rate is greater than the steady-state rate because the average flow rate in pseudosteady-state flow is less than the rate at the well (since production is due to fluid expansion) while in steady state the rate is the same at all points in the reservoir. Thus the average rate is the same as the rate at the well. Since pressure drop is proportional to rate, this means that the total pressure drop necessary to obtain a particular well rate in pseudosteady state is less than the total pressure drop necessary to obtain the same well rate in steady state. Consequently, the original observation that a pseudosteady-state well will produce at a higher well rate for a particular total pressure drop than will a steady-state well with the same pressure drop is logical.

Do not let this confuse you. Although at the initial total pressure drop the pseudosteady-state well will produce more than the steady-state well, remember that early in the life of the pseudosteady-state well the well pressure will reach its lower limit because the pressure throughout the reservoir is declining continuously. When this occurs the total pressure drop will begin to decline and the production rate will begin to decline. On the other hand, in the steady-state well, the total pressure drop will remain essentially constant throughout the well life and the production rate will decline only when the displacing water or gas begins to be produced in the well.

REFERENCES

1. D. L. Katz et. al., *Handbook of Natural Gas Engineering,* New York, New York: McGraw Hill, 1959.
2. B. C. Craft and M. F. Hawkins, *Applied Petroleum Reservoir Engineering,* Englewood Cliffs, New Jersey: Prentice Hall, 1959.
3. A. F. Van Everdingen and W. Hurst, "The Application of the Laplace Transformation to Flow Problems in Reservoirs," *Trans.* AIME (1949), *186,* 305.
4. H. S. Carslaw and J. C. Jaeger, *Conduction of Heat in Solids,* Oxford: Clarendon Press, 1959.

5. Thomas D. Mueller and Paul A. Witherspoon, "Pressure Interference Effects Within Reservoirs and Aquifers," J. Pet. Tech. (April 1965) 471–474.

6. H. C. Slider, "A Simplified Method of Pressure Buildup Analysis for a Stabilized Well," J. Pet. Tech. (Sept. 71).

7. A. F. Van Everdingen, E. H. Timmerman, and J. J. McMahon, "Application of the Material Balance Equation to a Partial Water-Drive Reservoir," Trans. *AIME* (1953), *198*, 51.

8. C. V. Theis, "The Relationship Between the Lowering of Piezometric Surface and Rate and Duration of Discharge of Wells Using Ground-Water Storage," *Trans.*, AGW (1935) II, 519.

9. E. R. Brownscombe and F. Collins, "Pressure Distribution in Unsaturated Oil Reservoirs," *Trans.* AIME (1950), 189, 371.

10. Muskat, M., *The Flow of Homogeneous Fluids Through Porous Media*, McGraw-Hill Book Co., Inc., New York, 1937.

# Well-Pressure Behavior Analysis

One of the major problems of reservoir engineering is improving the capacity of wells to produce. This is an individual well problem as opposed to reservoir engineering problems that are concerned with the reservoir as a whole—controlling production to obtain maximum recovery or application of secondary-recovery techniques to improve recovery. The individual well problem is basically concerned with fluid-flow technology as opposed to general reservoir problems that must be concerned with material balance and fluid displacement as well as some fluid flow.

Well-pressure behavior analysis is a term that it is hoped will convey the inclusion of all of the well-testing procedures. However, gas-well deliverability tests are so specialized that they are treated in a separate chapter. But now we will be concerned with the interpretation of the productivity-index flow tests and the common erroneous assumptions that are made by the practicing engineer concerning such tests. It will present an entirely new approach to transient pressure behavior termed negative superposition, that greatly simplifies transient pressure analysis and emphasizes the shortcomings of many of the commonly used pressure buildup and drawdown methods. This chapter also indicates the most foolproof methods of determining reservoir boundaries and discusses the unlikely possibility that a method could completely describe a reservoir drainage shape from a transient pressure test. Finally the problem of making realistic production rate estimates from drill-stem-test data is considered.

PRODUCTIVITY-INDEX TESTS

Historically the petroleum engineer has expected the production rate from a well to be proportional to the pressure drawdown, $p_e - p_w$, or inversely proportional to the well pressure, p_w.[1] Furthermore, he has more or less expected the relationship between the drawdown and the rate to remain constant with time. These ideas have been used to predict the production rate under various operating conditions, for example pumping a well versus flowing it. They have also been used to predict the variation in the producing rate with time or reservoir pressure depletion. The knowledge of fluid flow covered in Chapters 1 and 2 will provide us with tools to investigate the validity of these ideas.

The drawdown relationship to the producing rate is known as the productivity index and is mathematically defined as,

$$J = \frac{q_{STB}}{(p_e - p_w)}.\tag{3.1}$$

J is the symbol preferred by the Society of Petroleum Engineers to denote the productivity index. Note from Equation 3.1 that the rate in this equation is stated in stock-tank barrels, contrary to the general use of the rate symbol in this text which uses the units as reservoir barrels per day unless specifically designated otherwise. This variation is simply a means of conforming to the general field definition of the productivity index.

The above stated observation that the straight-line relation between the rate and drawdown is independent of the magnitude of the rate or reservoir depletion is then the same as assuming that the productivity index does not change with the rate or time (depletion). We will see that this assumption is true under only very special conditions.

A well producing at a constant well pressure from a solution-gas-drive reservoir would be in unsteady-state flow, governed by the constant-pressure solution, until the entire drainage area is affected, at which time it would tend to enter pseudosteady state. The term "tend to enter" is used because the well would not be theoretically in steady state until a constant rate (not pressure) is maintained for the stabilization time. We know that during the constant-well-pressure unsteady-state phase of the production, the rate would decline somewhat as the producing time increased. However, after the entire drainage area is affected flow will come closer to pseudosteady state than to any other flow regime for which we have working equations. Since most of the well's life will be spent in a flow regime approximating pseudosteady state we should be predicting pseudosteady-state flow rates from the productivity-index data. Furthermore, it would be impossible to even consider the unsteady-state constant-pressure portion of the life as being applicable to the productivity index idea, since we know that the flow rate varies during the constant-pressure unsteady-state flow regime.

When we examine the pseudosteady-state radial equation from Chapter 2,

$$q_w = \frac{7.08kh(p_e - p_w)}{\mu\left(\ln\dfrac{r_e}{r_w} - \dfrac{1}{2}\right)},\tag{2.41}$$

we see that if we substitute $q_{STB}B_o$ for q_w and write the equation for oil flow we can solve for the productivity index, J.

$$J = \frac{q_{STB}}{p_e - p_w} = \frac{7.08k_oh}{B_o\mu_o\left(\ln\dfrac{r_e}{r_w} - \dfrac{1}{2}\right)} \qquad (3.2)$$

Similarly, a well producing from a strong water drive or gas-cap-drive reservoir that tends to reach steady-state flow will have its productivity index described by a steady-state radial-flow equation similar to Equation 2.3.

$$J = \frac{7.08k_oh}{B_o\mu_o \ln(r_e/r_w)} \qquad (3.3)$$

If the productivity index as defined in Equations 3.2 and 3.3 are to remain constant during a flow test several conditions must be met. For the equations to be applicable it is necessary that a particular flow rate be maintained until the producing time at this rate exceeds the stabilization time defined in Chapter 2 as

$$t_s = \frac{0.04\phi\mu cr_e^2}{k} \qquad (2.63)$$

In applying this equation, care should be exercised to make certain the effective compressibility is used, especially when substantial gas saturations exist in the reservoir. The effective compressibility is described under EFFECTIVE COMPRESSIBILITY in Chapter 2. The saturations would normally be calculated by material balance. An equation for the calculation of the oil and gas saturations can be found in the Material Balance Chapter. Actually the stabilization time necessary to reach steady-state flow is somewhat longer than that indicated by Equation 2.63.[3] However, no significant error will be introduced by using Equation 2.63 for this purpose since the flow regime will be very near complete steady state at this time.

If producing times for each rate are not in excess of the stabilization time we could not expect to obtain an accurate productivity index. A straight-line relationship between the rate and drawdown may still be obtained due to insensitivity of the method or due to an "isochronal" effect if about the same producing times are used for each flow rate. The isochronal effect is discussed in Chapter 4. The effect is roughly one of affecting about the same amount of the reservoir at each rate so that the final drainage radius at each rate is about the same. This provides a basis for a straight-line plot of drawdown versus rate but this is not indicative of the productivity index and use of the value will result in predictions that are over optimistic.

Another factor that can keep the measured productivity index from being constant during a test is the effect of any change in the gas saturation on the permeability to oil.[4] As has been seen many times, the bulk of the pressure drop in the radial-flow system is right around the well bore. Consequently, this is the area where the largest change in the gas saturation occurs. Unfortunately this is also the area where a change in the effective permeability to oil will have the biggest effect on the average permeability of the drainage system. Since oil is the wetting phase relative to gas, small changes in the gas saturation can cause large changes in the relative permeability and the corresponding effective permeability. This reduction in permeability can be observed even when the gas saturation is less than the equilibrium gas saturation so that no free gas flows and there is no effect on the produced gas-oil ratio.

Many engineers erroneously believe that they can ascertain a substantial change in the gas saturation around the well bore by observing the produced gas-oil ratio. However, most of the produced gas originates in the reservoir, well away from the area where the gas saturation increases around the well bore. Consequently, although the permeability to gas increases with the increase in gas saturation around the well bore, there is no gas to replace this mobile gas around the well bore except gas coming out of solution as the oil passes through the low pressures near the well. Thus, there is no effect on the produced gas-oil ratio because the source of the produced gas-oil ratio, the area well removed from the well bore, does not change pressure significantly during a well test encompassing a few days at the most.

One other difficulty that often prevents the productivity index from being constant during a test is Darcy's basic equation. Remember that it was emphasized in Chapter 1 that the Darcy equation, which is the basis for the steady state and pseudosteady-state equations, is not valid for turbulent flow. This does not normally place a limitation on the applicability of equations for the flow of liquids when the flow rate is less than possibly 50 BPD/ft of sand. However, the engineer should constantly keep in mind that turbulence can cause a deviation in the productivity index and that the 50 BPD/ft of sand rule of thumb is just a guide line. Significant turbulence can occur at much smaller rates if the right combination of permeability, porosity, and pore-size distribution is encountered. In viscous or streamline flow the pressure drop is proportional to the mass flow rate as indicated in Darcy-based equations and the productivity index. In turbulent flow the pressure drop is roughly proportional to the square of the mass flow rate.

Insufficient producing times at each rate, a change in the gas saturation surrounding the well bore, or turbulence may prevent productivity-index tests from indicating a constant productivity index. These factors may also cause a variation in the productivity index when it is measured at different times or stages of depletion. The sufficiency of the producing time at each rate is not particularly associated with the stage of depletion except that the stabilization time will increase almost proportionately with the increase in the average gas saturation in the reservoir. Consequently, the higher the reservoir gas saturation, the more the engineer should be concerned about the stabilization time. Since in solution-gas-drive reservoirs the gas saturation increases as the reservoir depletes we can also emphasize that more consideration should be given to the stabilization time as the well life increases.

Turbulence, which as noted above seldom makes problems during oil well tests, will be even less likely to cause difficulties during the later life of a well when the total production capacity may have declined considerably due to the decline in the average reservoir pressure.

The same change in gas saturation around the well bore that may keep the productivity index from being a constant during testing may have an even more pronounced effect due to normal pressure depletion of the reservoir. Whereas during testing there may be a change in saturation around the well bore due to pressure drawdown, during depletion of a solution-gas-drive reservoir we are certain to have a change in the gas saturation, not only at the well bore, but throughout the reservoir. This change in gas saturation and pressure will change not only the mobility, k_o/μ_o, but also the oil formation volume factor, B_o. Consequently, if we can predict the change in the saturation and reservoir pressure we can predict the change in the productivity index by using a ratio of the mobilities and oil formation volume factors.[5]

$$\frac{J_1}{J_2} = \frac{(K_o/B_o\mu_o)_1}{(K_o/B_o\mu_o)_2} \tag{3.4}$$

Recommended procedure for interpreting productivity-index data. Many engineers simply use the flow rates and corresponding well pressures to calculate the productivity index at each rate to see if there is any systematic change in the values. However, to give as much statistical strength as possible to the interpretation, it is recommended that a plot of rate versus well pressure be constructed as illustrated in Fig. 3.1. The straight-line portion of the data may be considered as representative of the productivity index of this well. The productivity index can be calculated by evaluating the slope of the straight-line portion of the data and recognizing that the productivity index is this slope.

Note also that extrapolation of the curve to a q of zero will yield the pressure at the external boundary, p_e. This can be shown mathematically from the steady state and unsteady-state equations but physically it simply means that at a rate of zero the well pressure will be equal to the external pressure since no flow is taking place to make the pressures different.

Equation 3.4 can be used to predict the productivity index at any

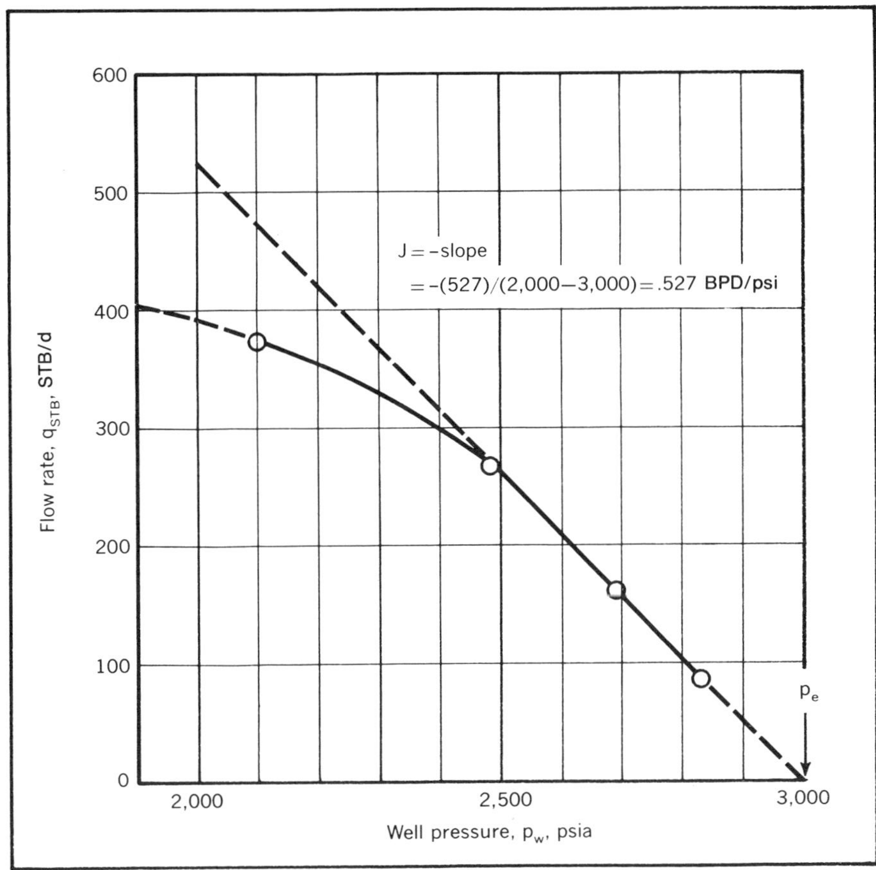

FIG. 3.1   Example productivity index.

future state of reservoir depletion. Equations 3.2 or 3.3 can be used to calculate kh/μ, kh, or the permeability, k, depending on the engineer's confidence in the independent evaluation of the thickness and viscosity. These calculated values represent the average values for the drainage area. They can be compared with the undamaged values for the well, if this is known, to determine the damage ratio. The damage ratio (DR) is a very simple expression that relates the actual production rate with the production rate that would be obtained if the well did not have any damage or improvement around the well bore. The ratio is evaluated as the undamaged (und) rate divided by the actual (act) flow rate which means that a DR > 1.0 indicates a damaged formation around the well bore and a DR < 1.0 indicates improved permeability around the well bore. Since the flow rates are proportional to the transmissibilities the damage ratio is equal to the ratio of the transmissibilities and the ratio of the permeabilities.

$$DR = \frac{q_{und}}{q_{act}} = \frac{(k_o h/\mu_o)_{und}}{(k_o h/\mu_o)_{act}} = (k_o)_{und}/(k_o)_{act} \tag{3.5}$$

The damage ratio has a big advantage over the Δp_{skin} or skin factor in dealing with managers or supervisory engineers because the DR is so much more readily understood and explained. Since damage ratios are so closely related to production rates it is easy to talk about the dollar value of getting rid of the damage in a well by treatment.

The productivity index can be useful in determining wells that have not been properly completed or that have sustained damage such as paraffin, cave, or some mechanical problem. A comparison of the productivity indices obtained at different times will indicate all of these problems except the damage or improvement of the original well completion. If a measured productivity index has an unexpected decline then one of the indicated problems should be investigated. A comparison of the productivity indices run on different wells in a particular reservoir at about the same stage of depletion should also indicate the wells that may have experienced unusual difficulties or damage during completion. Since the productivity indices may vary from well to well due to the variation in thickness of the reservoir, it is helpful to normalize the indices by dividing each by the thickness of the well. This is defined as the specific productivity index, J_s, and is

$$J_s = \frac{J}{h} = \frac{q_{STB}}{(p_e - p_w)h} \tag{3.6}$$

The productivity index could also be normalized for known variations in the permeability, drainage radius, or drainage geometry in much the same way.

When you must predict rates in the range of flow tests that fall off the straight-line plot of rate versus well pressure be sure to make your prediction on the extrapolated curved line. This can not be done with great accuracy for drawdowns much greater than those measured, but it will still obviously give much more accurate results than the straight-line extrapolation.

It is strongly recommended that you make every effort to determine the stabilization time for the well tested rather than to simply assume that your flow periods are sufficiently long. The most practical means of doing this for a solution-gas-drive reservoir is to determine the porosity, compressibility, and external radius squared product, $(\phi c r_e^2)$, by using the pseudosteady-state definition (Chapter 2) of the change in pressure with time.

$$(\Delta p/\Delta t)_{pseudo} = \frac{1.79q}{\phi h c r_e^2} \tag{2.41}$$

The change in pressure with time can be determined from previous pressure surveys or it can even be approximated from the change in surface tubing pressures if the well has produced at about the same rate and gas-oil ratio during the interim. Once the $\phi c r_e^2$ product has been determined it can be used in the stabilized time Equation 2.63 previously cited in this chapter to check the stabilization time.

It was previously noted that a productivity-index type of test is often used to predict the rate of production obtainable from a flowing well if that well is placed on pump. These tests are also often used to predict the results of putting a high-capacity submergible centrifugal pump in a well that is producing large amounts of water or in sizing pumps for waterflood production. In these cases the production rate must include the produced water.

Once the capacity of a reservoir has been determined by a productivity-index test it must be matched with the capacity of some lift system. The simplest approach to this problem is to assume that a pump can maintain a bottom-hole pressure close to atmospheric pressure and predict the production rate under these conditions. Or, that a flowing well will have a flowing bottom-hole pressure equal to a hydrostatic column of stock-tank oil extending from the bottom of the hole to the surface. This roughly assumes that the lightening effect of gas coming out of solution in the tubing is equal to the pressure drop due to flow through the tubing. This obviously is a crude approximation of the bottom-hole pressure that will be a function of the gas-oil ratio, the production rate, the tubing size, and numerous other factors.

Obviously, these methods leave much to be desired in the way of accuracy and do not pretend to be applicable to gas lift, plunger lift, free pumps, etc. A comprehensive treatment of these lifting problems is beyond the scope of this book. The reader is referred to *Principles of Oil Well Production* by Nind, *Gas Lift Theory and Practice* by Brown or some similar book.

To make certain you understand the principles of the productivity index tests work the following problem and check your solution against Appendix C.

Problem No. 3.1: Productivity Index Evaluation

A. The flow-test data plotted in Fig. 3.1A was obtained from a fractured well in a solution-gas-drive reservoir. This well has produced for 1 month. It drains about 40 acres and has a well radius of $1/_3$ ft. The undamaged permeability to oil as determined from a DST pressure buildup is 20.5 md and B_o is 1.25. The net thickness is 14 ft and the reservoir oil viscosity is 1.2 cp. Plot the productivity index vs. p_w. What is the damage ratio based on the straight-line portion of Fig. 3.1A data? Estimate the maximum pumping rate. What would be the maximum pumping rate indicated if the test rates were all less than 270 b/d?

B. It is estimated that production of 35,000 STB of oil from this reservoir will result in a pressure at the well drainage boundary of 2,800 psia and a reser-

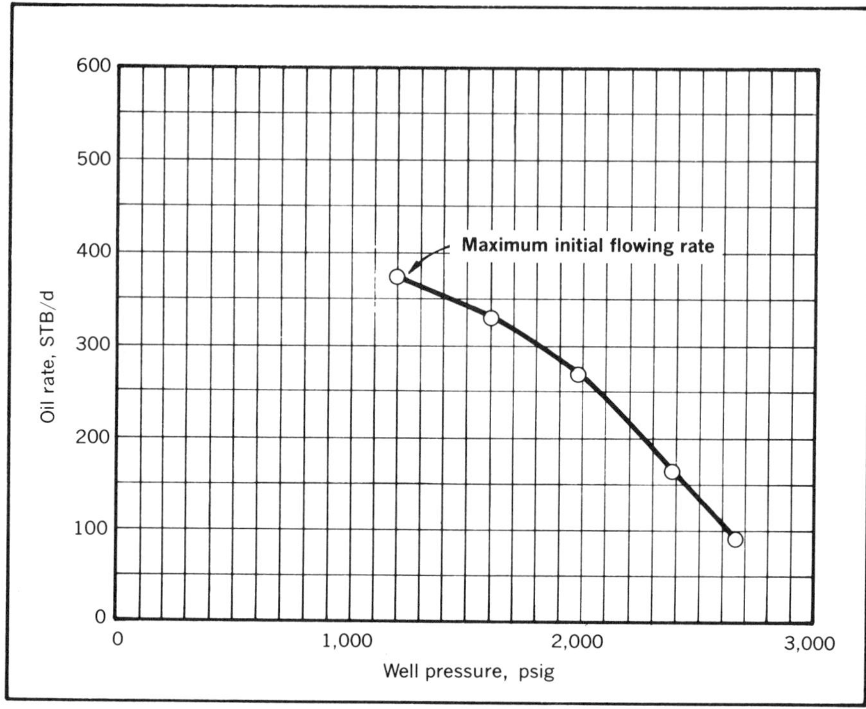

FIG. 3.1A Producing rate vs BHP for Problem 3.1.

voir gas saturation of 5%. At this time k_{ro} will be 0.5, k_{rg} will be 0.0, μ_o will be 1.1 cp, and B_o will be 1.15. Use k_{ro} in A as 1.0. Plot the P.I. vs. p_w at this time. Estimate the maximum flowing and pumping rates.

C. P.I. tests are being planned for the well at the time described in (B). The flowing tubing-head pressure has been declining at a uniform rate of about 3 psi/d while producing at 75 STB/d. Find the product, $\phi c r_e^2$. What stabilization times will be required?

PRESSURE BUILDUP ANALYSIS

Much has been written on pressure buildup analysis, possibly more than on any other specific subject in reservoir engineering. But because of confusion, it seems the engineer can seldom be certain that he is using the best or even a correct method of analysis. The Society of Petroleum Engineers Monograph No. 1 on pressure buildup analysis[6] did much to organize the published methods at the time of the preparation of the monograph. However, it still is the author's opinion that the monograph left a confusion of methods, the invitation to misapply some of the methods, and a dangerous set of "cookbook" forms that seem to imply that correct answers come from filling in forms. Probably 90% of the pressure buildup data can be analyzed using the same method of analysis regardless of whether the well is infinite acting, in pseudosteady state, or in steady state at the time of shutin.

Most pressure buildup methods claim to be able to determine the undamaged transmissibility, kh/μ (or some portion of it), some measure of the well damage, and the average reservoir pressure in the drainage area at the time of shutin. The transmissibility function can probably be evaluated with sufficient accuracy by about any method including many of those that are theoretically incorrect. The damage parameters are more difficult to determine accurately, and some of the methods for determining the average pressure are meaningless.

The most popular methods for determining the average pressure require an independent knowledge of the drainage-area shape and size (if you know the shape you must surely know the size), the average porosity, the average effective compressibility, and the cumulative production (rate and time). Using these values and a function of the buildup you can calculate the average pressure at time of shutin. However, a careful evaluation of the data required will show that there is really no need for the buildup data since the average pressure can be directly determined by material balance. This inability of the reservoir engineer to accurately evaluate average reservoir pressures is probably the biggest single reason for the many instances of "anomalous" reservoir behavior that is so often reported in the reservoir engineering literature. Because of the difficulty of determining the average reservoir pressure in a drainage area at the time of shutin this problem will be considered separately from the problem of determining the undamaged permeability and the amount of damage.

To save the experienced engineer from shock he is warned in advance that there is little use made of the Horner method in the methods developed and recommended in this chapter. As will be shown the Horner method is theoretically correct for an infinite-acting reservoir *only*. Except for gas wells, the infinite-acting portion of a well life is very small. One of the few times that infinite-acting behavior characterizes flow is during drill-stem tests, but in this case we will see that the pressure in the well on most drill-stem tests is increasing at the time of shutin while the Horner solution assumes the pressure is decreasing at the time of shutin.

Unchanging pressure at shutin. Pressure is not exactly unchanging, but we treat it that way if no appreciable or measureable change would have occurred during the shutin time if the well had not been shutin. This then will serve as our definition of "unchanging." The reader may wish to review the concept of superposition in Chapter 2. With this concept clearly in mind he will recognize the advantage of having a well whose pressure would not have changed during the time of shutin (if the well was not shutin) because we must superimpose the effect of the negative rate of shutting the well in upon the pressure that would have existed in the reservoir if the well had not been shutin.

The effect of a negative rate change is to cause a pressure increase. The effect is the same as the effect of injecting into the reservoir since an injection rate is the same as a negative production rate. Consequently, the pressure increase due to a negative rate, $-q$, acting for a time Δt, with the excessive (or additional) pressure drop due to damage around the well bore accounted for in a separate term, Δp_{skin}, is

$$\Delta p_q = \frac{0.141q\mu}{kh}\left(\frac{1}{2}\right)(\ln \Delta t_D + 0.809) + \Delta p_{skin} \qquad (3.7)$$

where the reduced time is

$$\Delta t_D = \frac{6.33k\Delta t}{\phi\mu c r_w^2} \qquad (3.8)$$

Equation 3.7 will be applicable only after Δt_D becomes greater than 100 and can be applied only until the effect of the shutin reaches the outer drainage boundary. On the basis of the well radius, the reduced time will generally become greater than 100 in a matter of seconds or minutes. Since the stabilization time is seldom less than a few hours, Equation 3.7 generally describes the pressure buildup for a long-enough period of time to permit an analysis.

Since the negative rate effect is superimposed on the pressure that would have existed in the reservoir if the $-q$ rate change had not occurred, we can state the well pressure during the time of shutin as

$$p_w = p_{wf} + \frac{0.141q\mu}{kh}\left(\frac{1}{2}\right)(\ln \Delta t_D + 0.809) + \Delta p_{skin} \qquad (3.9)$$

This equation may be clearer when it is presented graphically as in Fig. 3.2.

Equation 3.9 can be put in a more meaningful form if we substitute the expression of Equation 3.8 for Δt_D, convert from natural logs to logs with a base 10, and rearrange the equation to give,

$$p_w = p_{wf} + \frac{0.1625q\mu}{kh}\left(\log \frac{6.33k}{\phi\mu c r_w^2} + \frac{0.809}{2.3}\right) + \Delta p_{skin} + \frac{0.1625q\mu}{kh}\log \Delta t \qquad (3.10)$$

Recognizing that the first three terms of this equation do not change with time it can be written as,

$$p_w = \text{constant} + \frac{0.1625q\mu}{kh}\log \Delta t, \qquad (3.11)$$

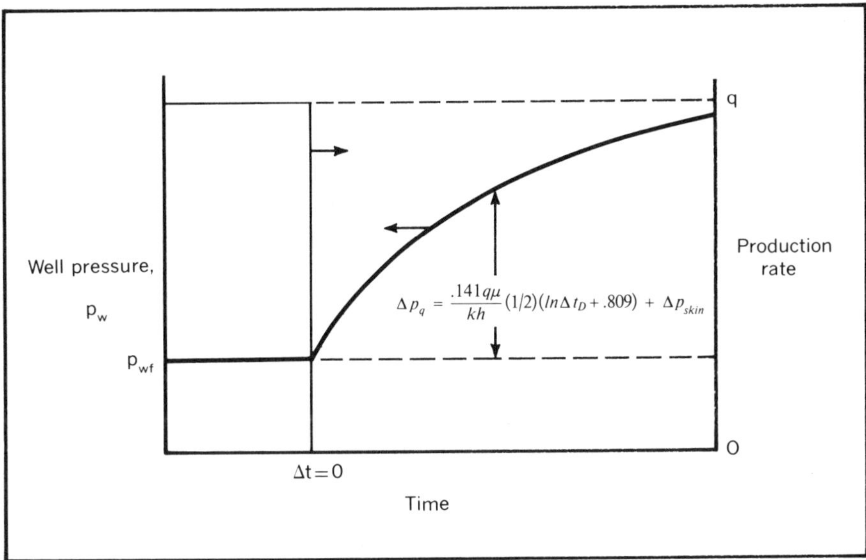

FIG. 3.2 Pressure buildup in a steady-state well or with an unchanging well pressure at shut-in.

to show that a plot of the well pressure, p_w, versus the log of the shutin time, log Δt, will result in a straight line whose slope will be

$$m = \frac{0.1625q\mu}{kh} \tag{3.12}$$

Such a plot is illustrated in Fig. 3.3. The shape of this plot is characteristic of pressure-buildup plots in general. The early portion of the curve does not fit the predicted straight line for several reasons. The biggest contributor to the nonideal early data is the fact that we have assumed that we were producing at a constant rate and that this rate was instantaneously reduced to zero. With the well being shutin at the surface (except on a drill-stem test) flow continues to take place across the well radius after flow has been stopped at the surface because the free gas in the tubing continues to be compressed with the increase in the well pressure. As the gas is compressed, oil continues to flow across the well radius to affect the reduction in the gas volume. This is called afterflow. When the gas is sufficiently compressed the flow rate approaches zero and the afterflow effect is essentially finished.

Since the flow rate is essentially zero in the vicinity of the well bore by the time the straight-line portion of the plot is reached, the slope of the straight-line portion is not affected by the damage around the well bore. During this period the damaged zone simply serves as a

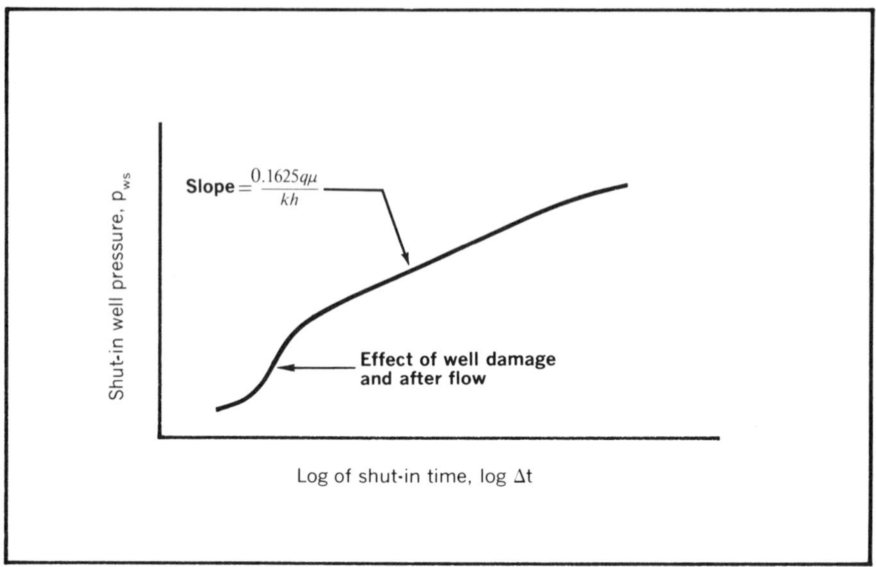

FIG. 3.3   Pressure buildup for a well with an unchanging flowing pressure at shut-in.

means of transmitting the pressure to the well bore where the pressure is measured. Consequently, the permeability, k, in the slope Equation 3.12, is the undamaged permeability. The same conclusion can be reached mathematically by examining Equation 3.10. The additional pressure drop in the damaged zone, Δp_{skin}, is a function of the rate, $-q$, and since this is constant during the shutin, the Δp_{skin} term remains constant and the slope, $0.1625q\mu/kh$, is unaffected by the damaged zone.

The early time nonlinear data are also contributed to by the damaged zone. In addition to the requirement that the rate at the well becomes zero before the straight-line relationship can be obtained, it is clear that the flow through the damaged zone must be essentially zero before the slope of the data can be a straight line reflecting the undamaged permeability of the formation.

Once the undamaged permeability has been determined from the slope of the plot, the additional pressure drop due to the damage around the well bore can be determined from Equation 3.9. This equation can be written as a function of the slope, m, as

$$p_w = p_{wf} + 0.867m\left(\frac{1}{2}\right)(\ln \Delta t_D + 0.809) + \Delta p_{skin} \qquad (3.13)$$

In using this equation it is necessary to determine the reduced time corresponding to the shut-in time. This means we must know the permeability, porosity, viscosity, effective compressibility, and well radius. The permeability divided by the viscosity can be determined from the slope, m, Equation 3.12, and the well radius is known. The other two constants ϕ and c, are generally more difficult to evaluate independently, particularly if a free gas saturation exists in the reservoir. The direct conventional calculation of these two terms from core or log analysis and published compressibilities, is recommended only as a last resort. If the reservoir is in pseudosteady state at time of shutin it is recommended that the terms be evaluated from the equation for the change of pressure with time during pseudosteady state.

$$(\Delta p / \Delta t)_{pseudo} = \frac{1.79q}{\phi h c r_e^2} \qquad (2.41)$$

for radial flow or

$$(\Delta p / \Delta t)_{pseudo} = \frac{5.615q}{\phi c V_b} \qquad (2.42)$$

for other flow geometries where V_b is the bulk volume in cubic feet. The terms, ϕc, may in some cases be evaluated through a knowledge of a nearby reservoir barrier. Later in this chapter we will learn that the distance to a barrier can be evaluated by noting the time when the straight-line portion of a buildup pressure plot deviates from the straight line. This time can be used with the distance to the barrier (use as r_e in the equation) to calculate ϕc from the time Equation, 2.63, derived in Chapter 2 and previously used in this chapter.

The term Δp_{skin} is generally sufficient for describing the damage in a particular well because it is possible to directly determine how much of an increase in the production rate will result when the damage is corrected or removed by treatment. However, if the engineer desires to compare the damage in wells in a particular reservoir, a more sophisticated term is necessary because he must normalize the pressure drops due to damage for the differences in the reservoir parameters between wells. Since the well damage is assumed to be very near the well bore we can safely assume that this zone is in steady state practically all of the time. This means that the pressure drop due to the skin would follow the steady-state equation for radial flow as derived in Chapter 1 and written in a slightly different form here.

$$(p_1 - p_2) = \frac{0.141q\mu}{kh} \ln(r_1/r_2) \qquad (1.12A)$$

This shows that the pressure drop due to the skin will be proportional to the group of terms, $0.141q\mu/kh$, regardless of how deep the damage (r_1) is. Since all of the factors in this group of terms may vary from well to well it is helpful to divide the additional pressure drop due to the skin by this group of terms to get a number that will give an indication of the relative damage or improvement in the permeability around a well bore.

$$S = \Delta p_{skin} \bigg/ \left(\frac{0.141q\mu}{kh}\right) \qquad (3.14)$$

S is defined as the skin factor. It is very useful in comparing damage between wells and finding the wells that have not been completed properly. Note that Equation 3.14 can also be written as a function of the pressure buildup slope, m, as defined in Equation 3.12.

$$S = \frac{\Delta p_{skin}}{0.867m} \qquad (3.15)$$

The constant 0.867 is simply a ratio of the constants 0.141 and 0.1625. A. F. Van Everdingen[7] is generally given credit for originating the concept of the skin effect and the skin factor.[6] He simply included the skin factor as a part of the constant-rate solution as a function similar to the pressure function. This would mean that Equation 2.16 described in detail in Chapter 2 could be written with the skin factor included as,

$$p_{w,t} = p_i - \frac{0.141q\mu}{kh}(p_{td} + S) \qquad (3.16)$$

From this equation it is easy to see that the additional pressure drop that defined in Equation 3.14 as Δp_{skin}.

Equations 3.7, 3.11, 3.12, 3.13, 3.14, and 3.15 can all be used to analyze a pressure buildup from an "unchanging" well pressure regardless of whether the well is infinite acting, in pseudosteady state, or steady state at time of shutin. These equations can be used to evaluate pressure fall-off tests, as in injection wells, as well as buildups. Also, by substituting the change in rate, Δq, for the rate, q, in the equations they can all be used for analysis of pressure tests where the rate is simply reduced rather than being shutin. These are known as two-rate tests. Thus, these same simple equations can probably be used to analyze 90% of the buildup and fall-off data. These same equations can be used to analyze drawdown data by substituting p_i for p_{wf}.

Two restrictions to the application of these equations should be again noted although they were cited previously: (1) The flowing well pressure at time of shutin must be sufficiently constant that a change in pressure during the ensuing shut-in time would be insignificant if the well were not shutin, and (2) The shut-in time, Δt, must be less than the stabilization time based on the distance to the nearest boundary.

The above methods of analysis can only be used when the well pressure at time of shutin is "unchanging." If the well is in steady state at time of shutin there is no problem in determining that the well pressure is "unchanging." However, in the case of wells in pseudosteady state or wells that are infinite acting at time of shutin, the problem of determining if the well pressure can be treated as "unchanging" is more difficult.

The pseudosteady-state case is probably the easiest of these to analyze. If the constant change in pressure with time has been established from past pressure surveys then it simply becomes a matter of multiplying this change of pressure with time under pseudosteady state by the maximum shut-in time to see how large a change in the well pressure would have occurred during the shut-in time if the well had not been shutin. If such pressure observations have not been made it is then necessary to estimate the change in pressure with time under pseudosteady-state conditions by Equations 2.41 or 2.42. Thus, determining the error introduced by the "unchanging" assumption for a well in pseudosteady state at shutin is reasonably simple.

Determining the error introduced by the "unchanging" assumption for a well that is infinite acting at shutin is somewhat more difficult. We can get an excellent evaluation of the change in pressure that would have occurred if the well had not been shut-in by comparing the theoretical value of $(p_w')_{min}$, the pressure that would have existed at the end of the shut-in period if the well had not been shut-in, with the theoretical flowing pressure at shutin, p_{wf}. Applying Equation 2.16 of Chapter 2, using a pressure function for an infinite-acting reservoir, and including the additional pressure drop due to the skin, we can write an expression for the flowing pressure at the producing time, t, at time of shut-in as

$$p_{wf} = p_i - \Delta p_{skin} - \frac{0.141 q\mu}{kh} \left(\frac{1}{2}\right)\left(\ln \frac{\eta t}{r_w^2} + 0.809\right)$$

The theoretical well pressure at the end of the shutin time that would have existed if the well had not been shutin for the time, Δt_{max}, would then be

$$(p_w')_{min} = p_i - \Delta p_{skin} - \frac{0.141 q\mu}{kh}\left(\frac{1}{2}\right)\left[\ln \frac{\eta(t + \Delta t_{max})}{r_w^2} + 0.809\right]$$

The difference between these two equations would then provide an estimate of the maximum change in the flowing pressure that would have existed if the well had not been shutin.

$$(\Delta p_w')_{max} = \frac{0.1625 q\mu}{kh} \log \frac{t + \Delta t_{max}}{t} \qquad (3.17)$$

This equation can then be used to determine the error that might be involved in assuming an "unchanging" well pressure for a well that is infinite acting at shutin.

If the change in the well pressure during the period of shutin would be significant if the well was not shutin, it is then necessary to use one of the more complex techniques of analysis described in the two sections that follow.

Problem 3.2: Pressure Buildup From an "Unchanging" Well Pressure

Fig. 3.4 is a plot of shutin pressure versus the log of shutin time for a well in Louisiana that was producing from a reservoir with a very active water drive. After 9 hr of shut-in time the pressure reached a constant value. The data listed below were determined for this well. The well was producing at a rate of 199 STB/d with an "unchanging" pressure at time of shut-in. Find the undamaged permeability. What is the (ϕc) product if the effect of the nearest drainage boundary 300 ft away is felt at the well after 7½ hr? What is the Δp_{skin} and skin factor, S, if this well had a flowing pressure of 702 psig? What would be the flow-rate increase if this Δp_{skin} was removed? Assume same p_w is maintained and the constant shutin pressure is the pressure at the external drainage boundary.

$$\mu = 2.0 \text{ cp}$$
$$r_w = \tfrac{1}{3} \text{ ft}$$
$$B_o = 1.15$$
$$h = 22 \text{ ft}$$

You may check your solution in Appendix C.

Finite acting at shutin.[8] If a well is finite acting at time of shutin and the pressure is changing too rapidly to permit the "unchanging" analysis described above, a fairly simple analysis procedure is still possible to obtain the undamaged permeability and an evaluation of the damage. The equation for the well pressure under these circum-

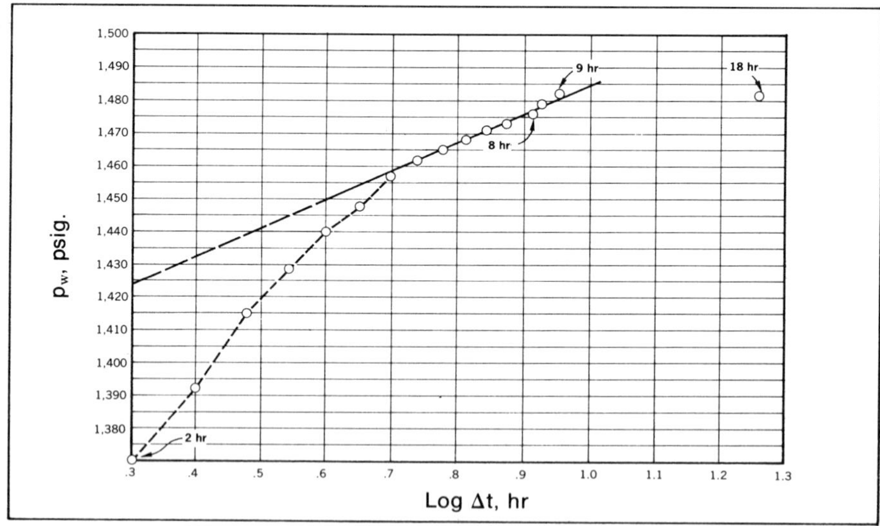

FIG. 3.4 Pressure buildup Problem 3.2 data.

stances can be derived in much the same way the well-pressure equation for the "unchanging" case was derived. We simply add the pressure increase caused by shutting the well in to the pressure that would have existed if the well had not been shut-in. The pressure, p_w', that would have existed if the well had not been shut-in is

$$p_w' = p_{wf} - (\Delta p/\Delta t)_{pseudo}\ \Delta t \qquad (3.18)$$

The pressure increase due to the negative rate introduced into the reservoir by shutting the well in, Δp_q, will still be governed by Equation 3.7. Adding this to the pressure indicated in Equation 3.18 we obtain,

$$p_w = p_{wf} - \left(\frac{\Delta p}{\Delta t}\right)_{pseudo}\ \Delta t + \frac{0.141q\mu}{kh}\left(\frac{1}{2}\right)(\ln\ \Delta t_D + 0.809) + \Delta p_{skin} \qquad (3.19)$$

The basis for this equation is illustrated graphically in Fig. 3.5.

Note from Equation 3.19 that there is no simple relationship between the well pressure and time. Both the second and third terms change with time. The change in the first term is negative and varies linearly with the shutin time, Δt, while the change in the third term is positive and is proportional to the change in the log of the shut-in time. Consequently, there can be no theoretically correct straight line between a plot of the well pressure and a log function of time nor can there be a straight-line plot between the well pressure and a linear function of time. For this reason we will choose to use a plot of Δp_q versus the log of the shut-in time, Δt, to determine the undamaged permeability and the extent of the damage in the well.

If we expand Equation 3.7 by expanding the Δt_D expression as in

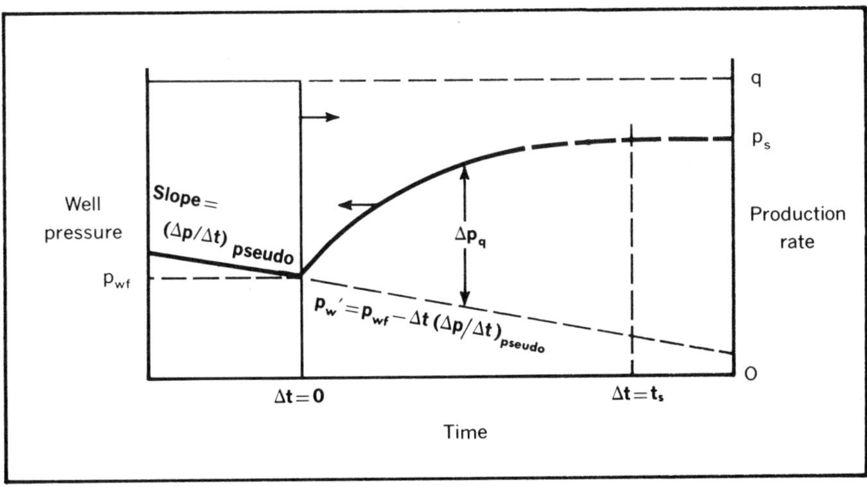

FIG. 3.5 Pressure buildup in a pseudosteady-state well.

Equation 3.8, converting from base e logs to base 10 logs, and rearranging the terms we obtain

$$\Delta p_q = \frac{0.1625q\mu}{kh}\left(\log\frac{6.33k}{\phi\mu cr_w^2} + \frac{0.809}{2.3}\right) + \Delta p_{skin} + \frac{0.1625q\mu}{kh}\log\Delta t$$

Note that only the last term changes with time so that we can write the equation as

$$\Delta p_q = \text{Constant} + \frac{0.1625q\mu}{kh}\log\Delta t. \qquad (3.20)$$

This clearly shows that a plot of Δp_q versus $\log\Delta t$ will give a straight line whose slope is again defined as in Equation 3.12. Furthermore, this slope can be used to calculate the undamaged permeability since the additional pressure drop due to damage is included in the Δp_{skin} term.

Once the permeability has been calculated, we can substitute a function of the slope, m, in Equation 3.7 to give an equation that can be used to calculate Δp_{skin}.

$$\Delta p_q = 0.867m\left(\frac{1}{2}\right)(\ln\Delta t_D + 0.809) + \Delta p_{skin} \qquad (3.21)$$

The skin factor can then be calculated from Δp_{skin} as described previously using Equation 3.15.

The data for the Δp_q versus $\log\Delta t$ plot can be obtained by extrapolating the pressure decline versus time plot and reading the Δp_q values for different Δt's from the plot. Alternatively, the Δp_q values can be calculated for a variety of shut-in times by substituting Δp_q for the last two terms of Equation 3.19 and rearranging the terms to obtain

$$\Delta p_q = p_w - p_{wf} + \Delta t\,(\Delta p/\Delta t)_{pseudo} \qquad (3.22)$$

A problem should help clarify the Δp_q plot type of solution.

Problem 3.3A: Pressure Buildup Analysis Of an Old Well

A well drilled in a field with uniform 40-acre spacing has produced at 280 STB/d for 10 days. Then the well is shutin for a pressure-buildup survey. In the days prior to shutin, the flowing tubing-head pressure declined about 24 psi/d. The gas-oil ratio has been constant during production. Calculate the 24-hr and 30-hr points as a check of the Fig. 3.6 data.

Assume the well was in pseudosteady state at time of shutin and find the product $(\phi\,c)$, the undamaged permeability, the approximate time when pseudosteady-state flow began, Δp_{skin}, and the skin factor, S.

Estimated Reservoir data:

Oil formation volume factor, B_o = 1.31
Oil viscosity, μ_o = 2.0 cp
Well radius, r_w = 0.333 ft
Net pay thickness, h   = 40 ft

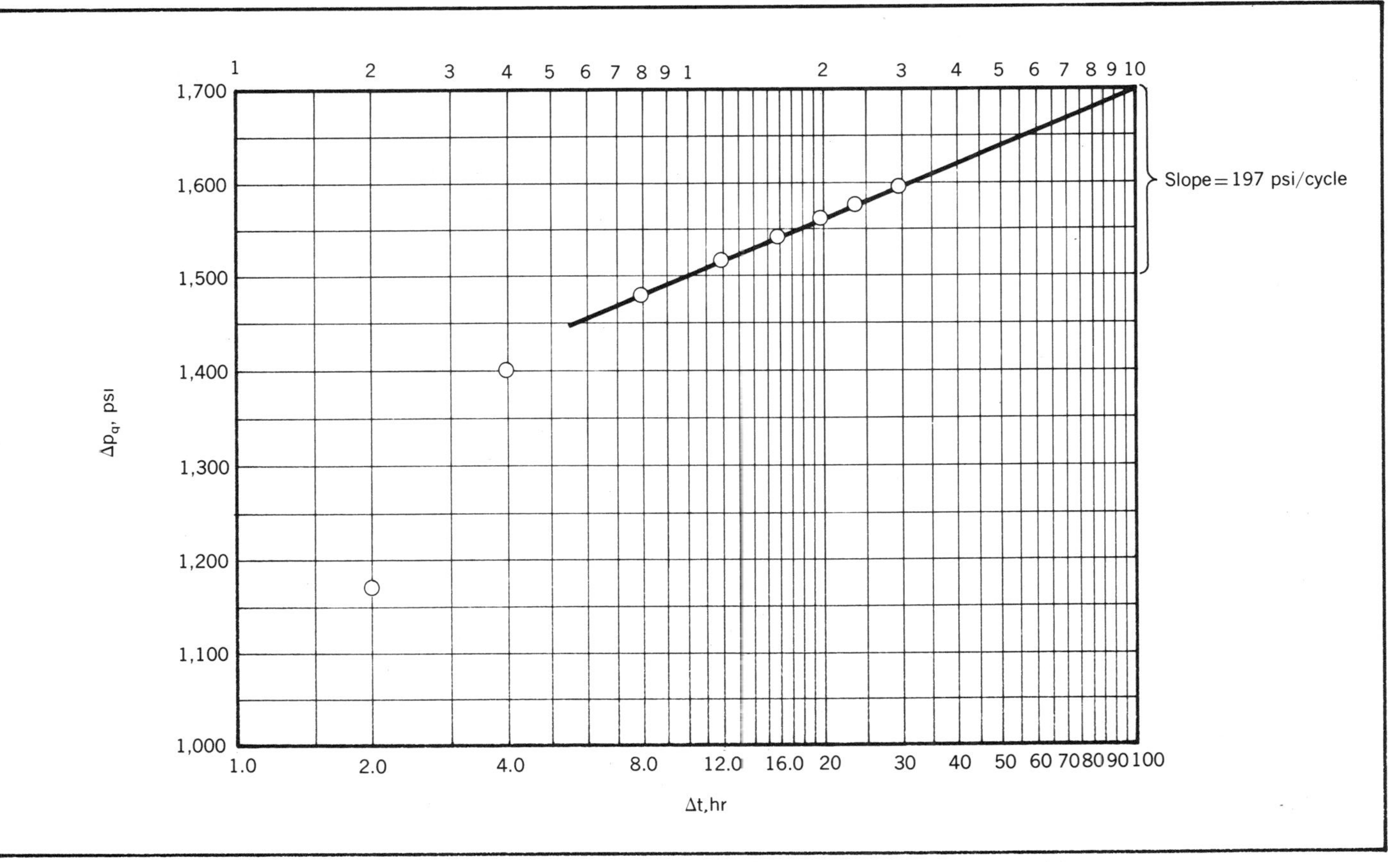

FIG. 3.6 Δp_q plot for Problem No. 3.3.

Pressure-survey data:

Time shutin, (hr)	Pressure, psia
0	1123
2	2291
4	2519
8	2592
12	2624
16	2647
20	2663
24	2675
30	2687

Check your solution against the one in Appendix C.

It is often convenient, for reasons described under the productivity index section of this chapter, to evaluate the damage as a damage ratio rather than as Δp_{skin} or the skin factor, S. To do this it is most convenient to derive an expression for the damage ratio defined as the undamaged permeability divided by the actual permeability. The undamaged permeability can be obtained from the Δp_q vs. Δt plot slope while the average or actual permeability can be calculated from the pseudosteady-state radial-flow equation using the flowing pressure at time of shutin.

$$DR = \frac{k_{und}}{k_{act}} = \frac{0.1625q\mu/mh}{0.141q\mu[\ln(r_e/r_w) - 0.5]/h(p_e - p_{wf})} \qquad (3.23)$$

This reduces to

$$DR = \frac{p_e - p_{wf}}{0.867m[\ln(r_e/r_w) - 0.5]} \qquad (3.24)$$

To use this equation it is necessary to determine the pressure at the outer boundary, p_e. We will see later that we can determine the average pressure for the drainage area of a well in pseudosteady-state flow at the time of shutin by using a rather simple equation. The pressure at the external boundary can then be calculated by recognizing that there is a fixed relationship between the average or static pressure and the pressure at the outer boundary.

$$p_e - p_s = 0.217m \qquad (3.25)$$

The basis for this expression is derived in the Craft and Hawkins[9] text by calculating an average pressure,

$$p_{avg} = \frac{\Sigma p \, \Delta V}{V} \qquad (3.26)$$

Substituting for ΔV in the radial system $2\pi r \Delta r h \phi$ and for p according to the pseudosteady-state radial-flow Equation 2.50.

$$p = p_w + \frac{\mu q \left(\ln \dfrac{r}{r_w} - \dfrac{r^2}{2r_e} \right)}{7.08kh} \tag{3.27}$$

and integrating between the external and well limits Craft and Hawkins[9] obtain

$$p_{avg} = p_w + \frac{\mu q}{7.08kh} \left[\ln \left(\frac{r_e}{r_w} \right) - \frac{3}{4} \right] \tag{3.28}$$

Since pseudosteady state assumes a constant compressibility the volumetric weighted average pressure must equal the static pressure. Therefore, the expression of Equation 3.28 is the static pressure, p_s. The pressure at the external radius, p_e, during pseudosteady state can be obtained from Equation 2.51 as

$$p_e = p_w + \frac{\mu q}{7.08kh} \left[\ln \frac{r_e}{r_w} - \frac{1}{2} \right] \tag{3.29}$$

When Equation 3.29 is subtracted from Equation 3.28 we obtain

$$p_e - p_s = \frac{0.141q\mu}{kh} \left(\frac{1}{4} \right) \tag{3.30}$$

When $0.867m$ is substituted for $0.141q\mu/kh$ and the expression is simplified the result is Equation 3.25.

When using Equations 3.28 or 3.29 individually the engineer should remember that the expressions do not include the additional pressure drop due to the well damage, Δp_{skin}. Consequently any application of these equations should add this expression. When substracting the equations to obtain Equation 3.25 the skin damage term cancels out. The calculation of the damage ratio using the data of Problem 3.3A will be demonstrated later in this chapter under the discussion of determining static pressures at time of shutin.

The general technique for analyzing pressure buildup by using a Δp_q plot has been proposed here for a well that is in pseudosteady state at shutin. However, the same technique can be used for any well if the pressure behavior prior to shutin can be extrapolated. A semilog plot of pressure versus time will permit the extrapolation of an infinite-acting well pressure while the pressure extrapolation for a steady-state well is simply the flowing pressure at shutin.

Infinite-acting at shutin. If a well is infinite acting at time of shutin and the well pressure would have changed significantly during the shutin period if the well had not been shutin, the methods described in this section may be usable for the pressure buildup analysis. If the pressure in the well would not have changed significantly during shutin if the well had not been shutin, then the engineer may choose to use the analysis described under the section of this chapter on unchanging pressure at shutin.

The classical method of analyzing a pressure buildup for a well that is in an infinite-acting state at time of shutin is the Horner method. This method is derived in many different reservoir engineering texts but a derivation will be repeated here for the benefit of those engineers who are not familiar with the basis for the Horner method and possibly to add more understanding to the method for those who have had contact with it before.

The derivation used here will be very similar to that employed previously to derive the expression for the shutin pressure for a well with an "unchanging" pressure and for a well whose pressure is changing according to pseudosteady state at time of shutin. The basic idea is to add the pressure increase that occurs due to the negative rate, $-q$, (introduced when the well is shutin), to the pressure that would have existed if the well had not been shutin. The pressure increase due to the negative rate is the same expression used previously for the "unchanging" and pseudosteady-state derivations.

$$\Delta p_q = \frac{0.141q\mu}{kh}\left(\frac{1}{2}\right)(\ln \Delta t_D + 0.809) + \Delta p_{skin} \qquad (3.7)$$

As illustrated in Fig. 3.7 the well pressure during shutin will be the sum of the Δp_q described by Equation 3.7 and the pressure that would have existed if the well had not been shutin, p_w'. Since the well is infinite acting the pressure that would exist without shutin would simply be the well pressure resulting from the pressure drop according to the constant-rate solution to the radial diffusivity equation for an infinite-acting system. If we restrict this to the time when the reduced time is greater than 100 we can then use the log solution, and the well pressure without shutin is

$$p_w' = p_i - \frac{0.141q\mu}{kh}\left(\frac{1}{2}\right)[\ln(t_D + \Delta t_D) + 0.809] - \Delta p_{skin} \qquad (3.31)$$

The use of the reduced time $(t_D + \Delta t_D)$ is the result of the positive rate, q, being effective for the total time of production and shutin. When the increase in pressure due to the negative rate, Δp_q according to Equation 3.7, is added to the pressure that would have existed without the negative rate, p_w' according to Equation 3.31, we obtain the shutin well pressure.

$$p_w = p_i - \frac{0.141q\mu}{kh}\left(\frac{1}{2}\right)[\ln(t_D + \Delta t_D) - \ln \Delta t_D] \qquad (3.32)$$

Note that the Δp_{skin} term and 0.809 constant drop out of the equation due to the difference in signs in Equations 3.7 and 3.31. Now note that the difference in the two logs in the bracket is equal to the log of the ratio of the arguments, $(t_D + \Delta t_D)/\Delta t_D$. Furthermore the ratio of the re-

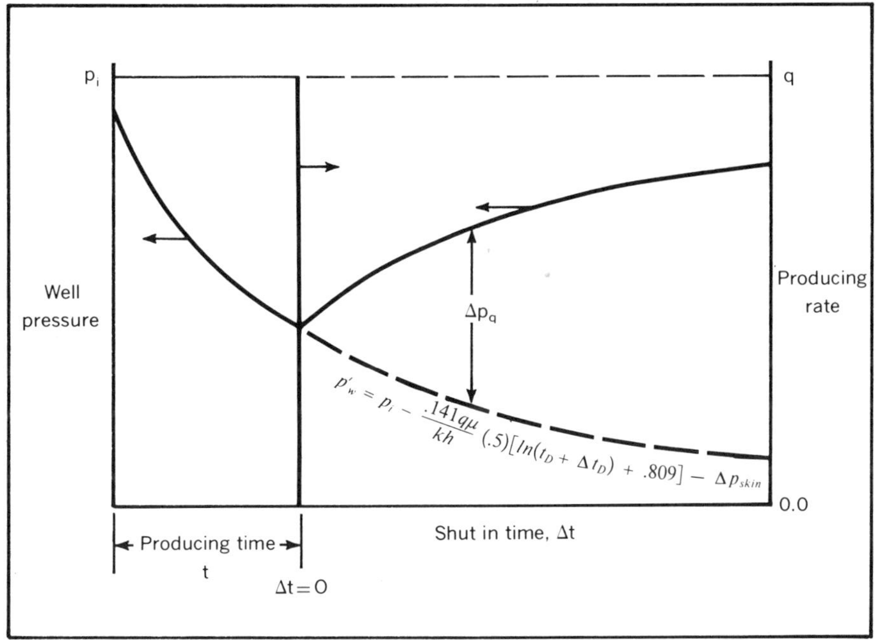

FIG. 3.7 Pressure buildup in an infinite-acting well.

duced times will be exactly the same as the ratio of the real times since the reduced time is simply the real time multiplied by the diffusivity constant and divided by the well radius squared. These constants cancel out of the ratio of the reduced times and we are simply left with the ratio of the real times. Equation 3.32 can then be written as

$$p_w = p_i - \frac{0.141q\mu}{kh}\left(\frac{1}{2}\right)\ln\left(\frac{t+\Delta t}{\Delta t}\right) \tag{3.33}$$

When this equation is written in terms of the log to the base 10 and all the numerical constants are combined we obtain the Horner equation,[10]

$$p_w = p_i - \frac{0.1625q\mu}{kh}\log\left(\frac{t+\Delta t}{\Delta t}\right) \tag{3.34}$$

This equation indicates that a plot of the shutin well pressure versus the log $[(t + \Delta t)/\Delta t]$ will give a straight line whose slope is 0.1625 $q\mu/kh$ which is the absolute slope of the other pressure buildup analysis plots we have examined and satisfies Equation 3.12. This is often referred to as a Horner plot.

An example of such a plot is illustrated in Fig. 3.8. Note that it exhibits the same early time afterflow—skin effect that have charac-

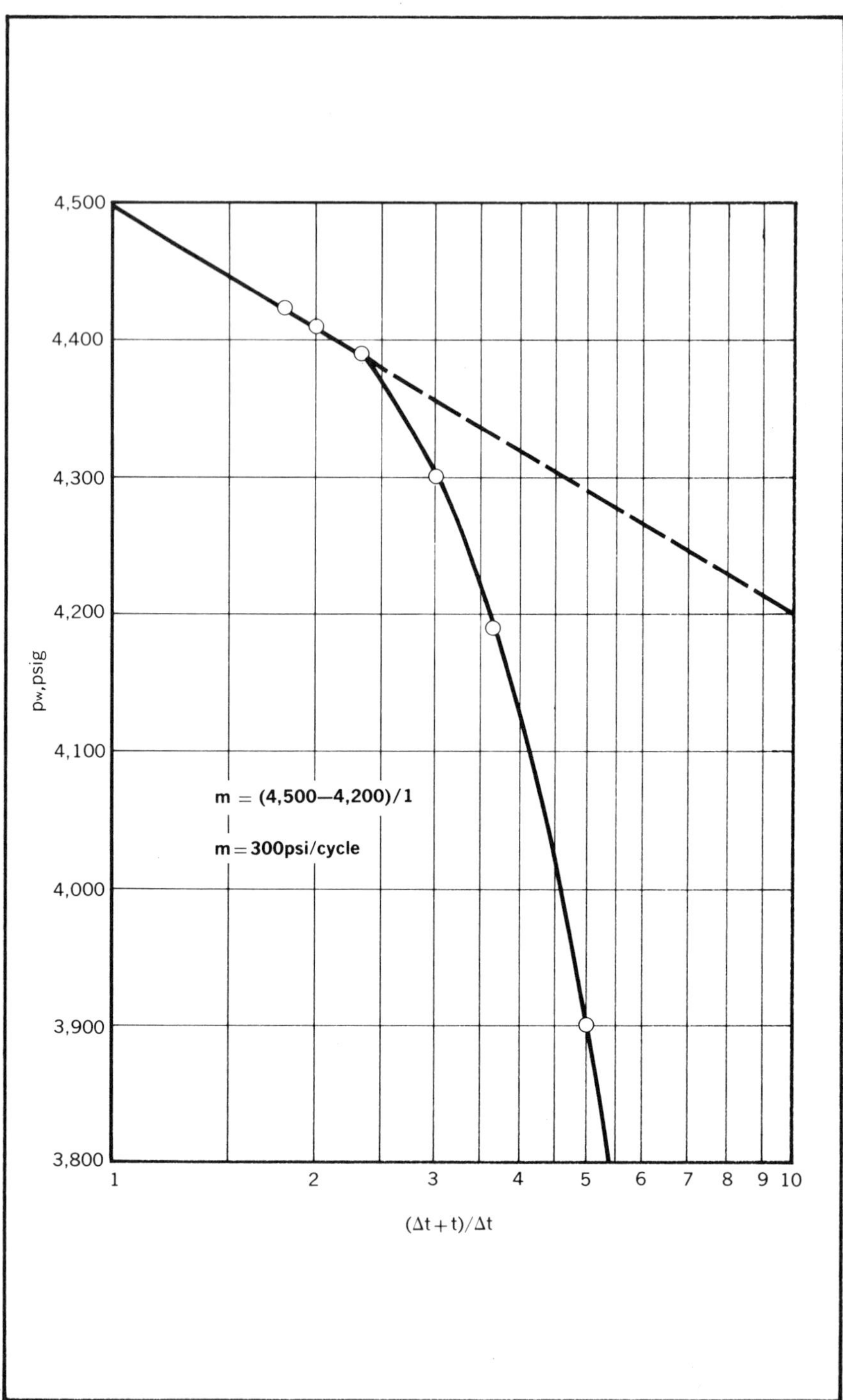

FIG. 3.8 An example Horner plot.

terized the other pressure buildup analysis plots we have previously examined. Since Fig. 3.8 represents the pressure-buildup data for a drill-stem test the early time anomaly includes one other nonideal characteristic that will be discussed under the section on drill-stem test interpretation.

Engineers are sometimes confused by the fact that the log of $\Delta t/(t + \Delta t)$ is often plotted rather than the log of $(t + \Delta t)/\Delta t$. When this is done the slope changes sign and the validity of this procedure can be verified by examining Equation 3.32. Note that we can change the sign of the second term of the right-hand side of this equation along with the sign of the two natural log expressions in the bracket without affecting the equality of the equation. Such changes would result in the sign of the second term of Equation 3.34 becoming positive and the time ratio being $\Delta t/(t + \Delta t)$.

The engineer might also be on the look out for reversed scales in a Horner plot. Many engineers prefer to have the shutin time increase to the right on a Horner plot and may actually reverse the time ratio scale to attain this quality.

Any analysis of the Horner plot to obtain the well damage must make use of the well pressure at time of shutin—the final flowing well pressure. This is simply the pressure equivalent of the constant-rate infinite-acting solution to the radial diffusivity equation. It may be simpler to obtain an expression for this flowing well pressure by examining Equation 3.31, the equation describing the well pressure that would be obtained if the well had not been shutin. This equation could be used to write an expression for the flowing well pressure at shutin by recognizing that this pressure, p_{wf}, is the same as p_w' for a reduced time of t_D and that $0.867m$ is $0.141q\mu/kh$.

$$p_{wf} = p_i - 0.867m\left(\frac{1}{2}\right)[\ln t_D + 0.809] - \Delta p_{skin} \qquad (3.35)$$

Assuming that the initial reservoir pressure, p_i, is known (and it will be shown that it can always be rather simply obtained), this equation can be used to determine Δp_{skin}. Then the skin factor, S, can be calculated as before from Equation 3.15. In using this technique the value of the mobility, k/μ, can be obtained from the slope,$-m$, for use in evaluating the reduced time. The reduced time should be based on the total producing time. This approach to evaluating the skin damage is simple and theoretically correct but it may place too much emphasis on the accuracy of the producing time. In applying the Horner method the assumption is normally made that the producing time is the cumulative production divided by the producing rate at time of shutin.

Possibly a better technique is to base the calculation of Δp_{skin} on the difference between the shut-in pressure at some particular time and the flowing well pressure at time of shut-in. A mathematical expression can be obtained for this pressure difference by recognizing from Equation 3.35 that the second and third terms on the right-hand side is the difference between the initial pressure and the flowing well pressure at time of shutin, $p_i - p_{wf}$, and that the second right-hand term of Equation 3.34 is the difference between the initial pressure and the well pressure at some shutin time Δt, $p_i - p_w$. When these two expressions are subtracted the initial pressure disappears and we obtain the equation for the difference between the flowing pressure at time of shut-in and the well pressure at some shut-in time, Δt.

$$p_w - p_{wf} = 0.867m\left(\frac{1}{2}\right)(\ln t_D + 0.809) + \Delta p_{skin} - m \log \left(\frac{t + \Delta t}{\Delta t}\right) \qquad (3.36)$$

This equation combines logs to the base e and logs to the base 10 but it provides a forthright and easily understood expression from which to calculate the pressure drop due to the well damage, Δp_{skin}. As before the skin factor, S, can be calculated from the Δp_{skin} by using Equation 3.15. Any pressure on the straight-line portion of the Horner plot can be used to calculate the additional pressure drop due to the well damage from Equation 3.36. The corresponding shut-in time must of course be used also. This equation is less sensitive to the producing time, t, since the first and last terms both contain the log of the producing time and have opposite signs.

Researchers and educators apparently decided that the practicing engineer was incapable of understanding and applying an equation such as 3.36 and needed a more specific calculation to be performed to calculate the skin factor, S. Consequently, an expression such as Equation 3.36 was expanded and "simplified" to provide an expression that could be used to calculate the skin factor directly.[11]

$$S = \frac{1.151(p_{1hr} - p_{wf})}{m} - 1.151 \log \left(\frac{q_{STB}B_o}{10.4mc\phi hr_w^2}\right) \qquad (3.37)$$

This equation can be obtained from Equation 3.36 by (1) writing the reduced time, t_D, as a function of the reservoir parameters, (2) writing Δp_{skin} as a function of the skin factor, S, (3) substituting a function of m for the mobility, k/μ, that appears in the reduced time term, (4) assuming that the log of t is equal to the log of $t + \Delta t$, (5) using the pressure at 1 hr as the well pressure and $\Delta t = \frac{1}{24}$ day, and (6) solving the resulting expression for the skin factor, S.

Assumption (4) above is the same as assuming that the pressure at time of shutin is "unchanging." Thus, Equation 3.37 can also be obtained from the equation for Δp_q, Equation 3.21. If the engineer still chooses to use Equation 3.37 to calculate the skin factor, S, he should be careful to read the pressure at 1 hr of shut-in from an extrapolation of the straight-line portion of the Horner plot if the straight-line portion does not extend through this shut-in time of 1 hr. The straight-line portion of many buildups do not encompass the 1-hr shut-in time.

In considering the use of the Horner method the engineer should remember that it is theoretically only applicable to a well that is infinite acting because the natural log solution used in the derivation is limited to infinite acting-reservoirs. Also remember that a well pressure plot versus any log function of time will not theoretically give a straight line for a finite-acting reservoir. Thirdly, bear in mind that there is no reason to attempt to apply the Horner method to finite-acting reservoirs since the methods described under the "unchanging" pressure and pseudosteady-state sections of this pressure buildup discussion are mechanically simpler to apply than even an incorrect application of the Horner method.

An example application of the Horner method will be given later.

Determining the average well drainage area pressure. As noted in the section on unchanging pressure at shut-in the undamaged permeability and a measure of the well damage may be determined using the same equations regardless of whether the well is in pseudosteady state, steady state, or is infinite acting at the time of the shut-in provided the well pressure qualifies as "unchanging." However, this is not true of the determination of average pressure that exists in the well drainage area at time of shut-in. To obtain this average pressure we must know what flow regime is governing flow at the time of shut-in. The methods used are different depending on whether the well at time of shut-in is infinite acting, in pseudosteady state, or in steady state.

We will first consider the Horner method. If the Horner semilog plot of well pressure versus $(t + \Delta t)/\Delta t$ is extrapolated to a value of $(t + \Delta t)/\Delta t$ of 1.0, this represents a shut-in time of infinity. Infinity is the only value of the shut-in time that would mathematically result in a time ratio of 1.0. At a time of infinity it is then logical to assume that the presssure recorded would be the average reservoir pressure. On the surface it appears that this concept may be at odds with the mathematics of the situation. If we examine Equation 3.34, the Horner equation, we see that a time ratio of 1.0 will result in the log of the time ratio being zero, which will make the term containing the log, zero, and will mean that the well pressure at this time will be equal to the initial pressure and not the average or static pressure. This would seem to represent an inconsistency until we recognize that the average reservoir pressure mathematically for an infinite-acting reservoir is the initial pressure. If a reservoir is truly infinite then the drawdown in pressure in any finite portion of the infinite reservoir would be insignificant and the average pressure would still be the initial pressure, p_i.

The practical significance of this discussion is that the well pressure on a Horner plot extrapolation to a time ratio of 1.0 is not the average reservoir pressure but is always the initial well pressure if the well is infinite acting, and theoretically the Horner plot does not apply to a well that is finite acting at time of shut-in. Although mathematically it appears irrefutable that the Horner equation does not apply to a finite-acting reservoir, engineers have always attempted to apply the method to finite-acting reservoirs. Some of the leading experts in this field still talk about the time limits that exist on the application of the Horner method to a finite-acting reservoir. They justify their approach by showing that computer manufactured and actual well data give a straight-line Horner plot over specified time ranges. The range of the apparent straight-line portion must largely be a function of the scale used for the plot. This approach to the analysis of a pressure buildup in a finite-acting reservoir would be admirable if there was no theoretically accurate or easy means of analyzing finite data but the methods previously described are both theoretically sound and simpler to apply than a Horner analysis.

The principal point to be made at this time is that engineers will probably continue to use a Horner plot for their finite well buildups. But they should be warned that the well pressure at a time ratio of 1.0 is not the average well pressure but approximates the initial pressure that would have been necessary for this well drainage area to still be infinite acting. Matthews, Brons and Hazebrook[12] recognized the fact that a Horner plot applied to a finite-acting well would not give either the intitial pressure or the average pressure when extrapolated to a time ratio of 1.0. Consequently, they coined the term, p^*, to identify this particular pressure. They recognized that p^* would be equal to the initial pressure if the reservoir is still infinite acting and they determined the relationship between p^* and the average reservoir pressure, p_s, that would exist if the well is finite acting at time of shut-in.

The relationship between p^* and the initial pressure, p_i, was shown to be a function of the reduced time and the shape of the drainage area. These relationships can be found in extensive plots in the SPE Monograph No. 1, Pressure Buildup Analysis. They are not included in this text because it does not appear they are needed for pressure buildup analysis.

To use the Matthews et. al. data it is necessary for the engineer to know the average compressibility for the reservoir to the point in time represented by the producing time at time of shut-in. He must also know the average porosity, the shape of the reservoir, and the thickness. With a knowledge of these reservoir properties he can generally calculate the average pressure by material balance.

$$p_i - p_s = \frac{5.615qt}{\phi c V_b} \tag{3.38}$$

In this equation the symbol V_b is the well drainage volume in cubic feet and the compressibility, c, is the average compressibility. It then appears that the Matthews, et. al. data has little if any advantage over a straight volumetric calculation of the average reservoir pressure, p_s. For an infinite-acting reservoir there could certainly be no advantage. Consequently, for an infinite-acting reservoir it is recommended that the engineer simply use Equation 3.38 to calculate the average pressure. It is recognized that this really represents no practical solution in most cases because the reservoir average pressure is generally needed for material balance calculations. Nevertheless, this is the only reasonable approach to the problem. If production has not affected the outer boundary of the drainage area it is illogical to expect a pressure buildup which does not affect the outer boundary to indicate the average pressure in the drainage area which has to be a function of the reservoir size.

If the well is in pseudosteady state at time of shut-in there are a couple of approaches that should yield a reasonably accurate average reservoir pressure. The preferred method is the use of the Odeh data[13] published in 1971 and presented here as Figs. 3.9 A and 3.9 B. Odeh apparently recognized the shortcomings of the Matthews et. al. data and solved the problem by relating the difference between the initial pressure and p^* to the difference between the initial pressure and the static pressure, p_s, with the magnitude of both normalized by dividing by the slope of the buildup curve, m. By handling the relationship in this way Odeh was able to avoid the need for a knowledge of the system volume and compressibility because they affect the two pressure differences equally. It is necessary to know the initial pressure to use the Odeh method whereas the Matthews, et. al. technique does not require a knowledge of p_i. Generally it is much easier to obtain an accurate initial pressure than it is to determine the data needed to utilize the Matthews, et. al. data.

In using Figs. 3.9A and 3.9B it is not generally necessary to make a full scale Horner plot. All that is needed is one point that is known to be on the straight-line portion. Generally a point about the middle of the data that forms the straight-line portion of the Δp_q or well-pressure plot versus shut-in time will give a point on the straight-line Horner plot. Once a point meeting these criteria has been determined, it is easy to calculate the p^* value. We know that the slope, m, of any of our build-up plots, are the same theoretical value and that the numerical value of the slope, m, will give the increase in the well pressure along the straight-line portion of the Horner plot for each log cycle of pressure change. Any log cycle of pressure change is equivalent to a change in the log of the time ratio of 1.0. For example the log cycle from 100 to 10 represents the difference between the log 100 and log 10 or $2 - 1 = 1$. Similarly the log cycle from 1,000 to 100 is log 1,000 $-$ log 100 or $3 - 2 = 1$, etc. Consequently, since we desire to extrapolate to a log value of zero we can directly calculate the well pressure, p^*, at this point.

$$p^* = p_w + m \log \left(\frac{t + \Delta t}{\Delta t} \right) \tag{3.39}$$

In using this equation the shut in time, Δt, must coincide with the pressure, p_w, from the straight-line portion of the Horner plot.

We previously identified p^* as the initial pressure that would have been necessary if a well was still to be infinite acting at the end of the producing time, t. A larger reservoir radius, r_e, would also be necessary.

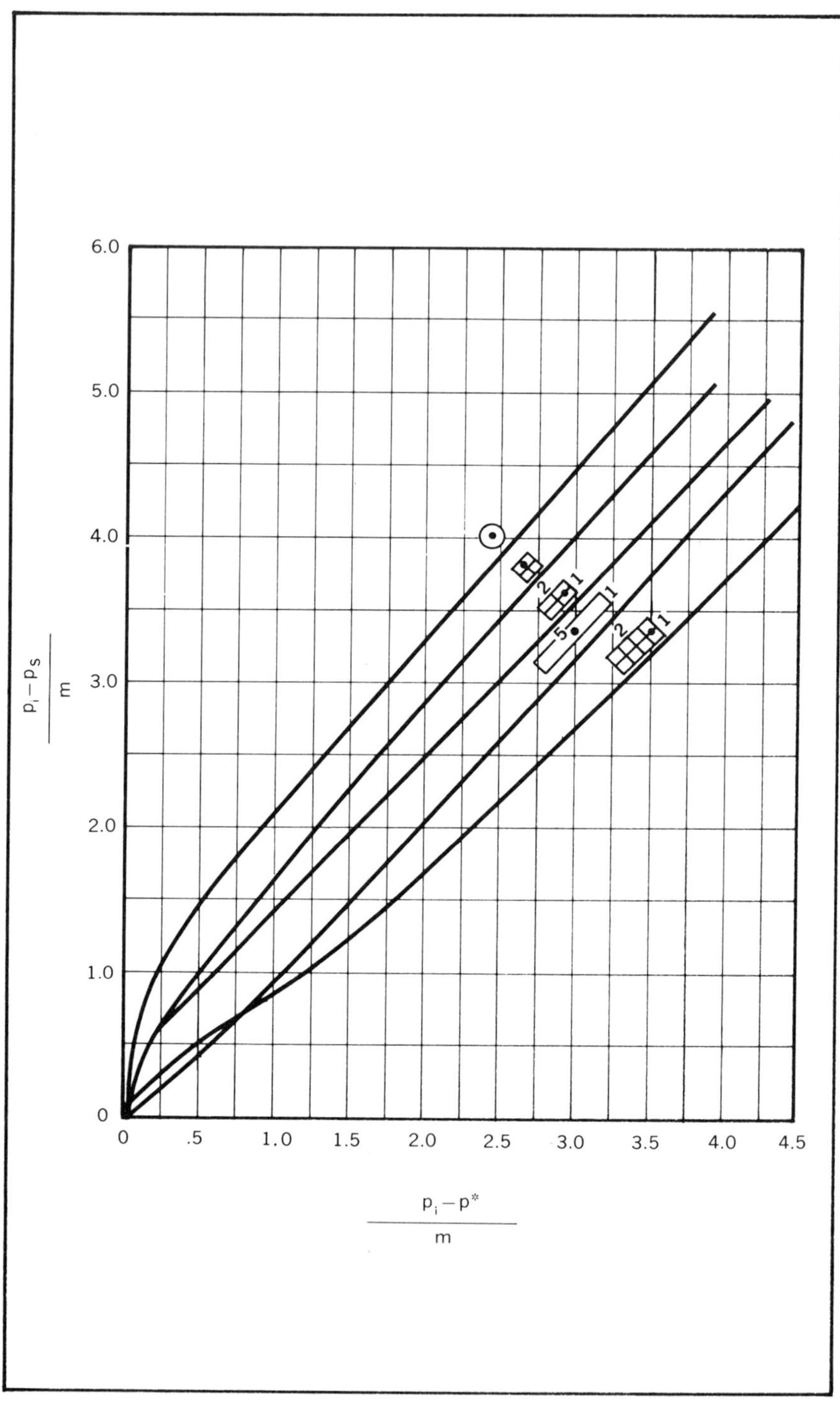

FIG. 3.9A Odeh data[9]—evaluation of static pressure from p*. (Odeh and Al-Hussainy, J. Pet. Tech.)

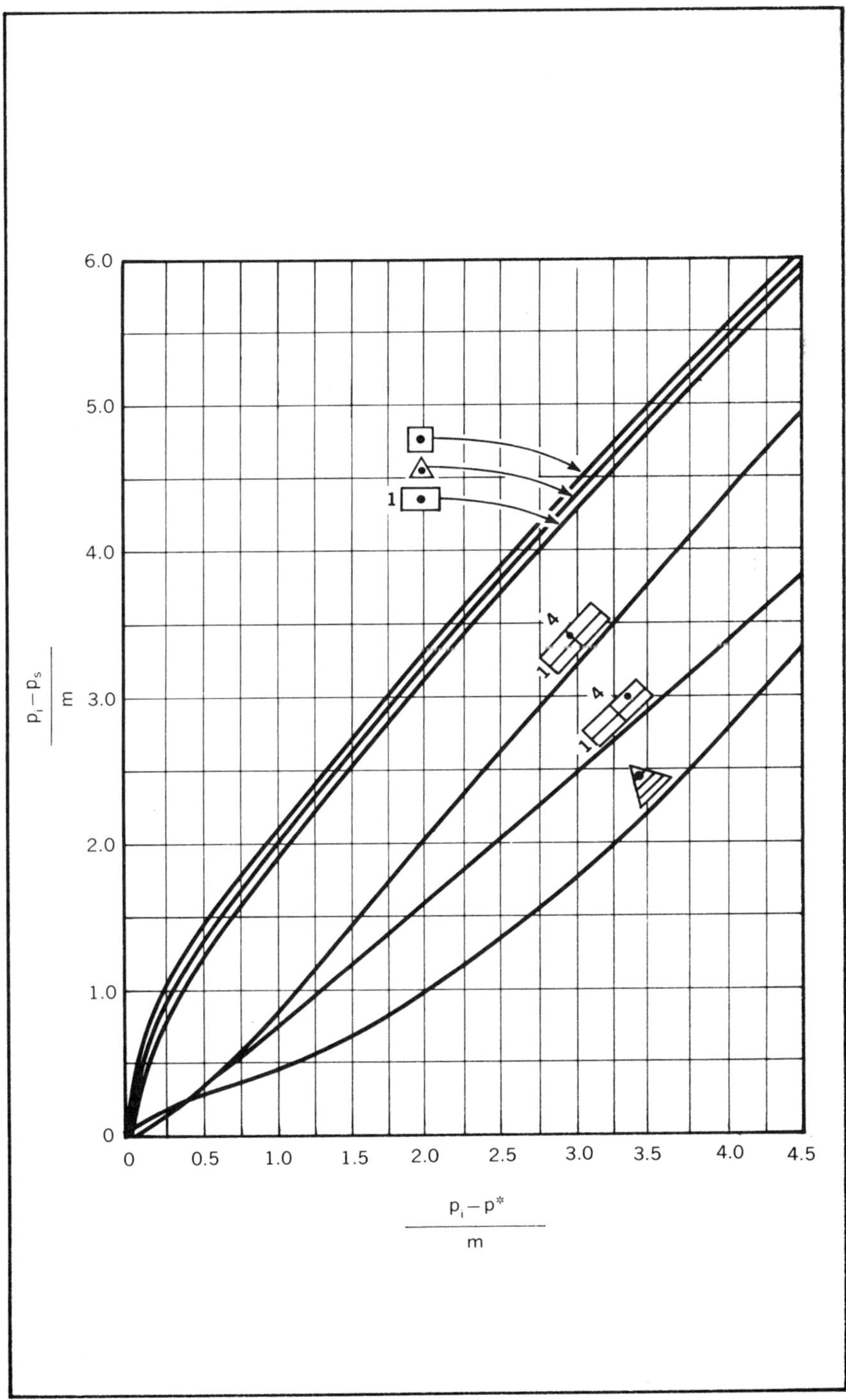

FIG. 3.9B Odeh data[9]—evaluation of static pressure from p*. (Odeh and Al-Hussainy, J. Pet. Tech.)

Thus, p^*, can also be back calculated from the equation for the flowing pressure at time of shut-in for an infinite-acting well. This means that p^* can be substituted for p_i in Equation 3.35 and when the expression is solved for p^* we obtain

$$p^* = p_{wf} + 0.867m\left(\frac{1}{2}\right)(\ln t_D + 0.809) + \Delta p_{skin} \qquad (3.40)$$

This equation is undoubtedly the most accurate equation theoretically for calculating the value of p^* to use in the Odeh data since it is not based on the erroneous assumption of a straight-line relationship for a Horner plot applied to a finite-acting reservoir. Nevertheless, there is much to be said in favor of the use of Equation 3.39 for this purpose. In applying Equation 3.40 it is necessary to use all of the Matthews et. al. properties we were trying to avoid to evaluate the reduced time. This error is generally magnified by the fact that the engineer must generally assume these same data in evaluating Δp_{skin} which must be determined before Equation 3.40 can be applied. Thus, Equation 3.39 appears to be the lesser of the evils.

Another method of determining the average pressure in the well-drainage area at time of shut-in in pseudosteady state can be used if the well is near the center of the drainage area. The equations are actually derived for radial flow but experience has shown that little error is involved if an equivalent radius is calculated for a well draining a square, a triangle, a hexagon or any similar regular area. In this method the well pressure that would be attained at a shut-in time equal to the stabilization time, $(\Delta t = t_s)$, is used as the basis for the static-pressure calculation. It will be shown that at this time the well pressure reaches the static pressure. It may be easiest to explain or understand this phenomena by writing an expression for the shut-in well pressure as a function of the initial pressure rather than as a function of the flowing pressure at time of shut-in as was done previously.

$$p_w = p_i - \frac{0.141q\mu}{kh}(PF)_{t+\Delta t} + \frac{0.141q\mu}{kh}(PF)_{\Delta t} \qquad (3.41)$$

No Δp_{skin} term appears in this equation because the additional pressure drop due to the positive producing rate, q, is equal and opposite in sign to the additional pressure increase due to the negative rate, $-q$. The equation indicates the pressure drop due to the positive rate acting in the reservoir for a time, $t + \Delta t$, and the pressure increase caused by the negative rate (caused by shutting the well in) acting for the time of shut-in, Δt. Now note that once the producing time becomes greater than the stabilization time, the change in the second term with time will be equal to the change in pressure with time under the pseudosteady-state flow regime, $(\Delta p/\Delta t)_{pseudo}$. Note also that once the shut-in time, Δt, exceeds the stabilization time, t_s, the change in this term with time will also be at a rate equal to the change in pressure with time under pseudosteady state, $(\Delta p/\Delta t)_{\text{pseudo}}$. Now if the second term of Equation 3.41 is decreasing at the same rate that the third term is increasing the well pressure will no longer be changing. When the well pressure reaches a state where it is no longer changing the pressure must surely be the static pressure.

Interference from other wells will generally prevent a direct measurement of the static pressure. Even if this subject well was the only well in the center of the reservoir the stabilization time would probably be too large in most cases for it to be practical to leave the well shut-in until the static pressure is reached. Consequently, it is generally necessary to obtain the static pressure by extrapolating the Δp_q plot to a time equal to the stabilization time, t_s. The resulting pressure term can then be used to calculate the corresponding well pressure by using the previously derived equation.

$$\Delta p_q = p_w - p_{wf} + \Delta t(\Delta p/\Delta t)_{\text{pseudo}} \tag{3.22}$$

In applying this expression the Δp_q will be evaluated at the stabilization time and could be appropriately subscripted. The well pressure will become the static pressure, p_s, and the shut-in time, Δt, will be the stabilization time, t_s. We can also write expressions for the stabilization time and the change in pressure with time under pseudosteady-state flow according to Equations 2.63 and 2.41. Both of these equations were derived in Chapter 2 but they have been previously used in this chapter. With these substitutions Equation 3.22 would become

$$p_s = p_{wf} - \frac{0.04\phi\mu cr_e^2}{k}\frac{1.79q}{\phi hcr_e^2} + (\Delta p_q)_{ts} \tag{3.42}$$

When like factors are canceled and numerical constants combined, the second term reduces to a function of $q\mu/kh$. A function of the pressure buildup slope, m, can then be substituted for this group of terms to obtain

$$p_s = p_{wf} - 0.439m + (\Delta p_q)_{ts} \tag{3.43}$$

This equation is somewhat misleading in that it may appear that you are able to evaluate the static pressure without a knowledge of the reservoir size and compressibility. However, note that Δp_q must be evaluated at the stabilization time and generally it is necessary to calculate this time from the stabilization time Equation 2.63. Thus, it appears that the difficulty of evaluating reservoir compressibility, porosity, and size is just the same here as it has been in most of the other methods of evaluating the average pressure. Nevertheless, it should be realized that the evaluation of the group of terms, ϕcr_e^2, for use in the stabilization-time equation is not as difficult as it is for evaluating this same group of terms for some other purposes. In this case we simply need the current value of the compressibility to determine the stabilization time areas. We need a well-life average compressibility if we are using the value in reduced time starting from the initial time and pressure such as would be the case in evaluating the reduced time to use the Matthews et. al. data or to calculate p* from the flowing well pressure at time of shut-in as in applying Equation 3.40.

If the engineer, in dealing with the field personnel, insists on the accurate recording of data as he should, he will have the decline in well pressure with time from successive bottom-hole pressure tests or from the decline in tubing-head pressures. These data will provide him with a good estimate of the change in pressure with time under pseudo-steady-state conditions, $(\Delta p/\Delta t)_{pseudo}$. From this value he can use the equation for this pseudosteady-state change of pressure with time to evaluate the current $\phi c r_e^2$ group of terms.

$$(\Delta p/\Delta t)_{pseudo} = \frac{1.79q}{\phi h c r_e^2} \qquad (2.41)$$

Note that this group, determined from recent pressure measurements, would be ideal for evaluating the current stabilization time but that it would not represent the average value that is needed to evaluate the reduced time starting from time zero and the initial reservoir pressure. What is needed to evaluate the group of terms for the latter purposes is the total drop in the average pressure from the initial time and the initial pressure to the time represented by the reduced time. If this pressure drop and the reservoir size were known the average compressibility could be calculated by material balance. However, in this particular instance the average pressure after some period of production, t, is what we desire as an answer and the solution would at best be trial and error.

Another method for determining the average pressure in the drainage area of a well in pseudosteady state at time of shut-in is based on the equation for the average pressure for radial pseudosteady-state flow derived previously as Equation 3.28 but listed here with the additional pressure drop due to well damage added.

$$p_s = p_w + \frac{q\mu}{7.08kh}\left(\ln \frac{r_e}{r_w} - \frac{3}{4}\right) + \Delta p_{skin} \qquad (3.44)$$

This equation may well be as accurate for determining the average pressure for a well near the center of a drainage area as any of the other methods cited. It certainly is much simpler to use and requires only a knowledge of the effective well radius in addition to the pressure buildup slope, m, and Δp_{skin} which we have shown can be determined independent of the average pressure. For determining the average or static well pressure, Equation 3.44 can be stated in a more convenient form.

$$p_s = p_{wf} + 0.867m\left(\ln \frac{r_e}{r_w} - \frac{3}{4}\right) + \Delta p_{skin} \qquad (3.45)$$

The average pressure in the drainage area of wells that are in steady state at time of shut-in can be found by using an equation similar to Equation 3.45 that applies to steady state rather than pseudosteady state.

$$p_s = p_{wf} + 0.867m \left(\ln \frac{r_e}{r_w} - \frac{1}{2} \right) + \Delta p_{skin} \qquad (3.46)$$

This equation is derived in an identical manner to the procedure followed in obtaining Equation 3.45 for pseudosteady state. An equation identical to the equation for the average pressure in pseudosteady-state flow, Equation 3.28, is also derived in the Craft and Hawkins text for steady-state flow except that the constant $^3/_4$ is replaced by the constant $^1/_2$. This difference is carried through to the final Equation 3.46 which differs from the similar pseudosteady-state Equation 3.45 to the extent of the difference in the same numerical constant.

Thus, to obtain the average pressure for the drainage area of a well that is in steady state we first determine the buildup slope, m, and Δp_{skin} as outlined under the unchanging method and use these values with the effective external well radius, r_e, to calculate the static pressure from Equation 3.46. Since steady-state flow generally exists only in a reservoir with a strong water drive and the average reservoir pressure is not generally as important in such a reservoir as it is in a depletion-type reservoir, there is not much consideration given in the literature to this problem. This may be fortunate since we need not be confused as to which method is to be used.

Steady-state reservoirs do present one common problem of pressure behavior misinterpretation. Many reservoirs with strong water drives are very permeable and porous. These reservoirs tend to have short stabilization times and thus often reach constant well pressure in a short time after shut-in. Most engineers erroneously interpret this constant pressure as the average pressure in the drainage area at time of shut-in. A little thought should convince the engineer that the constant pressure he is observing is simply the pressure near the outer drainage area of one of the surrounding wells. When a well is shut-in the area that has previously been drained by the shut-in well will soon come under drainage by the offset wells. As soon as this new drainage system has stabilized, the pressure distribution throughout the reservoir will become constant. The observed well pressure will become constant and simply represent one of the pressure points in the drainage system of an offset well. The only way a constant pressure in a well could represent the average pressure for the well's drainage area is if all the wells in the reservoir were shut-in.

In summary, the following methods are recommended for determining the average pressure in the drainage area of a well at time of shut-in.

(1) If the well is in steady state at time of shut-in and it is situated near the center of its drainage area, Equation 3.46 should be used to calculate the average pressure.

(2) If the well is in pseudosteady state at shut-in, the Odeh data of Figs. 3.9A and 3.9B should give the most accurate average pressures. The author cautions that at the time this chapter was being prepared the Odeh data had been used very little.

(3) If the well is in pseudosteady state at shut-in and is situated near the center of the drainage area, Equation 3.43 will give accurate results if the stabilization time is based on a field-measured compressibility-porosity product. If pressure data are not sufficient to permit recent past performance evaluation of the compressibility-porosity product, Equation 3.45 can be used to estimate the average pressure although this approach appears to be completely untested in practice.

(4) If a well is infinite acting at time of shut-in the average pressure can only be determined by material balance, Equation 3.38.

A problem will be used to illustrate the application of some of these methods. Since the calculation of the average pressure for steady state and infinite-acting wells seems to be straight forward (regardless of the accuracy) the problem will be limited to the evaluation of the average pressure for a pseudosteady-state well.

Problem 3.3B: Determining the Average Pressure in the Drainage Area of a Pseudosteady-State Well

In example 3.3A the pressure buildup for a well in pseudosteady state at shut-in was described. The tubing-head pressure decline just prior to shut-in was given as 24 psi/d and the solution of the problem established the slope, m, as 197, the Δp_{skin} as 415 psi, the stabilization time, t_s, as 3.65 days, and the undamaged permeability as 15.1 md. The Δp_q plot is shown in Fig. 3.6. The initial pressure in this drainage area when the well was drilled was 2,960 psia. Find average pressure for the 40 acres drained by this well and the damage ratio of the well.

The solution to this problem can be checked against the solution in Appendix C.

Two-rate build-up tests.   A two-rate build-up test uses the effect of the negative rate change, $(q_1 - q_2)$ to cause an increase in well pressure. This situation is illustrated in Fig. 3.10. An analysis of this pressure increase can provide a means of determining undamaged permeability, the additional pressure drop due to damage, and the average drainage-area pressure.

There are many problems associated with pressure build-up tests that can be avoided by simply reducing the producing rate and observing the increase in pressure without completely shutting the well in. The resulting improved characteristics of the build-up data are the result of

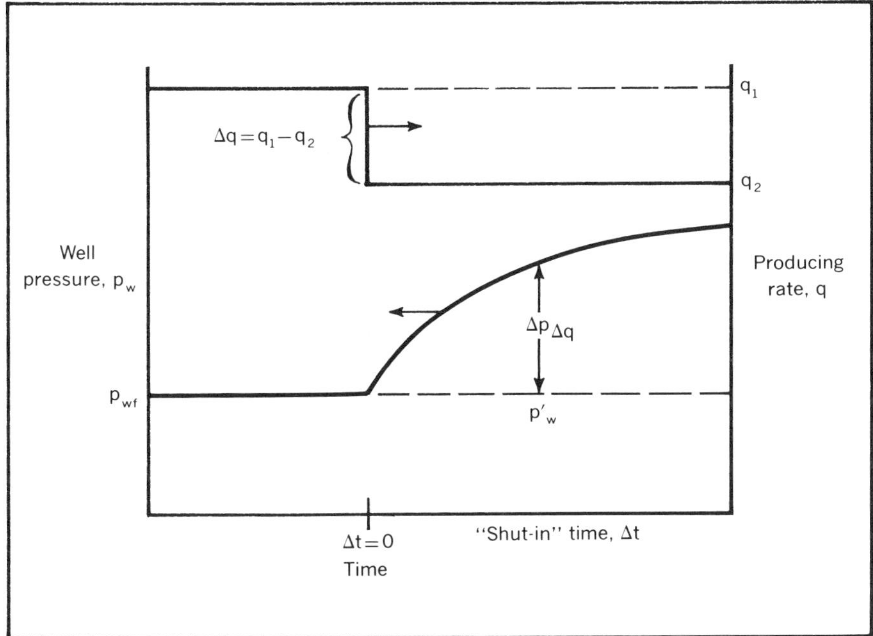

FIG. 3.10 Two rate buildup data.

a much more rapid stabilization of the rate change. The engineer should now be aware of the difficulty experienced normally in trying to obtain an instantaneous reduction in flow rate from a rate q to zero.

The difficulties resulting from the afterflow period were discussed at some length earlier.

Afterflow effects are the result mainly of the gas in the tubing and/or casing at time of shut-in (depending on whether a packer is run in the well or not). Clearly the worst situation that can exist for afterflow would be to have the producing system filled with low-pressure gas. This is the situation that would exist in a high-pressure gas reservoir producing at a high drawdown at time of shut-in such that the pressure in the producing system at shut-in is low. Pressure buildups conducted under such conditions often give afterflow effects that last for days and virtually prohibit the type of analyses previously discussed in this chapter.

Afterflow difficulties grade from this extreme to the ideal situation in which there is no gas in the producing system at shut-in and there is virtually no afterflow. If the flow rate from a well is simply reduced, the reduced rate will reach a steady value rather quickly as compared with a complete shut-in. Consequently, the resulting pressure increase is generally much easier to analyze. This is particularly true if the well pressure at the time the flow rate is reduced is unchanging as defined previously.

It will be recalled that the unchanging well pressure was defined as a pressure that would have changed an insignificant amount during the shut-in period if the well was not shut-in. In the case of the two-rate test we can say that the pressure is unchanging if the pressure would change an insignificant amount during the time the build-up data will be analyzed if the rate were not changed. Under these conditions we will see that the equations developed for the pressure buildup analysis from an unchanging well pressure can be used to analyze the two-rate test by simply using the rate change as the rate in the subject pressure buildup analysis equations.

Use of the two-rate type of pressure buildup test also may be used to overcome some of the other difficulties commonly encountered in pressure-buildup data. These situations will be discussed under the section of this chapter captioned, Pressure Buildup Anomalies. One such anomaly that is sometimes encountered is commonly termed inversion. What happens is that the redistribution due to gravity of the gas and oil trapped in the tubing at shut-in will often cause the recording of an abnormally high pressure early in the shut-in. This high pressure has in some cases been known to exceed the average shut-in reservoir pressure. As noted above the cause of this phenomena is more completely discussed in the Pressure Buildup Anomalies section; however, for our purposes we need to recognize that it is caused by the free gas that is more or less uniformly distributed throughout the oil in the tubing at time of shut-in, completely segregating following shut-in so that there exists in the well a column of oil containing no free gas and a column of gas over the oil column. The point is that simply reducing the flow rate rather than completely shutting the well in will avoid this complete separation of the liquid and gas and will thus avoid the inversion anomaly.

The two-rate test may also be used to avoid or minimize the difficulties caused by the existence of two or more beds of vastly different permeabilities or of a reservoir with a matrix permeability much different from the permeability of the fracture or vugular system.

It is impossible to provide the engineer with specific guide lines as to when a two-rate test should be employed rather than a conventional build-up test. Most competent reservoir engineers use the build-up test in all cases where good interpretable results are obtained and go to the two-rate test only when difficulties are encountered in running or interpreting the standard buildup. The reasons for this preference for the buildup is probably more a function of the history of the development of the technology than it is due to the advantages afforded by the build-up-type test. In other words the build-up type test has been in use much longer than the two-rate test; thus, engineers are more familiar with the test and use it in preference to the two-rate test which is not as widely understood.

One last practical advantage of the two-rate test should be noted before considering the specific analysis equations employed for a two-rate test. If there is concern about the "lost" production that will result from a build-up test you may want to go to the two-rate test. This is particularly true if the subject well has an excess producing capacity. It is obvious that reducing the producing rate will not result in as large a loss of production as will shutting the well in but note that with an excess producing capacity you may be able to produce the well at a higher-than-normal rate until it reaches steady state or pseudosteady state so that when the rate is reduced no net loss in production will result for that month. You can do the same sort of thing in preparing for a buildup but the quantitative possibilities are much more restrictive. This is no substitute for the engineer's ability to justify the cost of obtaining reservoir data from a profitability standpoint but the approach has been found helpful in obtaining permission for well tests.

In this section we will only consider methods that are applicable to wells that have unchanging well pressures. Much more complex methods are available in the literature (e.g. SPE Nomograph No.—Pressure Build-up Analysis)[6] that are applicable to the more general case. However, the need for a two-rate test arises from the fact that considerable amounts of gas are being produced and, as the amount of free gas in the reservoir increases, the tendency toward an unchanging well pressure increases. Consequently, the engineer will find that in almost all cases where two-rate tests are necessary an unchanging well pressure will exist in the reservoir and the simple methods described herein will suffice as a means of analysis.

Consider again the illustration depicting the conditions of a two-rate pressure build-up test, Fig. 3.10. This represents a well in which well pressure would not have changed significantly if the well flow rate had not been changed. The well pressure that would have existed without a change in rate is shown as p_w'. The pressure increase that is superimposed upon this pressure, p_w', is indicated as $\Delta p_{\Delta q}$ which is simply the increase in pressure caused by a negative producing rate change, $q_1 - q_2$, for a time Δt, the time the second rate has been effective. This is governed by the constant-rate infinite-acting solution to the radial diffusivity equation until Δt is greater than the stabilization time t_s (see Equation 2.63 derived in Chapter 2 and used previously in this Chapter). This is the same expression used to define the Δp_q expression, Equation 3.7, except that the change in rate, Δq, is substituted for the rate, q.

$$\Delta p_{\Delta q} = \frac{0.141 \, \Delta q \mu}{kh} \left(\frac{1}{2}\right) (\ln \Delta t_D + 0.809) + \Delta p_{skin} \qquad (3.47)$$

In this equation the reduced time, Δt_D, has the same value as that described in Equation 3.8. Equation 3.47 must be added to p_w to obtain an expression for the well pressure after the rate reduction. The well pressure at time of shut-in, p_{wf}, is p_w'. When Equation 3.47 is added to p_{wf} and the expression is expanded with all the terms that do not change with time grouped into a constant term (in the same manner that Equation 3.11 is derived from Equation 3.7), we find that the well pressure at some time after the rate change, Δt, is,

$$p_w = \text{constant} + \frac{0.1625\ \Delta q\mu}{kh}\ \log\ \Delta t \qquad (3.48)$$

This expression is the same as that of Equation 3.11 which describes the well pressure for a well with an unchanging well pressure at shut-in except that the change in rate, Δq, is substituted for the rate, q. Equation 3.48 shows that a plot of the well pressure, p_w, versus the log of the time since the rate change, will give a straight line whose slope is the familiar $0.1625\ \Delta q\ \mu/kh$. Whereas the rate, q, has been used previously in the build-up slope expressions we now must use the rate change, Δq.

$$m = \frac{0.1625\ \Delta q\mu}{kh} \qquad (3.49)$$

The permeability in this expression is the undamaged permeability since the additional pressure drop due to the damage around the well bore is included in the Δp_{skin} term.

To find the additional pressure drop due to the damage, Δp_{skin}, we can simply use Equation 3.47 with a function of m substituted for the group of terms, $0.141\ \Delta q\mu/kh$, to give

$$\Delta p_{\Delta q} = 0.867m\left(\frac{1}{2}\right)(\ln\ \Delta t_D + 0.809) + \Delta p_{skin} \qquad (3.50)$$

This equation is used by reading a value of $\Delta p_{\Delta q}$ from the data and using the corresponding time since the rate change, Δt, as the basis for calculating the reduced time, Δt_D, which is calculated according to Equation 3.8. As in other applications, the mobility, k/μ, can be evaluated from the slope, m, and it is advisable to determine the ϕc product from reservoir behavior if it is at all possible. The most popular method of determining ϕc by performance is to calculate it from the equation for the pseudosteady-state change in pressure with time, Equations 2.41 or 2.42 derived in Chapter 2 and previously used in this Chapter. The skin factor, S, can be calculated from Equation 3.15.

In applying Equation 3.50, the engineer should remember that the additional pressure drop, Δp_{skin}, represents the pressure drop due to only the rate change. Consequently, he may want to calculate the additional pressure drop due to damage at a normal producing rate. This can be done by first calculating the skin factor, S, and then using this factor in Equation 3.14 to calculate the additional pressure drop for any producing rate, q.

A problem should help clarify the calculating procedure.

Problem 3.4: Analysis of a Two-Rate Pressure Buildup

A well in the Vicksburg 8000-ft sand in Brooks County, Tx. stabilized at a rate of 78 STB/d for a week, at which time the bottom-hole pressure and surface pres-

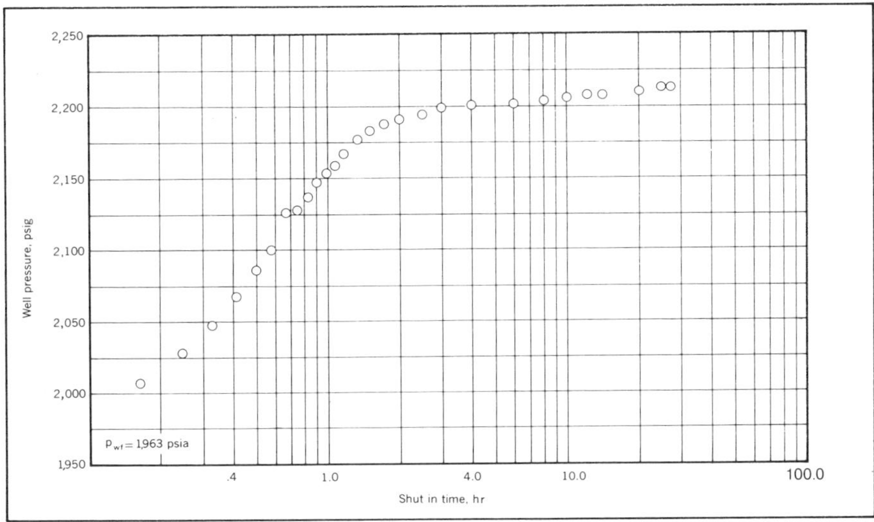

FIG. 3.11 Pressure data for two-rate buildup example problem.

sure appears to be unchanging. The well rate is then reduced to 64 STB/d with the pressure history indicated in Fig. 3.11. Find K_o, Δp_{skin}, S, and p_s if the following data apply to this well.

$c_e = 1.379 \times 10^{-4}$/psi (must be calculated considering the gas, oil, and water saturations)

$B_o = 1.322$

$\mu_o = 0.39$ cp

$h = 20$ ft

$\phi = 0.2$

$r_w = 0.265$ ft

$r_e = 1,490$ ft (calculated on basis of estimated drainage area)

Pressure falloff tests. Typical data for a pressure falloff test is shown in Fig. 3.12. A negative rate applies to injection; thus, falloff-test analysis applies to injection wells or for disposal purposes. Since injection wells generally reach the unchanging well pressure state quickly, the equations of the section dealing with pressure buildup for wells with an unchanging well pressure can be applied in most cases if care is taken to use the proper sign on the terms that are a function of the well rate. Some additional problems and some simplifications are typical of pressure falloff analysis.

Since we are normally dealing with liquid injection we find that afterflow effects are almost nonexistent in pressure falloffs unless there is trapped gas someplace in the injection system. Otherwise the very small compressibility of liquid lets the rate go to zero almost immediately after the injection well is shut-in at the surface. In fact, there is often an

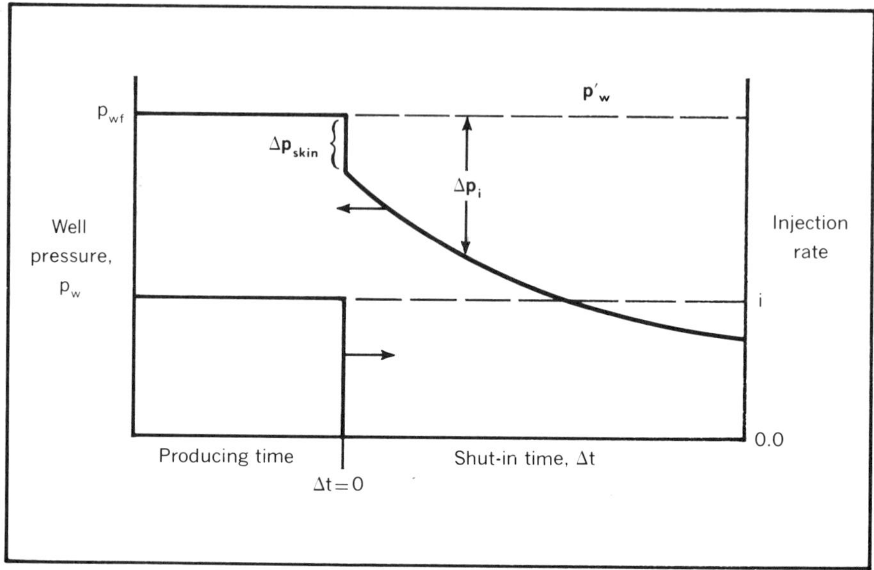

FIG. 3.12   Pressure falloff data.

instantaneous drop in the injection pressure that approximates the additional pressure drop due to well damage (Δp_{skin}), since the flow in the damaged zone also is abruptly terminated. This effect is analogous to having flow thru a small-diameter, long pipe that contains an orifice or other extreme flow restriction. If a valve just on the upstream side of the orifice is closed there will be an immediate drop in the pressure on the upstream side of the valve that will be equal to the flowing pressure drop across the orifice. Then the pressure will begin declining more slowly. In the case of the injection well the damaged zone is analogous to the orifice and the same pressure behavior occurs when the well is shut-in at the surface unless the damage radius of the system is substantial.

One of the problems that the engineer or operating personnel bring upon themselves, results from the attempt to use a surface pressure as a basis for the falloff analysis rather than running a pressure bomb to record the bottom-hole pressure. When this procedure is followed it is often found that the usable pressure record is of limited duration because the pressure at the surface falls to zero. Obviously this difficulty can be overcome by running a pressure bomb and recording the bottom-hole pressure as is normally done in other types of pressure-behavior tests.

This is not meant as an implication that all falloff tests should be run with a bottom-hole pressure bomb. If a reservoir is sufficiently tight and the injection pressure high enough to provide a usable pressure record, the engineer would be well advised to use the surface pressure and avoid the expense and always-present danger of running a bomb into the hole. The difference between the surface and bottom-hole pressure after shut-in without gas in the system will be constant and the change in the surface pressure with time will be the same as the change in the bottom-hole pressure with time. However, only experience will permit the engineer to be able to predict whether or not the surface pressure record will be of sufficient duration to permit an accurate analysis. When in doubt, run a bottom-hole pressure bomb and repeat the test. However, when the surface gauge pressure falls to zero, flow into the well will be resumed due to the static head of fluid in the well. If this flow rate is significant the fall off analysis will be complicated.

As noted above, we can obtain the equations for the analysis of a falloff test by following the same procedures developed in the section on pressure buildup analysis captioned, "Unchanging" Pressure at Shut-In. Referring again to Fig. 3.12, note that the well pressure during falloff will be the sum of the pressure, p_w', that would have existed without shut-in and the pressure decline caused by the positive rate change resulting from a discontinuation of the injection, Δp_i. The Δp_i equivalent of Equation 3.7 would then be

$$\Delta p_i = \frac{0.141 i \mu}{kh} \left(\frac{1}{2}\right) (\ln \Delta t_D + 0.809) + \Delta p_{skin} \qquad (3.51)$$

where i is the injection rate in barrels per day. In this equation the Δt basis for the reduced time Δt_D is the time the injection well has been shut-in. When this expression is added to p_w' which is p_{wf} for an unchanging injection pressure, we obtain the equivalent of Equation 3.9,

$$p_w = p_{wf} - \frac{0.141 i \mu}{kh} \left(\frac{1}{2}\right) (\ln \Delta t_D + 0.809) - \Delta p_{skin}. \qquad (3.52)$$

When this equation is expanded the equivalent of Equation 3.11 in the unchanging section is

$$p_w = \text{constant} - \frac{0.1625 i \mu}{kh} \log \Delta t \qquad (3.53)$$

This equation shows that the slope of the plot of well pressure, p_w, versus the log Δt is

$$m = -\frac{0.1625 i \mu}{kh} \qquad (3.54)$$

Once m has been evaluated a form of Equation 3.52 with 0.867m substituted for $0.141q\mu/kh$ can be used to evaluate Δp_{skin}.

$$p_w = p_{wf} - 0.867m \left(\frac{1}{2}\right) (\ln \Delta t_D + 0.809) - \Delta p_{skin} \qquad (3.55)$$

In using this equation the reduced time must be calculated by Equation 3.8 which means that k/μ must be determined from the slope, m, using Equation 3.54, and the porosity and compressibility must be determined independently. Fortunately this is not too difficult for an injection well because with only liquid flowing the determination of the effective compressibility from empirical data as shown in the appendix is relatively accurate.

Generally, the average pressure in the drainage area of an injection well is relatively unimportant since the engineer is much more concerned with displacement efficiencies under such conditions than he is with material balance and resulting saturations at various stages of injection. If the engineer does desire an average pressure for the drainage area it can be obtained with reasonable accuracy by using Equation 3.46 which is the average pressure for a radial steady-state flow system. When the equation is applied to a pattern flood remember that the drainage area of the injection well does not include the drainage area of the producing wells for the purposes of applying this equation. For example, when applied to an injection well in a 20-acre five-spot pattern the drainage area of the injection well would only be 10 acres. The effective external well radius would then be calculated on the basis of 10 acres.

The following problem calculations should help to explain the details of a falloff analysis.

Problem 3.5:   Pressure Falloff Analysis

Given the falloff pressure data of Fig. 3.13 and reservoir parameters listed below for an injection well in a five-spot pattern flood. Find the undamaged permeability, the pressure drop due to the well skin, and the skin factor, S. What is the maximum radius of investigation at a shut-in time of 15 min? What would be the injection rate if the skin factor is reduced to zero by acidizing? Assume no pressure loss in the tubing due to friction, water specific gravity of 1.0, the producing wells are kept pumped down (pressure is atmospheric), and Δp_{skin} at the producing wells is zero.

$$h = 50 \text{ ft}$$
$$c = 5 \times 10^{-6} \text{ psi}^{-1}$$
$$\mu = 1.0 \text{ cp} \qquad \text{Average sand depth} = 3,000 \text{ ft}$$
$$r_w = \tfrac{1}{4} \text{ ft} \qquad \phi = 20\%$$

Surface injection pressure at shut-in = 500 psig
Injection rate before shut-in   = 250 b/d

The danger of using too short a shut-in time as a basis for the pressure data analyzed is emphasized in Problem 3.5 where the calculated radius of investigation was very small. This problem is present in practically any type of well test but it is particularly dangerous in the case of pressure falloff due to the temptation to use surface recorded data and the fact that the surface pressure falls to zero and becomes unusable very quickly in many cases.

If the engineer experiences difficulties in obtaining a useable pressure fall off record he should consider using a constant rate pressure

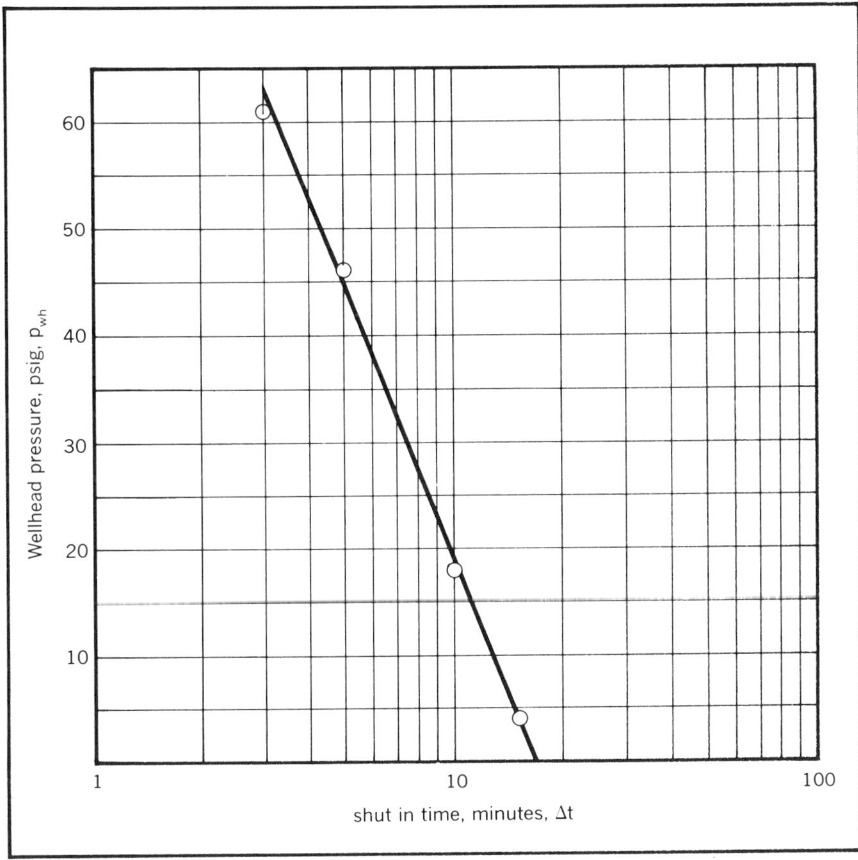

FIG. 3.13 Well pressure vs shut-in time for a falloff test date for problem 3.5.

build up similar to the constant rate pressure fall off test for producing wells. This is discussed later in this chapter.

Pressure buildup anomalies. Unexpected pressure behavior is often encountered in pressure testing of wells. We often refer to these unexpected data as anomalies. Since we believe we can explain and understand some of this data it may not be grammatically correct to refer to this behavior as anomalies but it does provide a convenient classification under which to discuss this type of phenomena.

Adoption of this definition of anomalies might mean that we should include afterflow and the effect of the skin damage in this section. However, since we have already discussed this matter in previous sections of this chapter the discussions will not be repeated here. Furthermore, the objective of this section will be to simply identify some of the anomalies without going to detailed explanations. The subject will be treated in this way here because great numbers of papers have been written on these problems (possibly many more than the subject warrants), and excellent, more detailed, discussions are readily available in the SPE Monograph No. 1—Pressure Buildup and Flow Tests in Wells. In this section we wish to make note of (1) the humping effect caused by the redistribution of the gas and oil following shut-in, (2) the effect of a fault or nearby boundary, (3) the effect of stratification with widely varying permeabilities, and (4) the effect of two widely different parallel permeabilities in a reservoir such as vugular and matrix permeability in limestone, or lateral increase or decrease in mobility as would be encountered with a gas cap or water drive.

A typical case of humping due to phase separation following shut-in is illustrated in Fig. 3.14. It has been shown theoretically and in laboratory experiments that such an anomaly is due to the phase redistribution in the tubing caused by the subsequent rise of the gas bubbles in the oil column following shut-in. Our intuition again serves us poorly because most engineers erroneously conclude that if the contents of a vessel remain the same the pressure on the bottom of the vessel must be constant regardless of the distribution of the phases in the vessel. Study of Fig. 3.14A will show the erroneous nature of this conclusion. One portion of

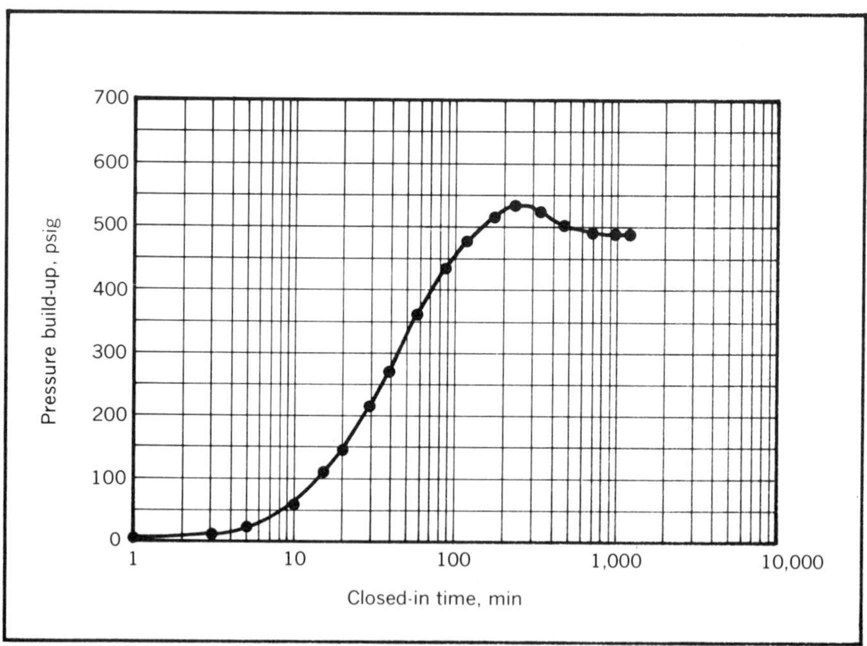

FIG. 3.14 Humping in a pressure buildup (after Mathews and Russell[5], SPE of AIME)

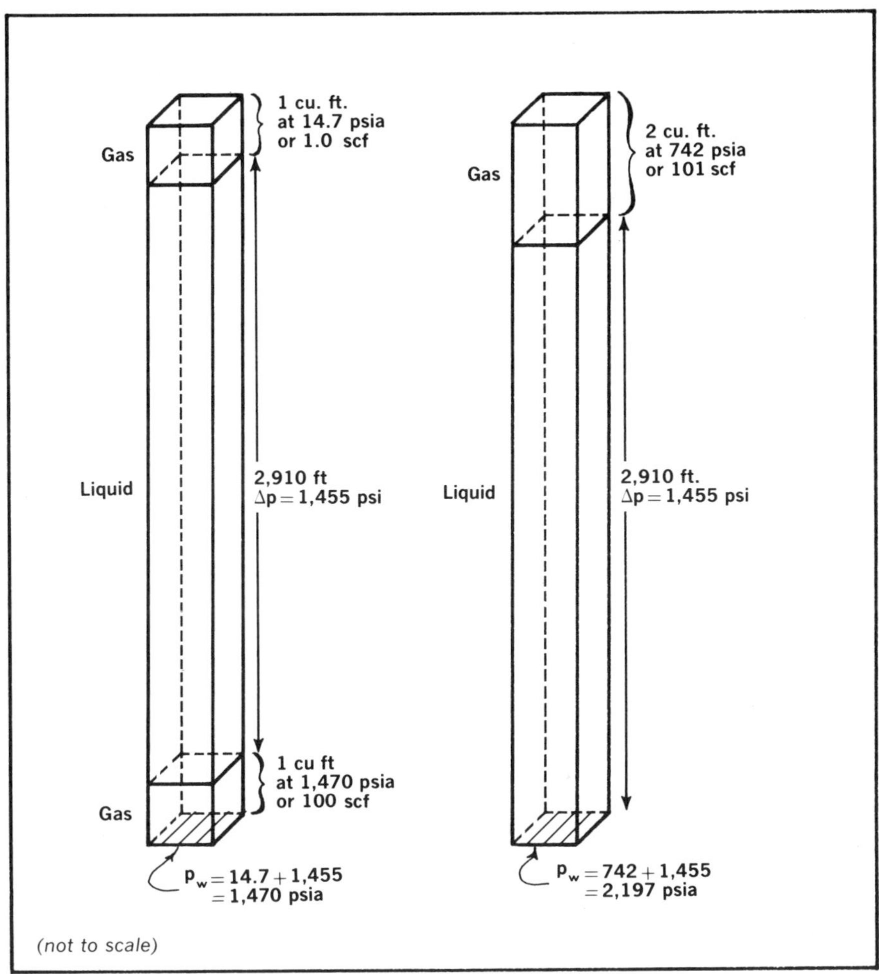

FIG. 3.14A   Effect of phase distribution on the bottom-hole pressure

the diagram indicates the bottom-hole pressure that would exist in this well if it contained 1.0 cu ft of air at atmospheric pressure above a 2,910-ft column of fluid with a pressure gradient of 0.5 psi/ft which has trapped beneath it at a pressure of 1,470 psia 1.0 cu ft of air. To simplify the calculations assume that the temperature is standard throughout the column and the cross-sectional area of the well is 1.0 sq ft.

Under these conditions the air in the bottom of the hole amounts to 100 scf so that the total mass of air in the well is 101 scf. Now if all of this air occupies the 2.0 cu ft at the top of the well, the resulting pressure would be 742 psia. When this is added to the pressure caused by the 2,910 ft of fluid, the bottom-hole pressure will be 2,197 psia, an increase of 727 psi. The lower bottom-hole pressure that follows the abnormally high pressure is due to the flow of fluid back into the shut-in well as a result of the abnormally high pressure. Consequently, it seems likely that the existence of the abnormally high pressure occurs much more often than we realize because a reservoir must be of a reasonably high permeability with a minimum of well damage to permit flow back into the reservoir. Unless flow back into the reservoir occurs we do not observe the abnormally high pressure due to the phase redistribution. The difficulties caused by phase redistribution can often be avoided by using a two-rate test as previously defined.

When a plane barrier or boundary exists in one direction from a well, the engineer should recognize that the pressure behavior of the well will be the same as the pressure behavior of two wells producing from an infinite-acting reservoir with identical producing-rate histories and a distance between the wells equal to twice the distance to the fault. (See Fig. 3.15). This phenomena was discussed under superposition in Chapter 2. Such a modeling of the actual case provides a drainage pattern in the infinite-acting model for each well that is identical to the actual well's flow pattern. That is, there will be no flow across the boundary in the actual case and there will be no flow across the line midway between the two infinite-acting wells in the infinite-acting model.

Now in the model consider that the wells have been producing long enough so that an unchanging well pressure at shut-in results. Then the shut-in well pressure is the flowing well pressure at shut-in plus the increase in the well pressure caused by the negative rate change of shutting in the real well plus the well pressure caused by the negative rate change of shutting in the image well.

$$p_w = p_{wf} + \Delta p_q + (\Delta p_q)_{image} \qquad (3.56)$$

In expanding this expression, the Δp_q term will be the same as defined by Equation 3.7 but the term representing the pressure increase

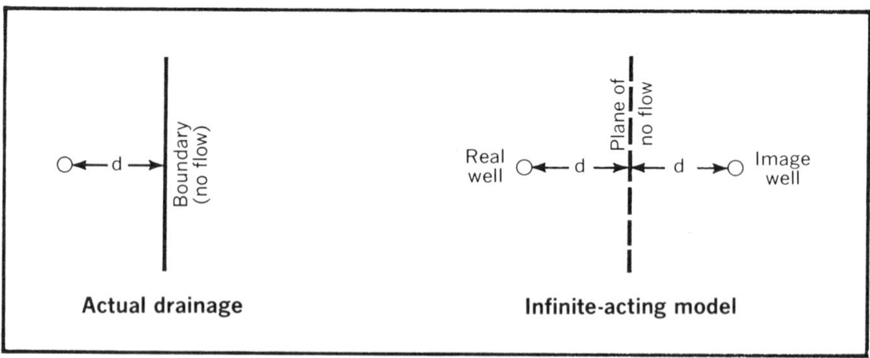

FIG. 3.15   Modeling a single plane flow boundary.

due to the negative rate change in the image well $(\Delta p_q)_{image}$, involves the calculation of the pressure drop at a radius equal to twice the distance to the fault. With this large a radius as a basis for the reduced time, Δt_D, there may be a substantial period of time before Δt_D is greater than 100 which is the arbitrary lower time limit we have been using for the application of the log equation in evaluating the pressure functions. Nevertheless, when Δt_D based on a radius twice the distance to the boundary is greater than 100 we can write Equation 3.56 as

$$p_w = p_{wf} + \frac{0.141q\mu}{kh}\left(\frac{1}{2}\right)(\ln \Delta t_D + 0.809) + \Delta p_{skin}$$
$$+ \frac{0.141q\mu}{kh}\left(\frac{1}{2}\right)[\ln (\Delta t_D)_{image} + 0.809] \tag{3.57}$$

Now note that we can expand both of the log of the reduced time terms by breaking them into the log of the diffusivity constant divided by the radius squared plus the log of the shut-in time, Δt. After doing this we will again use the technique of placing all the resulting terms that do not change with time into one constant term.

$$p_w = \text{constant} + \frac{0.141q\mu}{kh}\left(\frac{1}{2}\right)(\ln \Delta t + \ln \Delta t) \tag{3.58}$$

In this equation the two $\ln \Delta t$ terms are obtained from the two log of reduced time terms in Equation 3.57. Now when Equation 3.58 is written in terms of base 10 logs and a function of the normal build-up slope, m, is substituted for $q\mu/kh$ we obtain

$$p_w = \text{constant} + 2m \log \Delta t \tag{3.59}$$

This equation then shows that the ultimate effect of a plane boundary on an otherwise infinite-acting pressure shut-in effect is to double the normal pressure build-up slope, m. This effect was verified for an unchanging well pressure but the same procedure could be followed to show that the same effect will be experienced if the well is in pseudo-steady state or infinite acting at shut-in. However, note that to obtain the double slope it is necessary for the reduced shut-in time based on a radius of twice the distance to the fault to reach 100 before the shut-in time in days becomes greater than $r_e^2/4n$ where r_e is the distance to the next closest boundary. This represents the approximate time when the effects of the closest boundary will affect well pressure and at this time Δp_q will no longer be based on the log equation. The deviation from the build-up slope, m, can be used to calculate the distance to the closest boundary in many cases regardless of whether or not the slope doubles. This technique will be discussed fully in the section of this Chapter captioned, Reservoir Limit Tests. The point to be made at this time is that a boundary near the shut-in well will result in an increase in the slope of the build-up data which may double if there is sufficient difference in the distance to the nearest and next nearest drainage boundaries. An example is shown in Fig. 3.16.

Stratification of a reservoir does not generally result in as many anomalies in pressure buildup as might be expected. Chapter 10 of Monograph No. 1, referred to many times in this chapter, contains an excellent theoretical and practical discussion of the effects of reservoir heterogeneities on pressure behavior. From a practical standpoint we simply need to realize that stratified reservoirs will behave as would a homogeneous reservoir with the average characteristics of the stratified reservoir *if* there is unrestricted communication or permeability between the reservoir strata. However, if the strata are in communication only at the well bore they will act like two separate reservoirs being produced through a common producing system which is exactly what they will be.

In the latter case it should be realized that depletion of the two or more noncommunicating strata during production will be at constantly changing rates even though we maintain the total producing rate from a well as a constant value. This then means that at any particular time of shut-in the different strata will be at different stages of depletion. This will result in the strata being at different pressures and having different saturations. Under these conditions the resulting buildup is practically useless. Fig. 3.17 illustrates a buildup in a two layer reservoir. The early part of the buildup is probably affected most by the most permeable strata which has been more thoroughly depleted than the strata of lesser permeability and thus has a lesser reservoir pressure. After the flattening of the pressure history which probably reflects the magnitude of the pressure in the most permeable strata, note that the pressure starts to

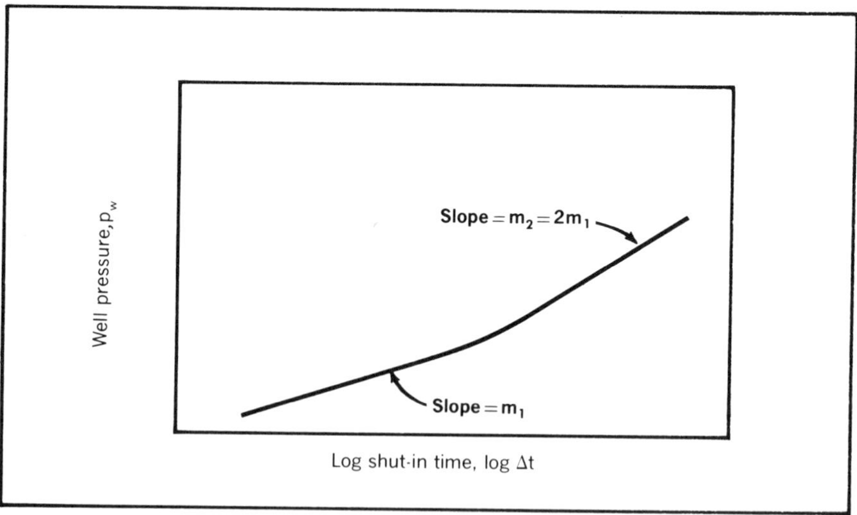

FIG. 3.16   Effects of a plane barrier on pressure buildup.

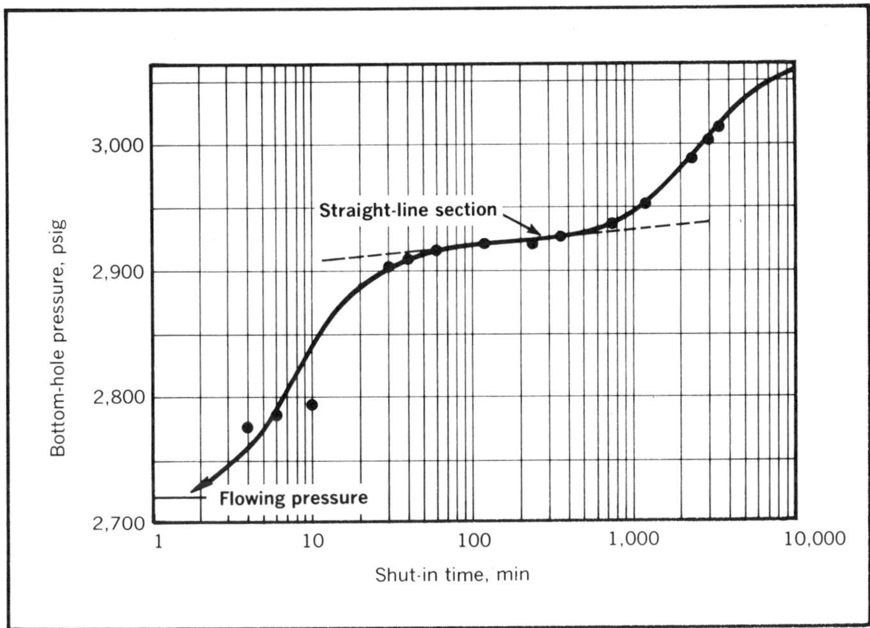

FIG. 3.17 Effects of a two-layer reservoir on pressure buildup[10] (after Mathews, J. Pet. Tech.)

increase more rapidly again. This buildup is probably due to the less permeable strata which has a higher reservoir pressure.

A reservoir that contains two vastly different pore structures such as a jointed or fractured reservoir or a vugular limestone may give pressure build-up characteristics similar to those of the buildup for a reservoir with two noncommunicating strata. Whether the characteristics exhibited resemble this or a regular homogeneous reservoir appears to be a function of the degree of porosity and permeability differences, the ratio of pore volumes represented by the two pore systems, and the degree of depletion.

Regardless of whether a build-up anomaly such as that of Fig. 3.17 is due to uncommunicating strata or two widely different pore structures in the reservoir, it has been found that using a two-rate flow test will often permit evaluation of average reservoir characteristics. This is probably due to the fact that by simply reducing the rate the flow remains in the same direction in the reservoir at all times whereas a rate of zero as is experienced with a shut-in may result in flow between the strata or different pore systems with the results that in some cases flow in one strata or system is taking place in one direction while in another strata or pore system the flow at the same instant is in the opposite direction. Regardless of the reason, experience has shown that a two-rate test will often give a pressure behavior that can be interpreted while a build-up test gives uninterpretable pressure data. Use of a constant rate pressure drawdown test provides similar advantages and will be discussed later in this chapter.

One other type of anomaly that may be encountered is the change in the build-up slope that may take place when the buildup begins to be affected by a part of the reservoir that has a decidedly different saturation. This is the sort of anomaly that is often observed in the presence of a water-oil or gas-oil contact. The engineer simply needs to be aware of the fact that the presence of a contact can cause such an anomaly. There does not appear to be any way that such an anomaly can be avoided. The engineer should simply limit his analysis to that portion of the buildup obtained before the change in saturation affects the build-up data.

RESERVOIR LIMIT TESTS

As has been noted previously under Pressure Buildup Anomalies and in Chapter 2 under Superposition, reservoir boundaries cause anomalies in pressure behavior. Since this is true and the engineer often needs an independent means of determining the size and shape of a reservoir it behooves him to find a means of interpreting the pressure behavior caused by a reservoir limit in such a way as to determine the size and shape of the limit. Many different methods have been devised and tried to obtain such an interpretation of pressure behavior. Many of these methods are theoretically sound but useless from a practical standpoint. Many misinterpretations of data have been caused by the engineer's inability or carelessness in using a quantitative evaluation. Many looked-for anomalies have been too small for a conventional bottom-hole pressure gauge to record. Some quantitative calculations would have shown this and avoided erroneous conclusions. Often the engineer has mistaken the anomaly caused by a change in gas saturation around the well bore as an anomaly caused by a reservoir boundary. These and other problems associated with reservoir limit or interference tests will be discussed in the section along with recommended methods of analysis.

The most useful test in an evaluation of reservoir limits is the drawdown test. The ideal means of running this test is to start producing a well in a reservoir at a constant rate with the pressure at the time production is begun being uniform throughout the reservoir. For most practical cases this situation would exist only at the time a reservoir is discovered, but other states of depletion could be used under certain conditions. In the preceding section, the effect of a plane boundary on a pressure buildup was considered. We will now consider the effect such a boundary would have on the flowing pressure of a well that is produced at a constant rate. As was pointed out in the pressure buildup discussion, a well a distance d from a plane reservoir boundary (as in Fig. 3.15) will have the same behavior as either of the wells a distance 2d apart in an infinite-acting reservoir if both produce at identical rates. This is true because there would be no flow across the line representing all the points equidistant from the two wells as shown in Fig. 3.15 and the flow geometry into the wells in the infinite-acting model would be exactly the same as the flow geometry into the actual well.

To analyze the pressure behavior in the infinite-acting model with both wells producing at a constant rate we must apply the constant-rate solutions to the radial diffusivity equation. The well pressure at any particular time will be equal to the initial pressure minus the pressure drop caused by the flow rate q at the well radius, minus the additional pressure drop due to the well damage, minus the pressure drop caused by the rate q in the image well at a radius of 2d.

$$p_w = p_i - \frac{0.141q\mu}{kh}(PF)_{real} - \Delta p_{skin} - \frac{0.141q\mu}{kh}(PF)_{image} \qquad (3.60)$$

The only way that we can evaluate the pressure drop caused by the flow rate q at a radius 2d is by using the Ei function solution since we do not know the flow rate at the radius 2d. Be careful that you do not become confused by the fact that in the infinite-acting model the real well is producing at a rate q at a distance 2d from the image well. We account for this flow in the second term of Equation 3.60 but this has nothing to do with finding the effect of the flow from the image well at a radius of 2d. In this case (the last term of Equation 3.60) we do not know the rate equivalent of the image well rate at a radius of 2d.

Since we can only use the Ei function solution when the reservoir is infinite acting, we can also use an infinite-acting pressure function for the real well. The log function equation can then be used to evaluate the pressure function for the real well because in this case we desire the pressure drop at the well radius where we know the flow rate. To use the log function it is, of course, necessary for the reduced-time basis for the function to be greater than 100, but this will normally occur in a matter of a few hours, minutes, or seconds since the reduced time is based on the small well radius which makes the reduced time increase very rapidly. Thus, using the log function for the real pressure function and the Ei function for the image pressure function we can expand Equation 3.60 to

$$p_w = p_i - \frac{0.141q\mu}{kh}\left(\frac{1}{2}\right)(\ln t_{D\ real} + 0.809) - \Delta p_{skin}$$

$$- \frac{0.141q\mu}{kh}\left(\frac{1}{2}\right)\left(Ei\ \frac{-1}{4t_{D\ image}}\right) \qquad (3.61)$$

An actual well pressure from a drawdown test when plotted against the log of the producing time will have the characteristics illustrated in Fig. 3.18 if there is a plane reservoir boundary nearby. Note that after

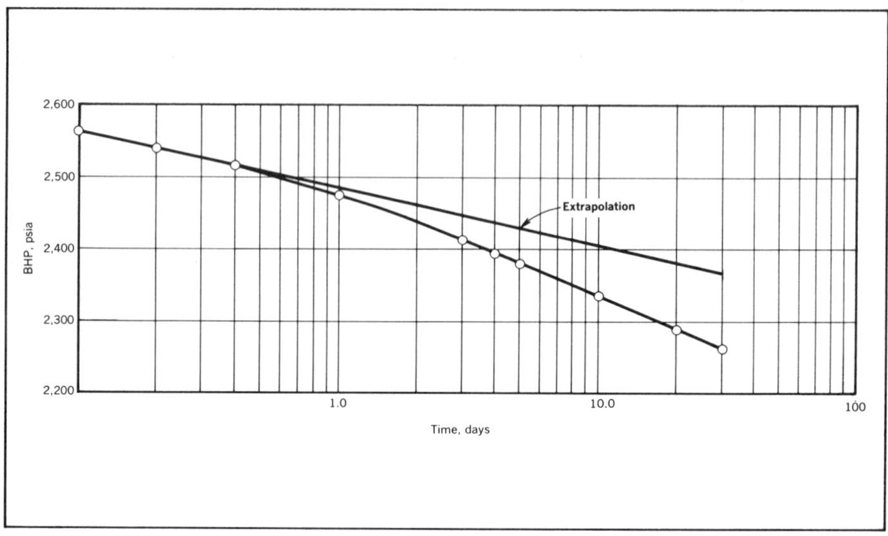

FIG. 3.18 Effect of a plane barrier on a drawdown.

a brief period corresponding to the time for $t_{D\ real}$ to exceed 100, there is a straight-line relationship between the well pressure and the log of the producing time. By examining Equation 3.61 we can see why this is true. The terms p_i and Δp_{skin} will not change with time so they will not affect the slope. Also remember that it takes some finite period of time for the production from the image well to affect the pressure at the radius 2d. Or another way of explaining this is to note from the plot of the Ei function versus the reduced time that the function is essentially zero until the reduced time exceeds about 0.1.

Now note that if we extend the early straight-line portion of this plot, the difference between this extrapolation, p_w', and the actual well pressure (after we have felt the effect of the reservoir boundary at the well) is equal to the last term of Equation 3.61.

$$p_w' - p_w = \frac{0.141q\mu}{kh}\left[\left(\frac{1}{2}\right)\left(-Ei\ \frac{-1}{4t_{D\ image}}\right)\right] \tag{3.62}$$

Also note that $t_{D\ image}$ is based on a radius of 2d.

$$t_{D\ image} = \frac{6.33kt}{\phi\mu c(2d)^2} \tag{3.63}$$

These equations and observations then give us a means of determining the distance to a reservoir boundary. If we plot the well pressure versus the log of the producing time as in Fig. 3.18, extrapolate the early straight-line portion, and read the difference between the extrapolated straight line and the actual recorded pressure at some noted producing time, we can substitute this value for $p_w' - p_w$ in Equation 3.62 and solve for the entire Ei function. Note that this includes everything in the brackets. Now using the plot of the Ei function versus reduced time, Fig. 2.8 in Chapter 2, we can find a corresponding reduced time, $t_{D\ image}$. Putting this reduced time in Equation 3.63 we can then calculate the distance to the boundary, d.

This calculating procedure can be made simpler by using the early straight-line plot to evaluate $(0.141q\mu/kh)$. Note that if we expand the reduced time in the p_w' expression of Fig. 3.18, separate time in days from the rest of the log term, and change the log to a base 10 log we can evaluate the slope of p_w' versus the log of time. We have derived similar slope expressions several times. In this case we find that the value of the slope is the same as the value we have encountered so many times previously.

$$m = \frac{0.1625q\mu}{kh} \tag{3.12}$$

Now we can use a function of the slope in Equation 3.62 to obtain

$$p_w' - p_w = 0.867m\left[\left(\frac{1}{2}\right)\left(-\text{Ei}\ \frac{-1}{4t_{D\ images}}\right)\right]. \tag{3.64}$$

The slope can also be used to evaluate the mobility, k/μ, to use in Equation 3.63 to calculate the distance d from the reduced time, $t_{D\ image}$. The details of this calculating procedure may be clarified by solving a problem and comparing the solution with the one in Appendix C.

Problem 3.6: Determining the Distance to a Reservoir Barrier from a Drawdown Test

Fig. 3.18 represents data from a constant-rate drawdown test on discovery well OSU-ChE No. 1. Geological evidence indicates the possibility of a nearby fault.

Determine (a) the reservoir mobility, k/μ, (b) the distance to the fault, and (c) Δp_{skin} and the skin factor, S.

Initial pressure = 2,800 psia
Production rate = 80 STB/d
B_o = 1.25 res bbl/STB
Effective compressibility = 10×10^{-6} psi^{-1}
Effective porosity = 0.20
Formation thickness = 10 ft
Saturation pressure = 1,800 psia
The well was drilled with a 6-in. bit.

Note that in the above example not only was the distance to the reservoir barrier calculated but the drawdown data were used to calculate the additional pressure drop due to well damage which in this case was negative because of permeability improvement around the well resulting from a well treatment. The important point to remember is that well damage can be evaluated from a drawdown test in much the same way that it can be determined from a build-up test. Although in this particular case the evaluation was performed on a test primarily designed to determine the distance to a reservoir barrier there is no reason why any drawdown test could not be used to determine well damage.

There have been many cases where drawdown data such as that in Fig. 3.18 have been erroneously interpreted to be the effect of a reservoir barrier when actually it was due to some other phenomena. One of the most common effects on a drawdown that resembles the effect of a barrier is a change in gas saturation around the well bore due to excessive drawdown. The lower pressures around the well bore cause a greater amount of gas to be liberated at this point in the reservoir and this increased gas saturation causes a corresponding decrease in the effective permeability to oil and the effect is similar to that caused by a reservoir barrier. Also note that interference from another producing well could cause an identical effect to that of a well barrier. It seems superfluous but the engineer should make certain he is not seeing the effect of another well rather than the effect of a barrier. The error has been repeated many times.

Due to the possibility of confusing a barrier effect with other effects in the reservoir, several safeguards have been proposed to minimize the possibility of such an error. Note from Equations 3.60 and 3.61 that when the reduced time base for the image well becomes greater than 100 that the Ei function can be replaced by the log equation. Equation 3.61 could then be written as

$$p_w = p_i - \frac{0.141q\mu}{kh}\left(\frac{1}{2}\right)(\ln t_{D\ real} + 0.809) - \Delta p_{skin}$$
$$-\frac{0.141q\mu}{kh}\left(\frac{1}{2}\right)(\ln t_{D\ image} + 0.809) \tag{3.65}$$

Now note that the change in the second term with time will be proportional to the change in the log of the producing time and the change in the value of the last term with time will be proportional to the change in the log of the producing time. Comparing this change in well pressure with time, with the change of the well pressure with time before the effects of the barrier is felt at the well, shows that the slope of the plot of well pressure versus the log of time is exactly double that experienced before the effect of the barrier was felt at the well. The more exact quantitative aspects of this situation may be better understood by referring back to the discussion of pressure buildup anomalies where the effect of a boundary on a pressure buildup slope was discussed.

Since the negative slope of the pressure plot versus the log of time will eventually double, some engineers have proposed that the technique of determining the distance to a barrier be limited to those data where the slope exactly doubles. There seems to be little question that following such a procedure would keep the engineer out of trouble as far as misinterpreting the effect of a barrier is concerned, but it would also mean that he would be missing many valid applications of drawdown data because the barrier must be very near the well in most cases if the slope is to double in any practical length of time. For example, in the above problem it was found that the reduced time for the image well was 3.0 when the produced time was 10 days. Consequently, if the reduced time is to be greater than 100, the requirement for the slope to double, the producing time would have to be about 330 days. In this length of time the reservoir would certainly be affected throughout its entirety and the reservoir would no longer be infinite acting. Thus, Equation 3.61 would no longer be applicable because the pressure functions included apply only to infinite-acting reservoirs.

The recommended technique for determining if the pressure behavior observed is actually due to a reservoir barrier or is caused by some other effect is to repeat the calculation of the distance to the barrier by using different times and corresponding producing times. All of these calculations should result in the same distance to the barrier within the accuracy of the measured and plotted data. For example, refer again to Problem 3.6 and note that at a producing time of 5 days, p_w' is 2,430 psia and the well pressure is 2,383 psia. From Equation 3.64 we calculate an Ei function value of 0.677 and from Fig. 2.8 the corresponding reduced time is 1.5. Putting the time of 5 days and $t_{D\ image}$ of 1.5 into Equation 3.63 we calculate a distance to the barrier of 229 ft which was the answer obtained in Problem 3.6 using time of 10 days. Such calculations can of course be repeated until the engineer is confident as to the nature of the pressure anomaly he is analyzing.

Note that if a second barrier exists in a reservoir such that the distance to the first barrier can be established from the pressure drawdown before the second barrier begins to affect the drawdown pressure, it may be possible to determine the distance to the second barrier. For example, consider the simplest case as in Fig. 3.19 where the distance to the nearest barrier is still d, and the distance to the next closest barrier is d_2. In order to have the same drainage and flow pattern in an infinite-acting reservoir it would be necessary to have four wells spaced as shown producing with the same rate histories. A similar boundary problem was considered in Chapter 2 under Superposition and if the engineer has difficulty with this modeling of the reservoir boundaries, particularly the need for the third image well, he might wish to restudy this section of Chapter 2.

Based on the pattern of Fig. 3.19 we can write an equation for the pressure in the real well at any time by using the Ei function equation to evaluate the pressure drops caused by the three image wells at radii of 2d, $2d_2$, and the distance of image well number 3 from the real well, which is the hypotenuse of the right triangle whose sides are 2d and $2d_2$.

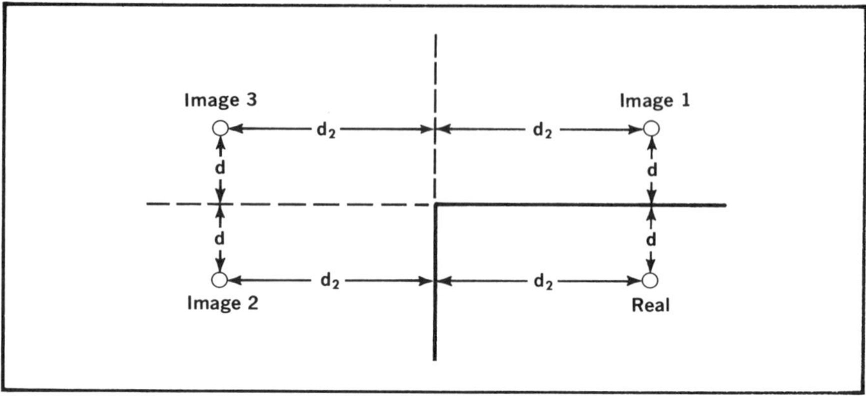

FIG. 3.19 Modeling two perpendicular boundaries.

$$p_w = p_i - 0.867m\left(\frac{1}{2}\right)(\ln t_{D\ real} + 0.809) - \Delta p_{skin}$$

$$- 0.867m\left(\frac{1}{2}\right)\left(-Ei\ \frac{-1}{4t_{D\ image\ 1}}\right) - 0.867m\ \left(\frac{1}{2}\right)\left(-Ei\ \frac{-1}{4t_{D\ image\ 2}}\right)$$

$$- 0.867m\left(\frac{1}{2}\right)\left(-Ei\ \frac{-1}{4t_{D\ image\ 3}}\right) \tag{3.66}$$

Now note that the image 1 term will become significant before the terms of the other image wells take on significant values. If this period of time is long enough it will be possible to determine the distance d to the nearest boundary as was done in Problem 3.6. Once the distance d to the nearest boundary has been determined, the term containing the reduced time for image well number 1 can be evaluated for any producing time desired. In such a case the term for image 2 will become significant before the term for image 3 is significant. Thus, $p_w' - p_w$ can be evaluated for sometime after the image 2 term becomes significant and before the image 3 term becomes significant and the distance to the second closest barrier can be calculated from the equation,

$$p_w' - p_w = 0.867m\left[\left(\frac{1}{2}\right)\left(-Ei\ \frac{-1}{4t_{D\ image\ 1}}\right)\right] - 0.867m\left[\left(\frac{1}{2}\right)\left(Ei\ \frac{-1}{4t_{D\ image\ 2}}\right)\right],$$

$$\tag{3.67}$$

and the equation for the reduced time, 3.63. As before, the reduced time for image 2 would be evaluated by first calculating the $(1/2)(-Ei\ldots)$ term by Equation 3.67 and then finding the corresponding reduced time from Fig. 2.8.

Careful consideration will show that this procedure could be followed for any number of barriers as long as they are sufficiently separated in distance so that each distance to a barrier can be determined before the next closest barrier affects the drawdown pressure. Note that the procedure could be further simplified if the relationship between the distances is such that the effect of the last boundary felt is governed by the log equation before the next closest boundary is felt at the well. In such a case the well pressure would again become a straight-line plot against the log of the producing time. This new straight line could be extrapolated and the difference between the extrapolation and the recorded well pressure could be equated to the appropriate Ei term as in Equation 3.64, so that the distance to the next nearest barrier could be determined as before.

Determining the distance to the two closest boundaries does not give the angle between the boundaries. Also note that it is impossible to carry the analysis further without knowing the angle between the two boundaries. If other than a right angle exists between the two, the modeling of an infinite-acting reservoir becomes very difficult. The angle between the two barriers can be determined by observing the relationship between the initial straight-line slope and the next straight-line slope. For example, in Equation 3.66, note that when the reduced time for the image 3 well is greater than 100 all four of the pressure-drop terms will be governed by the log equation and the slope of the well pressure versus log time plot will be four times the initial straight-line slope when only the term for image 1 is significant. This slope ratio of four is 360°/90° where the 90° is the angle between the two boundaries. Now suppose that the angle between the two boundaries is 180°. This is equivalent to having only one boundary and the well pressure is governed by Equation 3.65. Under these conditions, as was pointed out previously, the slope will eventually double. The ratio of the original and final slopes will be 2.0, or 360°/180°. It can be shown that this observed relationship between the intersecting angle and the slope ratios exists for any angle between the intersecting boundaries. Turning the observation around we find that once we have determined the ratio between the initial and final slopes (assuming that a final straight-line slope is reached), 360° can be divided by the ratio to give the angle between the two boundaries. If the slope ratio is 3.0, the angle between the boundaries is, 360°/3, or 120°.

The parenthetical phrase above is extremely important because, as was suggested previously, the laws of nature and chance are such that they seldom arrange the dimensions of a reservoir so that the effect of one boundary is so well separated from that of another that each will be governed by the log equation before it is confused with the next effect. It has been the author's experience that you can determine the distance to the closest boundary in most cases and that in only about 25% of the cases can you determine the distance to the next closest boundary with a reasonable amount of confidence. Furthermore, it is seldom that you can obtain any more information from the pressure behavior with the exception of the total reservoir size.

If the constant-rate production is continued until all of the reservoir has been affected, flow behavior will enter the pseudosteady-state regime and there will be a straight-line relationship established between a linear plot of the well pressure versus the producing time. The slope of this plot will be the change in pressure with time according to the pseudosteady-state flow equation.

$$(\Delta p / \Delta t)_{pseudo} = \frac{5.615q}{\phi c V_b} \qquad (2.33)$$

This equation, which was derived in Chapter 2, can be used to calculate the total reservoir volume, V_b in cubic feet. By noting the time when pseudosteady state begins, an estimate of the distance to the furthermost point in the reservoir can be obtained. The observed time is used in the stabilization time equation and r_e is calculated. This is the distance to the furthermost part of the reservoir.

$$t_s = \frac{0.04\phi\mu c r_e^2}{k} \qquad (2.53)$$

This technique sounds good from a theoretical standpoint but in practice it leaves much to be desired because of the difficulty in determining exactly when the plot of well pressure versus time becomes straight. Since the distance desired is proportional to the square root of the time, sizeable errors can result from the engineer's mechanical inability to accurately determine this critical time. It is recommended that this technique be used but only in a qualitative sense with a full realization of the possible inaccuracies involved.

Many engineers use a similar technique to determine the distance to the nearest boundary and other reservoir dimensions. They use the semilog plot of the well pressure versus time to determine when the effect of the nearest boundary is felt at the well. This time coincides with the departure of the plot from the original straight line. The subject time is then used in the stabilization time Equation 2.53, to calculate the distance r_e to the nearest boundary. This calculation has the same shortcomings noted for the calculation of the distance to the furthermost part of the reservoir plus some additional problems.

First there is the problem of accurately determining the time when the semilog plot deviates from the straight line. This is difficult enough when working with a linear plot but when the number desired is plotted as a log it is much more difficult to obtain an accurate value. In addition, the stabilization equation is at best an approximation which gives a built-in error of unknown magnitude to this technique. Also, some independent technique such as the doubling of the slope would have to be used to verify that you are actually dealing with a boundary effect and not with some other phenomena. Due to these shortcomings and the availability of the method proposed in this section it is recommended that the engineer avoid using the stabilization equation for determining reservoir dimensions except as a check of other methods and for estimating the distance to the furthermost part of the reservoir.

One of the biggest problems associated with reservoir limit and interference tests is the failure of the engineer to determine that the pressure anomaly he is seeking can actually be measured with the equipment he has available. Many engineers need to determine if two offset wells are in the same or different reservoirs. The problem sounds simple enough and the engineer tries to observe the effect of one well on the other. He may decide to double the producing rate in one well and monitor the bottom-hole pressure in the second to determine the effect of the double rate. After watching the pressure in the second well for several days he may see no effect and conclude that the wells are in separate reservoirs. However, subsequent development and pressure surveys may prove his conclusion erroneous. If this engineer had taken the time to qualitatively determine the magnitude of the pressure disturbance he was trying to observe he would quite likely have avoided the embarrassment of his erroneous conclusion by finding that the pressure change was too small for the equipment being used. This problem was discussed in Chapter 2 under the title Superposition and an example calculation was provided. The procedure simply involves estimating the minimum pressure difference you feel you can detect with confidence and then back calculating the amount of time required to obtain this pressure difference at the radius in the reservoir where the observations will be made. For details of this type calculation the engineer is referred to Problem 2.3B in Chapter 2.

INTERPRETING DRILL-STEM TEST DATA

The use of drill-stem tests to predict the performance and production potential of wells seems to have steadily declined in recent years although the author cannot cite any specific statistics on the subject. The apparent decline in the use of drill-stem tests is probably due to many things. Geophysical logging has been refined to the point that good producers can be predicted with confidence on the basis of logs alone. Furthermore, marginal producers are so reliant upon well-treating techniques, such as fracturing and acidizing, that cannot precede the DST (drill stem test), that the DST is looked upon as useless except in rare cases. However, it is the author's opinion that the DST companies' reluctance to provide the producer with production-rate estimates has also contributed immensely to the decline in the popularity of the DST as an evaluation tool. When the DST company does provide some sort of an analysis of the data, it is limited to determining the undamaged permeability and some abstract measure of the skin damage such as the skin factor or an infinite-acting damage ratio which is very misleading.

The producer wants and needs some kind of an estimate as to what the well will produce. Although, as will be discussed later, the technology we have available is not well suited to analyzing DST data, estimates of production rates can be made that are very helpful in predicting the ultimate behavior of a well.

Historically, the DST has been viewed as a means of determining whether or not a well would produce and not at what rate it would produce. As noted above, logging and well-treating developments have nearly relieved the DST of this function in modern drilling. However, a carefully interpreted DST can provide a useful estimate as to the rate of production expected from a well under various circumstances which cannot be provided by logging. Thus, the DST can be a primary factor in determining whether or not a completion is economically advisable.

Mechanically, a DST is simply a means of temporarily relieving the producing formation of the static mud pressure so that the reservoir fluid can flow into the well bore. Originally a DST consisted of running a dry string of drill pipe into the well, setting a packer to relieve the prospective producing formation of the static mud pressure, permitting the well to flow into the dry drill stem for a period of time, shutting the well in to obtain the reservoir pressure, and pulling the drill string with the produced fluid held in the drill pipe by a retaining valve. However, a modern DST can be run with several flow periods and pressure build-ups and the produced fluid can be circulated to the surface before the tool is pulled from the hole.

To predict the settled production rate of a well from the DST we must use the pressure buildup and fluid recovery data from the test. The drill-stem test provides a unique opportunity to attempt to directly measure the initial pressure in the well drainage area. Consequently, it is strongly recommended that the engineer make an attempt to obtain this direct initial pressure measurement. It is recognized that the engineer will not always be successful, but the short time invested in the attempt seems well worthwhile when it is realized that this is probably the only possibility for obtaining such a direct measurement.

The problem in directly measuring the initial pressure is that the formation in the vicinity of the well bore will generally be charged with some filtrate loss from the drilling mud and will have a pressure higher than the static reservoir pressure. Consequently, it is necessary to permit enough of the reservoir fluid to enter the well bore to permit relief of this abnormally high reservoir pressure before shutting the well in to record the initial pressure; otherwise the recorded pressure may be too high. On the other hand, if too much fluid is permitted to enter the well bore before shut-in, a regular pressure buildup will occur and the data will have to be extrapolated to the initial pressure using the Horner method. This then requires some experience and judgement on the part of the engineer since a tight reservoir should be permitted to produce for a longer time than the more permeable reservoir.

Generally, a flow period of between 5 and 20 min will be successful in obtaining a direct measurement of the initial pressure. The recorded bottom-hole pressure versus time will provide a good clue as to whether or not the hoped-for initial pressure has been accurately recorded. If the flow period is too short, the pressure will move abruptly to an unchanging pressure without any rounding of the pressure record. If the flow period is too long, there will be a long rounded pressure record but the recorded pressure will still be increasing after 10 or 15 min of shut-in. Ideally then, the record will be well rounded but will reach constant value in 5–10 min.

Fig. 3.20 shows a pressure record for a typical DST although procedures vary so widely we probably cannot identify any as typical. This is not a pressure record as the engineer normally receives it from the service company. Fig. 3.20 has increasing time from left to right and increasing pressure from the bottom to the top whereas an actual record will have at least one of these scales reversed and they will overlap with the same x-axis point representing two or even three times on the scale.

A to B on Fig. 3.20 indicates the pressure increase that occurs due to lowering the element into the column of mud in the hole. The pressure at B will be the total static mud pressure and can be checked against the mud weight to make certain the service company is using the correct calibration chart for this particular pressure bomb.

Point D on Fig. 3.20 is the reduced pressure that results from setting the packer element and opening the flow valve so that the formation is relieved of the static mud pressure. Point C is an abnormally high pressure, sometimes discernible, caused by the compressing of the mud trapped below the packer before the flow valve opens. The portion of the pressure chart at E illustrates the ideal smooth rounded initial

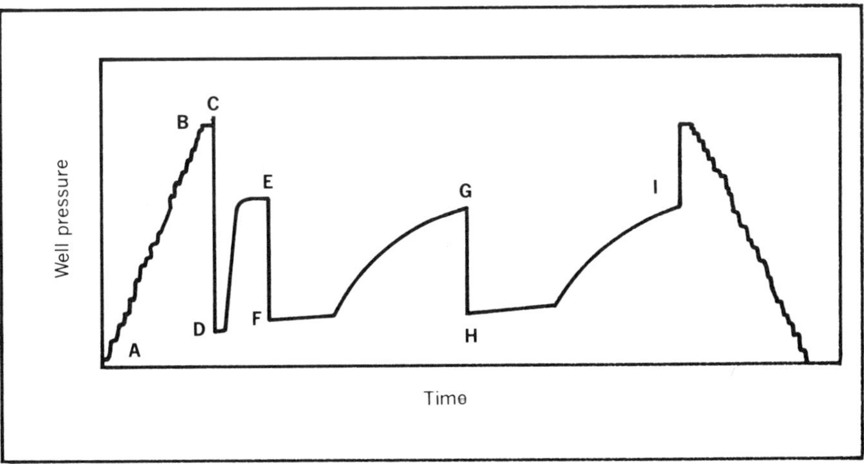

FIG. 3.20 Example DST pressure data.

buildup to a constant pressure that should represent the initial reservoir pressure. At F the well has been permitted to flow into the well bore. This represents the primary flow period and would normally cover 30 min to 2 hr although longer flow periods are not uncommon.

Following this initial flow period the initial pressure buildup data are recorded at G. This will probably be the most useful data for calculating the undamaged permeability and the additional pressure drop due to well damage. Following the initial pressure buildup this particular test employed an additional flow period at H and another buildup due to shut-in at I. More than one flow period and pressure buildup is normally used today in an effort to evaluate pressure depletion that may be occurring. This will be discussed later in this section. The remainder of the pressure record shows the effect of unseating the packer and removing the pressure element from the column of mud along with the drill pipe.

A drill-stem test represents one of the few times when the pressure behavior in a well can be relied upon to represent strictly infinite-acting behavior. Consequently, it is recommended that at the present time a Horner type of analysis as described in the part of this chapter dealing with, Pressure Buildup Analysis—Infinite Acting at Shut-In, be used to analyze the pressure buildup data from the DST. It will be emphasized later that the DST conditions do not exactly fit the Horner method but the apparent accuracy of the calculated results using the Horner method seems acceptable. In constructing the Horner plot it is necessary to determine the producing time, t. Normally it is recommended that in a Horner application the producing time be calculated by dividing the cumulative production by the rate at time of shut-in. However, in the application to DST data it is recommended that the producing time be used as the actual producing time and the average producing rate be calculated since we only know the total production (from recovery in the drill stem or flow to the surface). A useful relationship in this correction is that the drill pipe volume in barrels per 1000 feet is about equal to the inside diameter in inches squared.

Once the Horner plot has been constructed the slope can be used to determine the undamaged permeability. The rate in stock tank barrels would be determined as discussed above, the reservoir thickness can be determined from electric and/or radioactivity logs, and the fluid viscosity and oil formation volume factor can be determined from empirical data such as that found in the appendix. In evaluating the slope, consideration should be given to the use of the initial pressure (determined from the initial shut-in pressure) as a point plotted at $t + \Delta t/\Delta t = 1.0$. Due to the shortness of many of the DST buildup periods this additional control of the slope is often desirable. It is not suggested that the plot be forced through the initial pressure but it should be remembered that the plot should go through this point unless depletion has occurred or the initial pressure as determined from the initial shut-in is in error.

Once the slope and the undamaged permeability have been determined, the additional pressure drop due to the well damage can be determined from Equation 3.36.

$$p_w - p_{wf} = 0.867m\left(\frac{1}{2}\right)(\ln t_D + 0.809) + \Delta p_{skin} - m \log\left(\frac{t + \Delta t}{\Delta t}\right) \quad (3.36)$$

In evaluating the reduced time, t_D, based on the producing time prior to shut-in, the porosity can be determined from log analysis, and the compressibility can be determined more accurately from empirical data than is normally the case because there will generally be no gas saturation to deal with initially in an oil well, and no oil in a gas well. The bit size will normally suffice for the well radius.

In estimating the settled production rate the indicated productivity index using the flow rate and pressure drop, $p_i - p_w$, are meaningless. Some of the service companies refer to this value in predicting the settled rate, but since only a very small portion of the reservoir has been affected during a DST the so-called productivity index has no meaning. If this value is used to predict the flow rate under settled conditions and no skin damage, an extremely optimistic rate would be obtained.

It is recommended that the settled flow rate be based more or less on theoretical considerations and the parameters determined from the pressure buildup analysis. To use this approach the engineer must decide whether the well will reach steady state or pseudosteady state and what the drainage area will be. He must also estimate the skin factor that will be obtained by treating the well. When these estimates have been made he can then calculate the flowing or pumping rates that would be obtainable.

Unless there is good reason to believe otherwise, it is normally good practice to assume that pseudosteady-state flow will prevail. Now if we write the pseudosteady-state equation with the pressure drop corrected for Δp_{skin} and the external well pressure assumed equal to the initial pressure, p_i, we obtain

$$q_{STB} = \frac{7.08kh(p_i - p_w - \Delta p_{skin})}{B_o\,\mu\left[\ln(r_e/r_w) - \dfrac{1}{2}\right]} \quad (3.68)$$

This equation can be stated in a more convenient form by substituting a function of the slope, m, and the DST flow rate in stock tank barrels for $7.08kh/B_o\mu$.

$$m = \frac{q_{DST}B_o\mu}{7.08kh} \quad (3.69)$$

$$q_{STB} = \frac{q_{DST}(p_i - p_w - \Delta p_{skin})}{m\left[\ln(r_e/r_w) - \dfrac{1}{2}\right]} \quad (3.70)$$

This equation can be used to estimate production for a flowing well or a pumping well by simply adjusting the well pressure, p_w.

In estimating the well pressure, minimum errors are introduced by assuming that the pressure drop in the tubing due to friction is equal to the decrease in bottom-hole pressure caused by gas being liberated in the tubing as the oil moves to the surface. This is equivalent to calculating the difference between the bottom-hole pressure and the surface pressure as being the static pressure difference based on liquid alone.

$$p_w = 0.433\gamma_0 \, \Delta D + 15. \tag{3.71}$$

The constant, 15, in the equation is atmospheric pressure and ΔD is the depth in feet between the bottom-hole pressure datum and the top of the unsupported column of oil. If gauge pressure is desired the constant term should be omitted. If a well is flowing, ΔD will simply be the average depth of the producing formation but if the well is being pumped, ΔD will be the difference between the pump depth and the average depth of the producing formation.

One of the principal difficulties with a DST is that the flow periods may be so short that a very small amount of the reservoir affects the result of the test. This affords very poor statistical weight for the test. To obtain some idea as to how little of the formation is affected by a test, we use the term "radius of investigation," (which seems to have been borrowed from logging). We can simply use the stabilization time equation to calculate the radius of investigation, r_i.

$$t = \frac{0.04\phi\mu c r_i^2}{k} \tag{3.72}$$

In using this equation, the engineer must arrive at estimations of the porosity, ϕ, and compressibility, c. He should use the mobility, k/μ, that he obtains from the slope of the Horner plot. The porosity should be obtained from the electric or radioactivity logs and the effective compressibility can be accurately determined from empirical data since only one phase is normally present in the active pore volume during a DST. We will see later that the service companies, in an effort to oversimplify this calculation, provide average values that in some cases even include averages for the length of the flow period. It is the author's opinion that an engineer on the job should be able to obtain more accurate values for the porosity, compressibility, and flow time than is afforded by the use of a statistical average.

To make sure the details of a DST analysis are clear, work the following problem and check your solution against the one in Appendix C.

Problem 3.7: Analyzing DST Data

A drill-stem test recovers 7,500 ft of clean 40° API (S.G. 0.825) oil in an 80-min flow test from 20 ft of net pay at 9,985 to 10,015 ft using 4½-in. drill pipe (I.D.

3.83 in.) in 9.8 lb/gal mud. The following readings were obtained from the pressure recorder.

Hydrostatic mud pressure	5,080 psig
ICIP (Initial closed-in pressure)	4,500 psig
IFP (Initial flowing pressure)	250 psig
FFP (Final flowing pressure)	2,700 psig
(after 80-min open)	

Build-up pressure, psig	Shut-in time, min
3,460	10
3,900	20
4,190	30
4,300	40
4,390	60
4,410	80
4,422	100

Assume the following:
(1) The lightening of the oil by gas coming out of solution equals the friction loss in the tubing (a pessimistic assumption) during flow.
(2) $r_w = \frac{1}{4}$ ft (obtained from bit size).
(3) $c_{eff} = 5 \times 10^{-5}$/psi (a reasonable assumption).
(4) $\phi = 20\%$ (can be estimated from logs).
(5) $B_o = 1.75$ (can be estimated by assuming reservoir is initially at the saturation pressure).
(6) $\mu_o = 0.4$ cp (can be determined from empirical data by assuming $p_i = p_s$).
(7) The well can be treated to obtain $\Delta p_{skin} = 0$.

A. Calculate the Horner ratio for shut-in times of 60, 80, and 100 min. Determine the initial well pressure, p_i.
B. Check the hydrostatic mud pressure and determine the undamaged permeability.
C. Find the initial stabilized (pseudosteady state) flowing rate if the well drainage radius will be 745 ft (40 acre spacing).
D. Calculate Δp_{skin} and the skin factor, S.
E. Estimate the affected radius during the flow period.

The methods covered in this problem are highly recommended over those currently recommended by the leading DST companies. The leading DST companies recommend the use of a damage ratio calculation as a means of determining well damage. This parameter as previously discussed has many advantages over the use of the skin factor, S, and Δp_{skin} when properly applied. However, in applying the damage ratio concept to a DST it is necessary to evaluate the average permeability from an infinite-acting test which is somewhat meaningless since the damage ratio determined in this way is a function of the length of the flow period. In other words, in the presence of positive damage, the damage ratio continues to decrease as the length of the flow period increases. This can be readily understood if we look at the definition of the damage ratio as being the undamaged permeability divided by the average or damaged permeability. The undamaged permeability is fixed and is independent of the producing time. However, the average permeability which is an average of the reduced permeability and the undamaged permeability outside the damaged zone will continue to increase as the amount of affected reservoir increases. This increase in the average permeability results from the fact that the amount of damaged formation is fixed whereas the amount of undamaged formation in the flowing area continues to increase as the affected area increases. The continued increase in the average permeability results in a continued decrease in the damage ratio. In addition, the conventional DST does not provide a stabilized flow rate that can be corrected by the damage ratio.

In addition to the fact that the damage ratio is not a constant, the equations proposed by some for the calculation of the radius of investigation of a DST also incorporate average values for permeability, porosity, viscosity, compressibility, and length of flow test. This cannot be criticized as long as it is clear that the equations are based on average values. However, the ready acceptance of these equations by practicing engineers who are in a position to determine values for these assumed reservoir parameters with an accuracy much greater than that of an average value is at best confusing.

Use of the Horner method of analysis in analyzing the pressure buildup data of a DST leaves much to be desired. It is generally assumed that theoretically the Horner method fits the DST ideally because there is no question that the well behavior is infinite acting during the DST. However, if we review the derivation of the Horner equation we will see that it was derived for the case where well pressure was *declining* at time of shut-in and that if the well has not been shut-in the pressure would have continued to decline. The effect of the negative rate change superimposed upon this declining pressure base, p_w', resulted in the Horner equation. Contrast this with the actual situation that exists in most DST's. A DST that does not flow to the surface prior to shut-in will have a well pressure that is *increasing* at the time of shut-in. Consequently, the effect of the negative rate on the reservoir when the well is shut-in is not superimposed on a declining pressure base but is rather superimposed on an increasing pressure base.

This is probably why the buildup data for a DST takes so long to reach a straight line on a Horner plot. The effect of the increasing pressure at shut-in will decrease as the shut-in time increases and eventually a straight-line Horner plot results. Since the well is shut-in at the bottom of the hole on a DST, afterflow is negligible but it is only fair to note that most drill-stem tests are run in the presence of considerable well damage since the well has not yet been treated. This tends to aggravate the afterflow-well damage anomaly that delays the straight-line portion of the buildup.

It appears that some time in the future a calculating procedure will be available for the analysis of the DST buildup data that will account for the increasing well pressure at time of shut-in and will thus permit the utilization of much earlier shut-in pressures. In the meantime the methods illustrated in Problem 3.7 are strongly recommended.

A series of flow tests and shut-in periods have been used to evaluate pressure depletion of a reservoir due to the relatively small amount of fluid removed during a drill-stem test. Obviously, any measurable decline in the average reservoir pressure due to the small amount of fluid removed would mean that the reservoir is too small to justify completion costs of the well. The conclusion that pressure depletion is occurring should be reached only after repeated DST's so that the engineer is certain that the removal of some small amount of reservoir fluid results in a decrease in the average reservoir pressure. This analysis can in this instance be based on p^*, the pressure extrapolated to a time ratio of one since the decline in this value will be similar in this case to the decline in the average reservoir pressure. The note of caution—that pressure depletion is occurring—is probably unnecessary since it is generally impossible to convince an engineering supervisor that a well with a substantial initial rate of production should be plugged because the reservoir is to small to justify the completion costs even when conclusive test data are available. Much money has been wasted on completion costs that could have been avoided.

Accurate reservoir engineering analysis is impossible without good reservoir data and the conscientious reservoir engineer must be ready at all times to economically justify his need for data. In the case of a DST, accurate data will not generally add to the cost of the test. In this case it seems important to make it known that you want to do quantitative analysis of the data. Generally this is all that is required to have the service company take proper care to obtain the pressure data as accurately as possible. However, some degree of reliability will be added to the data if you ask for the date the pressure bomb was calibrated for the pressure and temperature ranges anticipated. As a rule of thumb, you might request the use of another bomb if the calibration date is more than 6 months old.

The engineer will find that the use of a blanked-off pressure recorder (a recorder outside the flow string) in addition to the recorder inside the flow string will help detect mechanical difficulties. Another rule of thumb the engineer might keep in mind in connection with a DST is— don't be hasty. Unless you anticipate difficulties in unseating the packer, make an attempt to record the initial reservoir pressure, use at least three flow periods, use a final flow period of sufficient length, and measure the gas, oil, and water production carefully even when the test appears to be noncommercial. The final flow period should be at least 2 hr if the well capacity is anticipated to be less than 10 md-ft and 30 min to 1 hr if the capacity is anticipated to be more than this value.

Side-wall fluid sampling devices are now in extensive use and provide an excellent means of determining water-oil and gas-oil contacts.

However, the pressure records obtained are not generally of sufficient accuracy to permit a quantitative interpretation.

REFERENCES

1. L. C. Uren, *Petroleum Production Engineering—Oil Field Exploitation*, McGraw-Hill Book Company, Inc.: New York and London, 1939.
2. Muskat, M., *The Flow of Homogeneous Fluids Through Porous Media*, McGraw-Hill Book Co., Inc., New York, 1937. (pg. 341).
3. A. Kumar, "Well Test Analysis for a Well in a Constant Pressure Square," SPE Paper #4054, San Antonio, Texas, October 1972.
4. T. E. W. Nind, *Principles of Oil Well Production*, McGraw-Hill Book Co., New York, 1964 (pg. 58).
5. C. Gatlin, *Petroleum Engineering—Drilling and Well Completions*, Prentice-Hall, Inc., Englewood Cliffs, N.J. 1960 (pg. 261).
6. C. S. Matthews and D. G. Russell, *Pressure Buildup and Flow Tests in Wells*, SPE Monograph No. 1, Society of Petroleum Engineers of AIME, New York, 1967.
7. A. F. VanEverdingen, "The Skin Effect and Its Influence on the Productive Capacity of a Well," Trans., AIME (1953) 198, 171–176.
8. H. C. Slider, "A Simplified Method of Pressure Buildup Analysis for a Stabilized Well," J. Pet. Tech. (Sept. 71).
9. B. C. Craft and M. F. Hawkins, *Applied Petroleum Reservoir Engineering*, Englewood Cliffs, N.J.: Prentice-Hall, Inc., 1959.
10. D. R. Horner, "Pressure Buildup in Wells," *Proc.*, Third World Congress, E. J. Brill, Leider (1951) *II*, 503.
11. API Mid-Continent District Study Committee on Completion Practices, "Selection and Evaluation of Well Completion Methods," *Drilling and Production Practice*, API (1955) pg. 421.
13. A. S. Odeh and R. Al-Hussainy, "A Method for Determining the Static Pressure of a Well from Buildup Data," *J. Pet. Tech.* (May, 1971) 621–624.
14. C. S. Matthews, "Analysis of Pressure Buildup and Flow Test Data," J. Pet. Tech. (Sept. 1961).

Gas Reservoir Engineering

Industry engineers have less experience with gas reservoir engineering than they have with reservoir engineering in general because it has only been in recent years that natural gas attained a price that commanded attention. Previously, no one worried much about how much gas they had in a reservoir or how long it would take to produce it. With the increase in the value of gas more attention was given to gas but total gas reserves and total deliverability became a problem only in very recent years when gas was found to be in short supply. Gas transmission companies have always had to know the total amount of gas and the rate at which it could be produced to be certain that they could show a profit on a particular pipeline. But even the transmission companies were not greatly concerned with sophisticated reservoir predictions in the past. They had so many prospects for production that they had to use only the most profitable, which could be determined accurately enough with very simple methods.

On the surface it would appear that gas reservoir engineering would be much simpler than oil reservoir engineering because there is no need for oil formation volume factors, gas-in-solution data, saturation data, etc. Also the volumetric behavior of gas seems to be much better defined than the behavior of a crude oil-solution gas system. However, the reservoir engineering technology was developed for oil reservoirs because at that time there was little profit to be made from the production of gas. Consequently, by the time there was a big demand for applying the previously derived reservoir engineering fundamentals to the production of gas, the reservoir engineering technology was developed to the point that it was not easy to modify it to the flow in gas reservoirs. This unfortunately is the situation in gas reservoir engineering today. We are taking technology developed originally for the production and analysis of oil reservoirs and are trying to apply it to gas reservoirs. We must do this by rather inexact mathematical methods.

The application would not present much of a problem except that the two fluids are so greatly different in their characteristics. Gas densities are very low compared to liquids. Gas viscosities may be only a small fraction of a liquid viscosity under similar conditions. This means that volumetric flow rates of gas may be 100 times the flow rate of liquid under similar circumstances. Thus, with gas we often have turbulent flow around the well bore whereas we seldom have this problem with oil. Since the Darcy equation was derived for streamline (viscous) flow we must in some way modify all of the basic flow equations to account for the effect of turbulence around the well bore.

In addition we must continue accounting for the gas deviation factor which is a function of both the temperature and pressure. Another difficulty is presented by the compressibility of the fluid. We find that the compressibility of oil is both small and practically constant whereas the compressibility of gas is about 100 times that of liquid and varies inversely with the pressure. In summary, we find that we must generally use equations designed for the flow of oil and apply them in a very inexact manner to the flow of gas by trying to adapt and modify them for the vast differences in the fluids.

NATURAL GAS PROPERTIES

The gas equation used to predict the behavior of gas is

$$pV = znRT \qquad (4.1)$$

In this equation the pressure, volume, and temperature are represented by the symbols, p, V, and T where the temperature is in absolute units. The number of moles of gas is n, R is the numerical constant that makes the equation correct for a particular set of units and z is the gas deviation factor. This equation represents a rather abstract expression for most engineers. It can be derived from basic considerations as is done in thermodynamics courses[1] but it still seems to remain somewhat ethereal to most engineers. Consequently, it is helpful to relate the expression to another equation that does seem to be readily understood. This more understandable equation relates the pressure, volume, and absolute temperature of a given mass of ideal gas at different conditions.

$$\frac{p_1V_1}{T_1} = \frac{p_2V_2}{T_2} \qquad (4.2)$$

This equation simply tells us that the volume of a gas is proportional to the absolute temperature and inversely proportional to the pressure. If we combine this equation with the observation that 1 lb mole of any gas at 60° F. and atmospheric pressure occupies 379 cu ft we can derive Equation 4.1. Let the left-hand side of Equation 4.2 represent standard conditions and the right-hand side reservoir or other conditions. Then the pressure and temperature at standard conditions would be 14.7 psia and 520° R. The volume will then be the number of moles, n, multiplied by 379 cu ft/mole. The right-hand side will remain as the general pressure, temperature, and volume.

$$\frac{(14.7)(n \times 379)}{520} = \frac{pV}{T} \tag{4.3}$$

$$pV = n \times 10.73T \tag{4.4}$$

$$pV = nRT \tag{4.5}$$

Equation 4.5 is the gas equation for an ideal gas. The gas constant R, is 10.73 for the units employed in this derivation. However, actual gases do not follow Equation 4.5 at high pressures such as we have in a reservoir. Solving for volume we obtain,

$$V = \frac{nRT}{p} \tag{4.6}$$

However, except at very low and very high pressures we find that this theoretical volume of gas according to Equations 4.5 and 4.6 is more than actually exists in practice. The gas is actually compressed more than we would have anticipated. To make Equations 4.5 and 4.6 accurate we then add a correction factor which accounts for the fact that the gas deviates from the theoretical prediction. This factor, z, is the gas-deviation factor. Its correct use is shown in the right-hand side of Equation 4.1.

Most practicing reservoir engineers are capable of evaluating a gas-deviation factor for a given set of conditions but most of them do not know whether or not this value is the most accurate one available. We will try to put this into proper perspective by noting the relative accuracy of the various methods of obtaining gas-deviation factors.

Evaluating z factors. On the surface it would seem that the most accurate values of z factors available for a particular gas would be those measured in the laboratory for that particular gas. However, most laboratory equipment used for this purpose is not sufficiently accurate to provide better data than that obtainable from empirical methods. A study of pure gases consisting of only one component (for example methane or ethane) showed that there was a well-established relationship between the z factors and the pressure, temperature, critical pressure, and critical temperature of the pure gas. This relationship takes the form of Fig. 4.0. More extensive z factor data are available in Appendix B.7. These figures consist of plots of the gas-deviation factor versus the reduced pressure, p_R, for various temperatures, T_R. The reduced pressure and temperature are related to the critical pressure, p_c, and critical temperature, T_c, as

$$p_R = \frac{p}{p_c} \quad \text{and} \quad T_R = \frac{T}{T_c} \tag{4.7}$$

The critical pressure and critical temperature are defined as the conditions above which liquid and vapor phases of the compound cannot be distinguished. This is illustrated by the phase diagram of Fig. 4.1. At temperatures below the critical temperature there is a particular

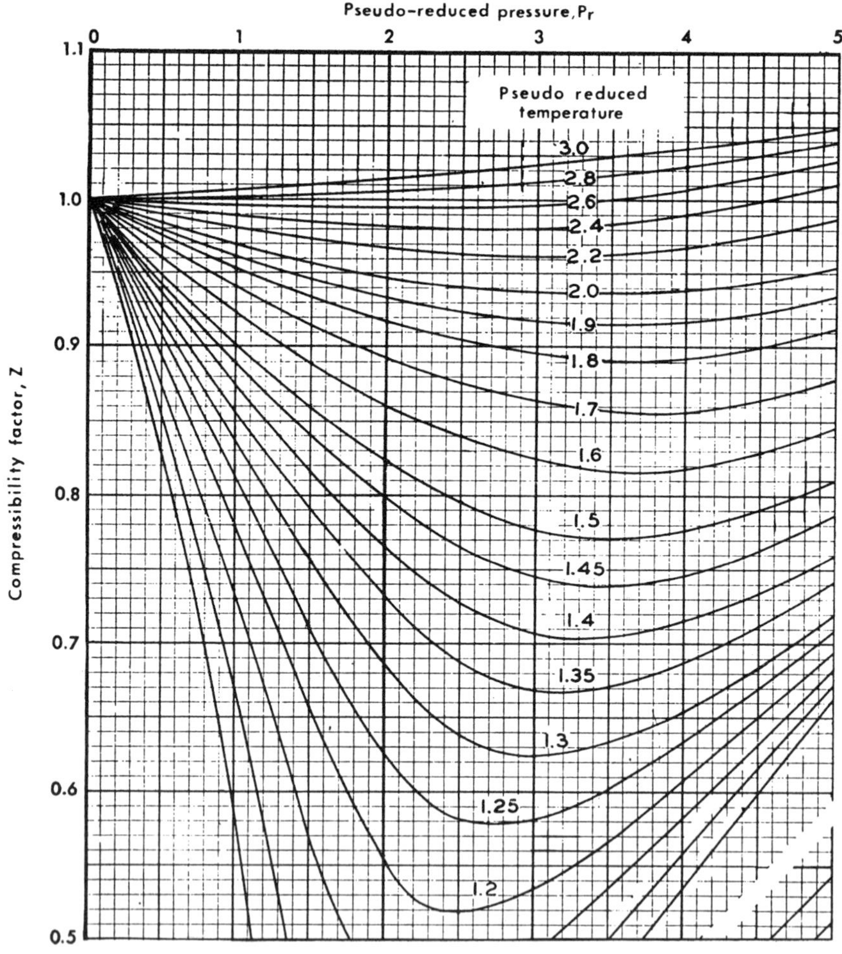

FIG. 4.0 Gas deviation factors.[2] (After Standing and Katz, *Trans. AIME.*)

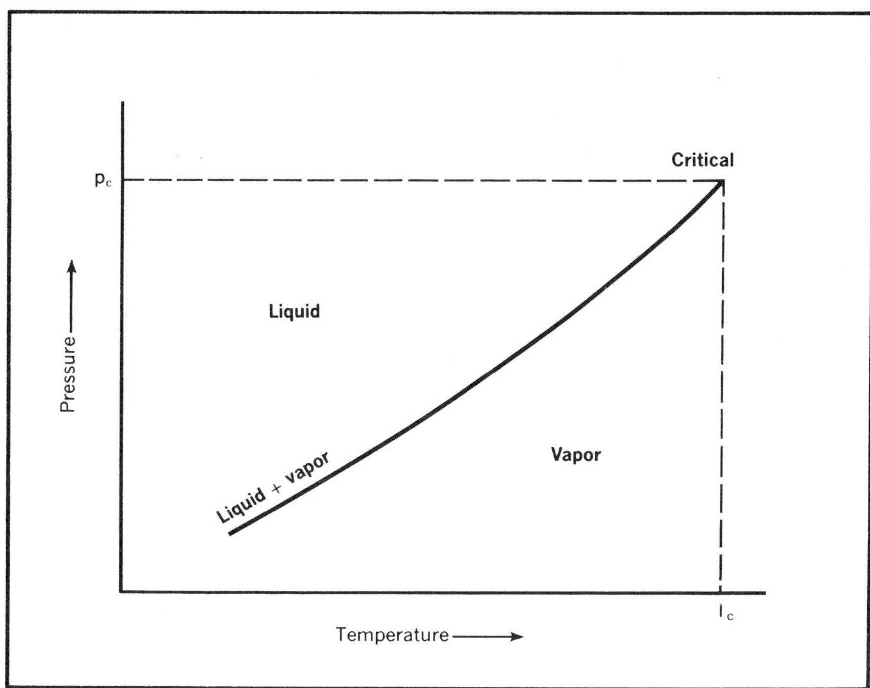

FIG. 4.1 Typical phase diagram for a single component system.[3] (After Burcik, John Wiley Publishers.)

pressure for each temperature at which liquid changes to gas or vice versa but at temperatures above the critical temperature the gas phase cannot be distinguished from the liquid phase regardless of the pressure. The same sort of statement can be made about the critical pressure and the corresponding temperatures.

For a gas that contains only one hydrocarbon component, data such as that of Fig. 4.0 can be used to determine the z factor at any particular pressure and temperature. The critical pressures and temperatures of various hydrocarbon compounds are listed in Table B11 in the Appendix.

Unfortunately, natural gases are a mixture of different hydrocarbons and consequently they do not have a true critical pressure and temperature as we have defined it. For example, if a natural gas is a mixture of methane, ethane, and propane and the pressure on the mixture in the liquid state is reduced isothermally until the first bubble of gas is formed, this would occur somewhere in the vicinity of the bubble point of the methane, the most volatile of the pure gases in the mixture. If the pressure decline was continued until the last drop of the mixture changed to gas this would occur somewhere near the bubble-point pressure of the propane, the least volatile of the gas components. Thus, all of the mixture would not change to gas at one pressure as is the case with a gas consisting of only one component. An example phase diagram for such a gas is illustrated in Fig. 4.2.

This means that the critical pressure and temperature terms lose much of their significance when applied to a gas that is a mixture of several different components as is the case with all natural gases. This then would mean that the z factor correlations of Fig. 4.0 could not be used for a mixture of gases except that it was found that use of a pseudo-critical pressure and temperature would permit the application of the z factor data to a mixture of gases as though the critical pressure and temperature were equal to the pseudocritical pressure and temperature. The pseudocritical pressure and temperature are mole-fraction-weighted critical pressures and temperatures. Using a subscript to denote the pseudo- part of the critical pressure and temperature, we may state the pseudocritical pressure and temperature as

$$_p P_c = \sum_{j=1}^{j=n} (MF_j \, p_{cj}) \tag{4.8}$$

$$_p T_c = \sum_{j=1}^{j=n} (MF_j \, T_{cj}) \tag{4.9}$$

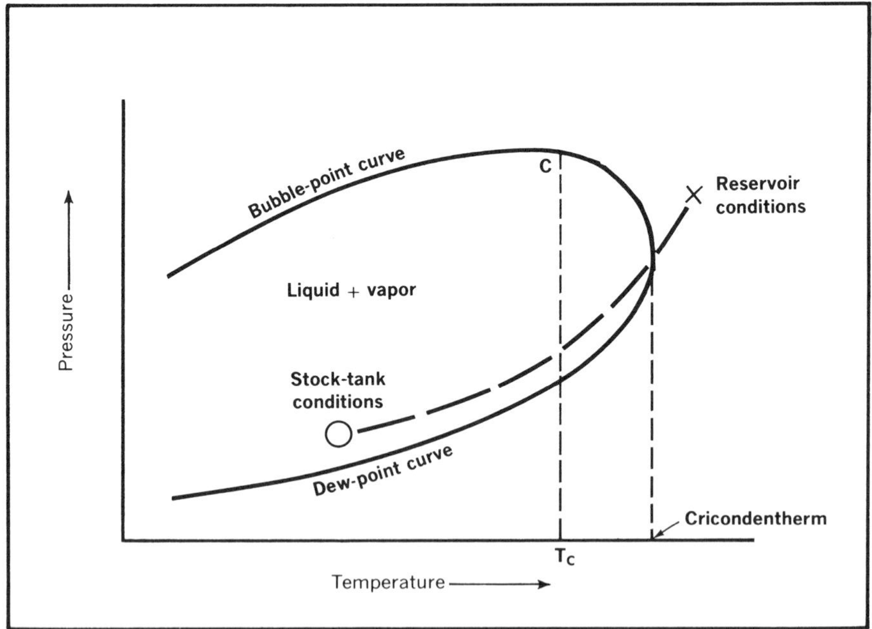

FIG. 4.2 Typical phase diagram for a multiple component system.[3] (After Burcik, John Wiley Publishers.)

The logical nature of Equations 4.8 and 4.9 can be recognized when we compare the equations with the manner used to calculate the molecular weight of a combination of compounds

$$MW = \sum_{j=1}^{j=n} MF_j \, MW_j \qquad (4.10)$$

The molecular weight, critical pressure, and critical temperature are all mole-fraction-weighted values.

The mole fraction of a gas is equal to the volume fraction at standard conditions and the n in these equations refers to the total number of components. The pseudocritical pressure and temperature should be evaluated according to these equations when possible but empirical methods are often necessary.

If gas is to be sold, there will always be a component analysis of that gas and it will then be a simple matter to calculate the pseudocritical values based on Equations 4.8 and 4.9. Care should be taken to include all of the material that was in the natural gas in the reservoir in calculating the pseudocritical values for application to a reservoir material balance. Specifically, the engineer should be careful to include the liquid condensate that is produced with the gas since this existed in the reservoir as gas. Thus, the gas-condensate ratio as well as the component analysis of the condensate and the gas must be used to determine the overall composition of the gas in the reservoir. This component composition must then be used in calculating the critical values of the reservoir gas and the corresponding gas-deviation factor.

When a component analysis is not available for a gas, empirical data are available that provide the pseudocritical pressure and temperature as functions of the specific gravity of the gas. These data are presented in Appendix B.8. To use the data you simply enter the plot with the specific gravity of the reservoir gas and read the pseudocritical pressure and temperature. To obtain the specific gravity of the gas, the specific gravity of the pipeline gas must be corrected for the condensate that is produced with the gas as was mentioned above. However, in this case there is little prospect for combining the compositions of the two materials into one composition of the reservoir material since we probably do not have the component analysis of the pipeline gas or we would not be using the specific gravity—critical value data. However, we can use the Standing correlation of Appendix B.12 to obtain the reservoir-gas gravity from the pipeline-gas gravity. In using this correlation note that the pipeline-gas gravity is called the trap-gas gravity and the well-fluid gravity refers to the reservoir-gas gravity. Furthermore, the correlation is based on the empirical relationship between the molecular weight and the specific gravity of the condensate stated as degrees API.

Referring again to the Appendix B.8. correlation of the gas gravity and the critical values, note that there are two correlation lines for both the critical pressure and critical temperature labeled "miscellaneous gases" and "condensate well fluids." Actually it will be found that the "miscellaneous gases" curves define the dry gases rather accurately. However, the gases that contain a considerable amount of condensate provide more of a problem in applying the data of Appendix B.8. For example a gas that has a condensate content of 10 bbl/MMcf of dry gas will have a much different pseudocritical pressure and temperature than a similar gas that contains 80 bbl/MMcf. Thus, the correlations leave much to be desired when applying them to gases that yield condensate. However, for gases that yield very little condensate the miscellaneous gases correlation provides reasonably accurate critical values. The engineer should also remember that CO_2 and H_2S severely limit the accuracy of the gas gravity-z factor correlations.

To summarize, we have two common methods of determining gas-deviation factors. The preferred method is to calculate pseudocritical pressures and temperatures by Equations 4.8 and 4.9 using the component analysis of the reservoir gas. Then use the calculated reduced pressure and temperature in Appendix B.7. to determine the z factors. The second method that should be used only when time or data are insufficient to permit use of the first method is to use the reservoir-gas gravity to obtain the pseudocritical pressure and temperature from Appendix B.8. Then evaluate the gas-deviation factor from Appendix B.7.

The engineer who has to do much work with gas reservoirs and has access to a digital computer will find it very helpful to have the gas-deviation factor data of Appendix B.7. in storage in the computer. Many gas reservoir engineering methods require iteration on the pressure with a corresponding change or iteration on the z factor. Thus it is very helpful to have the data of Appendix B.7. in storage and use a simple subroutine to interpolate the data to obtain the desired gas-deviation factor. The z factor data is digitized in Appendix table A-2 of the Katz et. al. *Handbook of Natural Gas Engineering.*[4]

Table 4.1 is a computer program designed to use data from a deck of cards with the z factor data punched in a 20F4.3 format. The program consists of two subroutines, ZREAD and ZCALC. The ZREAD subroutine simply reads the z factor data into the machine and the ZCALC subroutine furnishes the corresponding reduced pressures and temperatures and interpolates the data. Use of this program would not require permanent storage of the z factor data in the computer. Conversely, if the data could be stored permanently in the computer the ZREAD subroutine would not be needed. As is the case with all the computer programs in this text, this program is simply an example of the nature of the computer program that could be used for this purpose and is not meant to represent a final product. The program uses straight-line relationships between the pseudocritical pressure and gas gravity and the pseudocritical temperature and gas gravity. Appendix B.8. shows that this is not an accurate representation of the data. This approximation could be avoided by storing points off the relationship between the pseudocritical

Table 4.1

FORTRAN PROGRAM FOR STORAGE AND INTERPOLATION OF
GAS DEVIATION FACTORS

```fortran
      COMMON Z(297,20)
      CALL ZREAD
      READ(5,11) TEMP,PRESS,GRAV
   11 FORMAT(3F10.4)
      CALLZCALC(TEMP,PRESS,GRAV,ZOUT)
      WRITE(6,10)ZOUT
   10 FORMAT(E20.8)
      CALLEXIT
      END
      SURROUTINE ZREAD
      COMMON Z(297,20)
      DO 3 I=1,297
    3 READ(5,6) (Z(I,J),J=1,20)
    6 FORMAT(20F4.3)
      RETURN
      END
      SUBROUTINE ZCALC (TEMP,PRESS,GRAV,ZOUT)
      DIMENSION P(297), T(20)
      COMMON Z(297,20)
      P(1)=.2
      DO20I=2,297
   20 P(I)=P(I-1)+.05
      T(1)=1.05
      DO21 I=2,10
   21 T(I)=T(I-1)+.1
      DO24 I=16,20
   24 T(I)=T(I-1)+.2
      TC=173.+310.*GRAV
      PC=695.-40.*GRAV
      TR=TEMP/TC
      PR=PRESS/PC
      PP=(PR-.15)/.05
      M1=PP
      M2=M1+1
      DIF=PR-P(M1)
      DO30I=1,20
      IF(TR-T(I))31,31,30
   31 N1=I-1
      N2=I
      DIFI=TR-T(I-1)
      GOTO32
   30 CONTINUE
   32 CONTINUE
      PINT=DIF/.05
      TINT=DIFT/(T(N2)  T(N1))
      ZOUT=Z(M1,N1)+PINT*(Z(M2,N1)-Z(M1,N1))+TINT*(Z(M1,N2)
     -Z(M1,N1))
      RETURN
      END
```

ZREAD

ZCALC

pressure and temperature and the gas gravity and interpolating the data mathematically. An ideal program would also provide for the direct calculation of the pseudocritical pressure and temperature from a component analysis using Equations 4.8 and 4.9. Any practical program should certainly include an automatic shutdown of the program in case the reduced pressure or temperature exceeds the limit of the gas-deviation data of the program. These and other modifications and refinements could be easily built into a similar program by the engineer and the computer programmer.

Gas formation volume factor.   The engineer should already be familiar with the oil formation volume factor, the ratio of the volume of oil in the reservoir to the volume of oil that results when this oil is taken from reservoir conditions to stock-tank conditions. The gas formation volume factor is similar. It relates the volume of gas at reservoir conditions to the volume at standard conditions. The nature of the gas formation volume factor is somewhat different because both volumes in the ratio represent the same mass whereas in the oil formation volume factor the mass of the volume in the reservoir is different from the mass in the stock tank due to the liberation of gas that accompanies the change in conditions.

Since the gas formation volume factor simply relates the volume of gas at reservoir conditions to the volume of the same mass at standard conditions we can use Equation 4.1 to derive an expression for the gas formation volume factor. The units of the gas formation volume factor can vary but most engineers prefer to state this factor as reservoir barrels per standard cubic foot. Since 1 scf of any gas is equal to $1/379$ of a mole, we can substitute this into Equation 4.1 and solve for the volume at the reservoir pressure and temperature. Dividing the result by 5.615 (the number of cubic feet in a barrel), will change the volume to barrels and provide the gas formation volume factor, B_g, in reservoir barrels per standard cubic foot.

$$B_g = V/5.615 = \frac{z(1/379)(10.73)T}{5.615\ p} \tag{4.11}$$

$$B_g = \frac{0.00504zT}{p} \tag{4.12}$$

In applying this equation it is necessary to determine the gas-deviation factor, z, at the reservoir pressure and temperature, p and T, by the methods of the previous section.

Gas density.   Equation 4.1 can also be used to derive an expression for the gas density, in pounds per cubic foot. To do this it is necessary to write the number of moles of gas, n, as a function of the specific gravity of the gas, γ_g, and the gas weight, W. This is accomplished by recognizing that the number of moles of gas is the weight of the gas divided by the molecular weight of the gas and that the molecular weight of the gas can be stated as the specific gravity of the gas multiplied by the molecular weight of air. Thus,

$$n = \frac{W}{MW} \quad \text{or} \quad n = \frac{W}{29\gamma_g}$$

Substituting this expression for n in Equation 4.1 along with the gas constant for our system of units, we obtain

$$pV = z\left(\frac{W}{29\gamma_g}\right)(10.73)\,T \tag{4.13}$$

which, when solved for W/V gives the gas density, ρ_g, in pounds/cubic foot.

$$\rho_g = \frac{2.7\,\gamma_g p}{zT} \tag{4.14}$$

This equation can be used to calculate the gas density of any gas at any particular pressure and temperature after first evaluating the gas-deviation factor at those conditions.

Gas compressibility. Compressibility of gas is defined as the fractional change in the volume per unit pressure change.

$$c_g = -\frac{\Delta V/V}{\Delta p} \tag{4.15}$$

In practical units the compressibility is the fractional change in the volume per psi or since the fraction is dimensionless the unit of compressibility is simply 1/psi or psi^{-1}. When gas reservoir engineering technology was first being developed, most of the gas reservoirs being produced were low-pressure reservoirs in which the change in the gas-deviation factor with a change in pressure did not greatly affect the compressibility of the gas. Under these conditions it was sufficiently accurate to approximate the compressibility of the gas as being equal to the reciprocal of the pressure, 1/p. Consequently, many of the equations still being used in gas technology have had this expression substituted for the compressibility. However, the relatively high pressures of reservoirs being produced today precludes the approximation of the compressibility by this expression. We will see that the compressibility is only equal to the reciprocal of the pressure when an ideal gas is under consideration.

To find an accurate expression for gas compressibility, differentiate Equation 4.6 with respect to the pressure.

$$V = \frac{znRT}{p} \tag{4.6}$$

$$\frac{\Delta V}{\Delta p} = nRT \left[\frac{\Delta z}{p\Delta p} - \frac{z}{p^2} \right] \tag{4.16}$$

Now if this expression is substituted into Equation 4.15 along with the right-hand side of Equation 4.6 for the volume V, we obtain

$$c_g = -nRT \left[\frac{\Delta z}{p\Delta p} - \frac{z}{p^2} \right] \bigg/ \frac{znRT}{p} \tag{4.17}$$

When simplified, this equation gives an expression for the gas compressibility.

$$c_g = \frac{1}{p} - \frac{\Delta z/z}{\Delta p} \tag{4.18}$$

From this expression we can see that the compressibility will only be equal to the reciprocal of the pressure whenever the second term is zero. This means that z and/or $(\Delta z/\Delta p)$ has to be zero. Thus, the gas has to be an ideal gas for the compressibility to be equal to the reciprocal of the pressure.

If we made a plot of the pressure versus the gas-deviation factor for each gas we were considering we could simply use this plot to calculate the compressibility at any particular pressure. The slope of the tangent to the curve at the pressure of interest would be $\Delta z/\Delta p$ and the calculation could then be completed using Equation 4.18.

However, it is seldom necessary to prepare a plot of the pressure versus the gas-deviation factor so it is generally inconvenient to evaluate the gas compressibility in this way. However, we do have a plot of the gas-deviation factor versus the reduced pressure in Appendix B.7 or in Fig. 4.0. Consequently, we will change Equation 4.18 to use a slope $\Delta z/\Delta p_R$ instead of $\Delta z/\Delta p$. This is accomplished by substituting $p_R p_c$ for p in the equation.

$$c_g = \frac{1}{p_c p_R} - \frac{\Delta z/z}{p_c \Delta p_R} \tag{4.19}$$

In this form the slope of z versus p_R can be used as $\Delta z/\Delta p_R$ to calculate the compressibility. However, Trube performed these operations for the engineer and presented the results as a reduced compressibility plotted versus the reduced pressure where each curve represents a different reduced temperature. (see Appendix B.5) The reduced compressibility is defined as the gas compressibility multiplied by the critical pressure, $c_g p_c$. Thus, Equation 4.19 can be converted to a definition of the reduced compressibility by rearranging it to obtain

$$c_R = \frac{1}{p_R} - \frac{\Delta z}{z\Delta p_R} \tag{4.20}$$

Thus, to obtain the gas compressibility at a particular reservoir pressure and temperature, we determine the reduced pressure and temperature, evaluate the reduced compressibility from Appendix B.5 and calculate $c_g = c_R/p_c$.

Gas viscosity. By using a rolling-ball pressure viscosimeter in a temperature bath, the viscosity of a specific natural gas can probably be most accurately determined in the laboratory. However, the empirical data available for the purpose of determining the viscosity of gas is sufficiently accurate that it is seldom that an engineer finds it necessary to request a laboratory measurement of the viscosity. Considering the inaccuracy of some of the other data that are used in the same equations with the gas viscosity, the viscosity may well be the most accurate parameter even when it comes from empirical data. An example of such data is shown in Appendix B.14. These data can be used to determine the gas viscosity under various conditions within the data range if it is used carefully. More extensive correlations can be found in the Katz et. al. Gas Engineers Handbook.[4]

To demonstrate the application of the methods described in determining the characteristics of gas, work the problem that follows and compare your solution with the one in Appendix C.

Problem No. 4.1: Determining Gas Characteristics

To avoid repetitious calculations we will consider a hypothetical gas made up of 80% methane, 15% ethane and 5% normal butane by volume.

The gas is at 3,000 psia and 170° F.

A. Find the reduced pressure and temperature by using Equations 4.8 and 4.9 to obtain the pseudocritical pressure and temperature.
B. What is the gas-deviation factor?
C. Find the gas specific gravity.
D. What is the pseudoreduced pressure and temperature based on Appendix B.8.? What is the z factor?
E. What is the gas formation volume factor for this reservoir?
F. Find the gas density at reservoir conditions.
G. What is the compressibility of the gas? What is the compressibility according to $c = 1/p$?
H. Estimate the gas viscosity at reservoir conditions.
I. If 55° API gravity condensate was produced with this 0.7 gravity gas at a ratio of 40 bbl/MMcf, what would be the reservoir gas-deviation factor?

MATERIAL BALANCE

The material balance for a gas reservoir is undoubtedly the simplest material balance the reservoir engineer will encounter. Nevertheless, he must proceed cautiously in the application of the material balance if he is to avoid sizable errors. He should be aware that the original gas in the reservoir contains water vapor and condensate that is produced with the gas and that the gas material balance does not include automatic corrections for these factors. Thus, the gas produced must include the condensate and water vapor.

The engineer should also be aware of the fact that the early average reservoir pressures may be in error if the reservoir is of very low permeability and pressure buildup data are relied upon for determining average reservoir pressures. The calculation of water encroachment into a gas reservoir also presents special problems due to the long delay in the pressure drop reaching the original water-gas contact. Considerable error can be incurred by assuming that the decline in the pressure at the original water-gas contact is the same as the decline in the average pressure in the reservoir. These and other problems involved in the application of material balance to a gas reservoir will be considered in this section.

The gas material-balance equation.   A general discussion of material-balance technology will be provided in Chapter 5. This discussion of material balance of a gas reservoir may be looked upon as an introduction to the subject since it is concerned with the simplest case of material balance and the simplest material-balance equations.

Material-balance equations equate the volume of a mass of material at a particular pressure and temperature to the volume of the same mass of material at some different pressure and temperature. In a petroleum reservoir (including a gas reservoir) we define the original free-gas volume in the reservoir as G, and state this volume in standard cubic feet. Thus, the original reservoir volume of free gas is the product of G and the initial gas formation volume factor, B_{gi}, with the initial gas formation volume factor stated in reservoir barrels per standard cubic foot. If the gas reservoir is not subjected to a water drive and no interstitial water is produced, the reservoir volume occupied by the gas will remain constant for all practical purposes. Thus, the same initial reservoir volume can be stated in terms of the gas remaining in the reservoir at any particular time after some quantity of gas, G_p, has been produced. At this time the volume of gas remaining in the reservoir will be G-G_p standard cubic feet, and the reservoir volume will be (G-G_p)B_g barrels when the gas volume is converted to barrels at the current reservoir pressure. This means that the gas formation volume factor used to make this conversion must be based on the average reservoir pressure that exists after G_p standard cubic feet of gas has been produced. We can then equate the reservoir gas volume stated in these two different ways to obtain the material-balance equation for a gas reservoir that does not have a water drive and where no interstitial water is produced.

$$GB_{gi} = (G-G_p)B_g \qquad (4.21)$$

If a gas reservoir is subjected to a water drive, the gas volume in the reservoir will be reduced as the water encroaches into the original gas-bearing portion of the reservoir. The barrels of gas in the reservoir initially, GB_{gi}, less the barrels of gas in the reservoir at some time after a volume of gas, G_p, has been produced $(G\text{-}G_p)B_g$, will be equal to the barrels of water that has entered the original gas-bearing portion of the reservoir if none of the encroached water has been produced. If W_p barrels of the encroached water has been produced then the difference will be equal to the total number of barrels of water that has encroached into the original gas-bearing portion of the reservoir, W_e, less the produced water, W_p.

$$GB_{gi} = (G\text{-}G_p)B_g + (W_e - W_p) \qquad (4.22)$$

This equation assumes that the formation of liquid condensate in the reservoir and the evaporation of connate water causes insignificant errors in the equality. There are notable exceptions to the assumption that no liquid condensate forms in the reservoir but the assumption that the change in the gas reservoir volume due to water evaporation is negligible appears to be valid. The original gas in place can be calculated volumetrically and the water encroachment can be evaluated using the constant-pressure solution to the radial diffusivity equation. The evaluation of water encroachment will be discussed in a later section of this chapter. A lack of understanding of the time delay associated with the water encroachment into a gas reservoir appears to cause much difficulty in some segments of the gas-producing industry.

The gas production, G_p, to be used in Equation 4.22 should include the produced condensate and produced water that was in the original reservoir gas.

Material balance gas production. When dry gas at high pressure and temperature is removed to a low pressure and temperature some of the heavier hydrocarbons generally change to liquid or condensate. This is illustrated in Fig. 4.2 where the stock tank conditions are in a portion of the phase diagram where gas and liquid exist in equilibrium. The dotted line simply shows the possible pressure-temperature path the hydrocarbons might travel in going from far out in a reservoir to surface conditions. Since Equation 4.22 does not provide for production of anything except dry gas from the reservoir it is necessary to include the condensate production in the produced gas term, G_p, as equivalent dry gas.

Using Equation 4.1 we can derive an expression for the amount of gas in standard cubic feet that is the equivalent of 1 stock-tank barrel of condensate. This can be stated as a function of the specific gravity and molecular weight of the condensate. Since we want the volume at standard conditions, the pressure, temperature, and gas deviation factor are fixed at 14.7 psia, 520° R., and 1.0. The gas-deviation factor is 1.0 for all gases at standard conditions. Also, these units fix the gas constant, R, at 10.73. This leaves the number of moles, n, to be stated as a function of the specific gravity of the condensate, γ_L, and the molecular weight, MW. The weight of 1 bbl of fresh water is 350 lb so the weight of 1 bbl of a liquid whose specific gravity is γ_L will be $350\gamma_L$. Substituting these values into Equation 4.1 we can derive the expression for the gas equivalent of the condensate, GE, in standard cubic feet per stock-tank barrel.

$$(14.7)(V) = (1.0)\frac{(350\gamma_L)}{MW}(10.73)(520) \tag{4.23}$$

$$GE = V = 133,000\gamma_L/MW \tag{4.24}$$

For the greatest accuracy in determining the equivalent gas volume of the produced condensate this equation should be used and the molecular weight calculated from a component analysis or a laboratory measurement. However, the component analysis and/or laboratory molecular weight are seldom available for a condensate. In such cases the data of Appendix B.18 is very helpful. This table provides an empirical relationship between the condensate gravity in degrees API and the gas equivalent, GE, in standard cubic feet per stock-tank barrel. The table is based on the equations shown in the footnotes. These change degrees API to the specific gravity, γ_L; calculate the molecular weight, MW, from the specific gravity using an empirical equation; and calculate the gas equivalent, GE, from Equation 4.24.

We know that all petroleum reservoirs contain some water saturation because the pore volume was, sometime previously in geologic time, completely occupied by water. Then when the hydrocarbons migrated into the reservoir they were incapable of displacing all of the water. Hence, all petroleum reservoirs contain connate water even though it may be immobile for the entire producing life of the reservoir. The gas in a gas reservoir has been in the presence of this connate water for billions of years. Consequently, we can safely assume that the gas is initially saturated with water vapor in the reservoir. Since some of this water vapor will probably condense at surface conditions and most of the water vapor must be removed before the gas is metered for sale, the metered dry gas does not include the water vapor that was a part of the original gas in place in the reservoir. Consequently, we must correct the produced dry-gas volume for the water vapor before it is used in a material-balance calculation.

Appendix B.19 shows the water content of natural gas in the presence of liquid water at various pressures and temperatures. It can be used to determine the pounds of water in 1 MMscf of gas. The main portion of the graph gives the water content of freshwater but, in the lower right-hand corner, a graph of correction factors is included to account for the effect of dissolved solids on the water content of the gas. It is recommended that the original water-vapor correction be applied to all produced gas before the figures are used in material-balance calculations.

However, the exact handling of the water-vapor correction is not quite this simple. For one thing, the connate or interstitial water may become mobile as the reservoir pressure declines or in many cases the reservoir will exist in a transition zone (where the water saturation is greater than the irreducible water but less than 100%) and the interstitial water will be mobile from the moment of first production. Regardless of the reason, it remains that some of the interstitial water may be produced. Care must be taken to make certain that this produced liquid water is not assumed to be produced water vapor because it must be handled differently in the material balance. Produced water that was originally liquid in the reservoir should be included in the material-balance calculations as produced water, W_p, regardless of whether or not there is a water drive on the gas reservoir.

Examination of the gas material-balance Equation 4.22 will show that the equation will still hold even though there is no water encroachment. The effect of production of interstitial water will be one of increasing the reservoir volume of gas. If W_e in Equation 4.22 is zero the equation says that the original reservoir volume of gas, the left-hand side of the equation, is equal to the reservoir volume of gas after G_p standard cubic feet of gas has been removed, minus the produced interstitial water. Thus, the reservoir gas volume after production is equal to the original gas volume plus the interstitial water produced.

The interstitial water produced might be distinguished from the water vapor produced on the basis of the water salinity since water vapor will be fresh water and interstitial water produced as liquid will be brine. However, some of the interstitial water produced may actually have passed through the vapor state and thus be produced without its original salt content. Study of Appendix B.19 will explain this situation. Note that as the reservoir pressure decreases the equilibrium water vapor content increases. This means that part of the interstitial water can vaporize as the pressure declines. The evaporation of the interstitial water will make more space available for the reservoir gas and yet it may be produced as fresh water since it is produced in the vapor phase. Consequently, the author recommends that all of the produced water in excess of that originally in the produced gas in the vapor state in the reservoir be treated as W_p in Equation 4.22. This simply means that the engineer should calculate the amount of water vapor that was originally in the produced gas in the reservoir and this should be added to the metered dry gas, the equivalent gas of the produced condensate, and the vent gas from the condensate storage, to obtain the produced gas value, G_p.

$$G_p = \text{(Dry gas sold)} + \text{(Gas equivalent of condensate produced)}$$
$$+ \text{(Original water vapor in produced gas)}$$
$$+ \text{(Vent gas from condensate storage)} \tag{4.25}$$

The difference between the total produced water and the original water vapor in the produced gas must be treated as produced liquid water, W_p, in Equation 4.22. Note that this treatment still is not exact.

The difference between the produced water and the original water vapor content of the produced gas provides some correction for the additional pore volume available to the gas in the reservoir due to evaporation of the interstitial water that accompanies the decline in the reservoir pressure. However, this represents only part of the correction that is theoretically necessary as a result of this phenomena because all of the evaporated water is not produced but remains in the reservoir. This means that more space is available to the gas phase in the reservoir which is not accounted for in the material-balance equations discussed, but this additional pore space available is offset to a large degree by the fact that a larger gas phase mass is present in the reservoir than is indicated in the material-balance equations considered due to the evaporation of the interstitial water. Consequently, the net correction to the equations would be small. By accounting for the effect of the vaporized water produced we are accounting for the portion of the reservoir evaporation phenomena that is not offset by an additional gas mass in the reservoir.

The small equation error that remains could be taken into account except for the kinetics effect that accompanies the vaporization. It is uncertain just how long would be required for equilibrium to be obtained at each pressure. Gas production is continuous so that equilibrium cannot be reached at each pressure and the water is in the smallest pores which provides a very small interface between the gas and water. Consequently, the evaporation rate would be difficult to establish. Thus, it is recommended that Equation 4.22 be used in material-balance calculations. The gas production, G_p, should be calculated by Equation 4.25 and the water production, W_p, should include the produced water in excess of the original water vapor in the produced gas. To test your knowledge of these corrections work the following problem and then check your solution against the one in Appendix C.

Problem No. 4.2: Determining Gas Production for Use in Material Balance.

Calculate the total daily reservoir gas production, G_p, and water production, W_p, for material balance use from the following. (State G_p scfd. and W_p in BPD.)

> Separator gas production = 10.0 MMscfd.
>
> Condensate production = 150.0 STB/d
> Condensate gravity = 55° API (0.759 sp gr)
> Stock-tank gas production (vent gas) = 30 Mscfd.
>
> Freshwater production = 15 b/d
> Initial reservoir pressure = 4,000 psia
> Reservoir temperature = 220° F.
> Formation water salinity is 150,000 ppm
> The GE for water by equation 4.24 is 7390 SCF/Bbl.

Determining the reservoir pressure.   The most accurate means of determining the average reservoir pressure in a gas reservoir is to run a pressure bomb in a permanently shut-in well. In years past it was almost a certainty that some of the wells in a gas reservoir in good mechanical condition would be idle for long periods of time each year. These wells, during the idle periods would afford an ideal opportunity to directly measure the static pressure in a gas reservoir. However, the gas shortage of recent years has made it much more unlikely that such shut-in wells could be found in gas reservoirs.

Even when idle gas wells are available it is likely today that they will either be in poor mechanical condition (which accounts for their being idle) or the reservoirs will be at such a high pressure or so deep that the risk of running a pressure bomb in the well is believed to be excessive. When this occurs, or when running a pressure bomb in a well is believed unnecessary, bottom-hole pressure can be calculated from an observed static surface pressure. This can be done with sufficient accuracy if there are no fluids standing in the bottom of the hole and a calculating method is used that accounts for the variation in temperature and gas-deviation factors from the surface to the bottom of the hole. Two methods will be presented that provide such accuracy. One of these requires the use of a digital computer to perform the iterative calculations and the other is in the form of empirical data that has some limitations as to the maximum depths to be considered and the accuracy with which the charts can be read. We will also examine the basis of the equation generally recommended for use by state regulatory bodies, which requires the use of average temperatures and average z factors.

Static pressure differences for a column of gas are more difficult to evaluate than similar static pressure differences for a column of liquid because gas density is a function of pressure and the pressure in a gas column varies due to static pressure differences. The pressure gradient at any particular point of pressure and temperature in a gas column can be calculated from the density of the gas at those conditions. If the density is in pounds per cubic foot this is the same as a pressure gradient in pounds per square foot per foot of difference in depth. If we desire the pressure gradient in the more conventional psi per foot of depth we can simply divide the density in pounds per cubic foot by the number of square inches in a square foot, 144. Thus, we can obtain the pressure gradient in psi per foot from Equation 4.14 which defines the density in pounds per cubic foot.

$$\frac{\Delta p}{\Delta D} = \frac{\rho_g}{144} = \frac{2.7\gamma_g p}{zT} \Big/ 144 \qquad (4.26)$$

$$\frac{\Delta p}{\Delta D} = \frac{0.01875\gamma_g p}{zT} \qquad (4.27)$$

Equation 4.27 can then be used to determine the change in pressure over some small increment of depth by using the arithmetic average pressure as p in the equation.

$$\Delta p = \frac{0.01875\gamma_g p}{zT} \Delta D \qquad (4.28)$$

For hand calculations this equation would have to be assumed to be trial and error in its application. However, when a digital computer is used to make the calculation of a bottom-hole pressure from a surface pressure by applying Equation 4.28 to very small increments of depth, there is practically no limit on the minute size of the depth increments that can be used. A maximum increment size below which there is no significant increase in the accuracy of the calculations can be determined. A simple program for the calculation of the bottom-hole pressure from surface pressure, gas gravity, surface temperature, temperature gradient, critical pressure, critical temperature, and gas-deviation factor covering the anticipated range of reduced pressures and temperatures is shown in Table 4.2. This program is written in conversational language, PL 1. The program is designed to permit the submission of all data through a typewriter terminal. The procedure used is to simply

Table 4.2

COMPUTER PROGRAM FOR CALCULATING THE STATIC BHP OF A GAS WELL
(in PL-1, conversational language)

```
seq from 1 thru . . . from 1 thru . . .
  1.              PUT IMAGE(1)(head);
  2.      head:   IMAGE;
                  CALCULATION OF STATIC BHP IN A GAS WELL
  3.              GET LIST(sg,dx,psurf,tsurf,td,ttd,pc,tc,ntr,npr);
  4.              /*TEMPERATURES MUST BE IN DEGREES RANKIN*/;
  5.              /*  sg is gas specific gravity; dx is calculating interval desired;
                      psurf is surface pressure
  6.              /*  td is total depth; ttd is temperature at total depth; pc is critical
                      pressure;          */;
  7.              /*  tc is critical temperature; ntr is the number of tr values to be
                      entereed in program
  8.              /*  npr is the number of pr values to be entered in the program;
                      prntin is the desired pressure
interval*/;
  9.              /*  pr is the reduced pressure; tr is the reduced temperature*/;
 10.              ;
 11.              /*  prntin/dx and td/dx must be whole numbers   */;
 12.              GET LIST(prntin);
 13.              DECLARE pr(20) , tr(10) , z(10,20) ;
 14.              ;
 15.    loopl:    DO i=1 TO npr;
 16.              GET LIST(pr(i));
 17.              END loopl;
 18.    loop2:    DO j=1 TO ntr;
 19.              GET LIST(tr(j));
 20.              END loop2;
 21.    loop3:    DO i=1 TO npr;
```

Table 4.2 (*Continued*)

```
22.    loop4:   DO j=1 TO ntr;
23.             GET LIST(z(j,i));
24.             END loop4;
25.             END loop3;
26.             t=tsurf;
27.             p=psurf;
28.             ddd=0;
29.             dp=0;
30.             d=0;
31.             tgrad=(ttd-tsurf)/td;
32.             PUT IMAGE(1)(slip);
33.    slip:    IMAGE;
       Depth, ft.   Pressure, psia   Temperature, Rankine   Reduced temp.   Reduced
                    press.   z
34.    again:   ta=t+tgrad*(dx/2);
35.             tra=ta/tc;
36.             pra=(p+dp/2)/pc;
37.             i=0;
38.             i=i+1;
39.             h=i+1;
40.             IF pra>=pr(i)&pra<=pr(h) THEN GO TO tra
41.             IF h>npr THEN GO TO wrong;
42.             GO TO iinc;
43.    trac:    j=0;
44.    jay:     j=j+1;
45.             k=j+1;
46.             IF tra>=tr(j)&tra<=tr(k) THEN GO TO zprltc;
47.             GO TO jay;
48.    zprltc:  zprltr=z(j,i)+(tra-tr(j))/(tr(k)-tr(j))*(z(k,i)-z(j,i));
49.             zprhtr=z(j,h)+(tra-tr(j))/(tr(k)-tr(j))*(z(k,h)-z(j,h));
50.             zprtr=zprltr+(pra-pr(i))/(pr(h)-pr(i))*(zprhtr-zprltr);
51.             dp=.01875*sg*(p+dp/2)*(zprtr*ta);
52.             d=dx+d;
53.             t=t+tgrad*dx;
54.             p=p+dp;
55.             ddd=ddd+1;
56.             IF ddd=prntin/dx|d>=td THEN GO TO print;
57.             GO TO again;
58.    slider:  IMAGE;
       ------.--      -----.--     ---.--     --.--     --.--     -.---
59.             IF d<td THEN GO TO again;
60.    print:   PUT IMAGE(d,p,t,tra,pra,zprtr)(slider);
61.             ddd=0;
62.             IF d<td THEN GO TO again;
63.             STOP ;
64.    wrrng:   PUT IMAGE(1)(iml);
65.    iml:     IMAGE;
                            DATA ARE INSUFFICIENT
66.             STOP ;
```

calculate the pressure at the bottom of an increment by using the average temperature in the increment as representative of the entire increment and using the average pressure, p, for the calculation as being the pressure at the top of the increment plus half the pressure change, Δp, for the depth increment just above the subject increment. By starting the calculations at the top of the hole where the surface pressure is known it is then possible to use the pressure calculated for the bottom of each increment as the pressure at the top of the next lower increment and continue calculations until the depth at which the pressure is desired is reached.

As is the case with all the digital computer programs presented in this text, this is not a polished program but is simply presented to provide the engineer or programmer with the fundamental ideas so that a polished program can be developed if it is deemed advisable or necessary. If the program was used in a Fortran language with gas-deviation data in storage and a subroutine available to interpolate the data as previously defined, the program input and use would be greatly simplified.

The Texas Petroleum Research Committee published a booklet of nomographs, Bulletin No. 72, suitable for determining bottom-hole pressures. These nomographs evaluate static pressures to depths of about 15,000 ft with reasonable accuracy. Some data are presented for depths of as much as 24,000 ft. The nature of these data are illustrated in Figs. 4.3A through 4.3C for explanation purposes. Bulletin No. 72 can be obtained from the Texas Petroleum Research Committee, at the University of Texas, Austin, for $1.50 (1971 price). It will be obvious as we proceed with a discussion of these data that the full bulletin is necessary to obtain accurate calculations for most cases. The bulletin also provides a means of calculating bottom-hole flowing pressures from surface pressures but the author has not had occasion to check the accuracy of these data. The bulletin does not divulge the mathematical methods used to generate the data but random checks of the static pressures against the calculated values using the computer program of Table 4.2 show excellent agreement. The use of the bulletin to calculate flowing bottom-hole pressures will be discussed later in this chapter.

Basically the Bulletin 72 static-pressure data are simply plots of pressure versus depth (the almost straight lines) with each surface pressure represented by a particular line. The pressure scale is labeled "surface pressure" but it represents the surface pressure only at the surface depth which is "0" in the graphs. Each figure represents a particular gas gravity and thermal gradient and all of the basic charts are based on a surface temperature of 74°F., which is probably the mean average surface temperature for Texas. Since the surface temperature and temperature gradient are fixed, there is a fixed relationship between the depth and the temperature. This can be seen by noting that the horizontal axis has two scales, one of depth along the bottom of the graph and one of temperature along the top of the graph.

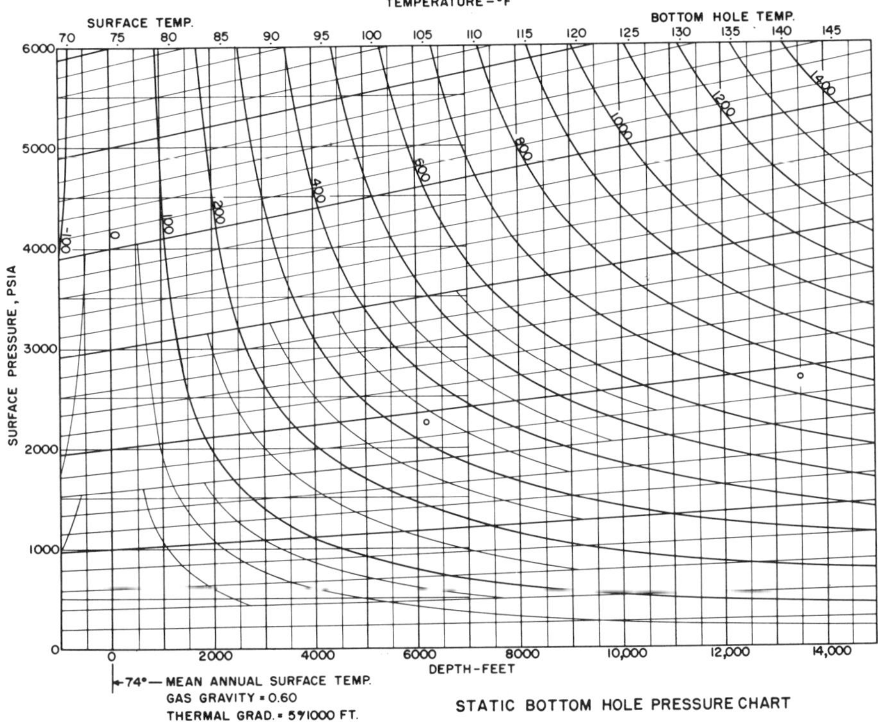

FIG. 4.3A Static bottom-hole-pressure nomograph. After Crawford and Fancher, TPRC Bulletin 72.

To illustrate the physical meaning of the data consider a simple example.

We will determine the pressure at a total depth of 6,500 ft if the surface temperature, surface pressure, gas gravity, and thermal gradient are 74°F., 5,000 psia, 0.60 sp gr, and 5°F./1,000 ft. Fig. 4.3A represents this combination of gas gravity of 0.6 and temperature gradient of 5°F./1,000 ft. The inclined straight line representing a surface pressure of 5,000 psia and temperature 74°F. is located by following the horizontal pressure line for 5,000 psia to a point where it represents a temperature of 74°F. Note that this also represents a depth of zero as read off the bottom scale. It will be found that a pressure-correlation line (inclined straight line) passes through the 5,000 psia-74°F. point. This is the line from which the pressure can be read at any depth. Reading this correlation line at a depth of 6,500 ft we will see that the pressure at this depth is somewhere near 5,700 psia.

With a scale of 1,000 psia per linear inch we would be hard pressed to read the pressure to an accuracy of more than possibly 50 psia directly

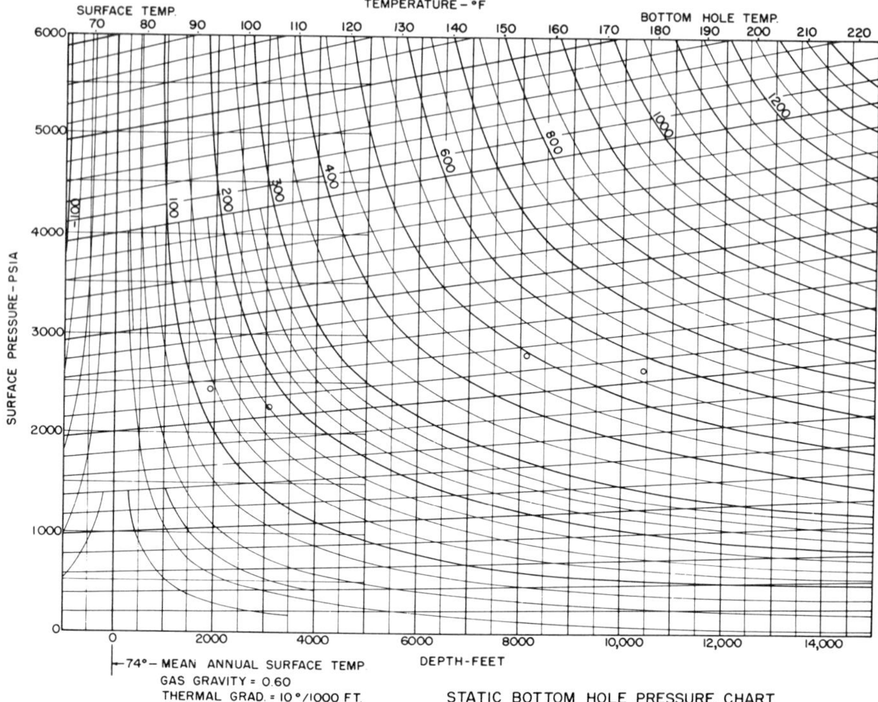

FIG. 4.3B Static bottom-hole-pressure nomograph. After Crawford and Fancher, TPRC Bulletin 72.

from the basic correlation. Consequently, the difference between the surface pressure and the pressure at any particular depth is also plotted. These are the "hyperbolic" appearing curves on the figures. In our particular example we see that a curve with a pressure difference of 700 psi passes exactly through the previously established point representing the pressure-depth correlation for a surface pressure of 5,000 psia (the inclined straight line) and a depth of 6,500 ft (bottom horizontal scale). Consequently, from the difference in pressure value alone we can determine that the pressure at a depth of 6,500 ft will be 5,000 psia (the surface pressure) plus the 700 psi pressure difference or 5,700 psia. By working with the pressure differences rather than the basic pressure-depth correlation the bottom hole pressure can be determined much more accurately. In our particular example it becomes a matter of reading a pressure scale representing about 200 psi per linear inch much more accurately than we can read the vertical pressure which represents a scale of about 1000 psia per linear inch.

Use of the Bulletin 72 nomographs becomes somewhat more confusing when we apply them to a situation where the average surface

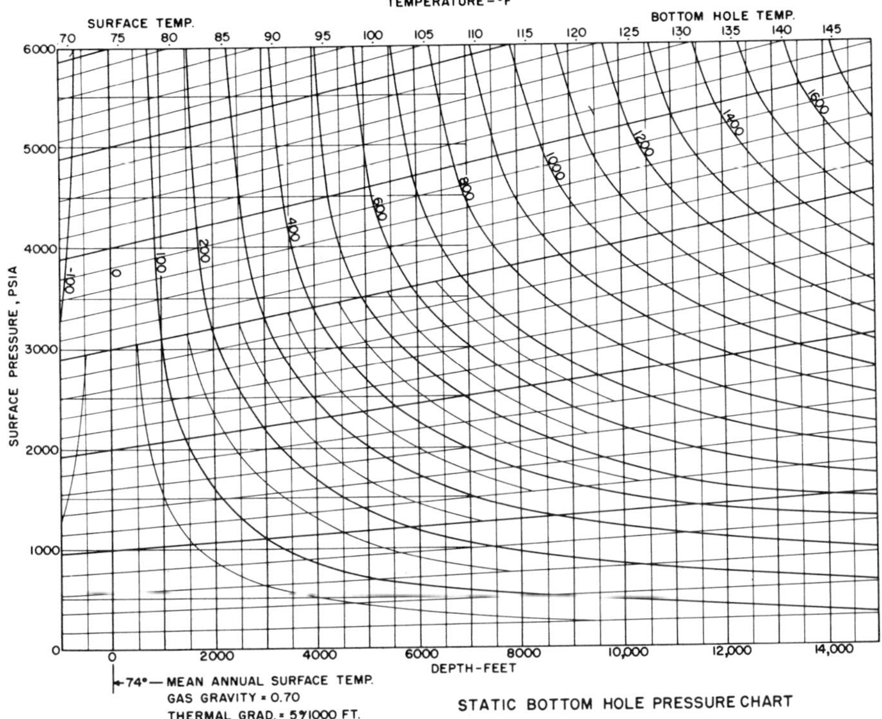

FIG. 4.3C Static bottom-hole-pressure nomograph. After Crawford and Fancher, TPRC Bulletin 72.

temperature is other than 74° F. which is of course infinitely more realistic. To demonstrate this procedure we will again consider the example discussed above except that we will change the mean annual surface temperature to 85° F.

We would of course still use Fig. 4.3A and we would follow the same procedure for establishing the pressure depth correlation line by finding the point represented by the surface pressure and surface temperature. However, note that at this point the solution differs somewhat since the point represented by a surface temperature of 85° F. does not correspond to a chart depth of zero as was the case when the surface temperature was 74° F.

In the new case the corresponding chart depth that represents the surface depth is about 2,200 ft. Consequently, to find the pressure at a depth of 6,500 ft we must read the chart at a chart depth of 2,200 plus 6,500 ft or 8,700 ft. If we read the pressure at this point established by the pressure-depth correlation line and a chart depth of 8,700 ft it is about 5,700 psia. This pressure must be established by extending a horizontal line through the established point back to the pressure scale since the pressure grid does not extend to this part of the graph.

To establish the pressure more accurately we should use the pressure difference lines as noted previously. In this example the pressure difference corresponding to the established correlation point for the chart depth of 8,700 ft should be about 920 psi. However, carefully note that this difference is the difference between the pressure at the chart depth of 8,700 ft and a chart depth of zero. Since we want the difference between the chart depth of 8,700 (actual depth 6,500) and the chart depth of 2,200 (actual depth of zero) we must correct the 920-psi difference for the difference between the chart depth of 2,200 ft and a chart depth of zero. This is done by reading the difference at the actual surface condition point which represents the pressure difference between the chart depth of 2,200 ft and the zero chart depth. This value is estimated as 230 psia. Thus the difference between the pressure at the chart depth of 8,700 ft (actual depth of 6,500) and the chart depth of 2,200 (actual depth of zero) is 920 minus 230 or 690 psi. With more careful interpolation of the data it may be possible to read the nomographs with an accuracy of plus or minus 5 psi.

Notice that double interpolation is not unusual when the Bulletin 72 data is used. It would be unusual for the gas gravity and temperature gradient to exactly fit the gas gravity and temperature gradient of one of the available charts in Bulletin 72. In order to test the reader's understanding of the Bulletin 72 nomographs he is invited to work the following problem and check his solution against the one in Appendix C.

Problem No. 4.3A Calculation of Static Bottom Hole Pressure from Surface Pressure Measurements

A reservoir contains a 0.68 sp gr dry gas at a temperature of 110° F. If the average depth of the formation is 3,676 ft, the average wellhead temperature 90° F., and the static surface pressure in the well is 2,715 psia, find the corresponding average reservoir pressure using the Bulletin 72 data.

The most often used method for calculating static bottom-hole pressure from surface pressure is probably the exponential form of Equation 4.28 based on an average gas-deviation factor and average temperature. This equation is obtained by treating the gas-deviation factor and temperature as constants and applying Equation 4.28 to all of the vertical increments, ΔD, from total depth to surface. If this is done and all equations are summed, we obtain

$$\sum_{p_1}^{p_2} \frac{\Delta p}{p} = \frac{0.01875\gamma_g}{z_{avg}T_{avg}} \sum_{0}^{D} \Delta D \tag{4.29}$$

$$\ln\left(\frac{p_2}{p_1}\right) = \frac{0.01875\gamma_g D}{z_{avg}T_{avg}} \tag{4.30}$$

Gas Reservoir Engineering

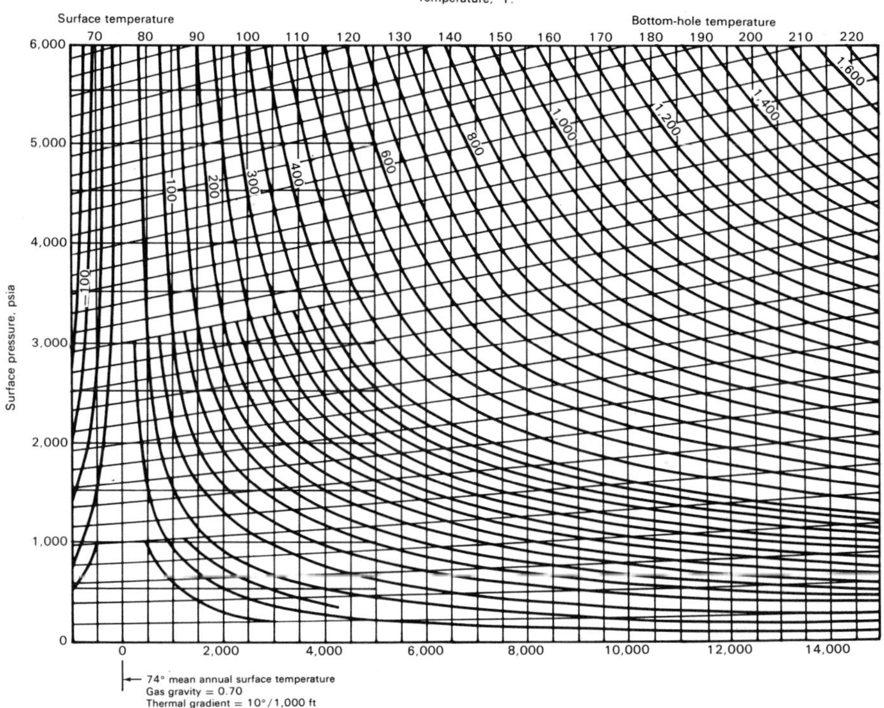

FIG. 4.3D Static bottom-hole-pressure chart. After Crawford and Fancher. TPRC bulletin 72.

In exponential form,

$$p_2 = p_1 e^{\left(\dfrac{.01875\,\gamma_g D}{z_{avg}T_{avg}}\right)} \tag{4.31}$$

Problem No. 4.3B Calculation of Static BHP from Surface Pressure Measurements

Rework No. 4.3A using Equation 4.31

When depths are shallow and pressures are relatively low, Equation 4.31 will give sufficiently accurate results for many purposes. However, the simplicity of the computer program necessary to make a much more accurate calculation and the availability of the Bulletin 72 data for most reservoir depths seem to make it unwise and unnecessary to use an equation that must assume an average gas-deviation factor and average temperature.

Use of Equation 4.31 is not as straightforward as it might appear even if we are willing to accept the inaccuracy of using an average gas-deviation factor and temperature. When an attempt is made to apply Equation 4.31 to calculate the bottom-hole pressure it will be found immediately that the pressure base for the average gas-deviation factor is not known. Since we are calculating the bottom-hole pressure we will not know the arithmetic average pressure until we have determined the bottom-hole pressure. By Equation 4.31 we can not evaluate the bottom-hole pressure until we know the average gas-deviation factor which requires a knowledge of the arithmetic average pressure. Thus, a trial-and-error application of Equation 4.31 is necessary. An average pressure is assumed, the gas-deviation factor is determined, the BHP is calculated, the arithmetic average pressure is calculated, the gas deviation is determined, etc., until there is no further change in the average pressure. In all fairness it should be noted that this trial-and-error is not as bad as it sounds. Normally the solution converges to an acceptable accuracy very quickly.

Graphical material balance. It is often useful to use a graphical form of the material-balance equation to analyze a gas reservoir's performance and predict its future behavior, especially if no water drive is present. Equation 4.21, the material-balance equation applicable in the absence of a water drive, can be placed in a form more useful for graphical interpretation by substituting for the gas formation volume factors according to Equation 4.12.

$$GB_{gi} = (G - G_p)\,B_g \tag{4.21}$$

$$G\left(\frac{0.00504 z_i T}{p_i}\right) = (G - G_p)\left(\frac{0.00504 z T}{p}\right) \tag{4.32}$$

When 0.00504T is cancelled from each side of this equation it can be solved for gas production.

$$G_p = G - \left(\frac{p}{z}\right)\left(\frac{Gz_i}{p_i}\right) \qquad (4.33)$$

Now note that a plot of the cumulative gas production, G_p, versus p/z should give a straight line that can be extrapolated to predict the cumulative production at any future average reservoir pressure, p. Further note that when p/z is zero, G_p will equal the original gas in place, G, and when the gas production, G_p, is zero p/z will equal p_i/z_i. These relationships are demonstrated graphically in Fig. 4.4.

Note from Equation 4.33 that if the gas in the reservoir was a perfect gas, such that the gas-deviation factor was always one, a straight-line plot would be obtained for cumulative production versus the pressure alone. It might seem unnecessary to make such a note but the practice still persists of simply plotting the pressure versus the cumulative production and expecting it to be a "relatively" straight line. Since there are no perfect gases in underground reservoirs, any straight lines obtained with such a plot are simply due to compensating errors.

Equation 4.33 and Fig. 4.4 appear to be deceivingly simple. However, even the apparently simple task of solving for the initial pressure, p_i, or the pressure obtained when a particular cumulative production is reached, presents difficulties. The engineer will note that only the ratio p/z or p_i/z_i can be readily obtained from the plot of the equation because the gas deviation factor, z, is a function of the pressure which is the unknown. If a computer is used that has the gas-deviation factors in

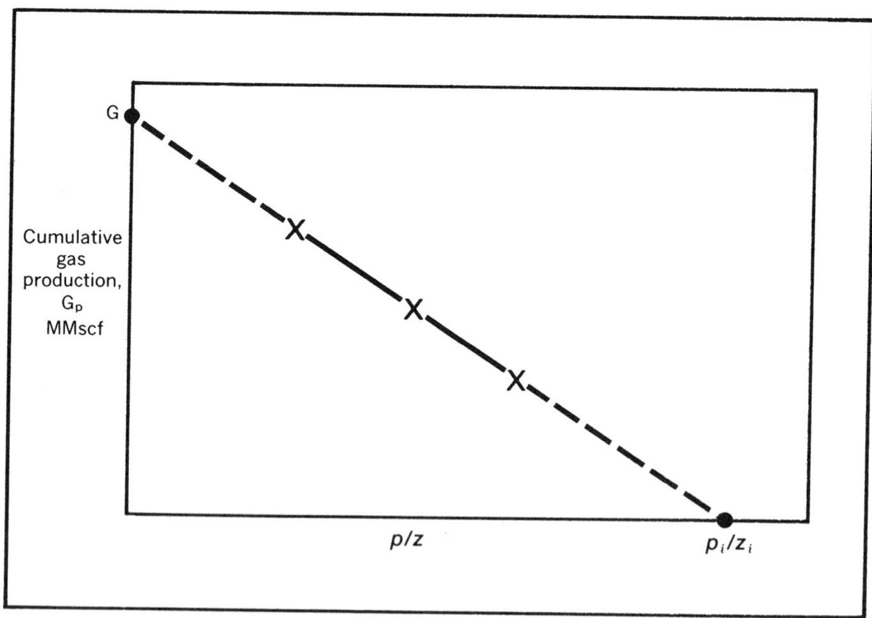

FIG. 4.4 Graphical material balance for a gas reservoir.

storage, the calculation of pressure presents no difficulty since the pressure can be determined by trial and error. However, if the engineer is performing hand calculations, the task of solving for the pressure is more difficult. In this latter case the engineer is encouraged to make use of the plot of (p_r/z) versus z for various reduced temperatures as presented in Appendix B.9. These data will facilitate the direct evaluation of the gas-deviation factor, z, from the ratio p_r/z, and the immediate calculation of the pressure by using z and the critical pressure without resorting to trial and error. The subject p_r/z versus z plot closely resembles the basic z, p_r, T_r correlation data presented in Appendix B.7.

Solution of the following problem will provide a check of your knowledge of the graphical gas material-balance technique. Your solution can be checked against the solution in Appendix C.

Problem No. 4.4 Application of Graphical Gas Material Balance Techniques

A dry gas reservoir has produced as follows:

Date	Cumulative production MMcf	Static reservoir pressure, psia
7–1–65	0	
7–1–66	1,800	3,461
9–1–67	3,900	3,370
10–1–68	5,850	3,209
11–1–69	9,450	3,029

Reservoir temperature = 100°F.
Gas gravity = 0.68

Determine the original reservoir pressure and original gas in place.

What will be the average reservoir pressure at the completion of a contract calling for delivery of 20 MMcfd for 5 years (in addition to the 9,450 MMscf produced to 11–1–69)?

Determining water encroachment. Water encroachment into a gas or an oil reservoir can be most accurately determined in almost all cases by using the constant-pressure solution to the radial diffusivity equation. Since the next section of this chapter deals with fluid flow in a gas reservoir the reader may wonder why this discussion is placed under the material balance section of this chapter. The problem of determining water encroachment would not be appropriate under fluid flow in a gas reservoir because the flow associated with water encroachment is actually taking place in the aquifer or water-bearing portion of the reservoir. Consequently, there is no basic difference between the calculation of water encroachment into a gas reservoir and water encroachment into an oil reservoir. However, the problem of calculating water encroachment into a gas reservoir has some peculiarities that should be recognized by the engineer.

If the engineer is not aware of the nature of unsteady-state flow, the general solutions that are available, and particularly the general constant-pressure solution to the radial diffusivity equation, he should go back to Chapter 2 and review this material. From a practical standpoint, the constant-rate solution permits the calculation of the cumulative flow of fluid across a particular radius in the reservoir where the pressure has been maintained at a constant value, p_r, which in turn has caused a constant pressure drop, $p_i - p_r$, to have been maintained at this radius for some period of time, t.

Mechanically, the engineer calculates the dimensionless time, t_D, that corresponds to the producing time, t, by using the equation derived in Chapter 2, modified slightly to apply to our application.

$$t_D = \frac{\eta t}{r^2} \qquad (2.10A)$$

When this equation is applied to calculate the cumulative flow of water across the original WOC (water-oil contact), the well radius, r, for the aquifer, is the internal radius of the aquifer or the external radius of the gas or oil reservoir. Once the dimensionless time, t_D, has been evaluated it is used to determine a corresponding Q_{tD} from Table 2.2 or 2.3 or Figs. 2.11A through 2.11D. If the reservoir is determined to be finite acting at the time of interest the dimensionless reservoir size, $r_{De} = r_e/r$, must be used to determine Q_{tD}. Once the function Q_{tD} has been evaluated the number of barrels of water that has flowed across the original WOC radius, r, can be calculated from another equation taken from Chapter 2.

$$Q = 1.12\phi hcr^2\, \Delta p\, Q_{tD} \qquad (2.29)$$

This equation could be used to calculate the water encroachment only if a constant pressure drop, $\Delta p = p_i - p_r$, was maintained for the entire period, t. However, we know that once the pressure at the original WOC begins to decline due to production of gas, the pressure at this radius will probably continue to decline if the gas reservoir is being produced at any rate approaching the capacity of the reservoir. Gas with its very low viscosity can flow much more rapidly than water. In other words water encroachment can not maintain a steady-state condition in the gas reservoir.

When the gas reservoir is not being produced at a rate approaching the reservoir capacity it is very likely that the production will be sporadic so that, in this case also, the pressure at the original WOC will vary with time. Consequently, Equation 2.29 must be applied by superposition due to the variation in the pressure at the original WOC. The technique recommended is to take the original WOC pressure history and approximate it with a sequence of finite pressure drops. The effective time for each pressure drop will then be the remaining time to the time of interest. For example, if a particular pressure drop is assumed to exist at a time of 1 year and we are calculating the cumulative water influx for a time of 18 months, the effective time to be considered for that particular pressure drop will be 6 months and an additional pressure drop that occurs at a period of 15 months would have an effective time period of 3 months. Once the finite pressure drops and the corresponding times that approximate the actual pressure history have been determined, it is possible to calculate the cumulative water encroachment due to each individual pressure drop acting for its appropriate time using Equation 2.29. The actual cumulative water influx at that time would then be the sum of all the individual Q's calculated.

Since the group of constants, $1.12\phi hcr^2$, appears in each individual calculation it is simpler to apply the calculations using the equation derived in Chapter 2,

$$Q = 1.12\phi hcr^2 \sum_{j=1}^{j=n} (\Delta p_j \, Q_{tDj}) \qquad\qquad (2.36)$$

Since the reservoir parameters in this equation apply specifically to the water-bearing portion of the reservoir they are normally evaluated only as a result of dry holes, and other errors. Since the accuracy of these data is always questionable, it is best to determine the group of terms, $1.12\phi hcr^2$, from the past reservoir performance. If we call the group of terms B, Equation 2.34 would become

$$Q = B \sum_{j=1}^{j=n} (\Delta p_j \, Q_{tDj}) \qquad\qquad (2.34A)$$

When this equation is applied to the history, material balance can be used to determine Q and thus everything is known except the constant, B, which can be calculated from the equation. Once the constant B is known, Equation 2.34A can be used to predict future water influx. You should not be unduly concerned with the fact that you do not know, with any great accuracy, the reservoir parameters necessary for the calculation of the dimensionless time, t_D. These parameters, the permeability, porosity, water viscosity, and effective compressibility, will not be known with any more accuracy than will the parameters that go to make up the constant, B. However, they will have a negligible affect on the accuracy of the predicted water encroachment because they are used in both the evaluation of B and the prediction of the water encroachment. Thus, compensating errors are introduced into the calculating procedure. Also, their effect is roughly proportional to the log of the values so that the inaccuracy is further reduced.

If sufficient history is available, the engineer may wish to make a plot of Q calculated from material balance versus the term $\sum_{j=1}^{j=n} (\Delta p_j \, Q_{tDj})$ and evaluate the slope which will be the constant B. (See Equation 2.34A). This will give more statistical weight to the evaluation of B, and will tend to minimize the importance of the early pressure data that is often inaccurate for various reasons.

Engineers generally assume that the change in pressure at the original WOC is the same as the change in average pressure in the gas or oil-bearing portion of the reservoir. When oil reservoirs are considered, this assumption seldom introduces any significant error. However, when calculating the water influx into a gas reservoir, the engineer should carefully evaluate the time before the pressure at the original WOC is affected. As previously noted, disturbances travel very slowly in many of the gas reservoirs; thus, a considerable amount of time may transpire before any substantial change in reservoir pressure occurs at the original WOC. Until a pressure drop occurs at this point in the reservoir, no water encroachment will take place. Consequently, care should be taken to estimate as accurately as possible the time when the first pressure drop occurs at the original WOC and this should represent the zero time for the calculation of the water influx.

Work the following problem to make certain you understand the method of calculating water encroachment. Compare your solution with the solution in Appendix C.

Problem No. 4.4A: Calculation of Water Encroachment in a Gas Reservoir

A gas reservoir with a strong water drive has the following pressure history:

Time, months	Average reservoir pressure, psia
0	2,500
6	2,450
12	2,356

Reservoir studies have shown:

(1) That the pressure at the original water-gas contact is 50 psi greater than the average pressure *after* initial conditions. (6 months of production is necessary to cause a significant pressure drop at the original WGC.)
(2) $t_D = 5.0$ when $t = 6$ months.
(3) The aquifer can be treated as being infinite.
(4) Material balance indicates that 192,000 bbl of water has encroached into the gas reservoir at the end of the first 12 months of production ($p_{avg} = 2,356$ psia)

Find the cumulative water encroachment after 18 months and after 24 months, if average reservoir pressures of 2,276 and 2,216 psia, respectively, are anticipated at the end of these periods.

You should note the difficulty of determining the average predicted reservoir pressure for the basis of the water influx calculation as in the above problem. Under the best conditions these pressures can only be determined by trial and error. The simplest situation is the one in which a known rate of production is to be maintained. As a place to start we might calculate the resulting reservoir pressure at the end of some period by assuming water encroachment will be zero during that period. Then based on the calculated average reservoir pressure we could calculate the water encroachment. This water encroachment would make the originally calculated reservoir pressure incorrect and this calculation could be repeated using the last calculated water encroachment. A new calculated pressure would result in a new calculated water encroachment which would change the pressure, etc. Each successive round of calculation will introduce smaller changes until the changes become insignificant. At this point the true value of the average pressure and water encroachment will have been reached.

If we assume maintenance of a maximum production rate, the solution becomes even more difficult because with each sequence of calculations the gas production during the subject period would change in accordance with the reservoir deliverability characteristics.

FLUID FLOW IN GAS RESERVOIRS

Reservoir fluid flow of gas presents additional problems above and beyond those concerned with the flow of liquid in the reservoir. These difficulties take several forms. What makes it most difficult is that the basic fluid-flow equations were derived for the flow of liquids at a time when the production of gas presented no problems since there was a plentiful supply of gas and its price was small compared to oil. Consequently, by the time the problems of gas production reached economic significance the technology of liquid flow was reasonably well established. Thus, it was natural for the engineer to simply attempt to adapt the liquid-flow technology to gas flow.

The difficulty with this approach is that the liquid-flow equations made use of several assumptions that simply were not valid for gas flow. Turbulence which seldom presents a problem in liquid flow presents a continual problem in gas flow into a well. Fluid compressibility, which is constant and very small for liquids, varies widely for gas and is quite large. In addition we are continually faced with the task of evaluating the gas-deviation factor to evaluate the flow rate. It would appear that due to these great differences between liquid and gas that gas flow equations should be derived from fundamentals to avoid the simplifying assumptions. Unfortunately, this was not done. Instead, the present state of the technology is to continue to use most of the equations derived for liquid flow with the inappropriate assumptions noted. Arbitrary and generally poorly defined limits on the applications tend to avoid the inherent errors caused by these inappropriate assumptions. Consequently, gas reservoir fluid flow is a very complex subject.

Darcy's equation with turbulence. We originally noted that the Darcy Equation was derived empirically for viscous flow and that it would not apply to turbulent conditions. Several attempts have been made to modify the Darcy Equation so it can be applied to turbulent flow. The pressure gradient for viscous flow stated in terms of Darcy Units would be,

$$\frac{\Delta p'}{\Delta x'} = \frac{\mu v'}{k} \tag{4.34}$$

It is recognized that the pressure drop in turbulent flow is proportional to the velocity squared. Many engineers and researchers have proposed and used various schemes to modify the Darcy equation to account for turbulence and other nonideal conditions. The form that we choose to use is probably no more accurate than some of the other schemes but we will see that it meshes well with the "additional-pressure-drop" idea introduced in discussing the skin effect. Consequently, it appears to be somewhat better suited for our reservoir fluid-flow technology than some of the other methods.

To account for the additional pressure drop due to turbulence, an additional term is added to Equation 4.34, to yield

$$\frac{\Delta p'}{\Delta x'} = \frac{\mu v'}{k} + \beta'\, \rho (v')^2 \tag{4.35}$$

In this equation ρ is the density in grams per cubic centimeter (or the specific gravity relative to water), and β is a porous-media constant describing the effect of turbulence in a particular porous media. Beta can be determined by appropriate measurements in the laboratory but it is more often evaluated from empirical data such as that presented in Appendix B. The engineer should note that these data present β in reciprocal feet rather than in reciprocal centimeters as is necessary for Equation 4.35.

The engineer is reminded that the velocity in the Darcy equation Equations 4.34 and 4.35 is an apparent, rather than an actual, pore velocity. Thus, in comparing Equation 4.35 with any equation that uses a velocity in a Reynolds number expression, the difference in the two velocities should be remembered.

Steady-state gas flow. For steady state to prevail in gas flow it is necessary for a constant mass flow rate to remain constant throughout the reservoir. We saw previously that a strong water drive helps liquid flow in a reservoir approach steady state. We saw that often the water enters the oil-bearing portion of the reservoir at the same rate that the oil is produced. However, steady-state flow of gas in the reservoir is very seldom if ever achieved today. Since the shortage of gas developed, practically all reservoirs are produced at near their peak production rate. Under these circumstances it is virtually impossible for water to enter a water-driven gas reservoir as rapidly as the gas is produced because of the difference in viscosities.

Nevertheless, the steady-state radial gas flow equations are often useful because the conditions around the well bore closely approximate steady state. Since the bulk of gas production comes from far out in the reservoir, the flow rate near the well bore is reasonably constant at all radii near the well bore. Since practically all of the pressure drop occurs near the well bore it is then possible to use the steady-state gas flow equation to estimate the effect of turbulence and well damage.

Equation 4.35 can be used to derive a radial-flow equation but it is quite complex due to the turbulence term. When the resulting equation is solved for the pressure drop we get

$$p_1{}^2 - p_2{}^2 = \frac{1.424\mu z_{avg}T_f q_g \ln(r_1/r_2)}{hk} + \frac{3.161\ (10^{-12})\ \beta\gamma q_g{}^2 z_{avg}T_f \left(\dfrac{1}{r_2} - \dfrac{1}{r_1}\right)}{h^2}$$

$$(4.36)$$

Examination of this equation shows that if the second term on the right-hand side is ignored, the equation is exactly the same as the steady-state flow equation that resulted when turbulence was ignored. The constant 1.424 is simply the reciprocal of the constant in the steady-state gas flow equation without turbulence, 0.703. This is true because Equation 4.36 is solved for the pressure drop rather than the flow rate that resulted in the constant 0.703.

Once we recognize that the first term of the right-hand side of Equation 4.36 is the $\Delta(p^2)$ value due to streamline or nonturbulent flow of gas we can then assume that the *additional* pressure drop caused by turbulence is represented by the second right-hand term. We could then use Equation 4.36 to estimate the additional pressure drop due to turbulence.

This steady-state equation can be used for this purpose even though we know that a well system is at a particular time in either unsteady state or pseudosteady state. This is because the conditions around the well bore closely approximate steady state even during unsteady state or pseudosteady-state flow and because all of the pressure drop due to turbulence occurs near the well bore. The latter point becomes clear when we recognize that turbulence is directly attributable to velocity and the velocity near the well bore is extremely high compared with the flow velocity further removed from the well bore.

We will see later that use of Equation 4.36 to estimate the additional pressure drop due to turbulence is extremely valuable because a standard pressure build-up analysis of a gas well results in a Δp_{skin} that includes turbulence as well as the damage or improvement around a well bore. Consequently, by using pressure buildup to determine the total additional pressure drop due to well damage and turbulence and then using Equation 4.36 to determine the additional pressure drop due to turbulence it is possible to differentiate between well damage which we may be able to repair and the effect of turbulence which we generally can do nothing about.

The modifier "generally" is used to qualify the conclusion that nothing can be done about the pressure drop due to turbulence because some believe that massive fracture treatments can change the pattern from essentially radial flow to something approaching linear flow. When this is done the velocity may be reduced enough to reduce the pressure drop due to turbulence. Solution of the following problem will help the reader understand the use of Equation 4.36 in differentiating between the pressure drop due to damage and the pressure drop due to turbulence. To find the pressure drop due to turbulence we simply calculate the well pressure that would result if there was no turbulence by using the first term of Equation 4.36 and subtract from that the well pressure calculated by using the entire equation.

Problem 4.5: Turbulent Effects in Gas Wells

A well is producing at a rate of 3.9 MMcfd before shut-in for a pressure build-up survey. The conventional build-up analysis indicates an undamaged permeability of 1.5 md, a "Δp_{skin}" of 1400 psia, and a pressure at the external drainage radius, an average of 550 ft away, of 4,583 psia.

(A) What would the flowing well pressure be if there was no well damage or turbulence?
(B) What would the well pressure be with turbulence and no well damage?
(C) What is the actual additional pressure drop caused by well damage and the actual additional pressure drop caused by turbulence?

$$
\begin{array}{ll}
r_w = 0.333 \text{ ft} & h = 30 \text{ ft} \\
\phi = 5.0\% & \gamma = 0.76 \\
\mu_g = 0.027 \text{ cp} & T_f = 712° \text{R.} \\
z = 0.97 \text{ (Assumed to simplify the problem. The} \\
\quad\quad\quad\quad \text{engineer must normally evaluate this at} \\
\quad\quad\quad\quad \text{the arithmetic average pressure.)}
\end{array}
$$

Although there is seldom any use for a steady-state linear gas flow equation it should be noted for completeness that Equation 4.35 can also be used to derive such an equation. The result is

$$
p_1{}^2 - p_2{}^2 = \frac{z_{avg}\, T_f \mu q_g L}{0.112 Ak} + \frac{1.254\,(10^{-10}) z_{avg} T_f q_g{}^2\, \beta L \gamma}{A^2} \tag{4.37}
$$

In this equation, as in Equation 4.36, the first term represents the $\Delta(p^2)$ that would occur in the absence of turbulence and the second term represents the additional $\Delta(p^2)$ caused by turbulence.

Pseudosteady-state gas flow. In applying the pseudosteady-state equations to gas, the engineer should remember that the equations represent true unsteady-state flow and they are derived for fluids with constant compressibility, whereas the compressibility of gas varies inversely with the pressure. Thus, the application of unsteady-state flow equations (including pseudosteady-state flow equations) to gas, lose their accuracy as the variation in pressure increases both radially and time-wise. This inaccuracy can be remedied by treating the reservoir as a sequence of different reservoirs in series and breaking the time interval, t, into smaller time segments.

In Chapter 2 it was shown that after a well had produced at a constant rate for a time in excess of the stabilization time defined as

$$t_s = 0.04 \ \phi\mu c \ r_e^2/k \tag{2.63}$$

the pressure at all points in the reservoir will begin declining at the rate

$$(\Delta p/\Delta t)_{pseudo} = 1.79 \ q/\phi h c_{sta} r_e^2 \tag{2.41}$$

It will be shown that it is necessary to evaluate the compressibility at the static pressure, hence; the symbol, c_{sta}, is used. Furthermore, it was shown previously that at this time the following rate equation stated in terms of the external radius and the pressure at this radius would prevail.

$$q_w = \frac{7.08 \ kh}{\mu} \ \frac{(p_e - p_w)}{\mu[\ln(r_0/r_w) - 0.5]} \tag{2.51}$$

When we think in terms of applying these equations to the flow of gas, remember that the gas compressibility is proportional to the reciprocal of the pressure. Consequently, Equation 2.41 cannot be exact for gas since the pressure varies with the radius and the compressibility varies with the pressure. Thus, the change in pressure with time for gas cannot be the same throughout the reservoir. Nevertheless, we can show by material balance that Equation 2.41 does indicate the average change in pressure with time if the compressibility, c, is evaluated at the average (static) pressure and we recognize the rate, q, as being the volumetric rate in barrels per day at the same average pressure.

In spite of these limitations we find that when the pseudosteady-state equations are converted to gas flow rates stated in Mscfd and a modification for turbulence is introduced, the resulting equation provides a reasonable description of the behavior of a gas reservoir after it has produced at a constant rate for a period of time governed by Equation 2.63 where the compressibility is based on the pressure at the external boundary during time t_s.

To convert Equation 2.51 to a gas flow rate we substitute for the rate, q, stated in reservoir barrels per day, a function of the gas flow rate, q_g, stated in Mscfd according to Equation 1.18 of Chapter 1. Since the conversion must include some pressure base the arithmetic average pressure $(p_e + p_w)/2$ is used.

$$\frac{5.04 q_g z_{avg} T_f}{(p_e + p_w)/2} = \frac{7.08 \ kh(p_e - p_w)}{[\ln(r_e/r_w) - 0.5]} \tag{4.38}$$

When this equation is solved for the flow rate, q_g, we obtain

$$q_g = \frac{0.703\, kh(p_e^2 - p_w^2)}{\mu z_{avg} T_f [\ln(r_e/r_w) - 0.5]} \qquad (4.39)$$

Remember that the Darcy equation used to derive Equation 2.51 and hence Equation 4.38 does not hold for turbulent flow. Consequently, Equation 4.39 must be modified to account for turbulence if it is to give reasonably accurate answers. This correction can be carried out empirically by the addition of a fractional exponent to the $p_e^2 - p_w^2$ term. In effect, the fractional exponent reduces the pressure drop necessary to provide a particular flow rate. Equation 4.39 then becomes

$$q_g = \frac{0.703\, kh(p_e^2 - p_w^2)^n}{\mu z_{avg} T_f [\ln(r_e/r_w) - 0.5]} \qquad (4.40)$$

This equation is often written with the constant 0.5 combined with the ln term as

$$q_g = \frac{0.703 kh(p_e^2 - p_w^2)^n}{\mu z_{avg} T_f \ln(0.606 r_e/r_w)} \qquad (4.41)$$

This equation is often called the gas deliverability equation and it is used as the basis for conventional deliverability tests. The turbulence constant is evaluated empirically from the flow tests and probably accounts for many nonideal conditions in addition to turbulence. For example, our equations assume isothermal flow, yet we know that the large pressure drops that accompany the radial gas flow and the attendant expansion of the gas has a cooling effect that is measurable. In fact, temperature surveys in the well bore are used to indicate which portions of a formation are producing gas. Where gas is being produced, large temperature anomalies exist in the well bore.

In very tight formations the turbulence constant may account for the change in the formation permeability with the change in the reservoir pressure. It may also account for the change in the gas viscosity and gas-deviation factors that accompany any pressure change. We will see that due to the empirical manner in which the constant is measured it may tend to correct the prediction for any or all of these factors. In fact some "experts" suggest that these "other" factors may be the dominating effect rather than the turbulence. It should be specifically noted that the constant n does not account for well damage. The way Equations 4.38 and 4.41 are written, the permeability, k, represents the average permeability which includes the skin damage. If you prefer to write the equations with undamaged permeability this can be done by adding the skin factor, S, to the log expression so that in Equation 4.41 [ln $0.606(r_e/r_w)$] would be replaced by [ln $(0.606 r_e/r_w) + S$]. Since in pseudosteady state the outer boundary is always fixed, it is simpler to work with the average permeability. This is not possible with transient unsteady-state flow because the drainage radius is continually increasing and the average permeability would be constantly changing.

Conventional back-pressure tests. The prevailing method of testing a gas well for predicting gas producing rate under various conditions is to flow the well at a particular rate until a "constant" well pressure is obtained. Then flow the well at another rate until a new "constant" pressure is reached, and continue this procedure until flowing well pressures have been recorded for a variety of rates. If three or four rates are used, the test may be referred to as a three-point or four-point test. If run and interpreted properly such tests may be used to predict deliverability at various drawdown pressures, at various states of depletion, and for a variety of spacing conditions. However, it is often impossible or very impractical to run a conventional back pressure test properly due to the time involved.

Note that since Equation 4.41 is basically a pseudosteady-state equation it is necessary for each flow rate to be maintained long enough to affect the entire well drainage area. This means that the producing time must be greater than the stabilization time, t_s, as defined in Equation 2.63. Until the entire drainage area is affected the drainage radius of a particular rate is continuing to increase and Equation 4.41 is not yet applicable.

However, if the test is run properly, a straight-line plot will be obtained when the log of the producing rate is plotted versus the log of the pressure drawdown term, $p_e^2 - p_w^2$. This can be shown by taking the log of both sides of Equation 4.41 to obtain

$$\log q_g = \log \frac{0.703kh}{\mu z_{avg}T_f \ln(0.606r_e/r_w)} + n \log (p_e^2 - p_w^2) \qquad (4.42)$$

The right-hand side of the equation is grouped in this way to show that the first term contains factors that can generally be treated as constants. Then we could write the equation as

$$\log q_g = \log C + n \log (p_e^2 - p_w^2) \qquad (4.43)$$

where

$$C = \frac{0.703kh}{\mu z_{avg}T_f \ln(0.606r_e/r_w)} \qquad (4.44)$$

Thus, it can be seen that a plot of the log of q_g versus log of $(p_e^2 - p_w^2)$ will provide a straight line whose slope is n if the terms in the constant C can be treated as constants. During a particular flow test the arithmetic average reservoir pressure, the basis for evaluation of μ and z, probably will not change sufficiently to affect the straight-line characteristics of the plot. However, pressure depletion during the production history may affect the average pressure to such an extent that the viscosity and gas-deviation factor will be changed enough to cause the straight line to shift.

Fig. 4.5 illustrates a deliverability curve. Experienced engineers should note that this plot may be reversed from the plot they normally use or the one required by their state regulatory body. The author prefers to plot the deliverability data in this way so that the slope is directly measurable as n.

Note that if the skin factor, S, and the undamaged permeability had been used in Equation 4.41, the skin factor would become part of the constant, C, and the straight-line quality of the deliverability plot would still be maintained. To illustrate the application of the deliverability curve to practical problems work the following problem and check your solution against the one in Appendix C.

Problem 4.6: Conventional Gas Well Back-pressure Test

The reservoir of Problem 4.4 totals 4,500 acres and is being produced by two wells when back pressure tests are conducted on one of the wells with the following results:

Stabilized BHP, psia	Producing rate, MMscfd
2,800	0
2,670	1.8
2,590	2.7
2,500	3.6
2,425	4.5

Reservoir Characteristics:

Permeability $= 74$ md
Porosity $= 15\%$
Gas viscosity $= 0.021$ cp (based on average pressure)
$c = 3.57 \times 10^{-4}$ (based on static pressure)

A. What must the BHP be in this well to produce at a rate of 5 MMscfd when p_e has declined to 2,000 psia if the constant C of equation 4.43 is not significantly affected?

B. How long must each rate be maintained to reach stabilized conditions during the test?

Throughout the consideration of the practical problems of gas reservoir engineering the engineer will find the continual problem of what pressure base to use for evaluating reservoir parameters that are considered constant mathematically but that actually vary with the pressure. In Problem 4.5 this situation was emphasized in the given data. It was noted that the compressibility was based on the static or volume-weighted average pressure while the viscosity was based on the arithmetically averaged pressure.

To see the logic behind this difference in the pressure bases we must

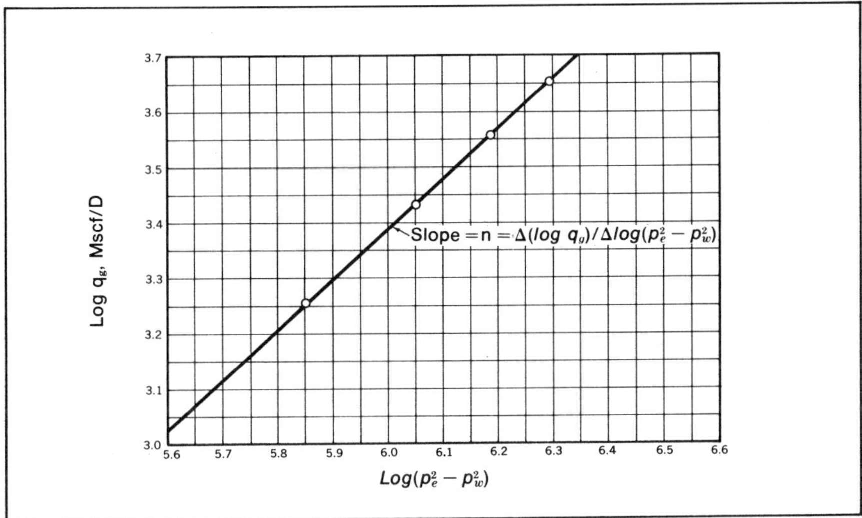

FIG. 4.5 Deliverability curve from stabilized drawdown data.

consider the basic purpose of each parameter. If we go back to the derivation of the radial diffusivity equation we will find the first introduction of the compressibility that eventually ends up in Equation 2.61. In the radial diffusivity derivation, the compressibility was introduced as a material-balance parameter in determining the change in pressure with radius according to material balance. Consequently, the average compressibility for a reservoir should be based on the reservoir static pressure since this is the basis for material balance in the reservoir.

On the other hand, the viscosity average has a different significance. Since pressure in radial flow varies radially, and viscosity is a function of pressure, the viscosity will also vary radially. This means that flow is taking place through a series of radial segments with different flow characteristics due at least in part to the change in viscosity. Thus, if we choose to use one viscosity in an equation representing the total geometry from the external radius to the well radius we must average the viscosities as in Chapter 1 under the section on Approximating Complex Geometries. When this is done we find that, to obtain single parameters to provide the correct total pressure drop in a gas system, we must evaluate the flow parameters at the arithmetic average pressure for either radial or linear flow.

Well spacing effects. Now consider the effect of well spacing on the deliverability curve. When the well spacing is changed, the constant C of Equation 4.43 and 4.44 is changed. However, the slope of the deliverability curve will remain the same. This assumes that during the testing of a well there is no significant change in the product of gas viscosity and gas-deviation factor. Since the slope is not affected by the change in spacing it is then possible to simply shift the deliverability curve vertically to obtain deliverability curves for different spacings. This vertical shift is proportional to the change in the log of the constant C. Note that if only the spacing of the effective drainage radius, r_e, is affected the change in C would be inversely proportional to the reciprocal of the ratio of the $\ln(0.606r_e/r_w)$ values. However, if a change in the state of depletion accompanies the change in spacing we should also account for the change in the viscosity and the gas-deviation factor at the same time. Thus, an expression for the change in the constant, C, that accompanies the change in spacing or depletion can be written.

$$\frac{C_2}{C_1} = \frac{\mu_1 z_{avg1} \ln(0.606r_{e1}/r_w)}{\mu_2 z_{avg2} \ln(0.606r_{e2}/r_w)} \tag{4.45}$$

Using this equation, a test run at one well spacing and state of depletion, (arithmetic average testing pressure), can be used to predict the deliverability at other well spacings and states of depletion. To do this it is not necessary to know all of the reservoir parameters in the expression for C, Equation 4.34. The constant C can most accurately be evaluated by calculating it from the observed deliverability curve. This can be done by first calculating the constant n from the slope and then reading the values for $p_e^2 - p_w^2$ and q_g corresponding to any point on the deliverability curve. When these values are put in Equation 4.43 along with the value of n, the constant C_1 can be calculated. Once C_1 is known, C_2 can be calculated by Equation 4.45. Using C_2 for C in Equation 4.43 and any assumed value for q_g a corresponding value of $p_e^2 - p_w^2$ can be calculated. This will be one point on the new deliverability curve which can be constructed parallel to the original deliverability curve.

Work the following problem and check your solution against the one in Appendix C to see if you understand this correction procedure.

Problem 4.7: Adjustment of a Gas Deliverability Curve for A Change in Well Spacing and/or Depletion

The deliverability curve resulting from Problem No. 4.6 is Fig. 4.5 which was obtained with a drainage area of 2,250 acres and a pressure at the external boundary, p_e, of 2,800 psia. In Problem No. 4.4 we found that the static reservoir pressure would be 1,568 psia at completion of the proposed contract. If additional wells are drilled so that a total of 5 equally spaced wells drain the 4,500-acre reservoir when the contract of Problem No. 4.4 is completed (i.e. $p_s =$ 1,568 psia) what must the bottom-hole pressure be for the well to produce its share, $(1/5)$, of the 20MMcfd required by the contract? Note that Problems 4.4 and 4.6 apply to the same reservoir. Well radius is .25 ft.

The engineer should carefully remember that the above analysis of a conventional back-pressure test is based on the assumption that stabilized or pseudosteady-state flow was obtained at each rate. The fact that the data obtained appears to give a straight-line deliverability curve does not mean that the flow rates have stabilized at each rate. If each flow period is about the same length it can be shown that the deliverability curve does closely approximate a straight line. In fact it is possible for the data to give an exact straight line. This situation will be better understood after the discussion of isochronal testing.

The engineer is urged to make a rough calculation of the stabilization time, t_s, by Equation 2.63 using approximations of the necessary reservoir parameters to ascertain that his producing times are sufficient. The author has seen data from very tight, widely spaced gas wells in the midwest that indicate stabilization times in excess of 1.5 years. While this is not considered typical, it does indicate the magnitude of the gas-well testing problem. When stabilization times are excessive, it is necessary to use some other type of gas-well test to predict behavior. For this purpose the technology of infinite-acting unsteady-state flow is necessary.

The radial diffusivity equation for gas. It may seem illogical to discuss the pseudosteady-state flow of gas before we discuss the more general case of unsteady state. However, the equations for pseudosteady-state flow are much simpler than those for general unsteady-state flow and the application of these equations in conventional gas-well testing is much simpler than the concepts of testing under infinite-acting behavior (isochronal testing). Thus, it appears easier to understand the material when it is presented in this sequence.

As is the case with fluid flow in general, the equations generally used for the flow of gas are very similar to those governing the flow of liquids. The radial diffusivity equation for the flow of a slightly compressible fluid such as a liquid was derived in Chapter 2 as

$$\frac{\Delta p}{\Delta t} = \left(\frac{6.33k}{\phi \mu c}\right)\left[\frac{1}{r}\frac{\Delta p}{\Delta r} + \frac{\Delta(\Delta p/\Delta r)}{\Delta r}\right] \qquad (2.7)$$

In this equation the group of constants, $6.33k/\phi\mu c$, is the diffusivity constant, η.

A similar equation can be derived in much the same manner for gas flow except mass flow rates must be used rather than volumetric flow rates. The derivation of this equation can be found in the Katz et al. *Gas Engineer's Handbook.*[4]

$$\frac{\Delta(p^2)}{\Delta t} = \left(\frac{6.33k}{\phi \mu c}\right)\left[\frac{1}{r}\frac{\Delta(p^2)}{\Delta r} + \frac{\Delta[\Delta(p^2)/\Delta r]}{\Delta r}\right] \qquad (4.46)$$

This gas diffusivity equation for radial flow differs from the equation applicable to gas only to the extent that the pressure, p, in Equation 2.7 is in every case replaced by the pressure squared, p^2. Consequently, a dimensionless solution treatment of the radial gas diffusivity equation will result in the same numerical answers as those obtainable for a dimensionless solution of the liquid radial diffusivity equation. Thus, any dimensionless solutions for the liquid radial diffusivity equation can also be applied to gas providing that they are converted to appropriate gas units. Instead of the general solutions being in terms of pressure they will be in terms of the pressure squared. Thus, the constant-rate solutions will permit a calculation of the change in p^2 and the constant-pressure solutions will be based on a constant change in p^2.

We should strongly emphasize that Equation 4.46 treats the gas compressibility as a constant. The engineer sometimes gets the impression that since the equation is derived using mass units rather than volumetric units, as was the case with Equation 2.7, that the variation in compressibility is accounted for. However, examination of Equation 4.46 shows that the compressibility is still treated as a constant. As a matter of fact either radial diffusivity equation could be applied to the flow of gas with the same resulting solutions. It would simply mean that the conversion equations applied to Equation 2.7 could be based on the volumetric flow rate at the arithmetic average pressure. This would mean that a separate calculation of this volumetric flow rate at the average pressure would have to precede the application of the equation that is based on this volumetric flow rate.

Since the radial diffusivity equation for gas assumes the compressibility of gas a constant, we must of course be careful that the compressibility variation throughout the reservoir is "small." Actually, we will see that, from a practical point of view, this limitation is not severe because most of our practical applications are for infinite-acting reservoirs where it is not difficult to determine the mathematical average compressibility. A little thought will convince us that for any reservoir that is still infinite acting, mathematically the reservoir outer boundary is infinite, and thus the average pressure is the pressure in the unaffected portion of the reservoir which is the initial pressure, p_i. Consequently, the average compressibility, c, will be based on the initial pressure. We will find that the infinite-acting solutions to the diffusivity equation give accurate answers when applied to gas reservoirs and compared with more exact numerical computer solutions. When the reservoir is no longer infinite acting we can treat it as a pseudosteady-state system and obtain adequate accuracy.

When the unsteady-state gas-flow equations of this text are compared with similar gas-flow equations from other sources, the engineer should be aware that the gas compressibility of this text may be replaced with the reciprocal of the pressure in other sources. This substitution was widely made when the gas reservoir engineering technology was developing several years ago because at that time all of the commercially productive reservoirs were low pressure reservoirs and the approximation that the gas compressibility equalled the reciprocal of the pressure was approximately correct. The gas compressibility does equal the reciprocal of the pressure for an ideal gas. However, as we produce higher-pressured gas reservoirs we reach the point where this approximation introduces errors that may approach 40%.

Unsteady-state gas flow—constant-pressure solution. We will first consider the constant-pressure solution to the radial diffusivity equation. The engineer should recall that this is a solution for a well in the center of a circular reservoir whose pressure is initially uniform throughout and in which a constant pressure is maintained from time zero throughout the producing life. The general solution is in the form of cumulative production combined with a group of reservoir constants, Q_{tD}, versus time combined with a different group of reservoir constants, t_D. The relationship between Q_{tD} and t_D is shown in Table 2.2 of Chapter 2. These reduced terms are defined in Chapter 2 as

$$t_{DW} = \frac{\eta t}{r_w^2} \qquad (2.10)$$

$$\eta = \frac{6.33k}{\phi \mu c} \qquad (2.8)$$

$$Q = 1.12 \, \phi h c r_w^2 \, \Delta p_w Q_{tD} \qquad (2.29A)$$

Equation 2.29A is a slightly modified form of Equation 2.29. The subscripts w have been added to make it clear that the radius for Δp and t_D in Equation 2.29A are the same. The cumulative flow, Q, in Equation 2.29A is stated in reservoir barrels. If we are to apply this equation to gas, it would be convenient to state the rate in Mscfd. This can be done by substituting the equivalent of Q stated as a function of Q_{Mscf} which is the cumulative gas flow in thousands of standard cubic feet. An expression for the relationship between reservoir barrels and thousands of scf was derived in Chapter 1 and applied to rates. Modified slightly to make it applicable to cumulative volumes we obtain

$$Q = \frac{5.04 Q_{Mscf} T_f z_{avg}}{p} \qquad (1.18A)$$

From this expression we see that some pressure base must be used to convert from reservoir barrels to Mscf. Since the Q_{tD} functions can be applied with reasonable accuracy only to infinite-acting gas reservoirs and the well pressure in an infinite-acting reservoir varies from the initial pressure, p_i, to the constant well pressure, p_w, we will use the arithmetic average pressure for the volume conversion. This is in line with the previously noted observation that for steady-state gas flow the total pressure drop is proportional to the gas flow rate stated at the arithmetic average pressure. When $(p_i + p_w)/2$ is substituted in Equation 1.18A for p and the resulting expression is substituted for Q in Equation 2.29A we can show that

$$Q_{Mscf} = \frac{0.111 \, \phi h r_w^2 \, c(p_i^2 - p_w^2) Q_{tD}}{z_{avg} T_f} \qquad (4.47)$$

This equation has not been used to any great extent by practicing engineers but it appears to háve considerable potential for use. After we have completed our consideration of the problem of predicting stabilized flow rates we will see that a reasonably accurate prediction of such rates can be made in most circumstances. However, as noted previously, many gas wells require considerable lengths of time before they reach stabilized flow. Equation 4.47 could be used to predict these flow rates prior to stabilization. In applying the equation be sure to note that the equation does not contain any correction for the additional pressure drop caused by well damage or permeability improvement, (negative Δp_{skin}), nor does it account for the additional pressure drop caused by turbulence. Consequently, the actual well pressure should be corrected for Δp_{skin} and turbulence before it is used in Equation 4.47. As noted previously, the compressibility should be evaluated at p_i which is the mathematical average reservoir pressure for an infinite-acting reservoir.

If it is desired to predict the producing rate under constant-pressure conditions prior to stabilization as described, it probably will be necessary to run a pressure buildup or constant-rate drawdown analysis to obtain the necessary reservoir parameters, especially the skin factors, Δp, caused by skin damage, and turbulence. Equation 4.47 provides excellent agreement with numerical computer solutions as long as the reservoir is infinite acting. Whenever the reservoir is no longer infinite acting, it can be assumed that it is in pseudosteady-state flow with the further cumulative production being calculated by these equations or material balance. Thus, the unsteady-state infinite-acting, constant-pressure solution together with pseudosteady-state flow should give an adequate prediction of cumulative gas flow versus time for constant-pressure conditions.

Unsteady-state gas flow—constant-rate solution. The constant-rate solution has found more practical application in gas flow applications than has the constant-pressure solution. As noted previously, the same numerical solutions apply to the constant-rate solution to the radial diffusivity equation as apply to the same case for liquid flow. Instead of using the solution to predict the change in pressure we can use the same pressure functions to predict the change in the square of the pressure, p_2. Consider the general pressure function definition as derived in Chapter 2 and rearranged here

$$(p_i - p_{r,t}) = \frac{0.141 q \mu}{kh}(PF) \qquad (2.14A)$$

In this equation, as was the case with practically all of the equations used in previous chapters, the rate, q, is stated in volumetric units of reservoir barrels per day. When we write this expression with the rate stated as a function of the mass flow rate in Mscfd. q_g, as indicated in Equation 1.18A, we obtain

$$(p_i - p_{r,t}) = \frac{0.141\mu}{kh}\left(\frac{5.04q_gT_fz_{avg}}{p}\right)(PF) \tag{4.48}$$

As was the case in deriving the expression to convert from Q_{tD} to Mscf above, it is necessary to determine some pressure base to use for the pressure, p, in Equation 4.48. Since the pressure will have varied from the initial pressure, p_i, to the pressure $p_{r,t}$ we will use the average pressure, $(p_i + p_{r,t})/2$, for this purpose. When this substitution is made in Equation 4.48 and the expression is solved for the square of the pressure at the radius, r, and time, t, we obtain

$$p_{r,t}^2 = p_i^2 - \frac{1.424q_g\mu z_{avg}T_f}{kh}(PF) \tag{4.49}$$

Equation 4.49 can also be written with the specific pressure functions, p_{tD}, or the Ei function expression substituted for the pressure function, PF, to obtain conversion equations similar to those used for liquid.

$$p_{r,t}^2 = p_i^2 - \frac{1.424q_g\mu z_{avg}T_f}{kh}(p_{tD}) \tag{4.50}$$

$$p_{r,t}^2 = p_i^2 - \frac{1.424q_g\mu z_{avg}T_f}{kh}\left[\left(\frac{1}{2}\right)\left(-Ei\frac{-1}{4t_D}\right)\right] \tag{4.51}$$

These equations have the same limitations as their counterparts for liquid flow. Equation 4.50 can only be used to calculate the pressure at a radius where the flow rate is known—for all practical purposes the well radius—and Equation 4.51 can only be used when the reservoir is infinite acting and the point-source solution is applicable, $\eta t/r_w^2 > 100$. We must also add a practical limit to Equation 4.50. Due principally to the invalid assumption that the fluid compressibility is constant, it has been found that the p_{tD} solution can be used with confidence only when the reservoir is infinite acting. This does not present a serious limitation because a finite-acting reservoir producing at a constant rate is in pseudo-steady-state flow and can be analyzed using these equations.

Equations 4.49, 4.50, and 4.51 covering the constant-rate case application to gas flow, do not include any provision for turbulence or well damage. These additional pressure drops are normally handled in a variety of ways depending on the approach that is most useful for a particular application. The skin factor, S, can always be added to the pressure function without changing most analysis procedures. However, many engineers prefer to include the additional pressure drop due to well damage in the turbulence effects as will be seen later. It is sometimes mechanically handy to treat the well damage in this manner but it is easy to see the inexact nature of such a mathematical treatment when we realize that the $\Delta(p^2)$ caused by turbulence is proportional to the square of the flow rate, q_g^2, while the $\Delta(p^2)$ caused by the well damage is proportional to the rate, q_g.

If we account for the well damage by adding the skin factor, S, to the pressure function, PF, and the term Bq_g^2 to account for the additional pressure drop due to turbulence, Equation 4.49 would become

$$p_{r,t}^2 = p_i^2 - \frac{1.424q_g\mu z_{avg}T_f}{kh}(PF + S) - Bq_g^2 \qquad (4.52)$$

Equations 4.50 and 4.51 could also be corrected in this way. The constant B in Equation 4.52 is simply a proportionality constant that must be evaluated by testing the well. Treating the additional $\Delta(p^2)$ due to turbulence as a steady-state effect in an unsteady-state equation introduces no significant errors because practically all of the turbulence occurs near the well bore where gas velocities are extremely high due to the very small cross-sectional area. Consequently, the turbulence pressure-drop term can be treated as a steady-state effect in much the same way that the pressure drop due to the well damage is treated as a steady-state effect.

Examination of Equation 4.36 which describes the steady-state flow of gas with turbulence included will also indicate the validity of assuming all of the additional pressure drop due to turbulence occurs at the well. The second term of Equation 4.36 contains the difference between the reciprocals of the well radius and external radius. Since the well radius is a fraction of a foot the limitation or affect of the reciprocal of the external radius on the accuracy of the equation is negligible. For example if the well radius is $\frac{1}{3}$ ft. (reciprocal 3.0) the reservoir outside a radius of 10 ft. (reciprocal 0.1) will affect the additional pressure drop due to turbulence about 3%.

Isochronal testing.   As noted previously, it is often difficult or impossible to test a gas well in the conventional back-pressure test manner to obtain usable data for predicting gas-well deliverability. When the stabilization time is excessive it is still possible to test the well under infinite-acting conditions and interpret the data to predict stabilized deliverability. This procedure is generally carried out in a test known as an isochronal test.

Ideally the procedure is started with the drainage area at static pressure throughout. The well is then produced for some short period of time, t^*, and the bottom-hole flowing pressure is noted. The well is shut-in until the static pressure is reached after which it is produced at a different rate for the same time period, t^*, and the resulting well pressure at this rate is noted. The well is shut-in until the static pressure is reached and is then produced at a third rate for a time t^* with the resulting well pressure noted. This procedure can be repeated to obtain well pressures resulting from flow at a variety of rates for a period of time, t^*.

When a log-log plot is prepared of q_g versus $p_e^2 - p_w^2$, a straight line is obtained. This plot will be shown to be the same as the stabilized deliverability curve of a particular drainage radius. The subject drainage radius is a function of the producing period, t^*, and the reservoir parameters. As shown for the conventional back-pressure test, once a stabilized deliverability curve has been obtained for one drainage radius it can be corrected to a stabilized deliverability curve for any other drainage radius desired. This testing procedure is called isochronal (equal-time) testing.

In effect, in isochronal testing we test the well under infinite-acting conditions and interpret the data for an equivalent stabilized system. To do this we determine the size reservoir that would give the same pressure drop term $(p_e^2 - p_w^2)^n$ as is experienced by an infinite-acting reservoir that has produced for a time t^*, $(p_i^2 - p_w^2)^n$. To obtain this relationship we will first solve Equation 4.41 for the pseudosteady-state pressure-drop term and add the skin factor, S.

$$(p_e^2 - p_w^2)^n = \frac{1.424 q_g \mu z_{avg} T_f [\ln(0.606 r_e/r_w) + S]}{kh} \tag{4.53}$$

Next consider Equation 4.50 written for the well pressure in an infinite-acting reservoir, solved for the pressure-drop term, and with the "turbulence" factor, n, and skin factor, S, added. Since the reservoir is infinite acting we can use the log equation for the pressure function, p_{tD}.

$$(p_i^2 - p_w^2)^n = \frac{1.424 q_g \mu z_{avg} T_f \left(\frac{1}{2}\right)(\ln t_D^* + 0.809 + 2S)}{kh} \tag{4.54}$$

The dimensionless time t_D^* is based on the isochronal producing time, t^*.

When we equate the right-hand sides of Equations 4.53 and 4.54 we obtain a relationship between the infinite-acting producing time and the external radius, r_e, of a stabilized or pseudosteady-state system that will give the same pressure drop term.

$$\frac{1}{2}(\ln t_D^* + 0.809) = \ln(0.606 r_e/r_w) \tag{4.55}$$

or

$$r_e = (38.5 k t^*/\phi \mu c)^{0.5} \tag{4.56}$$

This equation simply says that an infinite-acting well producing for a time t^* would have the same pressure-drop term as a similar well producing under pseudosteady-state conditions with an external drainage radius, r_e, as indicated in Equation, 4.56.

Note carefully that this is not a stabilization-time equation. Many engineers confuse Equation 4.56 with the stabilization-time equation since they are exactly the same except for the numerical constant. When we solve Equation 4.56 for the time we obtain

$$t^* = \frac{0.026 \; \phi\mu cr_e^2}{k} \tag{4.57}$$

Now compare the numerical constant of this equation, 0.026, with the numerical constant in the stabilization-time equation, 0.04. What this tells us is simply that the pressure-drop term for an infinite-acting reservoir is equal to the pressure-drop term for a pseudosteady-state system whose external boundary, r_e, satisfies Equation 4.56 before a system with the same radius, r_e, reaches pseudosteady-state conditions. This then means that the pressure at the external boundary, r_e, of a drainage system will have dropped prior to the time the system reaches pseudosteady state.

This is reasonable from a physical standpoint if we recognize that after the pressure at the outer boundary of a reservoir has been affected by production the pressure at the well continues to be infinite acting until the affect of the boundary travels back to the well through the reservoir. During this interim between the time when the pressure at the outer boundary is first affected and the time when the pressure at the well feels the effect of the outer boundary, the pressure at the outer boundary continues to decline. Consequently, it is perfectly logical for the pressure drop in a reservoir to be equal to the pressure drop that will ultimately be reached under pseudosteady state before the reservoir system actually enters pseudosteady state.

As indicated above, once a stabilized r_e is calculated by Equation 4.56 to represent the isochronal data, the data can be corrected to any desired reservoir drainage by using Equation 4.45 to shift the performance curve. Note that changing the drainage radius, r_e, simply shifts the performance curve, without changing the slope. Thus, if we establish the slope by an isochronal test we then need only one stabilized point to establish a new performance curve since the new curve can simply be drawn parallel to the previously established performance curve. Since any producing gas well must ultimately reach pseudosteady state, this stabilized point should always be used as the basis for positioning the stabilized performance curve for an existing well spacing. Equation 4.45 is still useful for predicting performance under various well spacings or states of depletion but its use in predicting the performance for a presently existing spacing should be limited to the time period necessary to obtain one stabilized point. This may be a substantial period when low permeability, low porosity, and wide spacing exist.

Work the following problem that demonstrates the use of isochronal data in predicting the stabilized deliverability of a gas reservoir, and compare your solution with the one in Appendix C.

Problem No. 4.8: Use of Isochronal Data

Isochronal back-pressure test data are obtained in the reservoir of Problem No. 4.4, with two wells producing. The tests were run by producing at 1,800 Mscfd for 60 min, shutting the well in until the pressure is again 2,800, producing at 2,700 Mscfd for 60 min, shutting the well in until the pressure is again 2,800, etc. Results were as follows:

q_g, Mscfd	BHP @ q_g after 60 min of production, psia
0	2,800
1,800	2,710
2,700	2,659
3,600	2,605
4,500	2,545

K = 74 md, c = 0.000357/psi, μ_g = 0.021 cp, ϕ = 0.15

A. What is the equivalent stabilized drainage radius for the isochronal data?
B. Calculate and plot the 1-hr isochronal curve and the stabilized back-pressure curve for a drainage area of 2,250 acres.
C. The well continued to produce at 4,500 Mscfd and reached a "constant" BHP of 2,425 psia. Plot the stabilized back-pressure curve based on this point.

One of the difficulties that arises in conducting an isochronal test is the problem of shutting the well in until it reaches static pressure before continuing the flow test at a new rate. Theoretically it might appear that this presents an insurmountable problem since we know that it would take a very long shut-in time before the well would approach the static pressure. Furthermore, this problem is worse when we have the greatest need for isochronal tests. To explain, remember that we need isochronal tests when the stabilization time is such that it becomes impractical to test under stabilized conditions. The same identical conditions that make the stabilization time too high to tolerate will have a similar effect on the shut-in time necessary to have the well pressure approach the static pressure. Consequently, we badly need some means of circumventing the necessity of shutting the well in until the pressure approaches the static pressure.

By using the negative superposition principle to obtain the $\Delta(p^2)$ value caused by a particular producing rate we can greatly shorten the necessary shut-in time. Fig. 4.6 illustrates the technique that can be used. It shows the pressure change that has resulted from flowing a well at a rate of 4 MMscfd for 1 hr and the resulting pressure buildup that results from the shut-in. Ideally we would leave the well shut-in until the pressure again reaches the initial pressure, 2,000 psia. The solid extrapolation indicates the shut-in pressure that would have been obtained if the new producing rate, 3 MMscfd, had not been initiated after 1 hr of shut-in. The dotted line shows the pressure drawdown that is obtained due to the new 3-MMscfd producing rate. Consequently, the

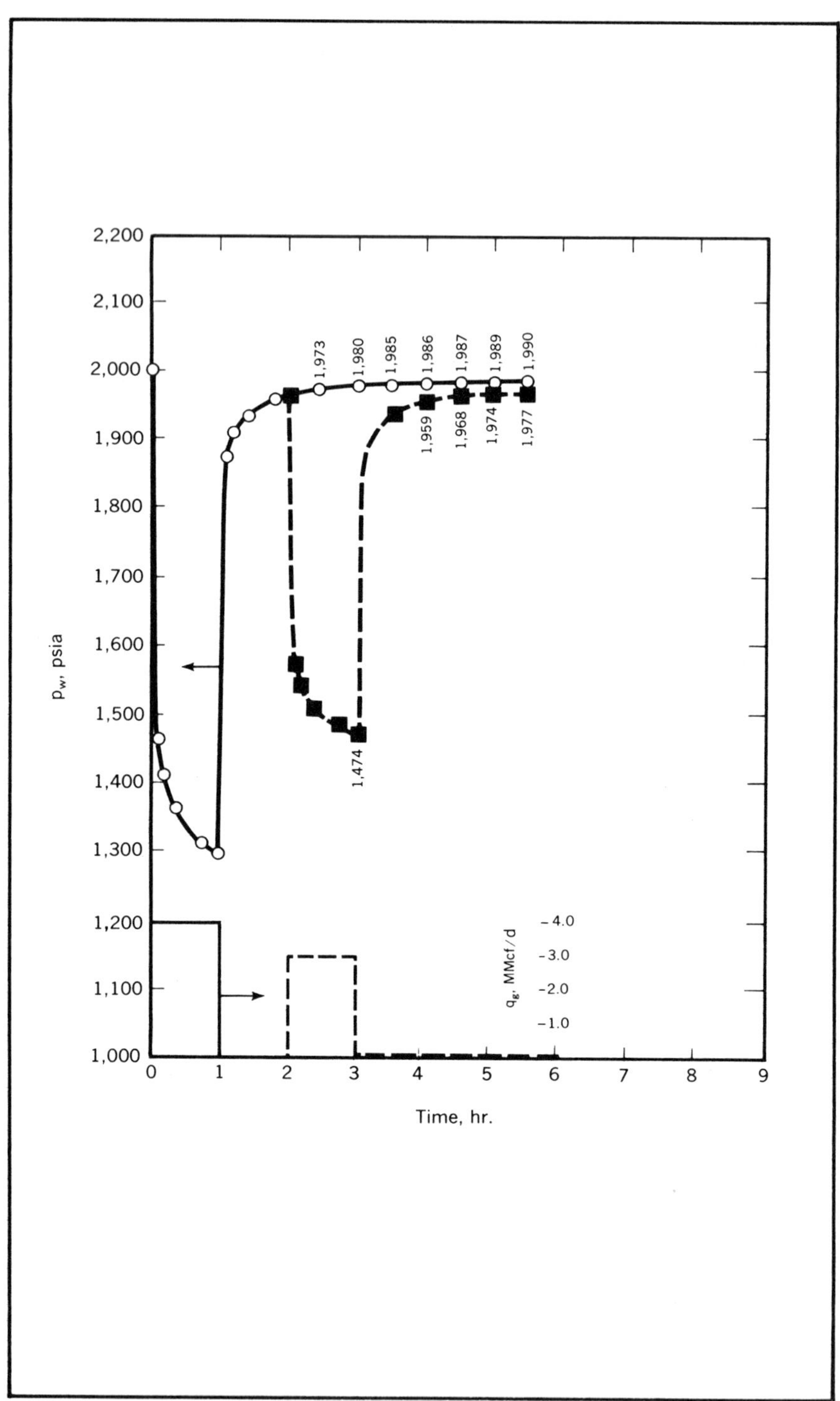

FIG. 4.6 Example of slow build-up to average pressure.[8] (After Slider, SPE paper.)

difference between the pressure that would have been obtained at the end of 3 hr if the 3-MMscfd rate had not been initiated, (1,980 psia), and the actual pressure obtained due to the 3-MMscfd rate, 1,474 psia, is the affect of the new rate alone.

Once the engineer understands this situation he can simply leave the well shut-in until he can predict with confidence what the pressure would be during the next flow period if the new flow rate was not initiated. Then the difference between the extrapolated pressure and the actual pressure at the end of the next flow period will be the desired isochronal pressure drop.

It must be noted at this point that superposition can only be applied to the portion of the $\Delta(p^2)$ that is caused by viscous flow and the well damage. In other words we can only use superposition to predict the $\Delta(p^2)$ value that excludes the additional pressure drop due to turbulence. This situation occurs because the additional $\Delta(p^2)$ caused by turbulence is proportional to the rate squared, q_g^2. If we write Equation 4.49 in a finite-difference form with the skin factor, S, added, we obtain the expression for the $\Delta(p^2)$ due to viscous flow alone.

$$\Delta(p^2)_{\text{VISC}} = \left[\frac{1.424\mu T_f z_{\text{avg}}(PF + S)}{kh}\right]\Delta q_g \qquad (4.58)$$

When this equation is applied by superposition to obtain the total $\Delta(p^2)$ caused by a series of rate changes we would obtain

$$\Delta(p^2)_{\text{VISC}} = \frac{1.424\mu T_f z_{\text{avg}}}{kh}\sum_{j=1}^{j=n}(\Delta q_{gj})(PF_j + S) \qquad (4.59)$$

Once this pressure-drop term due to viscous flow has been calculated the total pressure-drop term can be determined by simply adding the $\Delta(p^2)$ due to turbulence at the final rate.

$$\Delta(p^2) = \Delta(p^2)_{\text{VISC}} + Bq_{gn}^2 \qquad (4.60)$$

Consequently, the p_w^2 value that would be obtained by extrapolating the shut-in well pressure prior to initiation of a new rate will not include any turbulence affect since the rate, q_g, is zero. Thus, when the recorded pressure squared at the new rate is subtracted from the extrapolated value, the $\Delta(p^2)$ difference will represent the pressure-drop term due to viscous flow and the turbulence at that particular rate.

For example, in the case of Fig. 4.6, the Δp^2 due to producing for 1 hr at 4,000 Mscfd and being shut-in for 2 hr would be

$$p_e^2 - (p')^2 = \frac{1.424\mu T_f z_{\text{avg}}}{kh}(4,000)(PF_{3\,hr} + S)$$

$$- \frac{1.424\mu T_f z_{\text{avg}}}{kh}(4,000)(PF_{2\,hr} + S) + 0.0$$

where the 0.0 is due to the fact that there is no $\Delta(p^2)$ due to turbulence since the last rate is 0.0. Applying Equations 4.59 and 4.60 to determine $\Delta(p^2)$ due to producing at 4,000 Mscfd for 1 hr. being shut in for 1 hr. and producing at 3,000 Mscfd for 1 hr would give

$$p_e^2 - p^2 = \frac{1.424\mu T_f z_{avg}}{kh}(4,000)(PF_{3\,hr} + S)$$

$$-\frac{1.424\mu T_f z_{avg}}{kh}(4,000)(PF_{2\,hr} + S)$$

$$+\frac{1.424\mu T_f z_{avg}}{kh}(3,000)(PF_{1\,hr} + S)$$

$$+ B(3,000)^2$$

When these two expressions are subtracted we obtain

$$(p_w{}')^2 - p^2 = \frac{1.424\mu T_f z_{avg}}{kh}(3,000)(PF_1\,hr + S)$$

$$+ B(3,000)^2$$

This will be recognized as the expression describing the $\Delta(p^2)$ caused by producing at 3,000 Mscfd for 1 hr.

Determining isochronal data from continuous-flow data. State regulatory bodies recommend or require flow-after-flow type of flow tests without intervening shut-in. Consequently, most gas producers have many such flow-test records in their files that do not represent stabilized data. Thus, it seems wise to cover methods of interpreting such data.

Flow-after-flow test data that did not stabilize can be interpreted in terms of stabilized performance data in most cases. This can be done by recognizing the theoretical relationship between the actual pressure drops measured in flow-after-flow tests and the desired pressure drops. The pressure drop term, Δp^2, caused by a sequence of rates in an infinite-acting reservoir can be obtained by combining Equations 4.59 and 4.60 with the log equation substituted for the pressure function, PF.

$$(p_i^2 - p_w^2)_{act} - Bq_{gn}{}^2 = \frac{1.424\mu T_f z_{avg}}{kh} \sum_{j=1}^{j=n} (\Delta q_{gj})[0.5(\ln t_{Dj} + 0.809) + S] \quad (4.61)$$

However, the pressure term which we desire is the pressure drop caused by the last rate acting for the isochronal time, t^*.

$$(p_i^2 - p_w^2)_{qgn} - Bq_{gn}{}^2 = \frac{1.424\mu T_f z_{avg}}{kh} q_{gn}[0.5(\ln t_D{}^* + 0.809) + S] \quad (4.62)$$

Now dividing Equation 4.62 by Equation 4.61 we would obtain

$$\frac{(p_i^2 - p_w^2)_{qgn} - Bq_{gn}^2}{(p_i^2 - p_w^2)_{act} - Bq_{gn}^2} = \frac{q_{gn}[0.5(\ln t_D^* + 0.809) + S]}{\sum\limits_{j=1}^{j=n}(\Delta q_{gj})[0.5(\ln t_{Dj} + 0.809) + S]} \qquad (4.63)$$

To put the effect of the well damage as measured by the skin factor, S, in its proper perspective, we will rewrite Equation 4.63 with the pressure drop due to damage written as a separate term.

$$\frac{(p_i^2 - p_w^2)_{qgn} - Bq_{gn}^2}{(p_i^2 - p_w^2)_{act} - Bq_{gn}^2} = \frac{q_{gn}[0.5(\ln t_D^* + 0.809)] + \Delta p_{skin}^2}{\sum\limits_{j=1}^{j=n}(\Delta q_{gj})[0.5(\ln t_{Dj} + 0.809)] + \Delta p_{skin}^2} \qquad (4.64)$$

Now note from this equation that if the two pressure-drop terms in the left-hand side of the equation are nearly the same, the turbulence terms, Bq_{gn}^2, which is the same in both the numerator and denominator, will have little effect on the ratio. Similarly the pressure drop due to the skin damage will be the same in the numerator and denominator of the right-hand side and their effect on that ratio will be negligible. This conclusion is also based on the assumption that the pressure drop due to turbulence and the skin will be small relative to the total pressure drop. Consequently, if these assumptions are met the equation simplifies to

$$\frac{(p_i^2 - p_w^2)_{qgn}}{(p_i^2 - p_w^2)_{act}} \approx \frac{q_{gn}[(\ln t_D^* + 0.809)]}{\sum\limits_{j=1}^{j=n}(\Delta q_{gj})[(\ln t_{Dj} + 0.809)]} \qquad (4.65)$$

This equation provides a convenient and useful technique for correcting conventional flow-after-flow test data (that did not stabilize) into isochronal data. By using the observed pressure, rates, and times, and assuming reservoir parameters necessary for the evaluation of the dimensionless times, it is possible to arrive at much more meaningful data than that represented by the original test data. Since the assumed reservoir parameters necessary for the evaluation of the dimensionless times appear in the numerator and denominator as logs, the accuracy of these parameters does not greatly affect the results of the calculations. Furthermore, no difficulty is generally encountered in meeting the assumptions that the ratio of the observed and desired pressure-drop terms is near 1.0 and the additional pressure drops due to turbulence and the skin effect are small compared to the total pressure-drop terms.

If enough pressure data are available for the first pressure drawdown (the pressure decline caused by the initial producing rate), the assumptions inherent in Equation 4.65 can be avoided. This is accomplished by evaluating the group of constants in the right-hand side of Equations 4.61 and 4.62, $\mu T_f z_{avg}/kh$, from the slope of the drawdown plot of p_w^2 versus the log of producing time. The theoretical basis for this can be seen when we rearrange Equation 4.62 and write it for a general producing time rather than for t^*.

$$p_w^2 = p_i^2 - \frac{1.424\mu T_f z_{avg}(0.5)(2.3)}{kh}\,[\log t + \log(\eta/r_w^2) + 0.809 + (S/1.15)]$$

$$- Bq_g^2 \tag{4.66}$$

Note that only the well pressure and the log t multiplied by the group of constants varies with time. Since production is at a constant rate the additional pressure drops caused by the skin and turbulence remain constant. Consequently, we can write this equation with all the constant terms grouped together.

$$p_w^2 = \text{constant} - (1.638\mu T_f z_{avg} q_g/kh)(\log t) \tag{4.67}$$

This then shows that the semilog plot of the well pressure squared versus the producing time gives a slope which is

$$m = (1.638\mu T_f z_{avg} q_g/kh) \tag{4.68}$$

Once this group of terms has been evaluated, Equations 4.61 and 4.62 can be subtracted and rearranged to obtain

$$(p_i^2 - p_w^2)_{qgn} = (p_i^2 - p_w^2)_{act} - (1.428/1.638)(m'/q_g')\sum_{j=1}^{j=n}[(\Delta q_{gj}\log t_j)$$

$$- q_{gn}\log t^*] \tag{4.69}$$

Thus, by using the initial drawdown data to evaluate m, Equation 4.69 can be used to find the isochronal data without having to rely upon the assumptions of Equation 4.65.

Work the following problem and check your solution against the Appendix solution to test your understanding of the methods presented for correcting flow-after-flow type data to isochronal data.

Problem No. 4.9: Determining Approximate Isochronal Data from Conventional Drawdown Data

A well in the reservoir of Problem No. 4.8 is tested by flowing it at a rate of 1,800 Mscfd for 60 min. flowing at 2,700 Mscfd for 60 min. (no intervening shut-in), flowing at 3,600 Mscfd for 60 min., and flowing at 4,500 Mscfd for 60 min. This is an old well test with no additional pressure data available.

Total testing time, min	Rate during preceding 60 min, Mscfd	BHP, psia
0	0	2,800
60	1,800	2,710
120	2,700	2,653
180	3,600	2,596
240	4,500	2,532

Determine the approximate 1-hr isochronal test data and prepare a log-log plot of these data.

Turbulence and skin evaluations from continuous flow data. A major drawback to the above methods of correcting flow-after-flow data to isochronal data is that the testing procedure need not be related in any way to the isochronal data being calculated. Study of Equation 4.65 indicates this clearly. Note that from the basic flow-test data any isochronal rate and time data could be used in the equation and the pressure changes calculated. The final total flow rate for each measured pressure must be the same as the isochronal flow rate calculated for the equations to be valid. Otherwise the additional pressure drop due to turbulence would not be the same and cancel out. However, turbulence terms could be added to the equation to account for this difference. The point is that there is little relationship between the pressures measured during the flow-after-flow test and the calculated isochronal performance. When these calculations are studied in detail the engineer will see that the most direct measurement made in a flow-after-flow test is the effect of the last rate change and not the effect of the total rate that prevails at a particular time. By negative superposition we can directly measure the affect on the well pressure of a particular rate change. That is, by extrapolating the well pressure decline established prior to a particular rate change, and subtracting the well pressure resulting after the well rate change, the effect of the rate change on the well pressure at that particular rate change can be determined by difference.

Consequently, if turbulence was negligible we could run a test at a variety of rates chosen in such a sequence that they provide a variety of rate changes such as those in Fig. 4.7 and the resulting pressure data

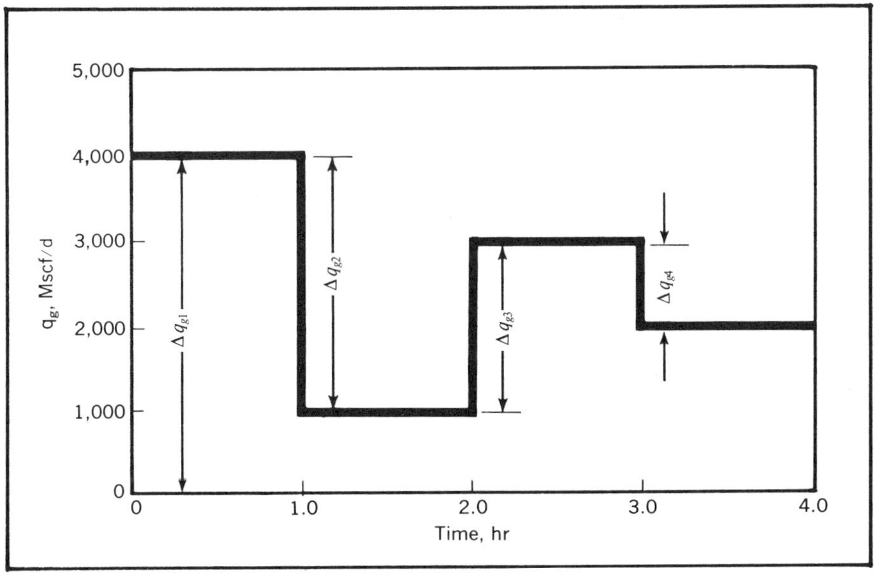

FIG. 4.7 Rate history for pressure history of Fig. 4.8.[8] (After Slider, SPE paper.)

could be interpreted with pressures extrapolated as in Fig. 4.8. The resulting differences between the extrapolated and measured pressure squared when plotted against the change in the rate would provide an isochronal data plot as indicated in Fig. 4.9. However, carefully note that this is true only if the turbulence effect is negligible. When the turbulence effect is not negligible the resulting data can be used to evaluate the additional pressure drop due to turbulence and the data can be corrected to a $\Delta(p^2)_{visc}$ plot. This plot can then be used to predict the well pressure at any production rate.

When turbulence is significant, a rate change at one rate level will not result in the same $\Delta(p_w^2)$ that would be caused by the same rate change at some other rate level. Study of Equation 4.61 will indicate that the change in the square of the well pressure due to a single rate change is

$$\Delta(p_w^2) = \frac{1.424\mu T_f z_{avg}}{kh} \Delta q_g [0.5(\ln t_D + 0.809) + S] + B \Delta(q_g^2) \quad (4.70)$$

The first term is proportional to the rate change and would be the same at any rate level but the turbulence term is proportional to the change in the square of the rate and thus varies with the magnitude of the rate at which the rate change is initiated. If the turbulence effect is actually negligible then the ratio of the change in the well pressure squared and the rate change, $\Delta(p_w^2)/\Delta q_g$, will be constant. Furthermore, we will see that when this ratio varies, a plot of the ratio versus the sum of the before-and-after rates gives a straight line whose slope is the turbulence constant, B.

To show this, first note that a rate change from q_{g1} to q_{g2} is related to the change in the squares as

$$q_{g1}^2 - q_{g2}^2 = (q_{g1} - q_{g2})(q_{g1} + q_{g2}) \quad (4.71)$$

or

$$\Delta(q_{g2}) = (\Delta q_g)(q_{g1} + q_{g2}) \quad (4.72)$$

By substituting the right-hand side of this equation into Equation 4.70 and dividing through by the change in the rate, Δq_g, we obtain

$$\frac{\Delta(p_w^2)}{\Delta q_g} = \frac{1.424\mu T_f z_{avg}}{kh} [0.5(\ln t_D + 0.809) + S] B(q_{g1} + q_{g2}) \quad (4.73)$$

Now if the time for each flow period is the same, the first term on the right-hand side will be the same for all rate changes and a plot of the ratio of the left-hand side versus the sum of the rates before and after the rate change, $q_{g1} + q_{g2}$, will result in a straight line whose slope is the turbulence constant, B.

Note that a sequence of high and low rates such as those of Fig. 4.7 does not lend itself to an accurate evaluation of the turbulence con-

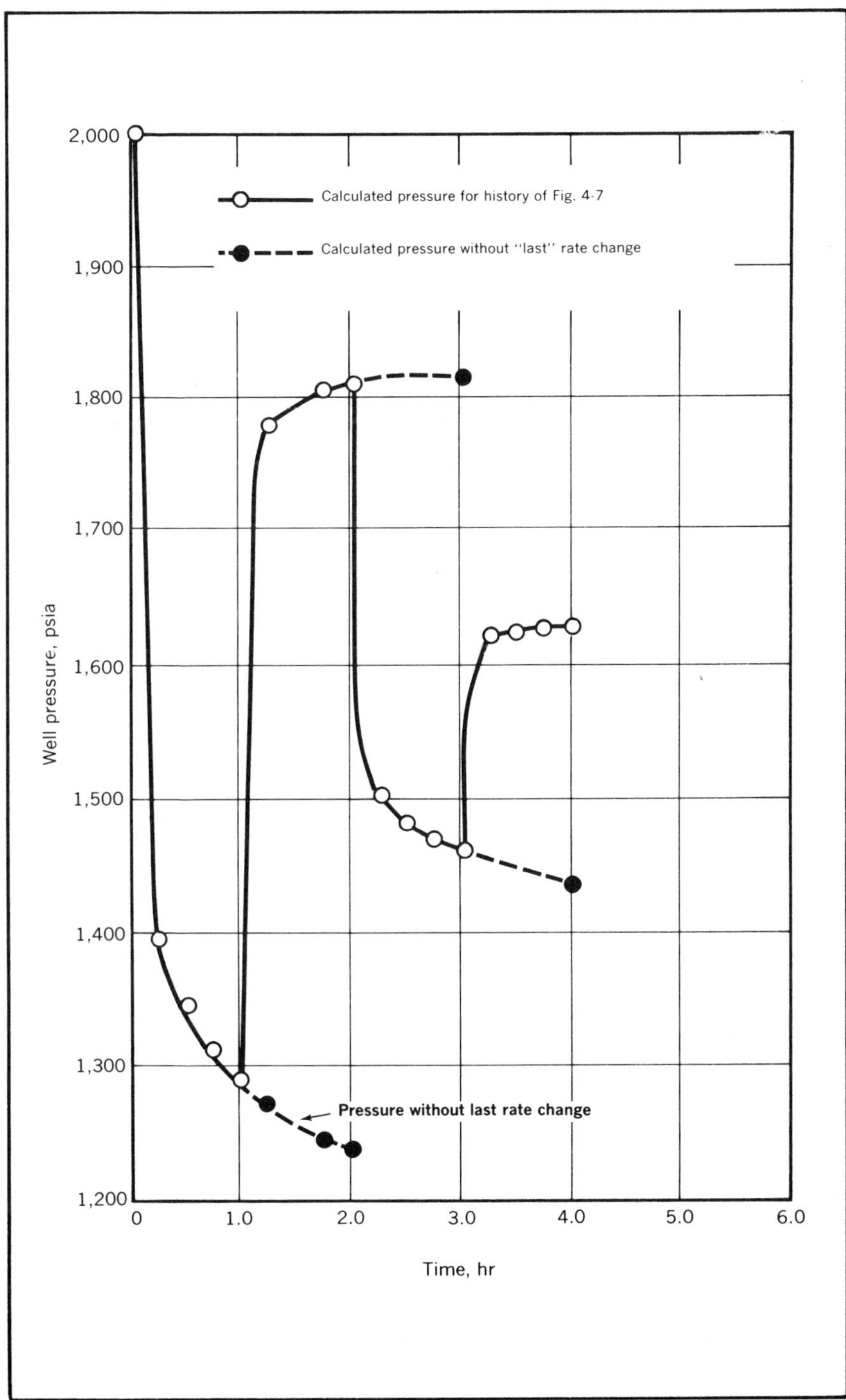

FIG. 4.8 Pressure history for rate history of Fig. 4.9.[8] (After Slider, SPE paper.)

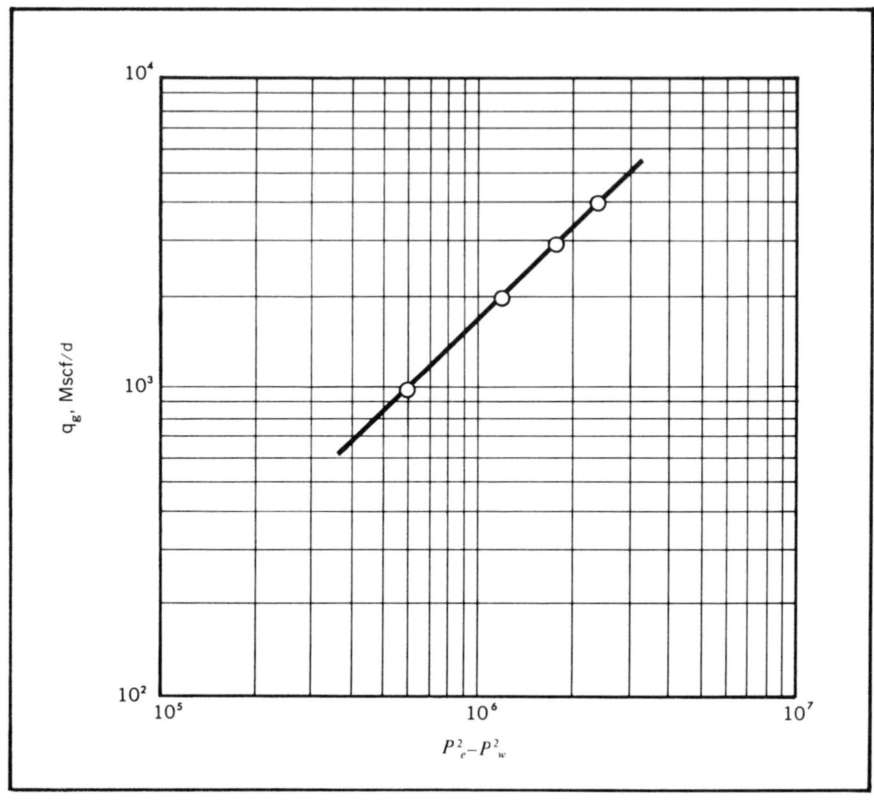

FIG. 4.9 One hour isochronal data from negative superposition interpretation of data of Fig. 4.8.[8] (After Slider, SPE paper.)

stant because the sum of the before-and-after rates varies very little when high and low rates are used in sequence. (e.g. $q_{g1} + q_{g2}$ for the four flow periods of Fig. 4.7 is 4, 5, 4, 5.) Consequently, it is recommended that the sequence of rates be designed so that a small rate change is attempted at low and high rates. To accomplish this and still obtain a wide variety of rate changes we might use a sequence of rates such as that exhibited in Fig. 4.10. With approximately the same rate change of 1.0, used at a high (5 to 4) and low (0 to 1) level, the high and low test values could be substituted into Equation 4.73 and the simultaneous equations solved to obtain a direct calculation of the turbulence constant, B.

$$B = \frac{[\Delta(p_w^2)/\Delta q_g]_{\text{HIGH}} - [\Delta(p_w^2)/\Delta q_g]_{\text{LOW}}}{(q_{g1} + q_{g2})_{\text{HIGH}} - (q_{g1} + q_{g2})_{\text{LOW}}} \tag{4.74}$$

If the effect of the turbulence is significant, it is then necessary to correct each observed $\Delta(p^2)$ value to a $\Delta(p^2)_{\text{visc}}$ value using Equation 4.60

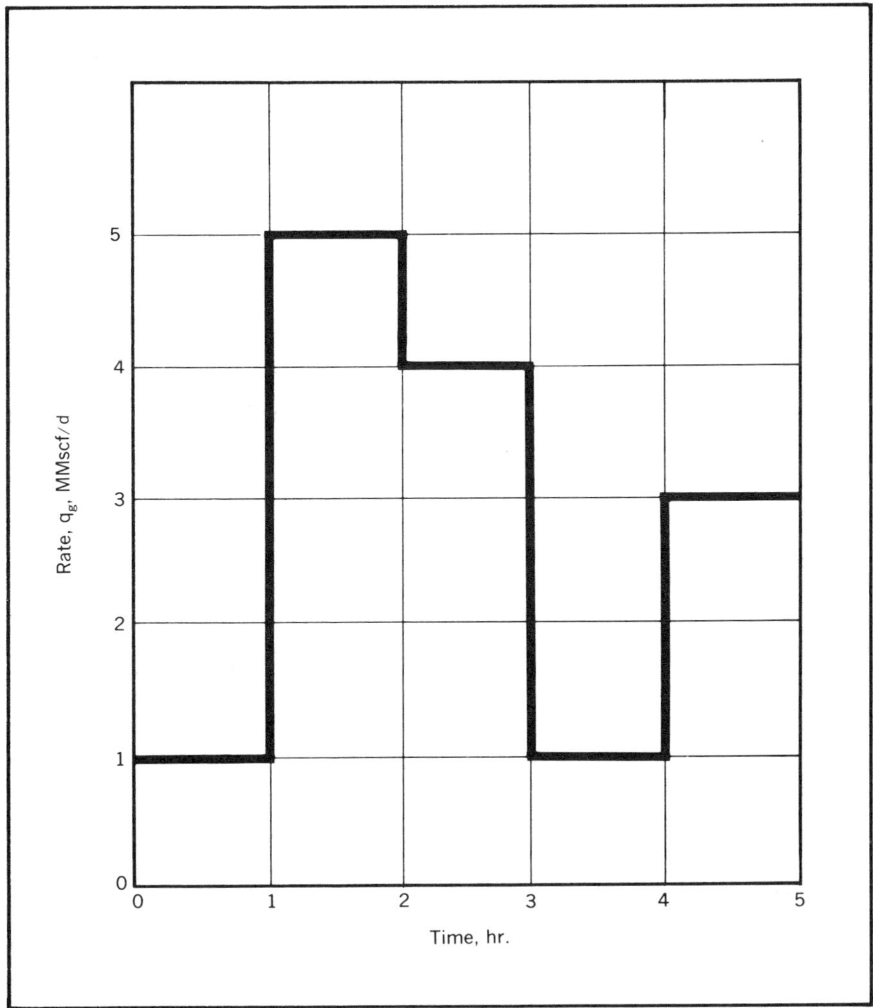

FIG. 4.10   Rate sequence for turbulence evaluation.

and the evaluated constant, B. The engineer will often find that the turbulence effect is negligible and thus he can use the simplified analysis of flow-after-flow type of deliverability tests.[5] However, when turbulence effects are significant, the approach outlined can be used without much additional difficulty to predict the deliverability at any flow rate.

Once the turbulence constant, B, has been evaluated, the skin factor, S, and the corresponding pressure drop due to the well damage can be evaluated. Probably the simplest way to obtain the skin factor is to apply Equations 4.66, and 4.68 to the initial drawdown data. A plot of the well pressure squared versus the log of the producing time at the initial producing rate will give a straight line whose slope will permit the evaluation of the group of constants in Equation 4.68. Substitution of these values, the turbulence constant, B, and corresponding values of the well pressure and producing time into Equation 4.66 will permit the evaluation of the skin factor, S.

Also note that the plot of the well pressure squared versus the log of the original producing time is generally necessary to obtain an accurate extrapolation of the pressure behavior during the first flow period for use in the evaluation of flow-after-flow data by the negative superposition method. During subsequent flow periods the change in pressure with time is small enough to permit a reasonably accurate extrapolation but this generally is not the case for the initial flow period.

The engineer should work the following problem in which turbulence is significant and compare his solution with the solution in Appendix C to make certain he understands the handling of turbulence in flow-after-flow data.

Problem No. 4.10. Determining Isochronal Data from Continuous Flow Test Data by Negative Superposition

A well in a reservoir has been shut-in for a long enough period of time so that the well pressure is "unchanging." The well is then subjected to a sequence of flow rates designed to permit an easy analysis of the data without requiring shut-in periods between rates. The following rate sequence was used in testing the well:

Time period, hr	q_g, Mscfd
0–1	1,000
1–2	5,000
2–3	4,000
3–4	1,000
4–5	3,000

The resulting well pressures were:

Time, hr	p_w, psia
0.0	2,000
0.1	1,905
0.2	1,900
0.3	1,897
0.4	1,895
0.6	1,892
0.8	1,890
0.9	1,889
1.0	1,888
1.4	1,178

Time, hr	p_w, psia
1.6	1,157
1.8	1.142
1.9	1,135
2.0	1,130
2.4	1,362
2.6	1,360
2.8	1,357
2.9	1,356
3.0	1,354
3.4	1,835
3.6	1,842
3.8	1,847
3.9	1,849
4.0	1,850
5.0	1,858

$\phi = 20\%$; $T = 700° R.$; $h = 5$ ft; $z = 0.96$; $c = 0.000357$/psi

Construct a stabilized deliverability curve for this well if its drainage area is 640 acres and r_w is 2.25 in.

Pressure buildup in gas wells. It is the opinion of the author that the general objectives of pressure buildup in a gas reservoir can be achieved by other methods more readily and more accurately. The deliverability of a gas reservoir can be predicted in a much more direct fashion through flow testing as previously described. The evaluation of the undamaged well capacity, kh, and the well damage as measured with the skin factor, S, can be determined with much less difficulty by using a drawdown test. Nevertheless, many engineers believe that the most direct method of predicting reservoir behavior is through the use of pressure buildup analysis. Consequently, those analysis methods most useful for gas wells will be described for those readers who subscribe to this opinion.

Much of the reservoir engineering community's enchantment with pressure buildup analysis as a means of predicting performance in a gas reservoir is probably due to disenchantment with drawdown tests. Since this disenchantment is believed to be due principally to a lack of understanding of drawdown data interpretation, it is believed that the reader who has carefully used the preceding material in this text will be convinced that pressure buildup testing in a gas reservoir is an unnecessary evil. Nevertheless, the author believes that the methods that follow will best serve the purposes of those engineers who remain convinced that pressure buildup analysis in gas reservoirs is desirable.

Essentially the same equations that were discussed under pressure buildup in Chapter 3 can be used for pressure buildup in gas reservoirs. In addition to the problems of turbulence, the gas-deviation factor, and the wide variation in compressibility that we have learned to live with previously in this chapter, pressure buildup in gas wells presents some additional problems that are more or less unique to this particular part of our reservoir-engineering technology. We will see that a constant reservoir rate prior to shut-in is practically nonexistent for a gas well and the same applies to constant rate drawdown. Also the after-flow effects often make pressure buildup data unusable for extremely long periods of time following shut-in. One plus factor for buildup in gas reservoirs will also be noted. Due to the same high compressibility that causes many of our problems with gas flow analysis, reservoir disturbances will be infinite acting for much longer periods of time than will similar disturbances in liquid-filled reservoirs. This means that the infinite-acting constant-rate solutions to the radial diffusivity equation that are so useful in transient-pressure analysis will be effective for much longer periods of time when we are working with a gas reservoir. The time during which a disturbance in a gas reservoir will be infinite acting can be best investigated quantitatively by examining the stabilization time equation which was derived in Chapter 2.

$$t_s = 0.04 \ \phi\mu c r_e^2/k \tag{2.63}$$

The compressibility may be a factor of 100 greater than that of an oil reservoir. In addition, gas wells are characterized by very wide spacing which tends to make the external drainage radius very large and permeabilities may be extremely small. All of these factors tend to make the stabilization time for a gas reservoir very large. For completeness we should note that the porosity and viscosity of gas reservoirs are characteristically low which would tend to reduce the stabilization time.

Since conditions change slowly in a gas reservoir practically all pressure buildup data for gas reservoirs (except that obtained during a drill-stem test) can be analyzed by the "unchanging" method described for oil and water reservoirs in Chapter 3.

The basic idea is that the flowing pressure at time of shut-in would not change significantly during the time of shut-in if the well had not been shut-in. Consequently, the pressure increase caused by the negative rate change of shutting in the well would be superimposed upon the flowing pressure at time of shut-in, p_{wf}. Thus, for the flow of liquid the well pressure would be

$$p_w = p_{wf} + \frac{0.141q\mu}{kh} (0.5)(\ln \Delta \ t_D + 0.809) + \Delta p_{skin} \tag{3.9}$$

This equation was derived in Chapter 3. When the equation is applied to gas we recognize that it is necessary to convert the gas flow rate normally stated in Mscfd to reservoir barrels per day at the well by using the following equation adapted from Chapter 1.

$$q = 5.04 q_g z_w T_w/p_w \tag{1.18}$$

Now note that a constant flow rate in Mscfd will not result in a constant reservoir flow rate at the well bore because the pressure at the well bore will be changing constantly. Although this is interesting and important when the well pressure is changing rapidly at time of shut-in (such as during a drill-stem test), reservoir conditions normally change so slowly that the well pressure and producing rate at shut-in are practically constant. Consequently, it is recommended that Equation 1.18 be used to calculate the rate, q, which can then be used in making an "unchanging" pressure buildup analysis as outlined in Chapter 3.

Many engineers prefer to use a pressure squared form of equation as the basis for pressure buildup analysis in a gas well. This can be obtained by using Equation 4.70 to describe the change in pressure from the flowing well pressure, p_{wf}, at shut-in to some shut-in pressure at a shut-in time, Δt. An expression similar to Equation 3.9 is obtained

$$p_w{}^2 = p_{wf}{}^2 + \frac{1.639 q_g \mu z_{avg} T_f}{kh} \left[\log \Delta t + \log (\eta/r_w{}^2) + 0.809 + (S/1.15)\right] + B\, q_g{}^2$$

$$(4.75)$$

Mathematically this equation indicates that a plot of the well pressure squared versus the log of the shut-in time would give a straight line whose slope would be described by Equation 4.68 which was derived for the slope of a drawdown curve. Note that all the other terms including the pressure drop due to turbulence remain constant.

It appears to be mathematically impossible for the well pressure versus the log of the shut-in time and the well pressure squared versus the log of the shut-in time to both give a straight line. This is true since different assumptions have been made (liberties taken) in deriving Equations 3.9 and 4.75 and applying them to a gas well. Note that when we use Equations 3.9 and 1.18 we are assuming that the well pressure is constant whereas when we use Equation 4.75 we are assuming that the average well pressure is $(p_w + p_{wf})/2$. Obviously neither is correct.

Generally the engineer will find that the well-pressure plot will give the best straight line. However, for reservoirs with lower reservoir pressures, the engineer may find that the well pressure squared plot will be easier to interpret.

Either approach is difficult to apply to most gas-well pressure buildups because of afterflow. Afterflow is extremely severe in gas wells due to the high compressibility of the gas. As the pressure in the well bore increases following shut-in, the gas is compressed and more gas flows into the well bore to satisfy the reduced volume. Serious afterflow in gas wells may last for days and effectively prohibit the analysis of the pressure buildup by conventional methods.

If an engineer is determined to evaluate a gas reservoir through the use of pressure buildup it is strongly recommended that he obtain his pressure-buildup data by the two-rate method rather than by using a conventional shut-in. By simply reducing the flow rate and observing the increase in the well pressure due to the negative rate change he will avoid much of the afterflow difficulty. Generally, a new rate of production can be established in a relatively short period of time.

The analysis of a two-rate test is exactly the same as the analysis of a shut-in buildup except that the rate used in all of the equations would be the rate change rather than the rate at time of shut-in. The engineer will be faced with the same choice of methods as described above but he will largely avoid the difficulty of extremely long periods of afterflow.

Several papers have been presented in recent years on the effect of the reservoir pressure on the permeability. We know that a change in reservoir pressure causes a small change in the pore volume of a reservoir. Consequently, it should not be surprising that the small change in the pore volume or porosity can have an affect on the permeability of a formation as shown by recent investigations.[6] The affect is only important for gas reservoirs since economic gas-producing rates can be obtained from tight reservoirs due to the low viscosity of gas. Consequently, it may be necessary to use somewhat different pressure-buildup methods when tests are run in tight reservoirs.

However, it appears that the "turbulence" factor, n, of the pressure deliverability equations, may include this variation in permeability with pressure. Vairog and Rhoades showed[6] that there tends to be a straight-line relationship between the log of $\Delta(p^2)$ and the log of the permeability to gas. It appears that such a relationship may be automatically included in the "n" factor and that the outlined methods of predicting deliverability based on flow tests using the deliverability equation will then automatically include the prediction of the variation of permeability with the reservoir pressure. In fact it was suggested by H. J. Ramey at the 1972 SPE Annual Fall Meeting that much of the so-called turbulence affect may be entirely due to the permeability variation affect and that the affect of turbulence per se may not exist in many cases. Regardless of whether this hypothesis ultimately proves to be correct or incorrect, it does appear that the recommended methods of predicting performance on the basis of flow tests and the deliverability equation will automatically include the affect of permeability variation with pressure.

EQUIPMENT CAPACITY LIMITATIONS ON DELIVERABILITY

Many reservoir engineers think only in terms of the capacity of the reservoir to produce when they are predicting the deliverability of a well.

We should never forget that the production must also pass through the tubing, separators, dehydrators, meter run, and flow line to the pipeline. Some pressure drop is associated with each one of these pieces of equipment and the pressure drop is a function of the flow rate. Consequently, we will find that in many cases the production rate is limited by the capacity of the equipment rather than the capacity of the reservoir to produce. When such a situation arises it may be possible to install larger diameter equipment but the point that is made here is that the capacity of a well to produce is a function of both the reservoir capacity and the equipment capacity.

This situation can be shown graphically by plotting the bottom-hole pressure versus the rate of flow for the reservoir assuming that no pressure drop occurs in the equipment. Then assume that you could remove all of the flow equipment from the well and run flow tests on it so that for some given pipeline pressure (upstream pressure) you could obtain a plot of the flow rate versus the entrance pressure into the system which would be the bottom-hole pressure. Now plot these two rate versus bottom-hole pressure curves on the same axes as illustrated in Fig. 4.11. The reservoir capacity curve represents a particular state of depletion or external reservoir pressure, p_e, and the equipment capacity curve represents a particular equipment set up and pipeline pressure. Now under these conditions note that as the bottom-hole pressure increases the rate of flow on the basis of the reservoir capacity decreases while an increase in the bottom-hole pressure actually results in an increase in the flow rate through the equipment. Consequently, at relatively low rates, the flow rate of a well may be limited by the capacity of the reservoir and at relatively high rates the flow rate of a well may be limited by the capacity of the flow equipment. In the latter case we could say that the reservoir will produce at a rate that exceeds the capacity of the equipment.

Note that for a particular set of equipment, pipeline pressure, and state of reservoir depletion there is some maximum rate that can be produced that is represented by the intersection of the two capacity curves. At this point the reservoir flow results in a bottom hole pressure that just matches the pressure drop needed for flow through the production equipment at this rate. At any other rate the capacity of the well to produce is limited by either the reservoir or the equipment capacity. If a reservoir engineer is to provide accurate predictions of deliverability under various conditions he must consider the capacity of the equipment.

Since equipment capacity generally falls into the responsibilities of a mechanical or production engineer we must often rely upon them for the actual equipment capacity curve upon which we base our predictions. However, we must frequently generate our own curves so it is necessary to discuss some of the techniques of obtaining them.

Equipment suppliers can generally supply you with the capacity of separators, dehydrators, and other pieces of equipment, so we will limit our discussion to methods of predicting the capacity of the tubular goods involved in the producing equipment.

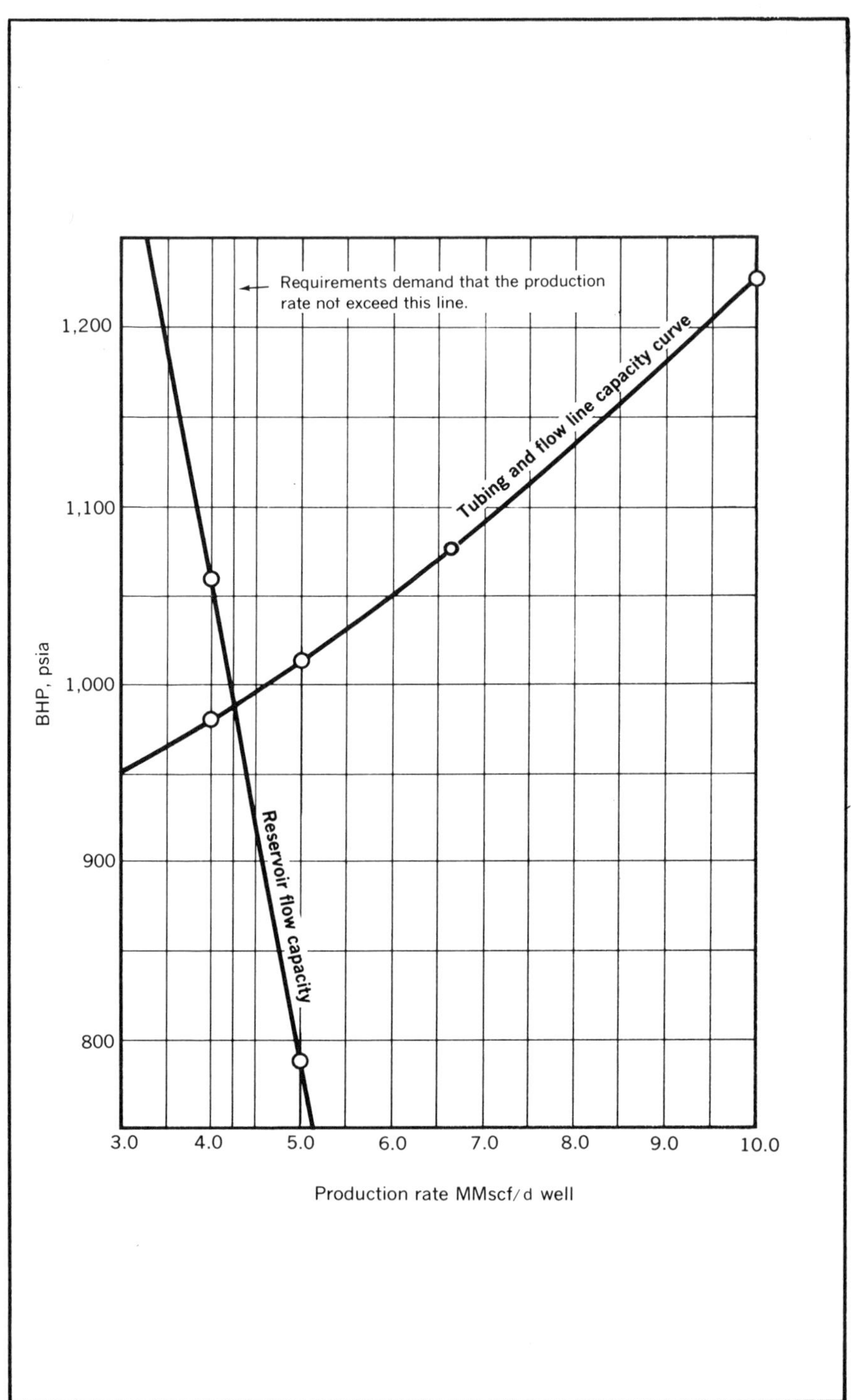

FIG. 4.11   Example relationship between reservoir and equipment capacity.

Flow-line capacity. Many different equations have been proposed for predicting the pressure drop in horizontal pipes at various flow rates. Each of these equations has a particular range of operating conditions under which they are most accurate. The choice of such equations to be used in a particular situation is extremely important if you are considering flow through many miles of pipeline. However, in our particular application where the flow line seldom will exceed 1–2 miles in length, the choice of equation becomes relatively unimportant. Consequently, we will base our discussion on the pipeline flow equation that is probably best known to engineers, the Weymouth equation.[7]

$$q_{scf/hr} = \frac{18.062 T_{SC}}{p_{SC}} \left[\frac{(p_1^2 - p_2^2)\, d_{in.}^{16/3}}{\gamma T_{avg} L_{miles} Z_{avg}} \right]^{(0.5)} \tag{4.76}$$

Careful note should be made of the units in this equation since they are unusual to the reservoir engineer, as indicated in the equation. The temperatures and pressures are in degrees Rankine and psia, respectively. The symbol, d_{in}, is the inside pipe diameter in inches and γ is the gas specific gravity relative to air.

Since we will use this equation to calculate the pressure, p_1, for a particular flow rate, the application must be by trial and error. This situation is apparent when we recognize that we must know the average pressure to evaluate the average gas-deviation factor, z_{avg}. This is the same situation that exists in solving many gas flow equations for a pressure. However, in this case, and particularly our specific application, the trial and error is generally a simple one. Since gas flow-line pressures are relatively low and the pressure drops are small it is seldom that more than two trials are necessary to solve the equation for p_1 with an acceptable degree of accuracy.

The Weymouth equation could be applied on an incremental basis to improve accuracy but again it is generally unnecessary to do so when the equation is applied to the typical low pressure, short flow line. The general form of the Weymouth equation can be derived analytically[4] but the friction factor included in the equation must be included in an empirical form.

Tubing or casing capacity. The problem of predicting the pressure drop through a vertical flow line, such as tubing or casing, is a difficult one. This can be seen qualitatively when the engineer realizes that Equation 4.76 and a static-pressure-difference equation, such as Equation 4.31, must be satisfied simultaneously under these conditions. Many different efforts have been made to obtain such equations. As an example, consider the equation developed by R. V. Smith.[4]

$$q_g = \frac{200 d^5 (p_2^2 - e^s\, p_1^2)\, s}{T_{avg} z_{avg}\, f\, \Delta X\, (e^s - 1)} \tag{4.77}$$

where

$$s = \frac{0.0375\gamma\Delta X}{T_{avg}z_{avg}}$$

(4.78)

In addition to the complexity of these equations note that it is necessary to independently evaluate the friction factor, f, which is a function of the flow rate, the depth of the well, the gas gravity, and the gas viscosity. Even with such a complex set of equations the solution is still trial and error since z_{avg} is a function of the pressure, which is unknown in our application. If an equation such as this is used through a computer program, or a program suggested by one of the multitide of computer analysis firms is used, the engineer should make certain that the results obtained are acceptable for the particular range of pressures, temperatures, depths, and gas gravities where it is applied. From a practical point of view such reliability can be determined only on a reservoir-to-reservoir basis. In other words the calculations should be compared with measured pressures for a particular reservoir before they are relied upon in that reservoir.

If an engineer must make predictions of flowing bottom-hole pres-

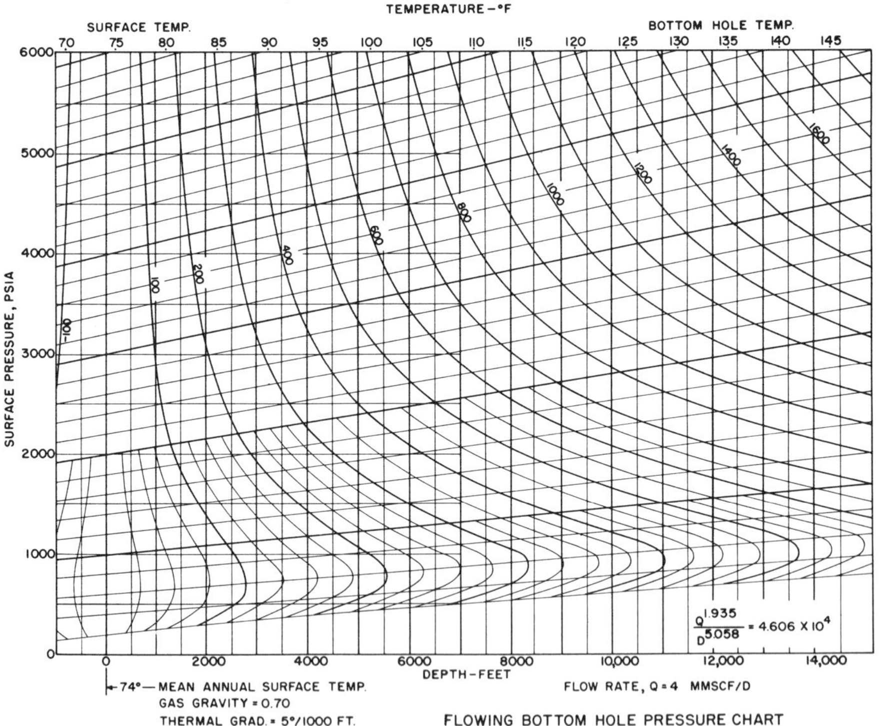

FIG. 4.12 Flowing bottom-hole-pressure nomograph for 4 MMSCF/D. (After Crawford and Fancher, TPRC Bulletin 72.)

sures without the benefit of a computer program the data of Bulletin 72 of the Texas Petroleum Research Committee is recommended. However, the same restriction is suggested for these data as was suggested for the computer programs.

The Bulletin 72 data were recommended previously as a means of determining the static bottom-hole pressure. The engineer may wish to return to this discussion under "Determining the Reservoir Pressure" in this chapter, to review the general nature of these data. In the case of the static pressure data each basic figure from Bulletin 72 represented a particular gas gravity and temperature gradient. The figures representing the flowing bottom-hole pressures have the additional parameter of flow rate fixed for each figure. Also, all of the flowing pressure drops used represent the flowing pressure drop through 2½-in. tubing. Bulletin 72 nomographs for flowing wells are illustrated by Fig. 4.12 through 4.15. These figures simply represent examples of the Bulletin 72 data. Any practical application would require the entire Bulletin.

The application technique to determine the pressure drop in 2½-in.

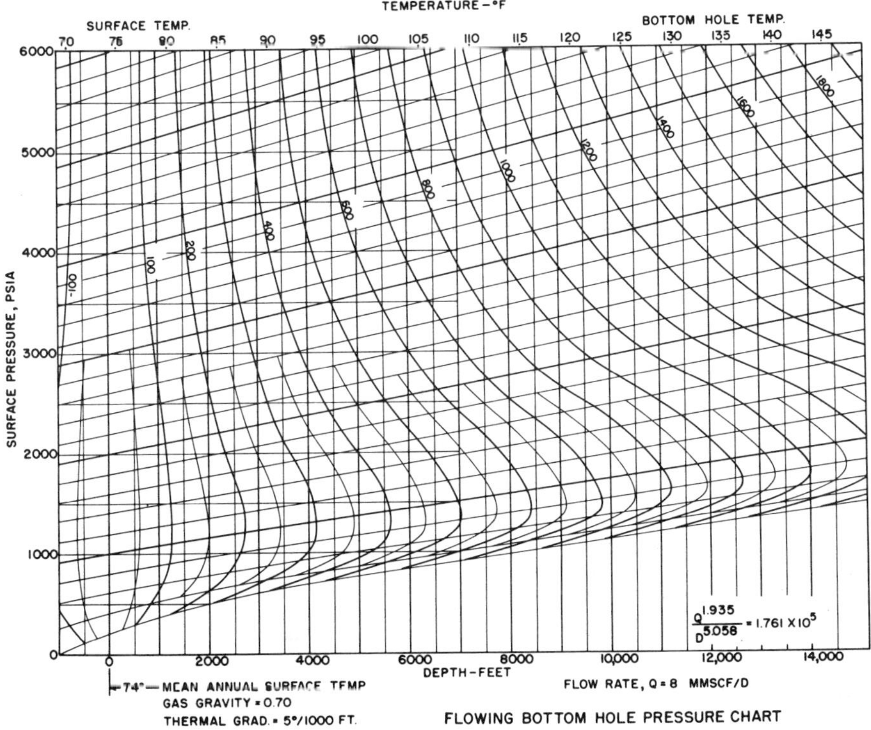

FIG. 4.13 Flowing bottom-hole-pressure nomograph for 8 MMSCF/D. (After Crawford and Fancher, TPRC Bulletin 72.)

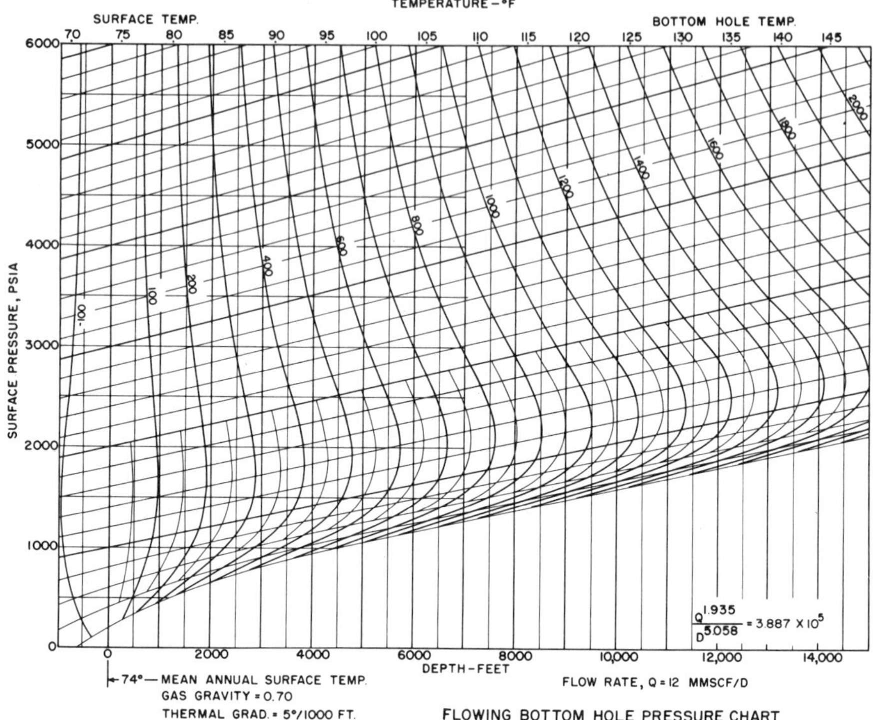

$$\frac{Q^{1.935}}{D^{5.058}} = 3.887 \times 10^5$$

FIG. 4.14 Flowing bottom-hole-pressure nomograph for 12 MMSCF/D. (After Crawford and Fancher, TPRC Bulletin 72.)

tubing is exactly the same as the technique used to determine the pressure difference in a static column of gas except that an interpolation must be made for three parameters instead of two. Interpolations will be necessary for the gas gravity, temperature gradient, and the flow rate. In addition, if the flowing pressure drop is required in some tubing or casing other than $2\frac{1}{2}$-in., it is necessary to first determine the equivalent flow rate for $2\frac{1}{2}$-in. tubing before making the calculations. In other words it is necessary to determine what flow rate would give the same pressure drop in $2\frac{1}{2}$-in. tubing as that being experienced in the subject tubing or casing size at the subject flow rate. These equivalent $2\frac{1}{2}$-in. tubing rates can be determined from Fig. 4.16.

Using the methods described above for predicting the pressure drop in the surface flow line and in the tubing or casing it is possible to predict the total pressure drop from the pipeline connection to the bottom of the well for different flow rates. The engineer should now use the Bulletin 72 data and the Weymouth equation to solve the following problem and check his solution against the one in Appendix C.

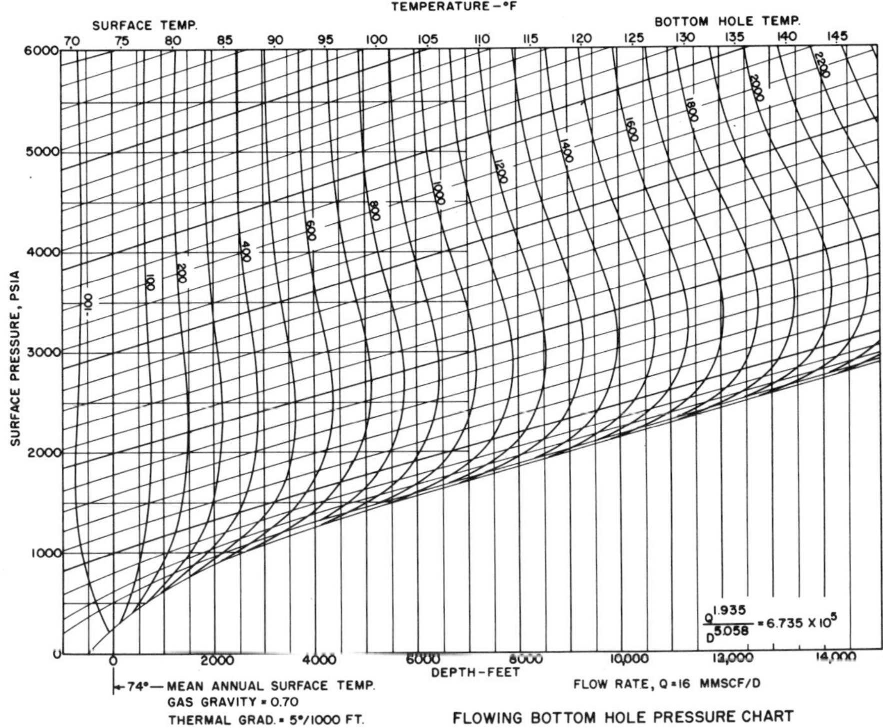

$$\frac{Q^{1.935}}{D^{5.058}} = 6.735 \times 10^5$$

FIG. 4.15 Flowing bottom-hole-pressure nomograph for 16 MMSCF/D. (After Crawford and Fancher, TPRC Bulletin 72.)

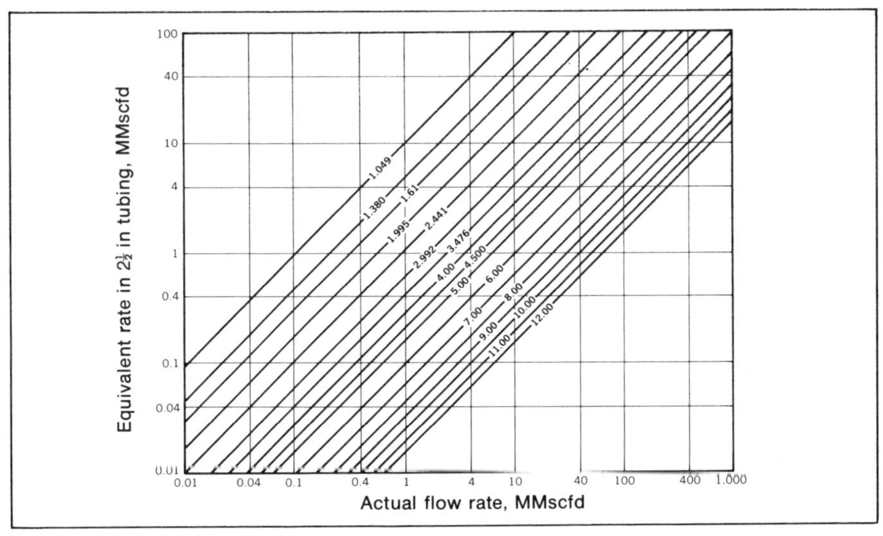

FIG. 4.16 Equivalent rate chart. (After Crawford and Fancher, TPRC Bulletin 72.)

Problem No. 4.11: Pressure Drops in The Producing System

The average well in the gas reservoir of Problems 4.4 and 4.6 is equipped with 3,700 ft of 3-in. tubing (I.D. = 2.992 in.) and a 1.0 mile 3.068-in. I.D. flow line to the pipeline. The gas temperature at the pipeline will be 70° F.; at the wellhead, 90° F.; and at the reservoir, 110° F.

Find the minimum BHP necessary to provide a pipeline delivery of:

(1) 4 MMcfd (2) 5 MMcfd (3) 6.67 MMcfd and (4) 10 MMcfd at 800 psia.

To simplify interpolations in calculating the Δp in the tubing, use the temperature gradient as 5°/1,000 ft and the gas gravity as 0.7. Also note that the static pressure data represents the BHP for a rate of 0.0.

PREDICTING RESERVOIR PERFORMANCE

To predict the production history of a reservoir it is necessary to consider the capacity of the reservoir to produce, the capacity of the equipment, and the state of depletion of the reservoir as predicted by material balance. Since the average reservoir pressure is a function of the previous production history and the producing rates are a function of the state of depletion, it is clear that all three parts of the reservoir prediction must be interrelated to achieve a prediction for the reservoir.

By using various testing procedures we learned that we could predict the capacity of a reservoir to produce under various conditions. This reservoir capacity can be represented as a plot of the bottom-hole pressure versus the producing rate for a particular state of depletion, (p_e). Such a curve was considered previously in Fig. 4.11. On the same figure is shown a similar capacity curve for the producing system. As the bottom-hole pressure declines, the flow rate through the reservoir will increase since a drop in bottom-hole pressure means an increase in the pressure drop in the reservoir. However, as the bottom-hole pressure declines the flow rate capacity through the tubing-flow line producing system declines because with a fixed pressure at the pipeline a decline in the bottom-hole pressure means that the pressure drop in the producing system declines.

When the capacity of the reservoir and the producing system are considered together, it is easily seen that the producing capacity of a well may be limited by the capacity of the producing system rather than by the capacity of the reservoir. For example, using the data of Fig. 4.11, consider what would happen if we attempted to flow the well at a bottom-hole pressure of 787 psia. Note that the reservoir alone without any back pressure or resistance from the producing system would produce at a rate of 5.0 MMscfd. However, this flow rate through the producing system requires a bottom-hole pressure of 1,015 psia. Consequently, with the particular tubing-flow line producing system represented by the capacity curve of Fig. 4.11 the well could not produce at this rate. In fact the well could not produce at any rate greater than 4.25

MMscfd. This is the rate represented by the intersection point of the two capacity curves. At this point the bottom-hole pressure of 988 psia will provide a flow rate through the reservoir of 4.25 MMscfd and will just be sufficiently high to provide the same flow rate through the producing system.

The engineer should bear in mind that Fig. 4.11 represents only one stage of depletion and one combination of tubing, separator, dehydrator, flow line, pipeline pressure, etc. Such a plot could be made much more general by adding other reservoir-capacity curves for different stages of depletion and a variety of producing-system-capacity curves representing different combinations of producing equipment. Such a family of capacity curves is illustrated schematically in Fig. 4.17.

As noted previously, the producing system capacity curve must be considered along with material balance and the deliverability curves to predict how a reservoir will perform under any given set of conditions. For example, suppose we needed to know how many wells to drill in a particular reservoir to fulfill some stated flow-rate contract from a reservoir for some stated period of time. From a rate standpoint the critical time will be the time at the end of the contract when the reservoir pressure has declined to a minimum under this contract. At this particular time we must be certain that a sufficient number of wells has been drilled to provide the required producing rate. As the number of wells is in-

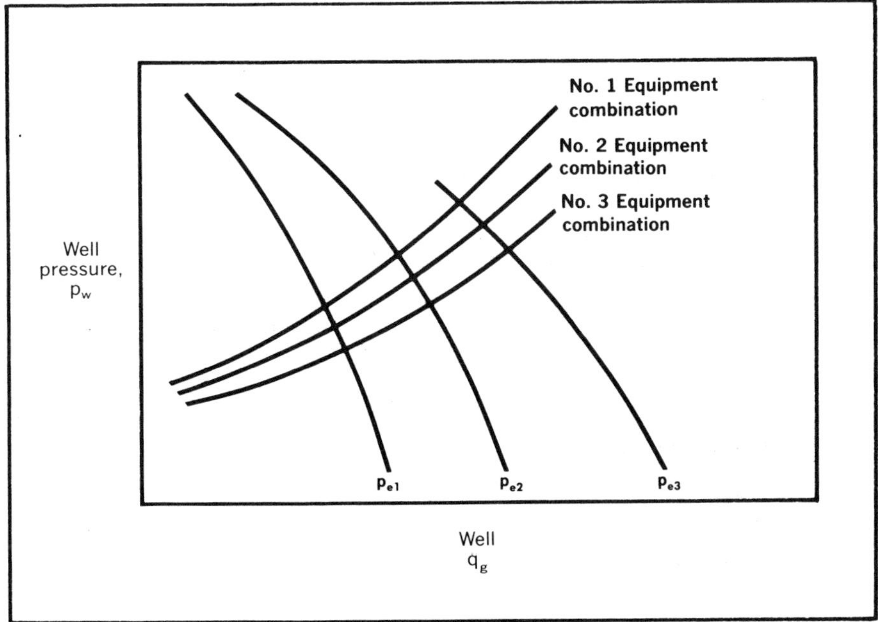

FIG. 4.17 Family of reservoir-equipment capacity curves.

creased the required rate per well will be reduced. Also remember that as the number of wells increases the production capacity of each well at any fixed state of depletion will be slightly increased because each well will be draining a lesser volume of the reservoir.

Consequently, this hypothetical problem must be solved by trial and error. With the state of depletion fixed by the contract length and total reservoir rate, we can determine the average pressure or the pressure at the external drainage boundary of each well. Now if we assume a number of wells we can use the basic deliverability curve and determine the deliverability curve for the resulting spacing. With the number of wells assumed we will, in effect, have assumed the rate per well since the contract fixes the total reservoir rate. Based on the per-well rate we can determine the bottom-hole pressure necessary to supply that rate from the deliverability curve. Then this bottom-hole pressure can be used together with the equipment-capacity curve to determine if the equipment capacity is sufficient to supply the required per-well rate at the subject bottom-hole pressure. If the equipment will not supply the rate a greater number of wells is considered. The number of wells must be bracketed before the engineer can be certain that he has determined the most economical solution to this problem.

This represents only one type of capacity problem. The more general problem today is probably concerned with the situation where a pipeline will prepare a contract for purchase of gas from a reservoir for some particular rate schedule that is not necessarily a constant rate and the engineer must determine the rate schedule and agreement terms at which his company can show the greatest profit. The more wells drilled the greater is the development cost but this will result in the quickest return of the revenue from the project. On the other hand, the fewer wells drilled the lower will be the development cost but the longer it will take to realize the profit from the project. Consequently, it is possible to optimize the number of wells drilled to maximize the deferred profit depending upon a particular company's financial situation and the profitability evaluation parameters the company believes are most realistic for them.

It is suggested that the reader work the following problem and check his solution against the solution of Appendix C.

Problem No. 4.12: Determining Gas Well Spacing for Completion of a Pipeline Purchasing Contract

The gas reservoir of Problem 4.4 and 4.6 must produce 20 MMcfd for the next 5 years (from 11-1-69) to meet contract requirements. If the physical equipment on each well is as described in Problem No. 4.11 and has the capacity indicated in Fig. 4.18A, how many equally spaced wells are needed to meet this contract?

Assume the p_e at the contract completion is 1568 psia as determined in Problem 4.4. Also use the stabilized performance curves in Fig. 4.18B.

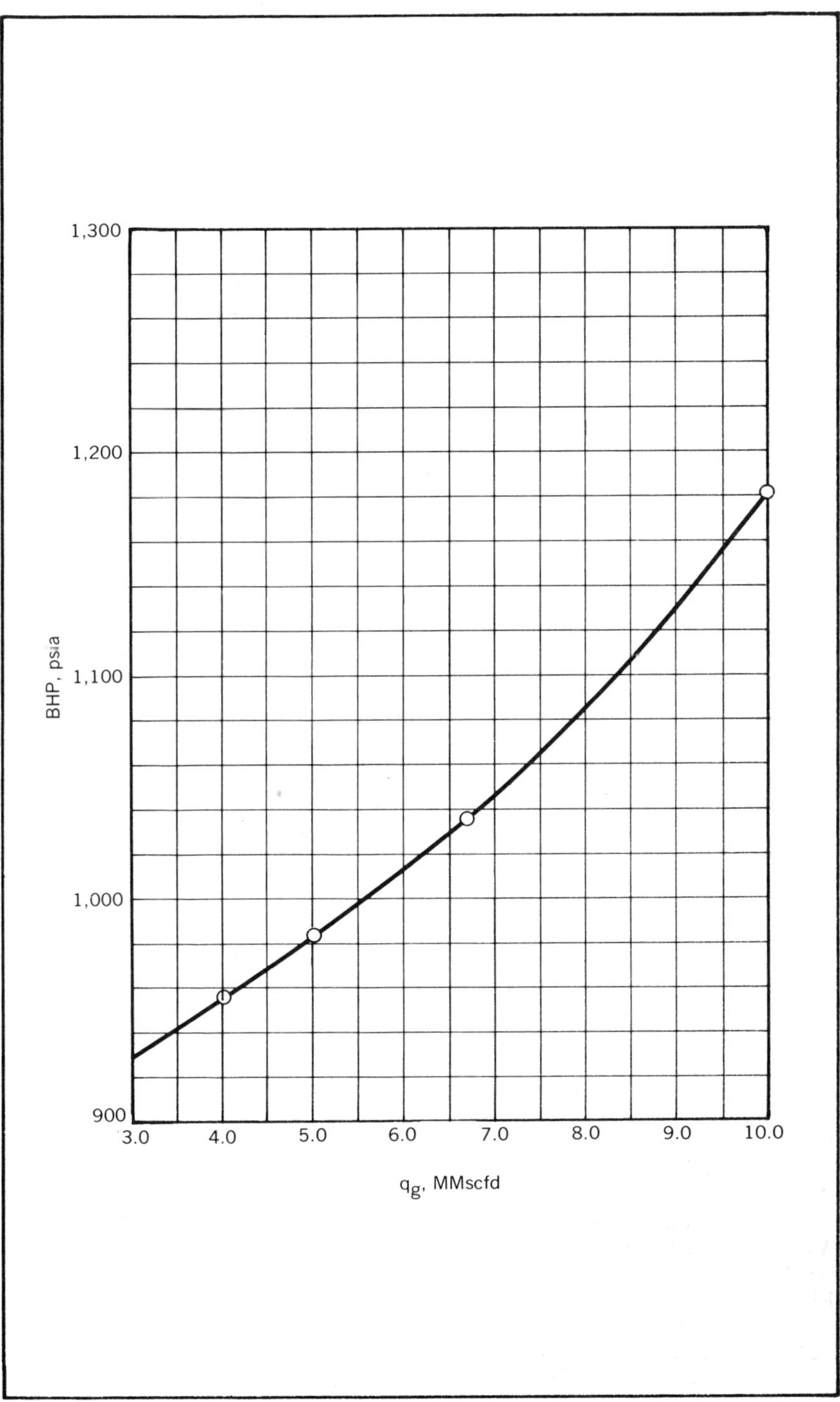

FIG. 4.18A Tubing and flow-line capacity—data for Problem 4.10.

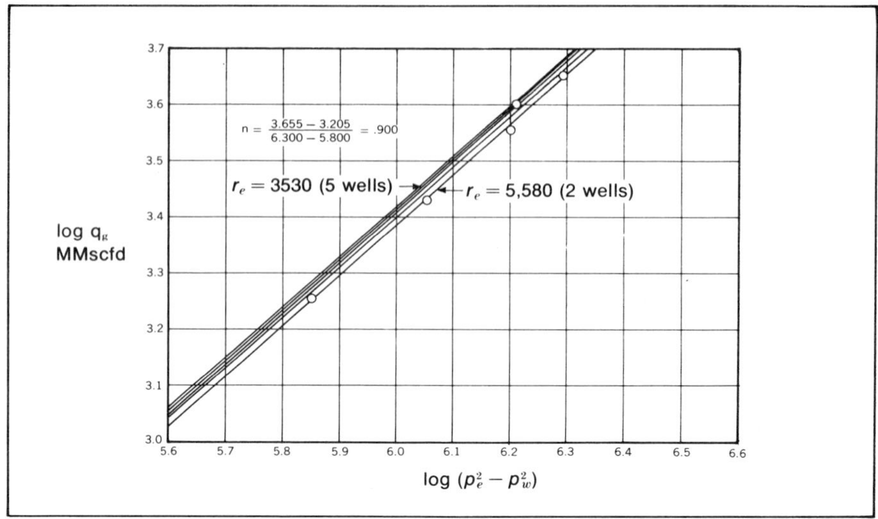

FIG. 4.18B Stabilized deliverability curves for different well spacings. (Results of Problem 4.7; data for Problem 4.12)

Since the calculations of this problem involve a trial-and-error procedure with a considerable amount of calculation in each trial it is advantageous to use a digital computer. A program is included as Tables 4.3B and 4.3C with symbols defined in Table 4.3A. To simplify the calculations this program uses the Smith equations for calculating the

Table 4.3A

A COMPUTER PROGRAM FOR DETERMINING THE
NUMBER OF WELLS TO DRILL IN A GAS RESERVOIR

(Basis for discussion — NOT recommended for use in this form)

EXPLANATION OF VARIABLES USED:

B(I)	= intercept of log q vs log ΔP^2 plot
F(I)	= friction factor data
FF(I)	= interpolated friction factor at specific flow rate
QT	= total daily flow rate = MMcfd
QG(I)	= effective flow rate for calculating friction factor
RE(I)	= reservoir outer radius for each well
PH(I)	= wellhead pressure, from horizontal pressure drop equation
FQG(I)	= effective flow rate data in friction factor data
I	= number of wells
PWT(I)	= BHP from Smith equation
PWR(I)	= BHP for reservoir for each flow rate
REI	= initial radius of investigation from isochronal testing
SLIP	= slope of line (average); log-log plot
SLOP	= slope of each individual line; log-log plot
K	= constant used to determine slope of log-log plot

Table 4.3A (*continuation*)

L	= constant used to determine slope of log-log plot
N	= constant used to determine slope of log-log plot
X	= variable used to determine slope of log-log plot
Y	= variable used to determine slope of log-log plot
BA	= intercept calculated from an individual data point
BB	= intercept calculated from an individual data point
BC	= intercept calculated from an individual data point
BD	= intercept calculated from an individual data point
BO	= average intercept of the preceding four intercepts
B(I)	= intercept calculated for each RE(I)
EX	= 1.1 SLIP
AREA	= 4,500 acres
Q(I)	= flow rate for each well = QT/WE(I)
WE(I)	= number of wells
QTEST(I)	= isochronal testing flow rate
BHP(I)	= corresponding pressures to QTEST(I)
PE	= 2,800 psi, used to get different intercepts

pressure drop in the tubing. This avoids the need to store the Bulletin 72 data and interpolate it in the program. The program simply calculates the bottom-hole pressure resulting from the flow of fluid at a particular rate and drainage volume and compares that with the bottom-hole pressure necessary to provide the same flow rate through the equipment.

Table 4.3B
A COMPUTER PROGRAM TO DETERMINE THE NUMBER OF WELLS
TO DRILL TO COMPLETE A GAS DELIVERABILITY CONTRACT[B]

```
PAGE 1
// JOB T
LOG DRIVE      CART SPEC      CART AVAIL      PHY DRIVE
   0000           0003           0003            0000

// FOR
*IOCS(2501READER,1403PRINTER)
*LISTSOURCEPROGRAM
      DIMENSION B(6),F(6),QG(6),WE(6),RE(6),Q(6),PH(6),FQG(6),FF(6),
     CQTEST(6),PWT(6),PWR(6)                   ,BHP(6)
      READ(8,2) Z,QT,AREA,PHI,PEI,TAW,TAP,PE
      READ(8,2) D,POUT,DIST,U,G,WD,RW,PERM
      WRITE(5,8)
    8 FORMAT(20X,17THE INPUT DATA IS,)
      DO3 J=1,5
      READ(8,4) QG(J),F(J),QTEST(J),BHP(J),WE(J)
      WRITE(5,9)QG(J),F(J),QTEST(J),BHP(J),WE(J)
    9 FORMAT(5F20.5)
    4 FORMAT(5F10.5)
    3 CONTINUE
    2 FORMAT(8F10.5)
```

Table 4.3B (continuation)

```
      REI=SQRT(38.5*PEI*PERM/(PHI*U*24.))
      RE(1)=SQRT(AREA*43560./(WE(1)* 3.14))
      SLIP=0.
      DO10 L=1,3
      K=L-1
      DO10 I=2,4
      IF(K-1)16,17,18
   17 IF(I-3)16,16,10
   18 IF(I-2)16,16,10
   16 N=I+1+K
      X=ALOG(PEI**2-BHP(      N)**2)-ALOG(PEI**2-BHP(I)**2)
      Y=ALOG(QTEST(        N))-ALOG(QTEST(I))
      SLOP=Y/X
      SLIP=SLIP+SLOP
   10 CONTINUE
      SLIP=SLIP/6.
      WRITE(5,21)SLIP
   21 FORMAT(//11H THE SLOPE=,F7.5)
      BA=QTEST(2)*1000./((PEI**2-BHP(2)**2)**SLIP)
      BB=QTEST(3)*1000./((PEI**2-BHP(3)**2)**SLIP)
      BC=QTEST(4)*1000./((PEI**2-BHP(4)**2)**SLIP)
      BD=QTEST(5)*1000./((PEI**2-BHP(5)**2)**SLIP)
      BO=(BA+BB+BC+BD)/4.
      B(1)=BO*ALOG(.606*REI/RW)/ALOG(.606*RE(1)/RW)
      EX=1./SLIP
      WRITE(5,6)
    6 FORMAT(//9X,12HFRICT. FCTR.,12X,10HRES. PRESS,7X,18HWELL
     HEAD PRESCSURE,5X, 9HEQUIP.BHP,3X,9HNO. WELLS,//)
      DO100 I=1,10
C     CALCULATING PW FOR RESERVOIR
      RE(I)=SQRT(AREA*43560./(WE(I)* 3.14))
      B(I)=B(1)*ALOG(.606*RE(1)/RW/ALOG(.606*RE(I)/RW)
      Q(I)=QT/WE(I)
      PM=(Q(I)/B(I))**EX
      PEE=PE**2
      IF(PEE-PM)100,100,61
   61 PWR(I)=SQRT(PEE-PM)
      CALCULATING P DROP FOR HORIZONTAL PIPE
      QH=QT/24.
      QH=(QT*14.7/(WE(I)*24.*18.062*520.))**2

      H=G*TAP*DIST*Z/D**5.3333
      PH(I)=SQRT(QH*H+POUT**2)
      CALCULATING P DROP FOR VERTICAL PIPE
      FQG(I)=QT*G/(WE(I)*U*1000.)
      IF(FQG(I)-QG(1))31,31,32
   32 IF(FQG(I)-QG(2))33,34,35
   35 IF(FQG(I)-QG(3))36,37,38
   38 IF(FQG(I)-QG(4))39,40,41
```

Table 4.3B (*continuation*)

```
41 IF(FQG(I)-QG(5))42,43,43
31 FF(I)=F(1)
   GOTO44
34 FF(I)=F(2)
   GOTO44
37 FF(I)=F(3)
   GOTO44
40 FF(I)=F(4)
   GOTO44
43 FF(I)=F(5)
   GOTO44
33 FF(I)=F(2)+(F(1)-F(2))*(QG(2)-FQG(I))/(QG(2)-QG(1))
   GOTO44
36 FF(I)=F(3)+(F(2)-F(3))*(QG(3)-FQG(I))/(QG(3)-QG(2))
   GOTO44
39 FF(I)=F(4)+(F(3)-F(4))*(QG(4)-FQG(I))/(QG(4)-QG(3))
   GOTO44
42 FF(I)=F(5)+(F(4)-F(5))*(QG(5)-FQG(I))/(QG(5)-QG(4))
44 S=.0375*G*WD/(TAW*Z)
   PWT(I)=SQRT((QT/(WE(I)*200000.))**2*G*TAW*Z*FF(I)*WD*(2.72**S-1)
   C/(D**5*S) | 2.72**S *PH(I)**2)
   WRITE(5,12) FF(I),PWR(I),PH(I),PWT(I),I
12 FORMAT(4F20.5,I9)
   IF(PWT(I)-PWR(I))51,51,100
100 CONTINUE
51 WRITE(5,7) I
 7 FORMAT(//30HTHE OPTIMUM NUMBER OF WELLS IS,I3)
   STOP
   END


FEATURES SUPPORTED
ICCS

ORE REQUIREMENTS FOR
COMMON          0 VARIABLES          248 PROGRAM 1334

END OF COMPILATION

/ XEQ
```

Table 4.3C

EXAMPLE DATA

THE INPUT IS

50000.00790	0.01630	0.00000	2800.00049	1.00000
100000.01580	0.01510	1800.00024	2710.00049	2.00000
200000.03161	0.01470	2700.00049	2659.00049	3.00000
300000.06323	0.01430	3600.00049	2605.00049	4.00000
400000.06323	0.01420	4500.00098	2545.00049	5.00000

Table 4.3C (continuation)

THE SLOPE=C.90418

FRICT. FCTR.	RES. PRESS	WELL HEAD PRESSURE	EQUIP. BHP	NO. WELLS
0.01463	105.29957	888.46582	1020.70739	3
0.01485	935.72119	850.89465	963.39013	4
0.01498	1171.67578	832.93090	935.51806	5

DETERMINING THE SIZE OF A GAS RESERVOIR

The problem of determining the size of a gas reservoir seems to be grossly misunderstood by managers and engineers in general. Given a few pressure measurements and a couple of days of production the average manager or engineering supervisor seems to think that their reservoir engineer should be able to tell them the drainage volume of the subject well. As a matter of fact if the engineer can tell the reservoir size from such meager tests the reservoir is probably so small that it will be uncommercial. In determining the size of a gas reservoir from test and pressure data there is no substitute for time.

One practical approach to this problem is for the reservoir engineer to determine the minimum production time necessary to calculate a reservoir size. This can be determined by using the stabilization time equation and the estimated reservoir parameters. Fig. 4.19 may be useful in this regard. Remember that this technique provides only a minimum reservoir size estimate. If the test is run improperly or the production rate is too small to result in interpretable pressure drops, the data may still be useless. However, the main point to be made is that there is no substitute for time and that a manager is simply misleading himself and making his reservoir engineer prognosticate when he demands an estimate of the reservoir size without adequate data.

The methods to be employed differ little from those used in determining the size of an oil reservoir. However, well spacing is so much more arbitrary in gas reservoirs that the problem seems to be encountered much more often.

The use of drawdown data to determine the distance to the nearest boundary in a reservoir was discussed extensively in Chapter 3 for oil reservoirs. The techniques apply in a similar way for gas reservoirs. Equations 4.66 and 4.67 give the well pressure resulting from production at a constant rate q_g for a time, t. A plot of p_w^2 versus the log t will give a straight line with a slope, m, that satisfies Equation 4.68 as long as the well pressure is infinite acting. However, once the nearest boundary is felt at the well the plot will no longer be a straight line and the difference between the extrapolation of the straight line and the actual pressure recorded will be

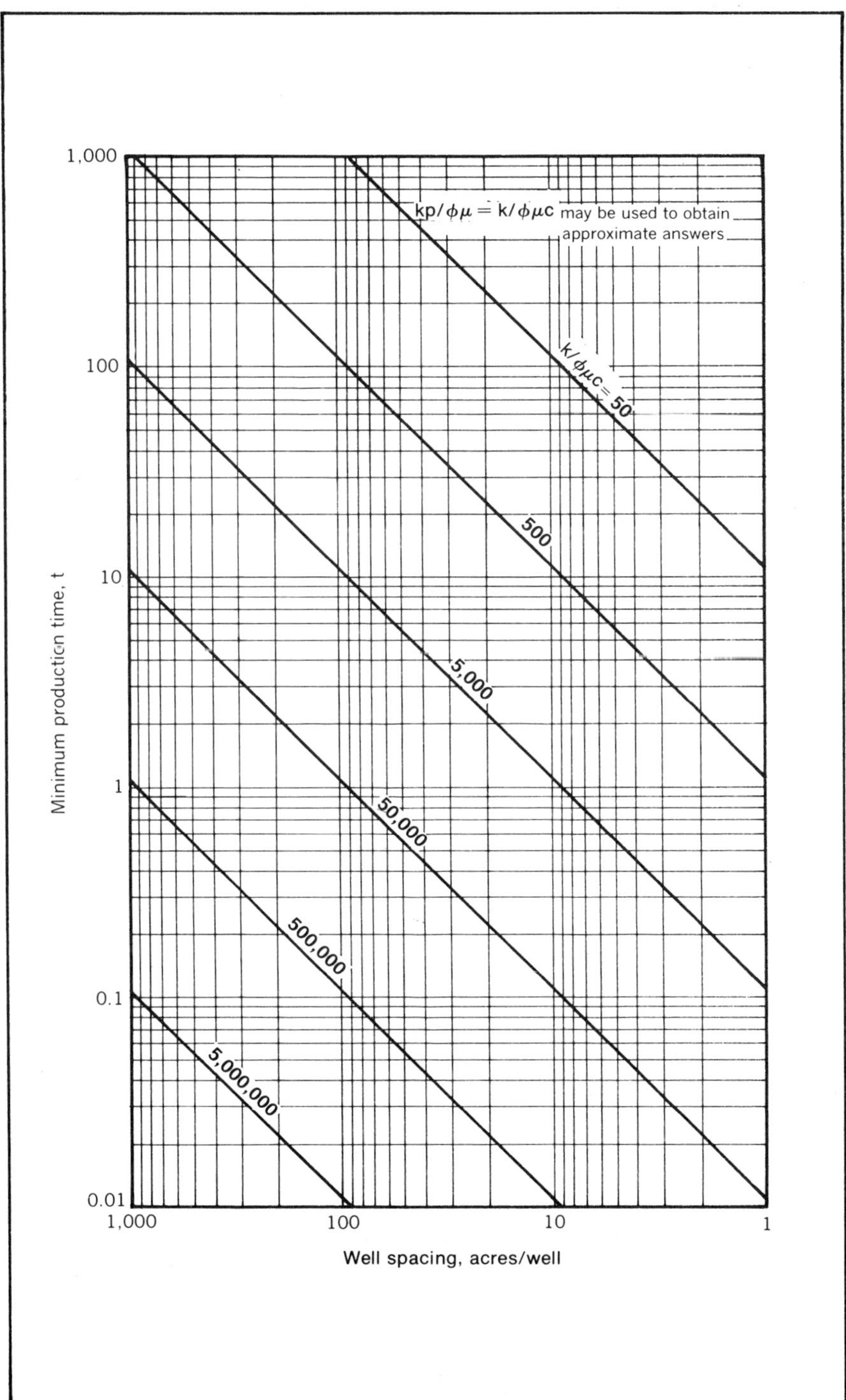

FIG. 4.19 Well spacing provable by a production time (based on $t = 0.04\phi\mu cr_e^2/k$)

$$(p_w')^2 - p_w^2 = \frac{1.424 q_g \mu z_{avg} T_f}{kh} (0.5)[-Ei(-1/4t_{Di})] \qquad (4.79)$$

The dimensionless time, t_{Di}, is a function of the distance to the closest reservoir boundary, d.

$$t_{Di} = (6.33k/\phi\mu c) \, t/(2d)^2 \qquad (3.63)$$

This relationship was derived in Chapter 3. To determine the distance to the nearest boundary the procedure is then to plot the square of the well pressure versus the log of the producing time, extrapolate the straight-line portion and read the extrapolated, $(p_w')^2$ and p_w^2 at some particular time, t. Then using Equation 4.79 the Ei function $0.5\left(-Ei\dfrac{-1}{4t_{Di}}\right)$, can be calculated, the corresponding t_{Di} can be determined from Fig. 2.8 of Chapter 2, and the distance, d, to the boundary can be calculated from Equation 3.63.

This technique and many other methods of determining reservoir limits are discussed in Chapter 3 along with some problems. The methods of this section can be used in gas wells to determine the distance to the next closest barrier, the angle between the barriers, the total reservoir pore volume, and obtain other ideas concerning the reservoir limits when appropriate conditions prevail.

In applying the techniques to a gas reservoir remember that the only difference is that a plot of p_w^2 versus the log of time must be used instead of a plot of p_w versus the log of time, and that the relationship is governed by Equations 4.66 and 4.67 rather than similar equations for the flow of liquids. Note that Equations 3.63 and 2.56 apply to both the flow of liquids and the flow of gas.

To test your knowledge of this section, and the one on pressure buildup in a gas reservoir, work the following problem and compare your solution with the one in Appendix C.

Problem No. 4.13: Evaluating a Pressure Buildup in a Gas Reservoir and Determining the Distance to a Reservoir Barrier

A discovery well in a gas reservoir is produced for a period of time during which 248 MMscf of 0.76 specific gravity gas is produced. The final rate is 3.9 MMscfd when the well is shut-in with the resulting pressure behavior as indicated in Fig. 4.20.

A. If the initial reservoir pressure was 4,990 psia, is this reservoir infinite acting at time of shut-in? (Find indicated p_i and compare with actual p_i).
B. What is the distance to the nearest boundary?
C. What is the gas pore volume if at time of shut-in the pressure was declining about 6 psi/day? (No water drive; assume $p_s = 4,600$).
D. Calculate the static pressure by material balance.

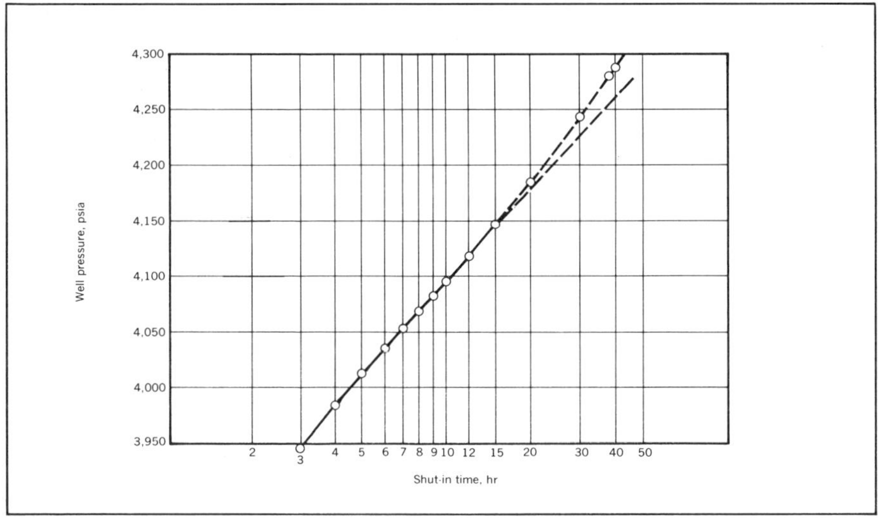

FIG. 4.20   Gas-well pressure buildup data for Problem 4.13.

Reservoir data

$\gamma\ = 0.76$	$p_i\ = 4{,}990$ psia	$r_w\ = 4$ in.
$\phi\ = 5.0\%$	$u_g\ = 0.027$ cp	$p_{wf} = 998$ psia
$T_f\ = 253°$ F.	$h\ = 30$ ft	$S_{wc} = 0.35$

REFERENCES

1. O. A. Househ, K. M. Watson, and R. A. Ragatz, *Chemical Process Principles*, New York, N.Y.: John Wiley and Sons, 1943.

2. Standing and Katz, *AIME TRANSACTIONS*, 1942.

3. E. J. Burcik, *Properties of Petroleum Reservoir Fluids*, New York, N.Y.: John Wiley and Sons, 1957.

4. D. L. Katz, et al., *Handbook of Natural Gas Engineering*, New York, N.Y.: McGraw-Hill, 1959.

5. C. S. Matthews and D. G. Russell, *Pressure Buildup and Flow Tests in Wells*, Dallas, Texas: SPE Monograph Volume No. 1, 1967.

6. J. Vairogs and V. M. Rhoades, "Pressure Tranient Tests in Formations Having Stress — Sensitive Permeability," SPE Paper No. 4050 presented at Annual Meeting, 1972.

7A. L. C. Uren, *Petroleum Production Engineering — Oil Field Exploitation*, New York, N.Y.: McGraw-Hill Book Company, Inc., 1934.

7B. Ballantyne, Wayne, A computer program design assignment from an undergraduate course, Ch.E. 743.01, Feb. 1969.

8. H. C. Slider, "Gas Well Testing Using Negative Superposition", SPE Paper No. 3875, Northern Plains Section Regional Meeting, Omaha, Nebraska, 1972.

Fluid Distribution and Frontal Displacement

In the first four chapters we discussed the reservoir without recognizing vertical variation in the distribution of reservoir fluids. We have at times referred to a water drive or gas drive but we have treated this phenomena as though the saturation changed abruptly in a vertical direction from the saturations of the oil-bearing portion of the reservoir to the saturations of a gas cap or a water-drive zone. We noted occasionally the variation in gas saturation laterally or radially due to the variation in pressure. However, we did not make note of the variation caused by the encroachment of gas-cap gas or water into the original oil-bearing portion of the reservoir. This does not mean that the methods of the first four chapters are impractical. It does mean that we must apply these methods to reservoir situations where they are not unduly affected by saturation variations either vertically or horizontally. A thorough understanding of this chapter on fluid distribution and frontal displacement will enable the engineer to intelligently and accurately apply the methods of the first four chapters.

In addition, the material of this chapter will specifically permit the engineer to determine the saturation distribution in the reservoir either vertically or laterally, initially or after producing the reservoir for some period of time. It will provide the fundamental understanding of the relationships between fluid saturation distribution, gravity forces causing segregation, and the differences between the viscous forces associated with the movement of the different fluids in the reservoir. These fundamentals can then be applied to the practical problems of evaluating the saturation distribution in the reservoir initially and during the course of any reservoir displacement.

The fundamentals are directly applicable to the problems of the dis-

264

placement of oil by a natural water drive or gas-cap drive, including coning, fingering, and hydrodynamic tilt of the water-oil or gas-oil contact. In addition, the frontal displacement fundamentals are directly applicable to most secondary recovery methods that employ a frontal displacement.

This chapter will begin with a discussion of the initial vertical distribution of the reservoir fluids and the means of evaluating this vertical saturation. It will stress the calculation of the saturation distribution from laboratory-measured capillary-pressure data. This technology will be presented not so much for its practical application as for its practical worth in providing an excellent physical understanding of the relationship between the microscopic saturation distribution, the pore-size distribution, wettability, and gravity segregation. Generally, the initial saturation distribution can be determined more readily and realistically from well logs but the complete physical understanding of the phenomena comes from the capillary pressure relationships.

Following the discussion of the initial saturation distribution, methods of determining the saturation distribution during a frontal displacement will be presented. These include the Buckley-Leverett and Welge methods. This material is presented in a general manner so that the equations can be applied to any immiscible fluid displacement in porous media including the displacement of gas by oil as encountered in waterflooding and other secondary processes where an oil bank is formed. The treatment in this chapter will also include those special cases where unbalanced viscous forces exceed the gravity forces and result in fingering, coning, and hydrodynamic tilt.

The fluid displacement treatment in this chapter does not apply to the internal gas drive that results from the formation of free gas from solution gas as the pressure declines in a solution-gas-drive reservoir. However, the fundamentals may be useful in evaluating the vertical movement of gas and oil in such a reservoir. This often results in the formation of a secondary gas cap.

INITIAL SATURATION DISTRIBUTION IN A RESERVOIR

Engineers often assume there is an abrupt change in saturation vertically when we consider movement from a water zone to an oil zone or from an oil zone to a gas cap. However, this is seldom the case. A more normal initial saturation distribution would be that depicted in Fig. 5.1. Here the saturations grade smoothly from 100% water in the water zone to an irreducible water saturation some vertical distance above the water zone. The total liquid saturation grades smoothly from 100% in the oil zone to 0.0 in the gas cap. This results in a similar gradual transition in the oil saturation and in the gas saturation if these values were plotted

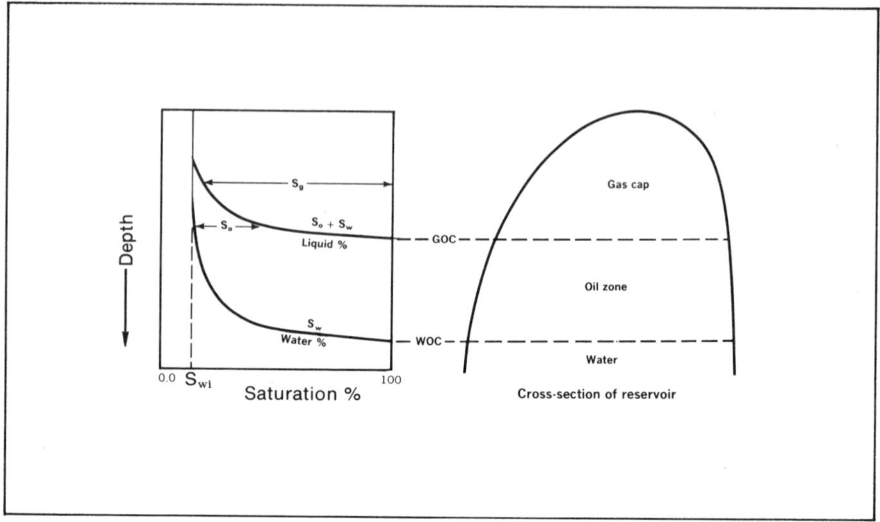

FIG. 5.1   Normal initial saturation distribution in a combination drive reservoir

separately. Consequently, the simplified diagram on the right-hand portion of Fig. 5.1 does not properly convey the actual saturation distribution that exists initially in the reservoir.

Fig. 5.1 serves as a definition of what is generally meant by a gas-oil contact or a water-oil contact. These terms mean different things to different engineers. For the purposes of clarity we will define the WOC (water-oil contact) as being the uppermost depth in the reservoir where a 100% water saturation exists. Note that this will also serve as a definition of a gas-water contact. Some engineers perfer to define the WOC and water-gas contact as the uppermost depth at which only water is produced. Due to relative permeability considerations these two depths are not the same.

Similarly, the GOC (gas-oil contact) is defined in this book as the minimum depth at which a 100% liquid (oil plus water) saturation exists in the reservoir. Again, some engineers prefer to define this as the greatest depth at which free gas is produced.

Determining the original WOC and GOC in a reservoir. Many different methods are used to determine the initial WOC and GOC in a reservoir. Most of these methods are more qualitative than quantitative to the extent that the engineer seldom attempts to distinguish between the 100% water level and the level at which 100% of the production is water. A similar statement could be made about the usual evaluation of the GOC. Obviously, a production test could then be used to determine the water-oil contact and the gas-oil contact.

A more sophisticated production test for the purposes of determining the WOC and GOC is the side-wall fluid sampler. This is a device that uses a packer arrangement together with one or two perforations to obtain a reservoir fluid sample from a small interval in the well. By taking a series of these samples it is possible to obtain a direct measurement of the WOC and GOC. The side-wall fluid sampler is a wire-line tool run during the drilling of a well before the casing is set. Interpretation of the data from these tests is complicated by the mud filtrate that has entered the formation and may constitute a large portion of the small (about 6 gal) sample that is taken but the interpretation of the data for the purposes of determining the WOC and GOC is generally conclusive.

Geophysical logs (electrical, radioactivity, etc.) are one of the most reliable methods of determining the WOC and GOC. These give the added attraction of providing the actual saturation distribution. However, as is the case with most log analysis there is generally some element of doubt associated with the log interpretation. Consequently, it is best to keep an open mind about the WOC and GOC levels and consider the evaluations determined from logs as being tentative conclusions subject to verification by flow testing.

Core analysis represents one of the most reliable methods of determining the WOC and GOC but the methods of interpreting these data to determine the WOC and GOC deserve some consideration. The engineer should remember that the saturation data obtained from a conventional core and conventional core analysis seldom represents the true saturations of the reservoir. If they do represent the actual saturations in the reservoir, that reservoir is probably so tight that it will be uncommercial to produce unless the formations are very thick.

When a conventional core is cut using a standard, diamond, or wire-line core barrel, the core is subjected to some degree of flushing by the drilling fluid or the drilling-fluid filtrate. Consequently, after cutting the core with the core still at reservoir pressure and temperature in the bottom of the hole, the core will generally contain mud filtrate, connate water, and the residual oil or gas left after flushing the core with the filtrate. While the core is being removed to the surface the pressure and temperature will be continually reduced until atmospheric conditions are reached. The solution gas in the oil left in the core after flushing with the mud filtrate will be liberated during the pressure reduction and will expand to force mud filtrate, oil, and possibly connate water from the core. Consequently, a core from the oil-bearing portion of the reservoir will contain gas, some small amount of oil, and a large amount of water which will be mud filtrate and connate water. Similarly a core from a gas zone will contain some gas, no oil, and a large amount of water, while a core from the water zone will contain almost 100% water and no oil. Thus, although the saturation values are of little use from a quantitative stand point, the change in the general characteristics of the saturation data from one zone to another provides a generally reliable means of evaluating the WOC and GOC.

The use of core analysis data to determine the WOC and GOC can be visualized by studying Fig. 5.2. Although this core has unusually low water saturations it is easy to pick the GOC at about 4,829 ft. It is also possible that a WOC exists at 4,849 ft although the change in the general saturations at this point are masked somewhat by the change in permeability and porosity. Unusually low water saturations as indicated in this core are sometimes caused by excessive weathering of the core (i.e. the core samples were not canned or in some other way sealed from the atmosphere quickly enough and much of the water evaporated).

Capillary-pressure data. When geophysical log interpretations can be relied upon these should be the basis for establishing the initial vertical saturation distribution in a reservoir. In many cases the accuracy of log interpretations is insufficient to permit the evaluation of the change in saturation with depth. In these cases it is often useful or necessary to evaluate the initial vertical saturation distribution from capillary-pressure data. A knowledge of this technique also provides the engineer with an excellent physical understanding of why there is a variation in the saturation vertically in the reservoir.

We will first examine the nature of capillary-pressure data. Naturally occurring porous media contain a wide variety of pore sizes. Most engineers are aware of the fact that a wetting fluid will rise in a small (capillary) tube due to the capillary attraction that exists between the fluid and the tube. The magnitude of the capillary rise is a function of the size of the tube, the angle formed between the wetting phase and the tube, and the interfacial tension that exists at the surface of the wetting phase.

Although all engineers have probably been previously exposed to the quantitative aspects of the relationship between capillary rise and these parameters during some elementary physics or science course, it seems wise to review this phenomena at this time.

In Fig. 5.3 a schematic core represented by three pore sizes is illustrated as being in a container of water with the water being attracted to a different level in each pore size. The capillary forces are indicated in the largest pore with the force, F, acting along the surface of the water which forms the angle, Θ, with the wall of the tube. The force is proportional to the energy required to maintain the interface between the gas and the liquid. This is called surface tension for a liquid-gas interface or interfacial tension for a liquid-liquid interface. Surface or interfacial tension has the units of force per unit of length. The length over which this force is being applied in a capillary tube is the circumference of a circle of radius r or $2\pi r$. Thus, the total capillary force is $2\pi r\sigma$ and the vertical force can be shown to be $2\pi r\sigma \cos \Theta$. When this is expressed as a pressure by dividing the vertical force by the cross-sectional area, πr^2, we obtain the expression for the capillary pressure.

$$P_c = 2\sigma \cos \Theta / r. \tag{5.1}$$

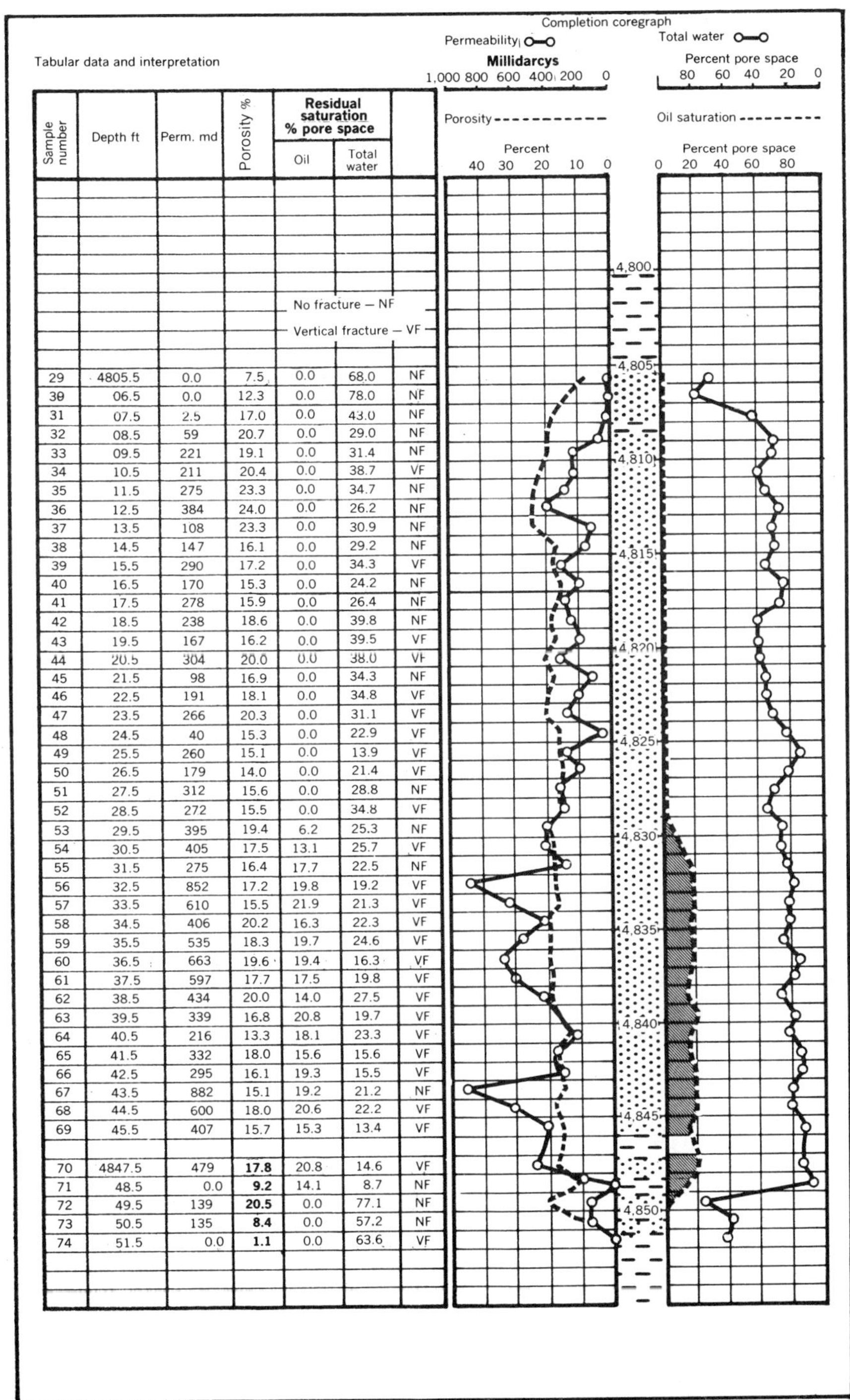

Tabular data and interpretation

Sample number	Depth ft	Perm. md	Porosity %	Residual saturation % pore space — Oil	Residual saturation % pore space — Total water	
					No fracture — NF	
					Vertical fracture — VF	
29	4805.5	0.0	7.5	0.0	68.0	NF
30	06.5	0.0	12.3	0.0	78.0	NF
31	07.5	2.5	17.0	0.0	43.0	NF
32	08.5	59	20.7	0.0	29.0	NF
33	09.5	221	19.1	0.0	31.4	NF
34	10.5	211	20.4	0.0	38.7	VF
35	11.5	275	23.3	0.0	34.7	NF
36	12.5	384	24.0	0.0	26.2	NF
37	13.5	108	23.3	0.0	30.9	NF
38	14.5	147	16.1	0.0	29.2	NF
39	15.5	290	17.2	0.0	34.3	VF
40	16.5	170	15.3	0.0	24.2	NF
41	17.5	278	15.9	0.0	26.4	NF
42	18.5	238	18.6	0.0	39.8	NF
43	19.5	167	16.2	0.0	39.5	VF
44	20.5	304	20.0	0.0	38.0	VF
45	21.5	98	16.9	0.0	34.3	NF
46	22.5	191	18.1	0.0	34.8	VF
47	23.5	266	20.3	0.0	31.1	VF
48	24.5	40	15.3	0.0	22.9	VF
49	25.5	260	15.1	0.0	13.9	VF
50	26.5	179	14.0	0.0	21.4	VF
51	27.5	312	15.6	0.0	28.8	NF
52	28.5	272	15.5	0.0	34.8	VF
53	29.5	395	19.4	6.2	25.3	NF
54	30.5	405	17.5	13.1	25.7	VF
55	31.5	275	16.4	17.7	22.5	NF
56	32.5	852	17.2	19.8	19.2	VF
57	33.5	610	15.5	21.9	21.3	VF
58	34.5	406	20.2	16.3	22.3	VF
59	35.5	535	18.3	19.7	24.6	VF
60	36.5	663	19.6	19.4	16.3	VF
61	37.5	597	17.7	17.5	19.8	VF
62	38.5	434	20.0	14.0	27.5	VF
63	39.5	339	16.8	20.8	19.7	VF
64	40.5	216	13.3	18.1	23.3	VF
65	41.5	332	18.0	15.6	15.6	VF
66	42.5	295	16.1	19.3	15.5	VF
67	43.5	882	15.1	19.2	21.2	NF
68	44.5	600	18.0	20.6	22.2	VF
69	45.5	407	15.7	15.3	13.4	VF
70	4847.5	479	**17.8**	20.8	14.6	VF
71	48.5	0.0	**9.2**	14.1	8.7	NF
72	49.5	139	**20.5**	0.0	77.1	NF
73	50.5	135	**8.4**	0.0	57.2	NF
74	51.5	0.0	**1.1**	0.0	63.6	VF

FIG. 5.2 Sample core analysis report (after Gatlin, *Petroleum Engineering*. Courtesy Baroid Petroleum Division, NL Industries, Inc.)

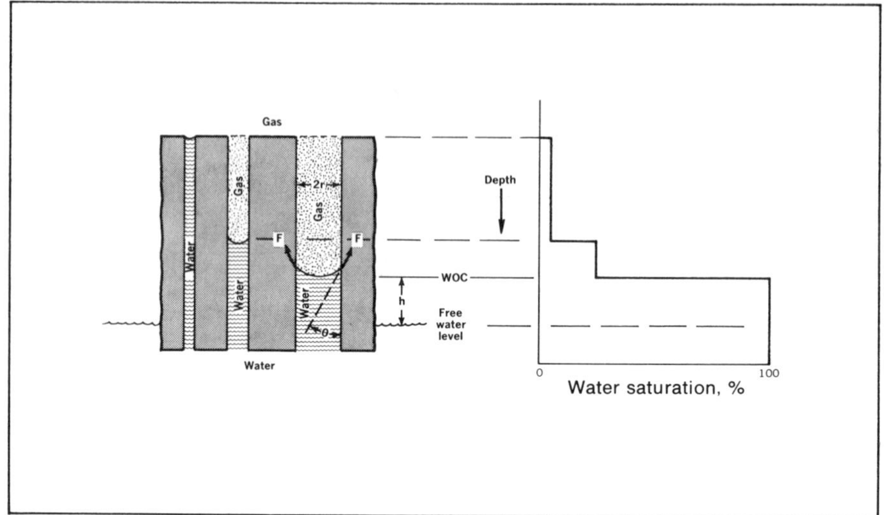

FIG. 5.3   Relationship between saturation distribution and pore-size distribution.

We can also state the capillary pressure, P_c, as a function of the unbalanced column of fluid of height, h, that it causes. We know that the pressure gradient for a column of water is 0.433 psi/ft and that the pressure gradient for a column of fluid of specific gravity, γ, is 0.433γ. Thus, the pressure equivalent of the unbalanced column of fluid will be equal to the capillary pressure.

$$P_c = 0.433(\Delta\gamma)h \tag{5.2}$$

The $\Delta\gamma$ in the equation is the difference in the specific gravities of the two fluids. If we equate Equations 5.1 and 5.2 we can show that the height of the capillary rise is proportional to the interfacial tension and the cosine of the wetting angle, and inversely proportional to the size of the capillary and the difference in the specific gravities of the wetting and nonwetting phases.

$$h \sim \frac{\sigma \cos \Theta}{r(\Delta\gamma)} \tag{5.3}$$

This expression is helpful in providing a qualitative knowledge of the effect of these parameters on the magnitude of the transition zone in the reservoir. The transition zone may be defined as the vertical thickness over which the saturations range from 100% water to irreducible water in the case of a WOC and from 100% liquid to an irreducible water saturation in the case of a GOC.

The general laboratory procedure used to measure the capillary pressure characteristics of a formation is to saturate the core sample with a wetting phase and then measure how much of the wetting phase is displaced from the sample when it is subjected to some given pressure of a nonwetting phase. To more exactly define this process we will assume that the schematic core of Fig. 5.3 is taken into the laboratory and saturated with water which we will assume is a wetting phase. We will further assume that the relative pore sizes and the fraction of the total pore volume represented by each pore size is as indicated in Fig. 5.4. Now imagine that we subject the sample to oil (the nonwetting phase) at the left end and increase the pressure on the oil until some water is displaced from the core. Note that the first displacement will take place when the oil pressure just exceeds the capillary pressure corresponding to the largest pore. In other words the capillary force will hold the water in the largest pore until the oil pressure is larger than the capillary pressure of the largest pore.

Also note that the amount of water displaced at that particular pressure will represent the pore volume of all pores of that particular size which, in this case, is 16/21 of the pore volume. Once this displacement has taken place the oil pressure will have to be increased to the next largest capillary pressure before any additional water will be displaced. Once this pressure is reached the water will be displaced from all of the pores of that particular size. After the additional 4/21 of the pore volume has been filled with oil the oil pressure will again have to be increased if additional water is to be displaced. But in our schematic core the smallest pore represents the pores of such small size that the capillary pressure is infinite and the water from these pores cannot be displaced. A plot of the displacement pressure versus the water displaced will then represent a plot of capillary pressure versus the percent of the pores with a capillary pressure greater than the subject capillary pressure.

Now remember that a reservoir rock contains a continuous variety of pore sizes from the largest to the smallest. Consequently, the capillary-pressure curve obtained for an actual reservoir rock will not have discontinuities such as those in the capillary-pressure curve of Fig. 5.4 which represents a hypothetical reservoir rock with only three pore sizes. Instead, the capillary-pressure curve of an actual reservoir rock will be a continuous smooth curve as shown in Fig. 5.5. Note that the capillary-pressure curve also can be calibrated to represent a plot of pore size versus the percent of pores whose pore size is less than the subject pore size since the interfacial tension and wetting angle of Equation 5.1 do not change during the test.

In the laboratory many different combinations of fluids have been used to measure capillary pressure. The most common combination is probably water and air. Some engineers prefer the use of mercury and air since equilibrium is reached almost immediately with these fluids while all other combinations require considerable lengths of time to reach equilibrium at a particular capillary pressure. However, other reservoir engineers contend that an air-mercury system is too different from the actual water-oil reservoir system to be representative of reser-

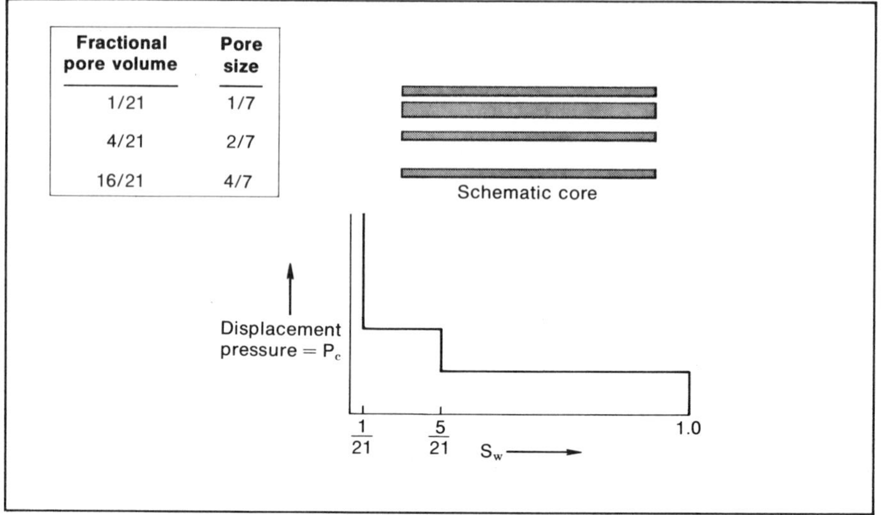

FIG. 5.4 Relationship between pore size distribution and a capillary-pressure curve.

voir capillary pressure. This writer believes that the additional accuracy gained by using a water-air or water-oil system in the laboratory is not justified when the much greater expense of running these tests is taken into consideration. Where both systems of tests have been run in a reservoir the data examined by the writer has seemed to correlate in an acceptable fashion.

Calculation of initial saturation distribution from capillary-pressure data. The general procedure for determining the saturation distribution in a reservoir from capillary-pressure data is to first take the family of laboratory capillary-pressure curves for a particular reservoir and interpret them in view of the general nature of the reservoir, to obtain one or more laboratory capillary-pressure curves to represent the reservoir. This will require the use of J functions, possibly permeability distribution data, along with a geologic understanding of the general nature of the reservoir. Next, it is necessary to convert the lab data to reservoir data by recognizing the difference between the characteristics of the fluids used in the laboratory and the actual reservoir fluids. Then the reservoir capillary-pressure curves can be interpreted in terms of a saturation distribution versus depth curve. In considering this general problem of predicting the saturation distribution from capillary-pressure data we will consider the technology in reverse order in the hope that it is more readily understood when it is presented in this way.

Once a capillary-pressure curve has been determined that is representative of the reservoir it can be directly interpreted to provide the saturation distribution versus depth. The first problem is to determine the depth of the free water level since capillary-pressures are measured

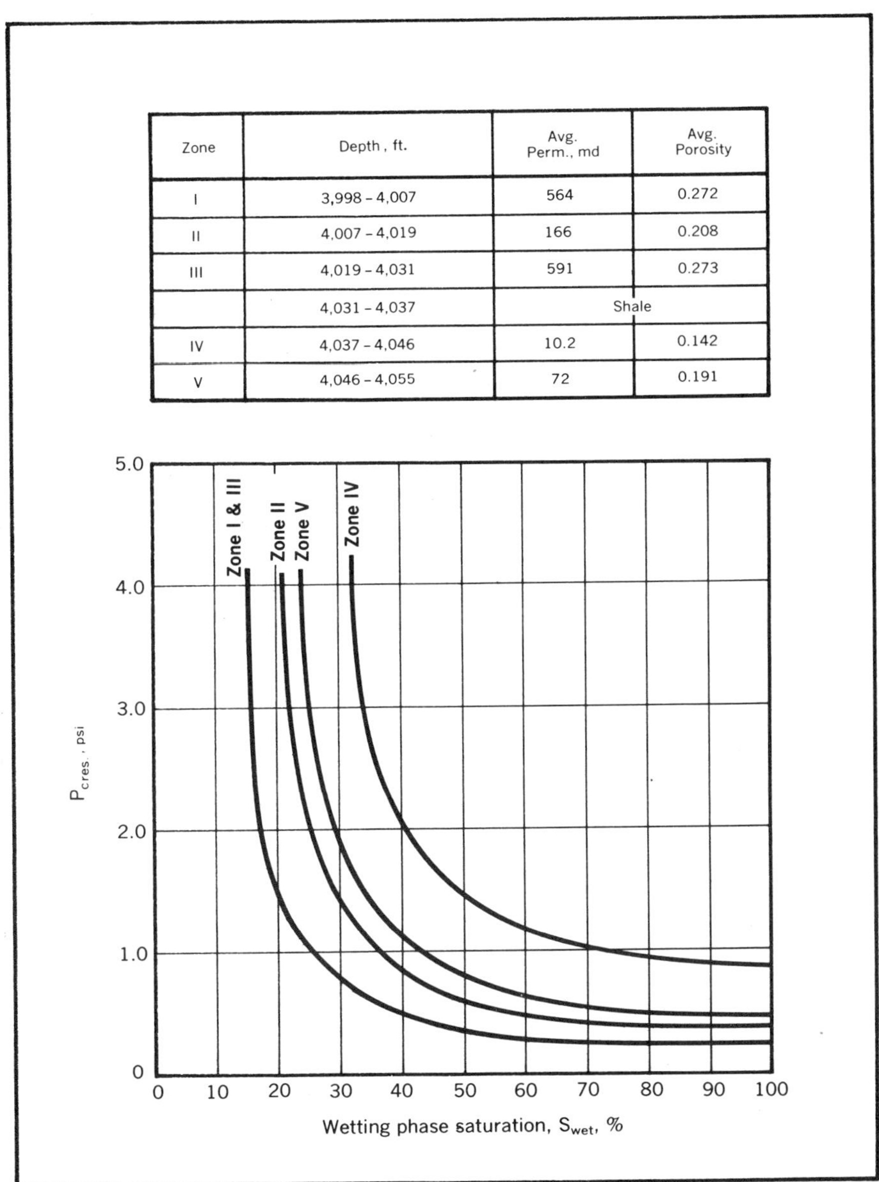

Zone	Depth , ft.	Avg. Perm., md	Avg. Porosity
I	3,998 – 4,007	564	0.272
II	4,007 – 4,019	166	0.208
III	4,019 – 4,031	591	0.273
	4,031 – 4,037	Shale	
IV	4,037 – 4,046	10.2	0.142
V	4,046 – 4,055	72	0.191

FIG. 5.5 Example capillary pressure curves for a stratified reservoir.

from the free water level. It was noted in Fig. 5.3 that there is a difference between the free water level and the minimum depth at which 100% water saturation exists. Note that this difference represents the capillary rise characteristic of the largest pore size in the reservoir. If this pore size is so large that there is no capillary rise in this size pore (as for example in vugular limestone) then the free water level and the 100% water saturation level will be the same. However, in most reservoirs these two levels are different because even the largest pore is small enough to cause some capillary rise. In very tight reservoirs the difference in these two levels may be many feet.

The difference can be obtained by reading the minimum capillary pressure that corresponds to the 100% water saturation on the capillary-pressure curve and converting this pressure to a height above the free water level by using Equation 5.2. This height can then be subtracted from the WOC (determined from well and test data as described above) to determine the free water level in the reservoir.

The engineer should carefully note that any discussion of the saturation above the free water level and the determination of the distance between the free water level and the 100% water level applies equally to the calculation of the saturation distribution in the gas cap above the 100% liquid saturation. The difference in this case would represent the height between the actual 100% liquid saturation level that exists in the reservoir and the 100% liquid saturation level that would exist if the reservoir contained a pore size that was so large that there was no capillary rise in the largest pores. Capillary-pressure would then be measured from this datum rather than from the free water level as is the case with the water-oil capillary-pressure data.

Once the free water level (or free 100% liquid level in the case of the gas cap) has been determined it is a simple matter to determine the saturation at any point above this level. The distance above the free water level, h, can be entered in Equation 5.2 and the corresponding reservoir capillary pressure can be calculated. The saturation corresponding to this capillary pressure can then be read from the reservoir capillary-pressure curve.

To test your understanding of these concepts work the following problem and compare your solution with the solution in Appendix C.

Problem No. 5.1: Determining the Static Saturation Distribution from Reservoir Capillary Pressure Data

Given the capillary-pressure curves of Fig. 5.5 for the various zones of the OSU Sand. Verify the plot of water saturation versus depth for this reservoir as depicted in Fig. 5.6. The 100% water saturation depth determined from core analysis is at 4,052.5 ft, the water density is 65.3 lb/ft^3 and the oil density is 56.2 lb/ft^3.

A well-stratified reservoir with much different permeabilities in the various strata can result in some unexpected saturation distributions. The capillary-pressure characteristics and zonation depicted in Fig. 5.5 and fluid characteristics enumerated in Problem No. 5.1 result in a saturation distribution in the stratified reservoir as shown in Fig. 5.6. Note

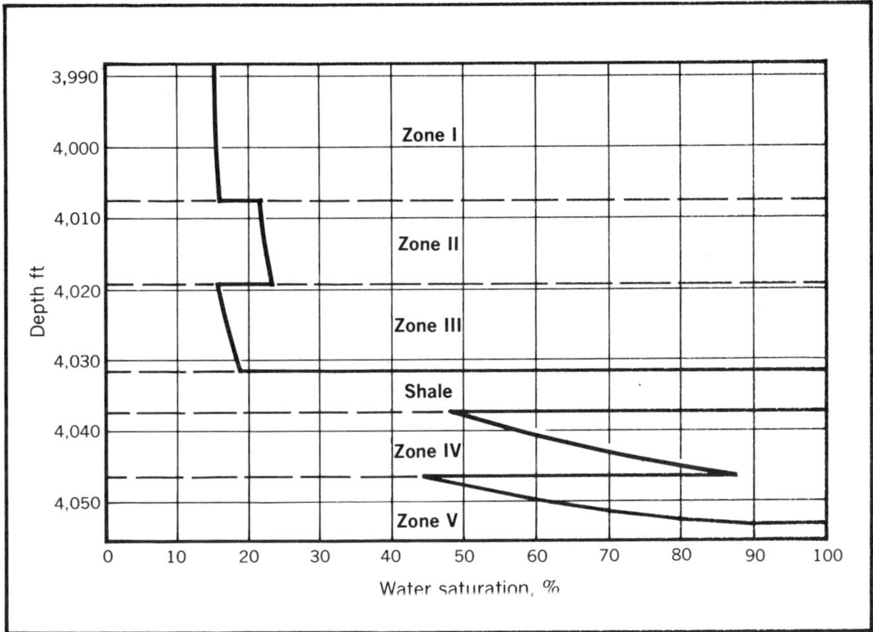

FIG. 5.6 Saturation distribution for reservoir with characteristics of Fig. 5.5.

that it is entirely possible to have substantially higher water saturations located vertically above lower water saturations. This phenomena may explain some of the "unexplainable" reports of cable tool drillers who sometimes note water production above oil productive zones and gas production below oil zones, in continuous reservoirs.

We should also note that differing reservoir characteristics at the free water level in a reservoir can cause a large difference in the vertical position of the WOC observed in various portions of a reservoir. The spread of the reservoir capillary pressures representing the 100% water saturation points in a suite of capillary-pressure curves such as Fig. 5.5 will give some insight into the variation in the observed WOC that may be expected in a particular reservoir.

Calculating reservoir capillary pressure from lab data. As discussed above, many different combinations of fluids are used in the laboratory for measuring the capillary pressure of a reservoir sample. Regardless of whether oil-water, water-air, or mercury-air fluid systems are used for the measurements, the laboratory fluids will not be the same as the fluids in the reservoir. Equation 5.1 clearly indicates that different fluid combinations will result in different capillary pressures because the interfacial tension, σ, and the wetting angle, Θ, vary with the nature of the fluids used. Thus, the capillary-pressure measurements from the laboratory must be corrected before they can be used in reservoir calculations.

If we know the interfacial tension and wetting angle for the fluids in the reservoir and the fluids used in the laboratory we can correct from the capillary pressures measured in the laboratory to capillary pressures in the reservoir. The mathematical basis for this calculation can be obtained by first using Equation 5.1 to write expressions for the laboratory capillary pressure and for the reservoir capillary pressure.

$$P_{c\ Lab} = \frac{2(\sigma \cos \Theta)_{Lab}}{r} \tag{5.4}$$

$$P_{c\ Res} = \frac{2(\sigma \cos \Theta)_{Res}}{r} \tag{5.5}$$

If we divide Equation 5.5 by Equation 5.4, and solve for the reservoir capillary pressure, we obtain the equation that can be used to correct laboratory data to reservoir data.[A]

$$P_{c\ Res} = P_{c\ Lab} \frac{(\sigma \cos \Theta)_{Res}}{(\sigma \cos \Theta)_{Lab}} \tag{5.6}$$

The practical use of Equation 5.6 presents some difficulties. The laboratory values of interfacial tension and the wetting angle are generally available or can be readily obtained by the lab. However, there is little data available for the empirical evaluation of the effective interfacial tension and wetting angle for the reservoir fluids at reservoir conditions. The measurement of the reservoir data in the laboratory is not generally possible because rather sophisticated equipment is necessary to measure the interfacial tension with gas in solution in the oil at a relatively high reservoir pressure and temperature. A smattering of data that can be used on an analogy basis can be found on page 130 of the *Handbook of Natural Gas Engineering*[1] by Katz et al.

Note that the units of the interfacial tension, σ, used in Equation 5.6 are immaterial since a ratio is employed.

One of the difficult problems that confronts the reservoir engineer in choosing capillary pressures to represent the reservoir is what we might call the zonation of the reservoir. The problem is one of characterizing the reservoir as far as the recognition of vertical changes in the reservoir permeability and porosity are concerned. We talk about a layered reservoir as being one in which, for example, the 100-md zone is at the top, the 50-md zone in the middle, and the 500-md zone is at the bottom of the reservoir. However, from well to well in this reservoir we may find one of the zones missing or worse, the sequence of permeabilities from top to bottom may be changed from low-middle-high to high-low-middle or middle-high-low. In many cases no effective zonation will exist but in others you may find an "average" zonation with wide swings around this "average" from well to well. There are some statistical methods of determining the best zonation to consider in a well, but the writer believes that, at this stage in the development of reservoir description methods, the reservoir engineer can arrive at his "best" zonation treatment by simply examining the data from well to well and using his best judgment as to how to group the reservoir permeability and porosity variations into zones with average values representing each zone.

To test your knowledge of the conversion of laboratory capillary pressures to reservoir data and familiarize yourself with the problem of zonation, work the following problem and compare your results with the solution in Appendix C.

Problem No. 5.2: Conversion of Laboratory Capillary-Pressure Data to Reservoir Capillary-Pressure Data

Given the laboratory water-air capillary-pressure curves of Fig. 5.7 and the core analysis for OSU No. 1 listed below. Verify the capillary-pressure data and zonation of Fig. 5.5 by calculating the reservoir capillary pressure for each zone at a 50% water saturation. The water-oil interfacial tension from this reservoir is estimated to be 28 dynes/cm and the wetting angle is 0°. Note that the core-analysis data are given as an average for every 3 ft to simplify the problem. Lab interfacial tension is 70 dynes/cm, and the lab wetting angle is 0.0°.

Depth	Permeability, md	Porosity, fraction
3,998–4,001	578	0.272
4,001–4,004	542	0.264
4,004–4,007	559	0.277
4,007–4,010	184	0.212
4,010–4,013	196	0.204
4,013–4,016	111	0.207
4,016–4,019	172	0.210
4,019–4,022	641	0.279
4,022–4,025	648	0.266
4,025–4,028	567	0.271
4,028–4,031	508	0.276
4,031–4,034	Shale	
4,034–4,037	Shale	
4,037–4,040	9	0.140
4,040–4,043	14	0.164
4,043–4,046	8	0.121
4,046–4,049	63	0.175
4,049–4,052	81	0.196
4,052–4,055	72	0.202

Averaging capillary-pressure data. It is seldom that the permeability and porosity of the reservoir samples whose capillary pressures are measured in the laboratory match the characteristics of the average permeability and porosity whose capillary pressure is desired. In most cases a suite of capillary pressures is run on a variety of reservoir samples

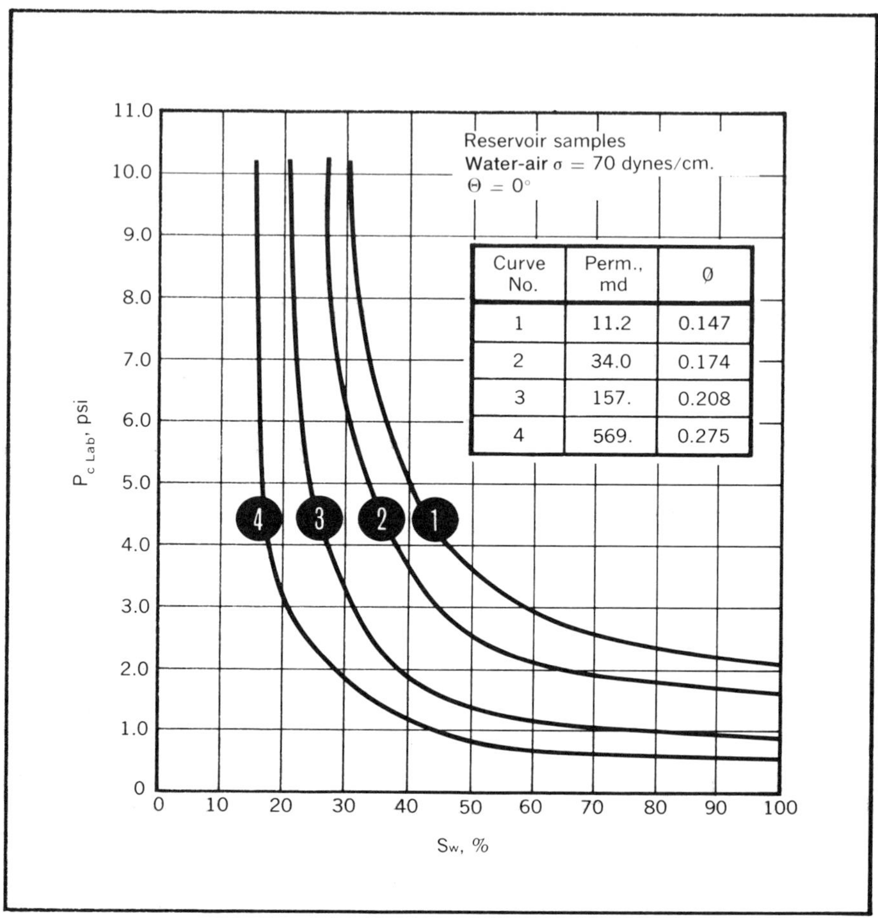

Curve No.	Perm., md	∅
1	11.2	0.147
2	34.0	0.174
3	157.	0.208
4	569.	0.275

FIG. 5.7　Capillary pressures of reservoir samples

with a wide range of permeabilities but none that correspond as closely to those desired as was the case in Problems 5.1 and 5.2. Consequently, the engineer is generally faced with a mass of capillary-pressure data, none of which is exactly what he wants. Furthermore, the method to be used in averaging the available data is not always apparent since the suite of capillary curves may cover the entire graph when they are all plotted on a single set of coordinates as in Fig. 5.7.

One of the classical methods of averaging capillary-pressure data is by using a J function. Sometimes the criteria necessary to make this technique very accurate are apparently met and the procedure is very useful, but other sets of data do not appear to conform at all to the assumptions of the method and the procedure is useless. Nevertheless, the procedure provides a good understanding of the relationship between the capillary-pressure characteristics and the permeability and porosity of the porous media and since it also is often a useful averaging technique it warrants our consideration.

The J function concept is based on the theoretical relationship between the permeability and porosity that can be derived by assuming that a porous media can be characterized as a bundle of capillary tubes that are not interconnected. The flow rate through one capillary tube according to Poisuelles equation[2] is,

$$q' = \frac{\pi (r')^4 \Delta p'}{8 \mu \, L'_{cap}} \tag{5.7}$$

where q' is the volumetric flow rate in Darcy units. Consequently, the flow rate through a number of capillary tubes, n, would be

$$q' = \frac{n \pi (r')^4 \Delta p'}{8 \mu \, L'_{cap}} \tag{5.8}$$

Now note that the porosity of a formation made up of a bundle of capillaries is

$$\phi = \frac{n \pi (r')^2}{A'} \tag{5.9}$$

and the permeability defined by the Darcy equation is

$$k = \frac{q' \, \mu L'_{core}}{A' \Delta p'} \tag{5.10}$$

Now if ϕA is substituted for $n\pi r^2$ and a function of the permeability, k, is substituted for q' in Equation 5.7, we can show that the permeability and porosity are related.

$$(r')^2 = (8k/\phi)(L'_{cap}/L'_{core}) \tag{5.11}$$

Note that the length of the capillary is recognized as being different from the core length. This becomes obvious when we recognize that a particular capillary path through a core is not a direct route from one face to another but that it actually tends to wind around in the porous media and follow a tortuous path. Consequently, this hypothetical ratio between the core length and the capillary length is often defined as the tortuosity.[3] If we assume that the tortuosity is constant for a particular reservoir and we combine this with the numerical constant of Equation 5.11 we can show that the radius is related to the permeability and porosity as

$$r = (constant)(k/\phi)^{0.5} \tag{5.12}$$

Substituting this definition of the radius into the definition of the capillary pressure, Equation 5.1, we obtain

$$P_c = 2\sigma \cos \Theta / (\text{constant})(k/\phi)^{0.5} \qquad (5.13)$$

When we solve this expression for the "constant," and incorporate the numerical constant of the equation, we obtain the J function.

$$J = \frac{P_c \, (k/\phi)^{0.5}}{\sigma \cos \Theta} \qquad (5.14)$$

The physical meaning of this expression is that by normalizing a capillary-pressure measurement for the differences in the permeabilities and porosities of the samples and the differences in the fluids used to measure the capillary pressure, a function of the capillary pressure is obtained that is independent of the permeability, porosity, interfacial tension, and the wetting angle. Furthermore, this function can be applied to a reservoir with any permeability and porosity that contains any combination of wetting and nonwetting fluids characterized by a particular interfacial tension, and wetting angle.

Consequently, this function of the capillary pressure can be calculated from capillary-pressure data and the characteristics of the samples and fluids used. One single curve will be obtained representing all of the capillary-pressure data for all of the samples measured. This means that a single J function curve not only represents many different permeabilities and porosities but the water-air, water-oil, and mercury-air laboratory systems are represented in one curve. Several sets of reservoir data have been observed by the writer that fit the J function theory. One such example is illustrated by the J function curve of Fig. 5.8.

However, keep in mind that the J function is based on the assumptions that the porous media acts like a bundle of capillaries and that the ratio between the length of the capillaries and the core length is a constant. Apparently, there are many reservoirs that do not fit these assumptions since a wide spread of data (that is almost useless for interpretation purposes) is often obtained when a J function plot is made for a particular reservoir. Another method of averaging capillary-pressure data will be presented but first it is suggested that the engineer work the following problem and check his solution against Appendix C to determine his knowledge of the J function concept.

Problem No. 5.3: Using the J Function to Average Capillary-Pressure Data

Given the capillary-pressure data of Fig. 5.7, verify the J function plot of Fig. 5.8 by calculating the J function at the 50% saturation point for each curve of Fig. 5.7. Using the J function curve of Fig. 5.8 calculate and plot a reservoir capillary pressure curve for Zone II of Fig. 5.5.

If the J function curve looks like a "buck-shot spread" or fails to give a well-defined relationship, then it is obvious that the reservoir with which you are working does not satisfy the assumptions inherent in the J function method. The engineer should then look to other averaging

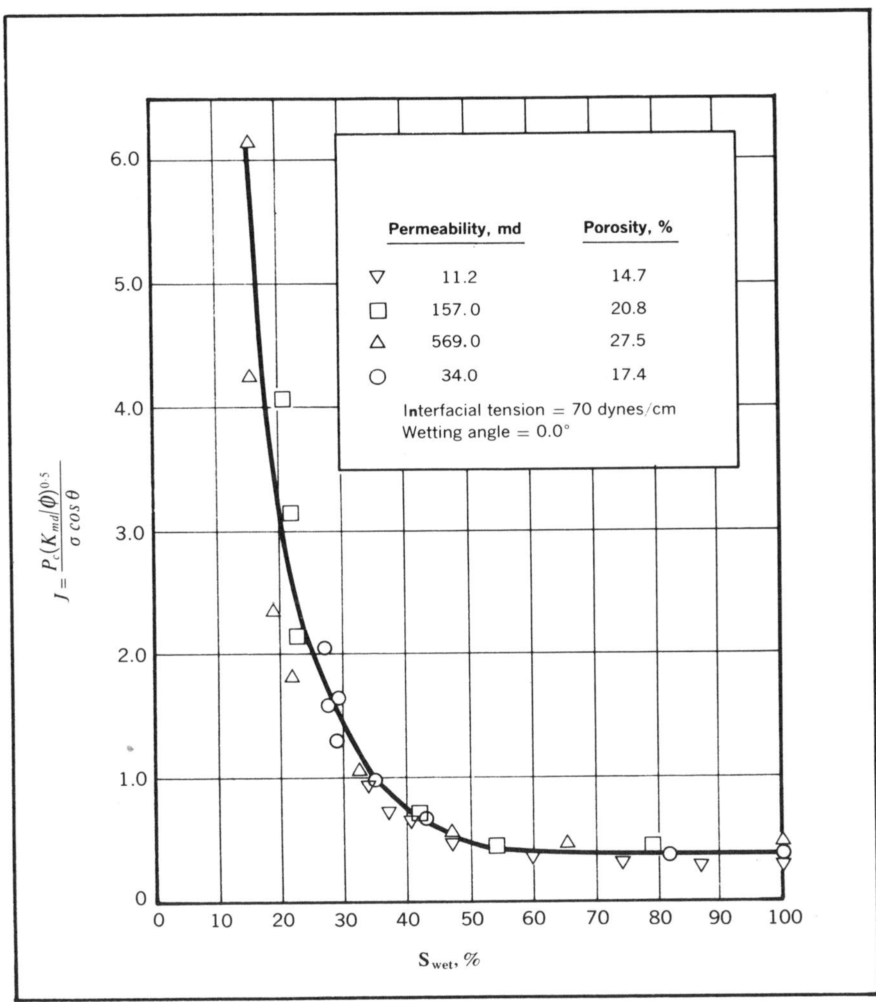

$$J = \frac{P_c (K_{md}/\phi)^{0.5}}{\sigma \cos \theta}$$

FIG. 5.8 A "J" function curve

techniques. One approach that often gives meaningful data is the method of weighting the capillary-pressure data on the basis of the statistical permeability distribution. In Chapter 1 the fact was recognized that even in the most uniform-looking natural porous media the measurement of permeability of small samples results in a wide range of measured permeabilities. Example data were presented as Fig. 1.1. Such data can be converted to a statistical form by converting the x-axis scale from "thickness" to "percent of reservoir." Then it is possible to determine the percentage of the reservoir that should be represented by each capillary-pressure sample on the basis of the permeability of that sample and the permeability distribution of the entire reservoir.

This simple technique can best be described by referring to some example data. Suppose that the statistical study of the permeability measurements for a reservoir indicates a permeability distribution such as that in Fig. 5.9 and that there is no indication of either an areal or vertical distribution of permeability. In other words we will assume that at any "point" in the reservoir all of the permeabilities exist and in the pro-

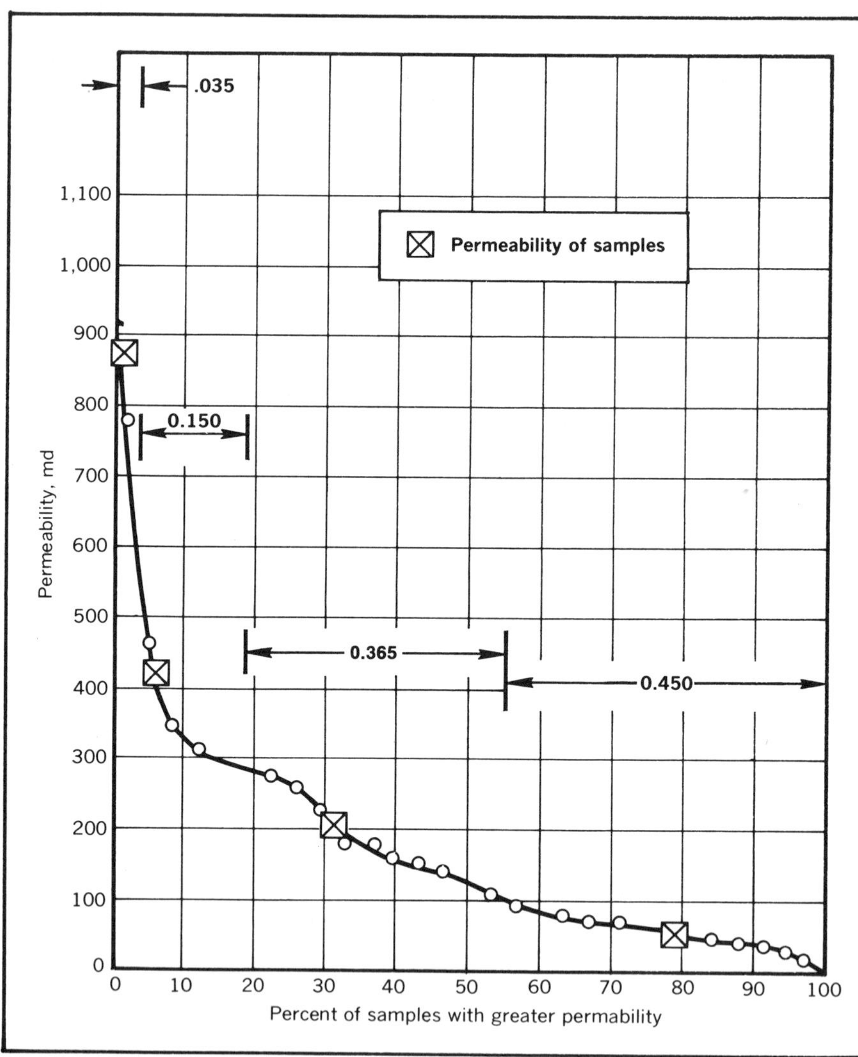

FIG. 5.9 Determining weighting factors from permeability distribution

portion indicated in Fig. 5.9. Now suppose that capillary pressures were run on only four core samples from this reservoir and that the permeabilities of these samples are as indicated on Fig. 5.9. Then it is logical to assume that at a particular level in the reservoir, represented by a particular capillary pressure, the average saturation could be determined by determining the saturation at this level (or capillary pressure) for each of the capillary-pressure samples. Then a weighted average could be calculated on the basis of the percent of the reservoir that can be considered to be represented by each sample.

For example, a study of the relationship between the permeability distribution of Fig. 5.9 and the permeabilities of the capillary samples, indicates that the sample permeabilities of 875, 420, 210, and 55.7 md represent 0.035, 0.150, 0.365, and 0.450 fractions of the reservoir. Consequently, the saturations for a particular capillary pressure or reservoir level should be weighted by these fractions.

RESERVOIR SATURATION DISTRIBUTION DURING DISPLACEMENT

In the above consideration of the initial saturation distribution we noted that an engineer's first assumption without any theoretical knowledge of the situation is that initial saturations are uniform throughout the water zone, oil zone, or gas cap of a reservoir. The same engineer would then assume that as production took place in a reservoir the gas cap would expand and/or the water would encroach and that the saturation in the invaded portions of the reservoir would be uniform. Such an assumption is often referred to as an assumption of "piston-like displacement." Just as uniform saturation distribution seldom exists in a reservoir, piston-like displacement seldom takes place in the reservoir.

What actually happens in the frontal displacement of oil by advancing gas or water is that a more or less uniform saturation distribution is formed with a discontinuity only at the furthermost advance of the gas or water. Our objective in this section will be to consider methods of determining the saturation distribution along the displacement streamlines. The concepts developed will be useful directly as a means of determining the saturation distribution in water-drive and gas-cap-drive reservoirs and will also provide useful general concepts concerning the displacement of one immiscible fluid by another in a porous media. These ideas should be particularly helpful in the consideration of secondary recovery.

We will first consider the general characteristics of the displacement and the general procedure to be followed in calculating the saturation distribution at any particular time. Once this procedure has been established the detailed steps in the analysis and some useful simplifying short cuts will be developed.

General characteristics of fluid displacement. To illustrate what happens when one fluid displaces another in the reservoir, consider a water-drive reservoir, Fig. 5.10, in which water is encroaching updip at a relatively slow rate as oil is produced near the top of the structure. Now consider the saturation in the horizontal slices of the reservoir as displacement proceeds. We will do this by identifying each horizontal slice as being some distance, X, from the initial minimum depth of the 100% water saturation with the distance, X, measured along the dip of the formation. When the initial saturation in these horizontal slices is plotted versus the X position of each slice we obtain a relationship such as shown in the left-hand side of Fig. 5.10.

Consider the nature of the saturation distribution after 1 year when a considerable amount of oil has been produced from this reservoir and a like amount of water has encroached into the initial oil-bearing portion of the reservoir. The resulting saturation distribution is shown in Fig. 5.11 along with the initial saturation distribution and the saturation distribution at later times of 2 and 3 years. This illustration has been generalized by using the general displacing fluid saturation symbol, S_d. Also the position of the original WOC is labeled as the "influx face" and the position of the nearest producer is termed the "producing face."

Notice that the only saturation discontinuties that exist are at the front or furthermost advance of the displacing fluid. Also note that the displacing-phase saturation is increasing at all times at almost all points in the reservoir behind the advancing front. In other words, the advancing front does not displace all of the mobile oil as it moves through the reservoir. Instead, it acts more like a very inefficient piston. The front of the displacing fluid corresponds to the first stroke of the inefficient piston which displaces some fraction of the mobile oil. As water continues to flow through the same pore volume it acts like successive piston strokes with some percentage of the mobile oil that is left being displaced until finally after many pore volumes have flowed through the same pore space (many piston strokes have taken place) all of the mobile oil has been displaced. The zone behind the displacing front is sometimes referred to as the "drag zone" which seems to be fairly descriptive of what takes place physically in this part of the displacement.

The engineer should be careful that he does not become confused as to the meaning of Fig. 5.11. A common error is to interpret this figure as indicating some vertical distribution of fluid with the y-axis mistaken to be the vertical distance so that there is water or displacing fluid underriding the oil. This of course is an erroneous impression. Remember that the diagram is simply a plot of the average saturation in horizontal slices of the reservoir plotted against the distance from the original 100% displacing phase saturation measured along the bedding plane.

We should note two general characteristics of fluid displacements in porous media that are clearly indicated in Fig. 5.11. First note that there is a saturation below which the saturation of the displaced (oil) phase

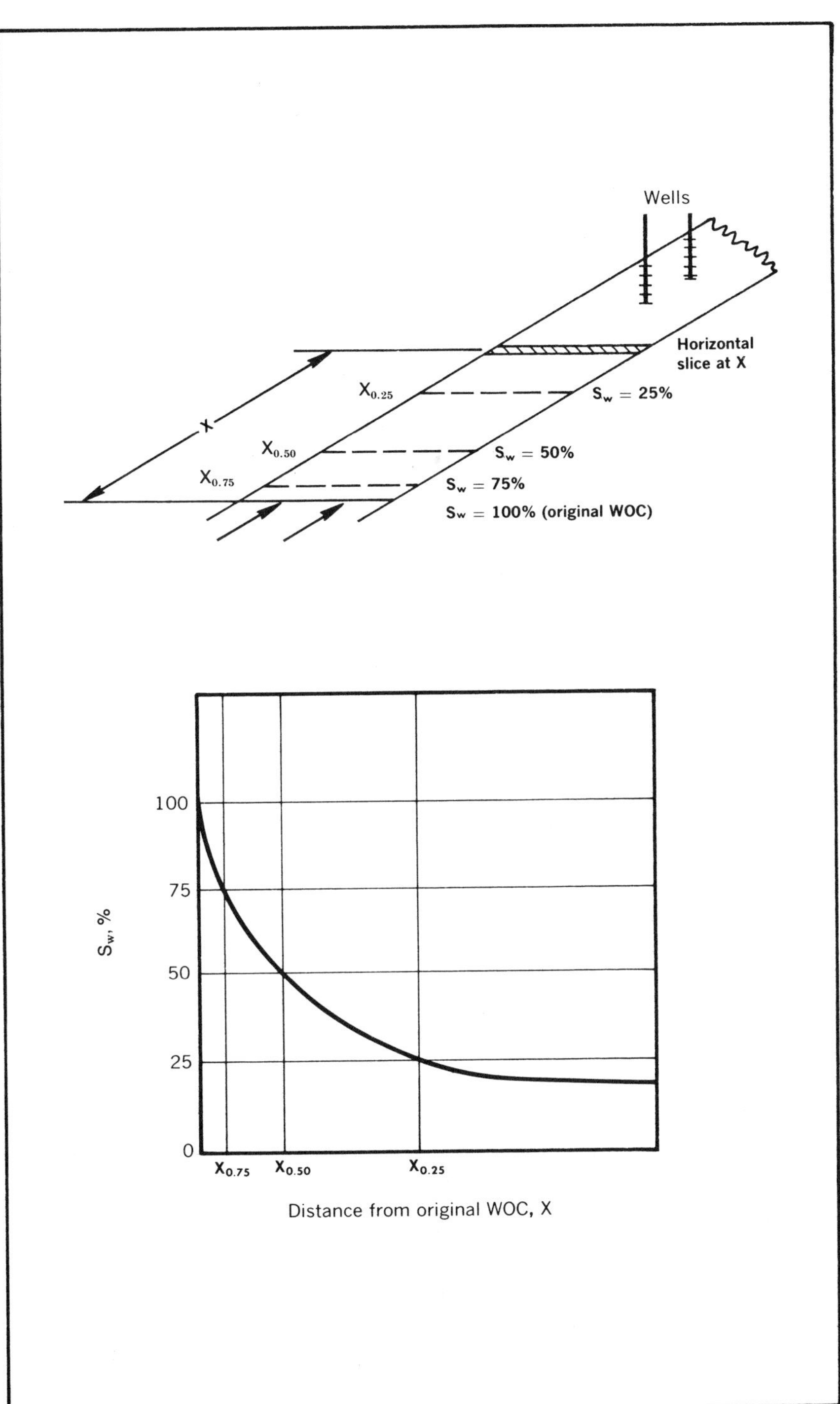

FIG. 5.10 Saturation profile in a water-drive reservoir.

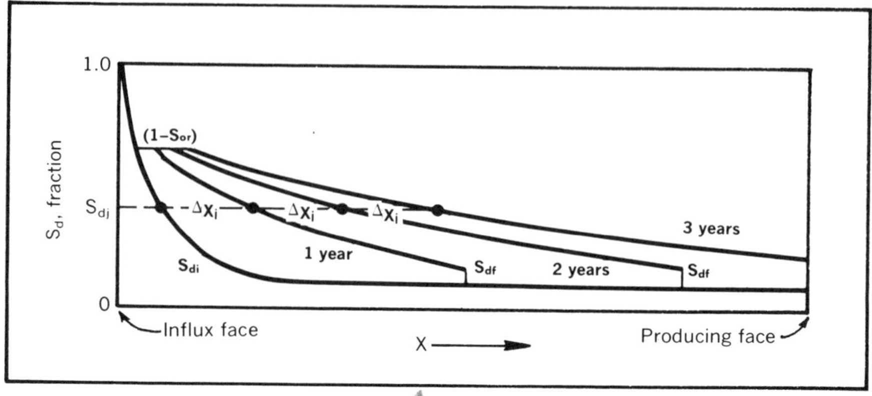

FIG. 5.11 Fluid-displacement characteristics with initial saturation distribution.

cannot be driven regardless of the amount of displacing fluid that passes through the porous media. This residual saturation is represented in Fig. 5.11 by the horizontal line labeled, $(1 - S_{or})$, which is the displacing fluid saturation equivalent of the residual oil saturation, S_{or}. Note that we refer to the displaced fluid generally as being oil since this is generally the situation. However, any immiscible fluid could be the displaced fluid. The technology can certainly be applied to the displacement of gas by water in a water-drive gas reservoir and we will apply the technology to the displacement of gas by oil when we discuss waterflooding and the formation of an oil bank in a later chapter.

The second general characteristic of immiscible displacement that we should note from Fig. 5.11 is that each saturation travels through the reservoir at some fixed velocity as long as the displacement rate, q_t, is constant. In other words if the process is the displacement of oil by a natural water drive, as was originally suggested, and the production and encroaching water rates were constant, the 50% water saturation would move through the reservoir at some constant rate and the 60% water saturation would move through the reservoir at a constant but different rate and the 65% water saturation would travel at a constant but still different rate, etc. This is illustrated in Fig. 5.11 by indicating that a particular saturation, S_{dj}, moves a distance ΔX_j through the reservoir during each year. Carefully note that during the first year this distance of movement is measured from the initial position of the S_{dj} saturation in the reservoir and is not measured from the 100% S_d saturation. This is pointed out specifically because such an error in application is very common.

We should also note in studying Fig. 5.11 that the frontal saturation does not remain constant when an initial saturation profile such as the one in Fig. 5.11 is encountered. This point is made to contrast with the situation that exists in Fig. 5.12 where the frontal saturation does remain constant. The difference in the conditions of Figs. 5.11 and 5.12 is that in Fig. 5.12 the saturation initially is uniform throughout. This situation can represent a waterflood being initiated in a reservoir that did not or does not have a natural water drive or gas-cap drive; or it could represent the situation that is approximated when the pores are so large in a reservoir that the transition zone can be treated as being of negligible thickness. In either case (and in many others) the initial saturation conditions can be treated as being uniform throughout the reservoir.

Note that having a uniform saturation throughout the reservoir initially actually means that an abrupt and very large discontinuity exists in the reservoir at the influx face. To the left of the influx face shown in Fig. 5.12, we know that the displacing-phase saturation is 100% and that to the right the saturation is, S_{di}, the initial displacing-fluid saturation that is uniform throughout the oil-bearing portion of the reservoir. This then also means that all of the S_d saturations from S_{di} to $(1 - S_{or})$ exist at the same point in the reservoir at the influx face.

Fig. 5.12 shows that the results of a displacing-fluid encroachment into a reservoir with a uniform initial saturation is very similar to the results of a displacing-phase encroachment into a reservoir that has a nonuniform saturation distribution initially. Note that there is still a lower limit on the oil saturation which cannot be further reduced regardless of how much of the displacing fluid travels through the reservoir, and that the velocity of a particular saturation is constant and different from the velocity of other saturations. However, it should be remembered that the initial position of all the saturations is the influx face in contrast to the situation in Fig. 5.11 in which each saturation has a different initial position.

There are general observations that should be made about Fig. 5.12 that are peculiar only to a displacement that occurs in a reservoir where the displacing-fluid saturation is uniform throughout the reservoir

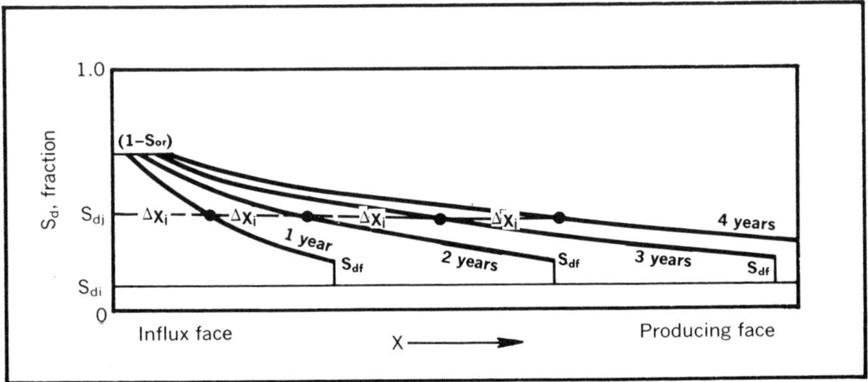

FIG. 5.12 Fluid-displacement characteristics, no initial saturation distribution.

initially. In this case the frontal saturation remains constant (as noted above) until it reaches the producing face. When this constant frontal saturation is coupled with the constant velocity of each saturation starting from the same point in the reservoir (the influx face) we can show that the average saturation behind the displacing front remains constant. This portion of the displacing-phase system is generally referred to as the displacing-fluid bank. For example, in a waterflood it is called the water bank. Although the displacing-fluid bank continues to increase in size as the displacement continues, remember that the average saturation in this bank remains constant until the front reaches the producing face if the initial displacing-fluid saturation was uniform throughout.

There is not always a displacing-fluid saturation in the reservoir prior to the initial displacing-fluid influx. For example, in a gas-cap drive when the gas begins encroaching into (expanding into) the oil-bearing portion of the reservoir, there may be no gas saturation present. On the other hand in a waterflood of a previously primary depleted oil reservoir there would be a uniform connate water saturation in the reservoir before water injection is initiated.

It may be helpful to list the previously noted characteristics of an immiscible displacement in porous media:

(1) There is a characteristic saturation of the displaced phase which cannot be further reduced regardless of the quantity of displacing phase that flows through the porous media.

(2) As long as the influx rate remains constant, each displacing-fluid saturation moves through the reservoir at a particular velocity that is characteristic of that particular saturation and different from the velocity of other saturations.

If the displacement takes place in a reservoir whose displacing-fluid saturation was initially uniform throughout the reservoir we also noted:

(3) The saturation at the displacing-fluid front remains constant until it reaches the producing face.

(4) The average saturation in the displacing-fluid bank remains constant until the front reaches the producing face.

General procedure for calculating saturation distribution during displacement. The displacing-fluid fraction of the total fluid flowing will vary with the distance behind the displacing-fluid front. This can be better understood by examining the nature of two-phase relative permeability with respect to the displaced and displacing fluids. For example, examine Fig. 5.13 which is the water-oil relative-permeability data for a water-displacement problem to be given later in this chapter. Note that when the saturation in the reservoir is 55% water and 45% oil the relative permeabilities to water and oil are the same. Consequently, if the viscosities of the two fluids were equal, the flow rates of the fluids would be equal—half water and half oil.

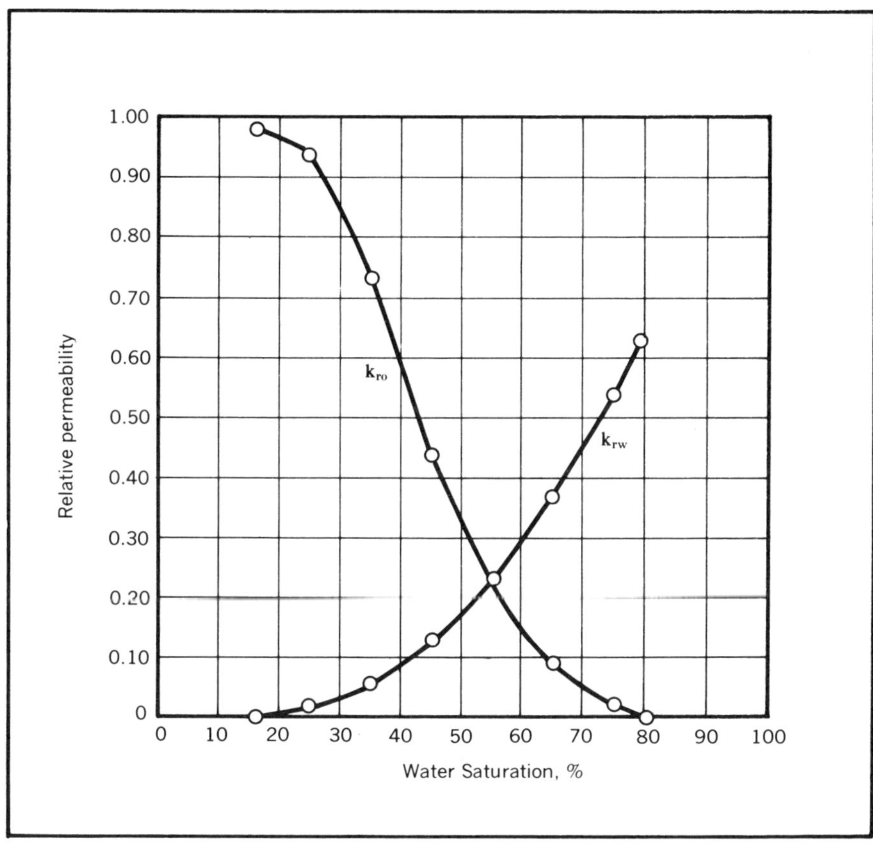

FIG. 5.13   Oil-water relative permeability data (used in Problem No. 5.3A)

Consider the situation at a point in the reservoir where the saturations are about 45% water and 55% oil. From the relative-permeability data note that the ratio of the permeabilities would be about three to one with the oil largest. Consequently, if we again assume the viscosities to be equal, the oil rate of flow would be about three times the rate of flow of the water and the fraction of water flowing would be about one-fourth of the total rate. Since the displacing-phase fraction of the total fluid flowing varies with the saturation, which in turn varies with the distance behind the displacing-phase front, the saturations behind the front must continually change with time, until the irreducible oil saturation is reached.

Consequently, the first step in calculating the saturation distribution during a displacement is to calculate the fractional flow curve based on the relative permeabilities, viscosities, and other reservoir parameters. Such a relationship will be a plot of the fraction of the total flow rate that is displacing fluid versus the saturation as illustrated in Fig. 5.14. This fractional flow curve can then be used to evaluate the change in the fractional flow with the change in saturation at any particular saturation ($\Delta f_d/\Delta S_d$). This value in turn is used in the Buckley-Leverett equation to calculate the position of a saturation in the reservoir at a particular time. The Buckley-Leverett equation for linear flow (which will be derived later) is

$$\Delta X_{Sdj} = \frac{5.615 q_t \, \Delta t}{\phi A} (\Delta f_d/\Delta S_d)_{Sdj} \tag{5.15}$$

The $\Delta f_d/\Delta S_d$ expression is the slope of the fractional flow curve at the saturation of interest, S_{dj}, and the ΔX is the distance this saturation moves during the time interval, Δt.

It would appear on the surface that the above procedure would result in the desired saturation distribution at a particular time because we would be able to simply calculate the position of any saturation in the reservoir at any time. However, the procedure and equations have a mathematical peculiarity that makes some further calculations necessary. Note that the fractional flow curve, Fig. 5.14, has slopes such that one slope represents more than one saturation. A mathematical interpretation of this means that a particular point in the reservoir could be represented by more than one average saturation. This represents a physical absurdity. Consequently, the engineer must have a means of determining just how much of the calculated saturation profile he should use. This is obtained by coupling the Buckley-Leverett calculation with a material-balance calculation and in effect it amounts to evaluating the frontal saturation.

To summarize, the general procedure for calculating the saturation distribution during a displacement is (1) calculate and plot the fractional flow curve, (2) use slopes evaluated graphically from the fractional flow curve, in Equation 5.15 to calculate the position of the various saturations at the time of interest, and (3) determine the frontal saturation at the time of interest.

These steps in the calculation will be considered separately.

Determining the fractional flow curve. The fractional flow, f_d, is simply defined as the fraction of the total fluid flow that is due to the flow of the displacing phase. If only two phases are flowing at a particular spot in the reservoir and the displaced phase is oil, we define the fractional flow as

$$f_d = \frac{q_d}{q_d + q_o} \tag{5.16}$$

Now suppose we substitute for the flow rates according to the Darcy equation originally introduced in Chapter 1,

$$q = -(1.127 kA/\mu)(\Delta p/\Delta X) \tag{1.3}$$

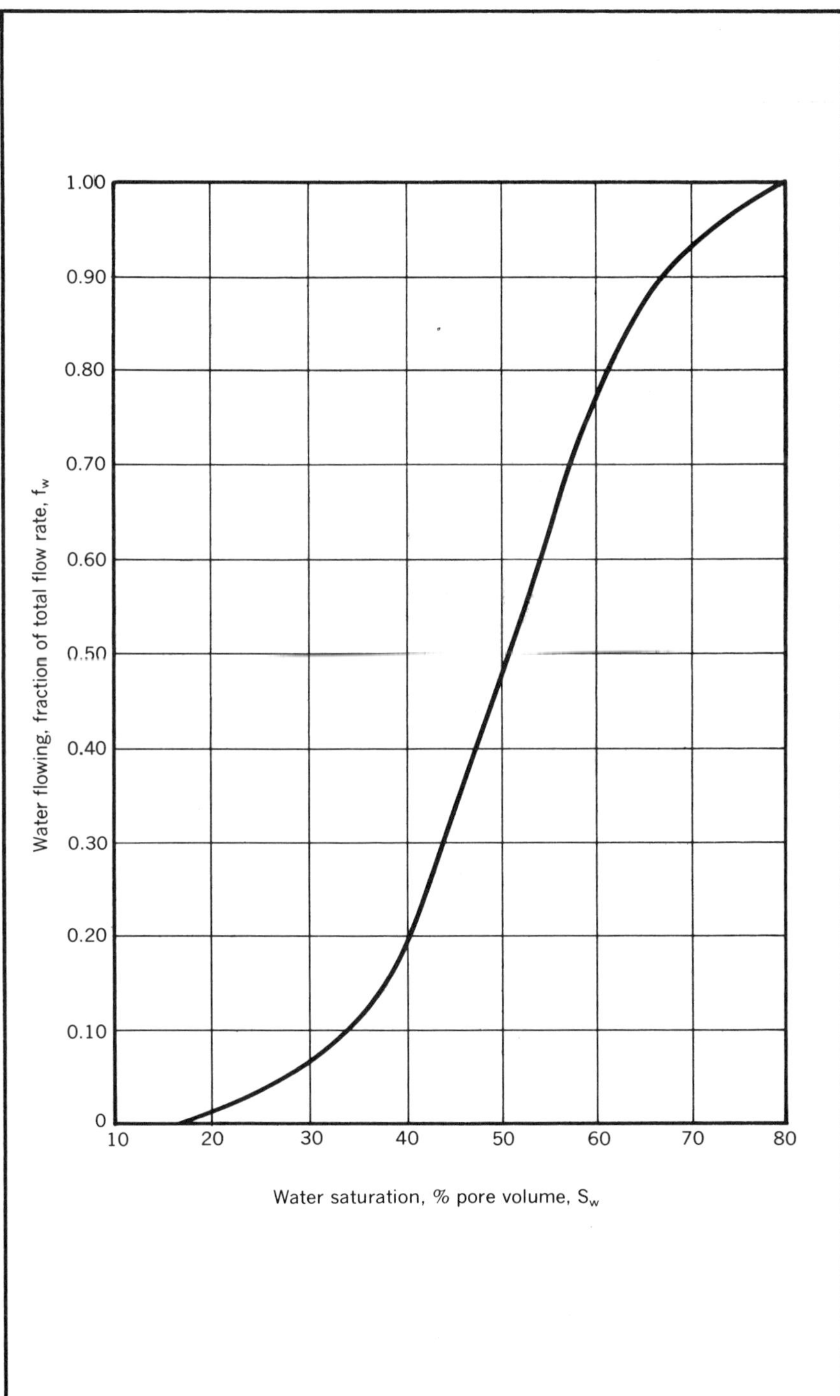

FIG. 5.14 Fractional flow curve.

If we assume that flow is horizontal and the capillary pressure can be neglected we could conclude that all of the parameters controlling the flow rate of each phase would be the same except for the permeability to and the viscosity of each phase. Substituting the resulting expressions into Equation 5.16 and simplifying we obtain an expression for the fractional horizontal flow of the displacing phase.

$$f_{dNO} = \frac{(1.127k_d A/\mu_d)(\Delta p/\Delta X)}{(1.127k_d A/\mu_d)(\Delta p/\Delta X) + (1.127k_o A/\mu_o)(\Delta p/\Delta X)} \quad (5.17)$$

$$f_{dNO} = \frac{1}{1 + (k_o/k_d)(\mu_d/\mu_o)} \quad (5.18)$$

The symbol f_{dNO} is used to denote that this fractional flow expression does not contain any gravity effects or as stated above, it applies to only horizontal flow or when the effect of nonhorizontal flow is negligible. When flow takes place in other than a horizontal direction we showed in Chapter 1 that Darcy's equation had to be modified to account for the angle between the direction of flow and the horizontal direction, α, if the total pressure gradient is used (rather than the pressure gradient due to flow that is represented in most flow equations).

$$q = (1.127kA/\mu)[(\Delta p/\Delta X)_{total} \pm 0.433\gamma \sin \alpha] \quad (1.52)$$

In this equation γ is the specific gravity of the flowing fluid relative to water because the numerical constant 0.433 is the vertical pressure gradient in fresh water. If the fractional flow with gravity is derived in the same way that the fractional flow without gravity was derived, except that Equation 1.52 (instead of Equation 1.3) is used to define the flow rates of the two fluids, the resulting expression is

$$f_d = \frac{1 - \dfrac{0.488k_{ro}KA|\Delta\gamma|\sin\alpha}{\mu_o q_t}}{1 + (k_o/k_d)(\mu_d/\mu_o)} \quad (5.19)$$

Note that this equation is simply a modification of Equation 5.18. The first term, 1.0 divided by the denominator, is f_{dNO} according to Equation 5.18 so that the "second term" is the modification necessary to have the equation apply to nonhorizontal flow.

Several items should be noted in connection with the application of Equation 5.19. Note that the gravity term contains the effective permeability to the oil stated as a function of the absolute permeability, k. This is done so that the application of the equation to any particular reservoir will permit the simplification of the equation to a form in which all the variables are a function of the saturation. In this regard it may be helpful to note that the ratio of the effective permeabilities, k_o/k_d, is the same as the ratio of the relative permeabilities, k_{ro}/k_{rd}. Also note that $\Delta\gamma$ in the equation, the difference between the specific gravities of the displaced and displacing fluids relative to water, has an absolute symbol around it. This comes about mathematically because of the way the writer prefers to define α, the acute angle between the direction of flow and the horizontal direction. This angle is the same as the conventional bed dip angle definition and is always taken as positive. It is simpler and less confusing to most engineers to treat the angle α as always being positive regardless of the direction of flow. Since this is a mathematical inconsistency, it is counteracted in Equation 5.19 by always considering the difference between the displaced and displacing phases as being positive. This is simply an artificial means of controlling the sign of the gravity term so that it will always be negative. This is normally encountered in practice. Water displaces oil or gas updip and has a greater density than either oil or gas so that the gravity tends to keep the fluids separated. Similarly, gas displaces oil downdip but has a density that is less than the oil so that it tends to keep the fluids separated by gravity. If you prefer a more rigorous approach to the definition of the formation dip and the specific gravity difference it is recommended that you check the Craft and Hawkins text, page 367.

To test your knowledge of the methods of calculating the fractional flow curve, work the following problem and check your solution against Appendix C.

Problem No. 5.4A: Calculation of a Fractional Flow Curve

A water-drive reservoir is of such size and shape that water encroachment to the first line of producers can be treated as linear flow. The water drive is sufficiently active that fluid flow is steady state. The withdrawal rate from the reservoir averages 2,830 reservoir BPD. Reservoir data are as follows:

Average formation dip	15.5°
Average "width" of reservoir	8,000 ft
Reservoir thickness	30 ft
Average cross-sectional area (A)	240,000 sq ft
Permeability	108 md
Connate water (irreducible water)	16%
Reservoir oil specific gravity	1.01
Viscosity	1.51 cp
Reservoir water specific gravity	1.05
Viscosity	0.83 cp

Relative-permeability data (plotted in Fig. 5.13)

S_w, %	k_{rw}	k_{ro}
79	0.63	0.00 (Critical)
75	0.54	0.02
65	0.37	0.09
55	0.23	0.23
45	0.13	0.44
35	0.06	0.73
25	0.02	0.94
16	0.00 (Critical)	0.98

Calculate the fractional flow for this reservoir corresponding to the saturations listed above in the relative-permeability data.

Equation 5.19 has some qualitative utility that may exceed its value for calculating the fractional flow curve. This equation provides an excellent means for determining the effect of various reservoir parameters on the displacement efficiency. The lower the fractional flow of the displacing fluid at a particular saturation, the higher the displacement efficiency, because the fraction of oil flowing must be higher. Note that if gravity does not have an effect on the fractional flow that the displacement efficiency will be a function of only the mobility ratio, $(k_d \mu_o / k_o \mu_d)$. The lower this mobility ratio, the higher will be the displacement efficiency. This we know intuitively or we can reach the same conclusion by studying Equation 5.18.

However, if gravity is affecting the fractional flow, all of the parameters in the second term of the denominator will have an effect on the efficiency. Since minimizing the fraction of displacing fluid flowing maximizes the fraction of oil flowing, maximizing the gravity term will maximize the displacement efficiency. Thus, increasing the permeability, specific gravity or density difference, formation dip, and the mobility of the oil, (k_{ro}/μ_o) will improve the displacement efficiency.

The flow rate and cross-sectional area are probably best thought of as a velocity. Equation 5.19 then shows that when gravity effects are significant the displacement efficiency is maximized by minimizing the velocity, q_t/A. It is also worthy of note that, according to this equation, there will be no gravity effect as long as flow is horizontal since the sine of 0° is zero.

The Buckley-Leverett equations. As noted previously, once the fractional-flow curve has been evaluated, the slopes from the curve can be used to calculate the position of a particular saturation at a particular time. The equation normally used for this purpose is Equation 5.15, the linear Buckley-Leverett equation.

This equation can be derived by evaluating the change in the displacing phase saturation in a Δx increment of the reservoir during some time period, Δt, and relating this expression to the change in the average fraction of displacing fluid flowing into and out of Δx during the time interval Δt. These relationships may be best visualized physically by referring to Fig. 5.15. Here it can be seen that Δt represents the length of time for a particular saturation in the reservoir to move a distance Δx. Now if we consider the change in average saturation during Δt to be ΔS_d, it is clear that the change in the number of barrels of the displacing fluid in the Δx segment during Δt will be the pore volume of the segment multiplied by the change in the saturation.

$$\Delta \text{ bbl of "d" in } \Delta x \text{ during time } \Delta t = (\Delta x A \phi / 5.615) \Delta S_d \qquad (5.20)$$

Also notice from Fig. 5.15 that the fraction of fluid flowing at a particular point in the reservoir changes as the saturation at that point

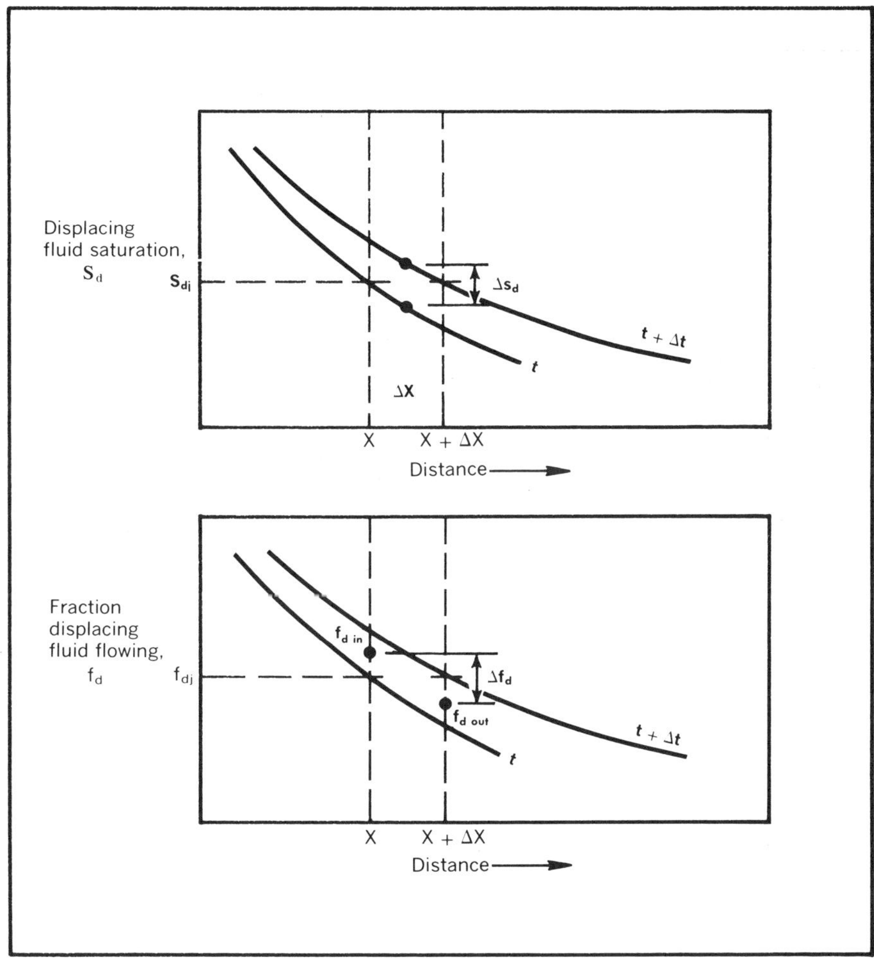

FIG. 5.15　Diagrams for Buckley-Leverett equation derivation.

changes, so that a fractional fluid flow plot at the beginning and ending
of the time interval would look as shown. Now note that the average
f_{dIN} multiplied by the cumulative total flow-in during the period Δt would
represent the total displacing fluid entering the Δx increment during
the period. Similarly, the average f_{dOUT} during the period multiplied by
the cumulative total flowout would be the displacing phase leaving
during the period. Consequently, the difference in the displacing fluid
in and the displacing fluid out will be the increase in the displacing
fluid in Δx during the period.

$$\Delta \text{ bbl of ``d'' in } \Delta x \text{ during time } \Delta t = q_t\,\Delta t \Delta f_d \qquad (5.21)$$

When Equations 5.20 and 5.21 are equated and solved for Δx we obtain Equation 5.15.

$$\Delta x_{Sdj} = (5.615\ q_t \Delta t / \phi A)(\Delta f_d / \Delta S_d)_{Sdj} \qquad (5.15)$$

As noted previously, this equation is used by evaluating the expression, $\Delta f_d / \Delta S_d$, graphically at the saturation of interest. By assuming different values for the saturation and calculating its position at a particular time it is possible to evaluate the saturation distribution at that time.

To test your understanding of the application of the Buckley-Leverett equation, work Problem 5.4B and compare your solution with the solution in Appendix C.

Problem No. 5.4B: Application of the Buckley-Leverett Equation

Fig. 5.16 gives the initial saturation distribution for the reservoir of Problem No. 5.4A and the calculated saturation distribution in this reservoir at the end of 0.5, 1.0, and 2.0 years. Verify these calculated saturation distributions by calculating the position of the 0.79, 0.75, and 0.70 water saturations at the times indicated. Use the previously verified fractional-flow curve of Fig. 5.14. Do not concern yourself with the evaluation of the frontal saturations. This is the next subject for consideration. The reservoir porosity is 21.5% and the average distance from the original WOC to the first line of producers is 350 ft.

Determination of the frontal saturation by material balance. Examination of Fig. 5.14, which is a typical fractional-flow curve, will show that it would be meaningless to evaluate the position of every

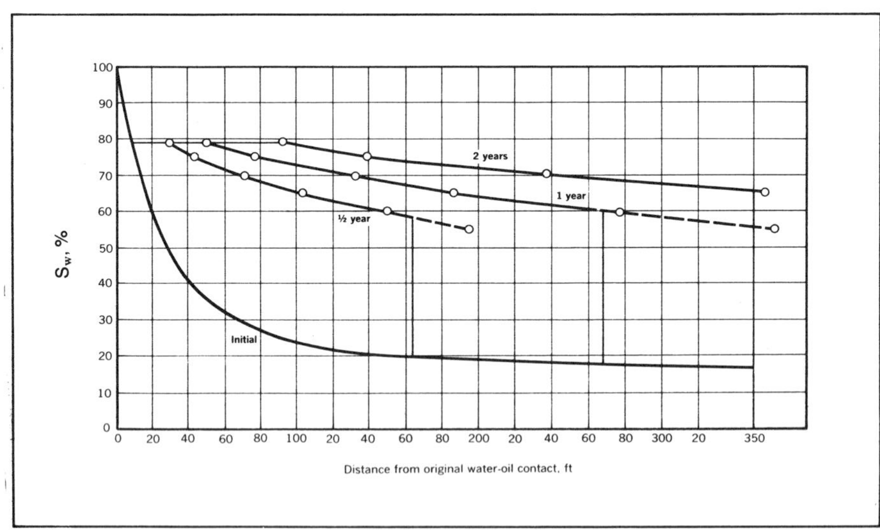

FIG. 5.16 Saturation distribution in a water-drive reservoir.

saturation represented on this curve. There is a duplication of slopes that would mean that the calculated reservoir position of the two different saturations with the same slope would be identical. Consequently, some method is required to determine how much of the calculated saturation profile is needed. This can be accomplished by material balance. We know that at any particular time of interest the total influx into the reservoir will be the influx rate multiplied by the time. Prior to breakthrough of the displacing front into the producing wells the total influx will be equal to the increased number of barrels of displacing fluid in the reservoir.

Initially we know that the number of cubic feet of displacing fluid in any particular Δx increment of the reservoir will be, $S_{di}\Delta x A\phi$. Thus, the increase in the number of cubic feet of the displacing phase in ΔX will be $A\phi\Delta X(S_d - S_{di})$. If we summed the increase in all segments from the influx face to the displacing fluid front we must obtain the total number of cubic feet of displacing fluid influx during the period of time, t.

$$5.615 q_t t = \phi A \sum_0^{X_f} (S_d - S_{di})\, \Delta X \tag{5.22}$$

This then fixes the frontal position as the X_f necessary to satisfy this equation. It is somewhat unusual because the unknown normally is not the limit of a summation term. The equation is generally solved graphically and the method is most easily explained by considering an example. In Fig. 5.17 is shown the initial saturation distribution for the reservoir of Problems 5.4A and B and the "apparent" saturation distribution calculated in Problem 5.4B and plotted in Fig. 5.16. The dotted lines on Fig. 5.17 are solution construction lines. The two curves are approximated with step functions. The objective, is to have the area under the step function be equal to the area under the curve. After this construction is completed we can begin numerically evaluating the summation terms of Equation 5.22, beginning with the area nearest the influx face. As we sum the $(S_d - S_{di})\Delta X$ terms we will approach a sum that satisfies Equation 5.22. Eventually we will find that a new term added to the summation will cause the summation to exceed the value necessary to satisfy the equation. In other words we will find

$$\Sigma(S_d - S_{di})\Delta X > \frac{5.615 q_t t}{\phi A} \tag{5.23}$$

When this occurs it will be necessary to adjust the ΔX for the last term to satisfy Equation 5.22. When this has been done the position of the front at this time will be fixed. Check your understanding of this technique by solving the following problem and checking your solution against Appendix C.

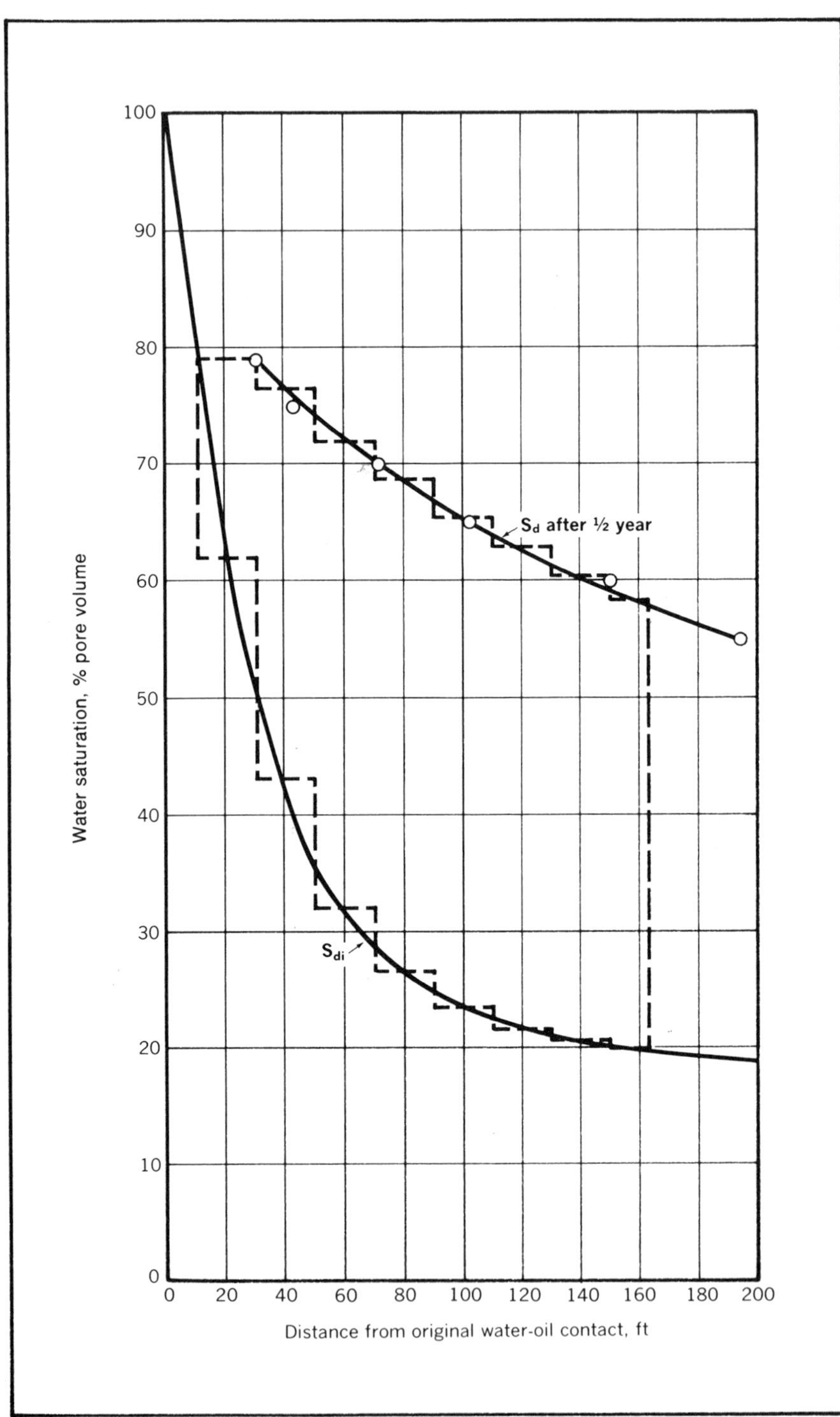

FIG. 5.17 Example graphical integration for Problem 5.4C.

Problem No. 5.4C: Evaluation of the Frontal Position by Material Balance

Use the "apparent" saturation profile for 0.5 year as shown in Fig. 5.16 and Fig. 5.17 to determine the frontal position at this time.

Note that the cumulative influx into the reservoir will also be equal to the total displacement of fluid from the subject portion of the reservoir since we have been working with a steady-state reservoir system. It may also be worthy of note that the displaced fluid less the production from the first line of producers will represent the fluid that is being moved into the reservoir above the first line of producers.

Graphical analysis of fluid displacement with uniform initial saturation distribution (Welge Method). Many years ago Henry Welge developed an intriguing graphical-mathematical method of analyzing saturation-distribution data. The method can theoretically be applied to reservoir displacements only if the initial saturation of the displacing fluid is uniform throughout the reservoir. Nevertheless, the method finds much use and, as is the case with most interesting reservoir engineering techniques, much misuse.

The Welge method is much easier to describe mechanically than it is to explain mathematically. The method centers around the graphical use of a fractional-flow curve. Some of the operations are indicated in Fig. 5.18. The saturation at the displacing-fluid front is determined by drawing a tangent to the fractional-flow curve through the point representing the initial S_{di} and f_{di}. The point of tangency fixes the displacing-fluid saturation at the front, S_{df}. Extension of the tangent line to $f_d = 1.0$ will provide the average displacing-fluid saturation in the displacing-fluid bank prior to breakthrough, $\overline{S}_d$. The reciprocal of the slope of the tangent will give the cumulative influx of the displacing fluid in pore volumes at time of breakthrough.

After breakthrough the cumulative influx and the average displacing-fluid saturation in the reservoir can be determined for any saturation at the producing face, S_{dpf}. To obtain these values, construct a tangent to the fractional-flow curve at some assumed value of the saturation identified as S_{dj} in Fig. 5.18. The reciprocal of the slope of this tangent will be the cumulative displacing-fluid influx in pore volumes. Where the tangent crosses the line represented by $f_d = 1.0$ we read the average displacing-fluid saturation in the reservoir when $S_{dpf} = S_{dj}$.

To understand the theoretical basis of the method, we will first recognize that the method of determining the frontal saturation by constructing a tangent to the fractional flow curve through the initial saturation, S_{di}, and the initial fractional flow value, f_{di}, is the same as saying mathematically that

$$\left(\frac{\Delta f_d}{\Delta S_d}\right)_{fr} = \frac{(f_{dfr} - f_{di})}{(S_{dfr} - S_{di})} \tag{5.24}$$

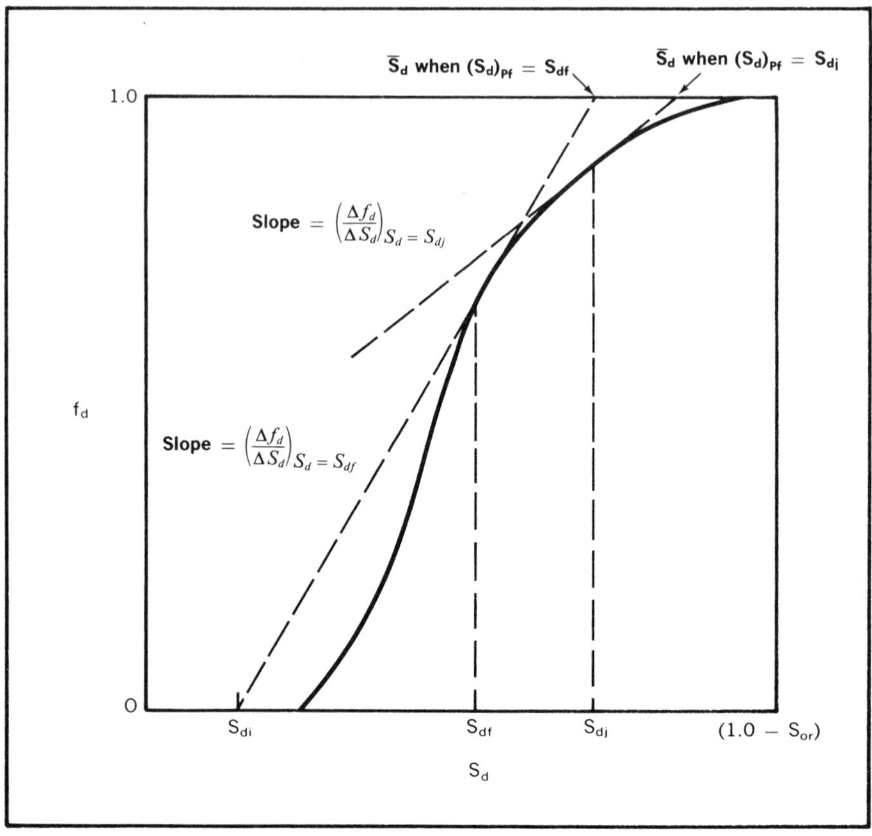

FIG. 5.18   Fractional flow curve (Welge analysis assumes no initial saturation distribution).

The graphical validity of this expression can be verified by studying Fig. 5.18. Once we have proven the mathematical validity of the expression we will have proven the Welge method of determining the frontal saturation.

It was noted previously that if the entire fractional-flow curve was used to calculate a saturation distribution a double-valued plot of saturations versus distance from the influx face would result over a portion of the plot. This is illustrated schematically in Fig. 5.19. In Equation 5.22 we showed that the cumulative displacing fluid influx would be proportional to the area on Fig. 5.19, representing the summation term, $\Sigma (S_d - S_{di})\Delta X$, in Equation 5.22, and areas B, C, and D in Fig. 5.19. Note that mathematically, based on the two-valued saturation curve, the cumulative influx can also be written as

$$5.615 q_t t = \phi A \sum_{(1-S_{or})}^{S_{di}} X \Delta S_d \qquad (5.25)$$

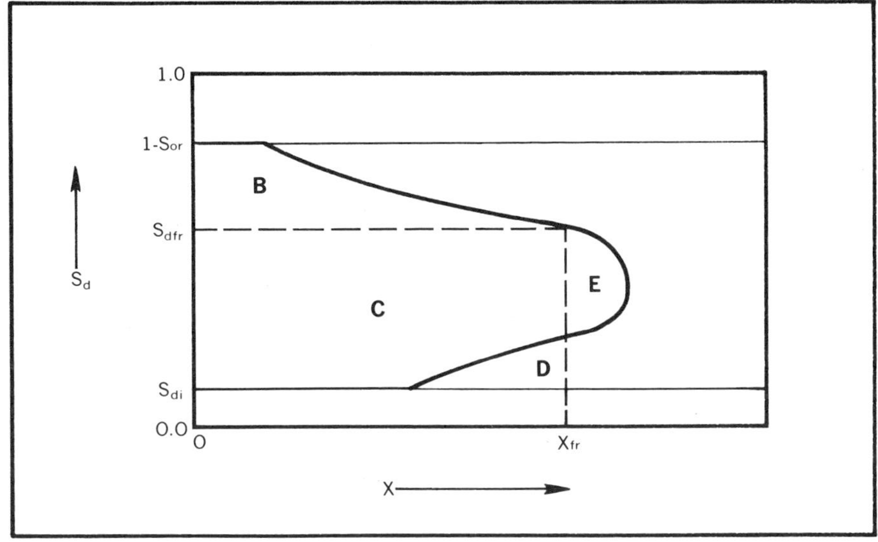

FIG. 5.19 Schematic of Buckley-Leverett S_d versus distance.

Note that the summation in this equation is then equal to areas B, C, and E in Fig. 5.19. Consequently, we see that $(B + C + D)$ must equal $(B + C + E)$ and thus area E must equal area D. Actually, this is the basis for one of the most common methods of determining the position of the front. The engineer simply determines by trial and error the position of the front that results in area E being equal to area D. This particular technique can be used whether or not the initial displacing-fluid saturation is uniform.

Continuing with the theoretical development of the Welge method, this means that area $C + E$ must equal area $C + D$ in Fig. 5.19. Mathematically, we can state this relationship as

$$\text{Area } (C + E) = \text{Area } (C + D) \tag{5.26}$$

$$\sum_{S_{di}}^{S_{dfr}} X\Delta S_d = X_{fr}(S_{dfr} - S_{di}) \tag{5.27}$$

Now remember that the X of the saturation-distribution curve is proportional to $\Delta f_d/\Delta S_d$ evaluated at the saturation represented by the subject X. (See Equation 5.15). Thus, with X's in all terms on both sides of Equation 5.27 the proportionality constant, $5.615 q_t \Delta t/\phi A$, cancels and the equation becomes

$$\sum_{(\Delta f_d/\Delta S_d)_{Sdi}}^{(\Delta f_d/\Delta S_d)_{fr}} \left(\frac{\Delta f_d}{\Delta S_d}\right)\Delta S_d = \left(\frac{\Delta f_d}{\Delta S_d}\right)_{fr}(S_{dfr} - S_{di}) \tag{5.28}$$

This simplifies to

$$f_{dfr} - f_{di} = \left(\frac{\Delta f_d}{\Delta S_d}\right)_{fr} (S_{dfr} - S_{di}) \qquad (5.29)$$

When this equation is solved for the fractional-flow curve slope at the front we obtain Equation 5.24 which, as previously stated, shows that the frontal saturation can be determined by drawing a tangent to the fractional-flow curve through the initial conditions of S_{di} and f_{di}. The frontal saturation will then be represented by the point of tangency.

By rearranging the Buckley-Leverett equation we will show that the reciprocal of the slope of the fractional-flow curve, when evaluated at the producing-face saturation after breakthrough, is equal to the cumulative displacing-fluid influx in pore volumes. Since we are evaluating the slope at the producing face the ΔX_{sdj} value in Equation 5.15 will be equal to the length of the linear system, and the pore volume in barrels, V_p, will be $\Delta X_{Sdj} A \, \phi/5.615$. Cumulative influx, Q_i, is $q_t \Delta t$. Now substituting into and rearranging Equation 5.15 we obtain an expression for the cumulative influx in pore volumes.

$$\left(\frac{Q_i}{V_p}\right) = \left(\frac{\Delta S_d}{\Delta f_d}\right)_{Pf} \qquad (5.30)$$

The last part of the graphical method is concerned with evaluating the average displacing-fluid saturation in the reservoir. This is done by constructing a tangent to the fractional-flow curve at the displacing-fluid saturation that exists at the producing face at a particular time and extending the tangent to the line representing $f_d = 1.0$. The saturation representing this intersection is the average displacing-fluid saturation in the reservoir at that time. Study of Fig. 5.18 will show that this graphical technique is equivalent to the equation,

$$\overline{S}_d = S_{dpf} + \frac{(1 - f_{dpf})}{(\Delta f_d/\Delta S_d)_{pf}}. \qquad (5.31)$$

It may be easier to note from Fig. 5.18 that the slope of the tangent to the fractional-flow curve at the saturation, S_{dj} is

$$\left(\frac{\Delta f_d}{\Delta S_d}\right)_j = \frac{(1 - f_{dj})}{(\overline{S}_d - S_{dj})} \qquad (5.32)$$

This equation can be rearranged to Equation 5.31 if we let conditions at the producing face, pf, be equal to the general conditions denoted as j. Thus, if we prove the theoretical validity of Equation 5.31 we will then have verified the graphical method as being theoretically correct.

The average saturation, S_d, for a saturation distribution after breakthrough as shown schematically in Fig. 5.20, is the area under the saturation curve divided by the length of the reservoir, L.

$$\overline{S}_d = \frac{\sum\limits_0^L S_d \Delta X}{L} = \frac{(1 - S_{or})X_{or} + \sum\limits_{X_{or}}^L S_d \Delta X}{L} \qquad (5.33)$$

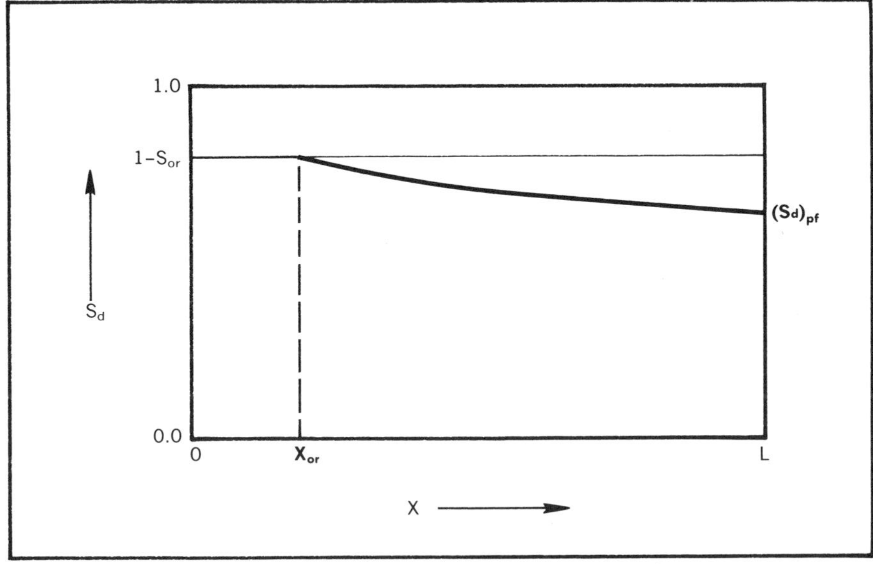

FIG. 5.20 Schematic of S_d versus distance, X, after breakthrough.

In this expression the area is broken into two sections to facilitate handling. Now note that each term in the numerator and each term in the denominator, L, contains an X value from the saturation-distribution curve. Since according to Equation 5.15 the X position of a saturation is proportional to the slope of the fractional flow curve at that saturation, the proportionality constant, 5.615 $q_t \, \Delta t/\phi A$ cancels and we can simply replace all of the X values in Equation 5.33 with the slope corresponding to that saturation.

$$\overline{S}_d = \frac{(1 - S_{or})(\Delta f_d/\Delta S_d)_{ro} + \sum_{ro}^{pf} S_d \, \Delta(\Delta f_d/\Delta S_d)}{(\Delta f_d/\Delta S_d)_{pf}} \qquad (5.34)$$

To simplify this equation to the desired pf form we transpose the summation term to one which is stated as a function of ΔS_d rather than as a function of $\Delta(\Delta f_d/\Delta S_d)$. Equation 5.34 then becomes

$$\overline{S}_d = \frac{(1 - S_{or})(\Delta f_d/\Delta S_d)_{ro} + S_{dpf}(\Delta f_d/\Delta S_d)_{pf} - S_{dro}(\Delta f_d/\Delta S_d)_{ro} - \sum_{ro}^{pf} (\Delta f_d/\Delta S_d)\Delta S_d}{(\Delta f_d/\Delta S_d)_{pf}} \qquad (5.35)$$

If the engineer is confused by the mathematical step from Equation 5.34 to 5.35 he might better recognize the pure math form of the summation term transformation as

$$\int u \, dz = uz - \int z \, du \qquad (5.36)$$

In our application of this equation u is S_d and z is $\Delta f_d/\Delta S_d$.

Note that $S_{dro} = (1 - S_{or})$ so that the terms in the numerator of Equation 5.29 that contain these expressions cancel out. Also the summation term is simply $f_{dpf} - f_{dro}$ and the fraction of displacing fluid flowing at the residual oil saturation, f_{dro}, is 1.0 since no residual oil can flow. Equation 5.35 can then be written as

$$\bar{S}_d = \frac{S_{dpf}(\Delta f_d/\Delta S_d)_{pf} - f_{dpf} + 1.0}{(\Delta f_d/\Delta S_d)_{pf}} \tag{5.37}$$

When this equation is rearranged it becomes Equation 5.31 which as noted previously is the mathematical equivalent of the mechanical procedure for drawing a tangent to the fractional-flow curve at the displacing-fluid saturation at the producing face, extending the tangent to the horizontal line representing $f_d = 1.0$, and reading the average displacing-fluid saturation in the reservoir as the saturation at the intersection.

The Welge graphical method of evaluating the saturation distribution during a displacement is a simple and fast technique when it can be used. Even when a substantial transition zone exists in the reservoir (i.e. the initial displacing-fluid saturation is not uniform) the engineer may wish to assume some average uniform initial saturation distribution to obtain a quick analysis before performing the time consuming but more exact type of analysis previously outlined.

The engineer should note that the fluid displaced from the reservoir is equal to the displacing fluid entering the reservoir. In other words, the increase in the average displacing-fluid saturation is equal to the decrease in the oil and gas saturation.

It is recommended that the reader work the following problem and check his solution against the solution in Appendix C in order to check his understanding of the graphical method.

Problem No. 5.5: Using the Welge Graphical Method to Calculate Displacement in a Waterflood

A line-drive waterflood is to be performed in a thin, homogeneous, horizontal reservoir with an initial gas saturation of 5%. The equilibrium gas saturation is 0.0. Assume all gas is displaced before the oil is displaced (an oil bank forms). The fractional-flow curve for this reservoir is shown in Fig. 5.21. The connate water saturation is 35%.

What will be the water saturation at the front prior to water breakthrough into the producing wells?

Estimate the oil displaced in pore volumes versus the cumulative water injection in pore volumes.

Up to this point our discussion of fluid displacement has been concerned entirely with the linear flow system. Generally the linear system seems to satisfy the fluid-displacement needs because natural encroach-

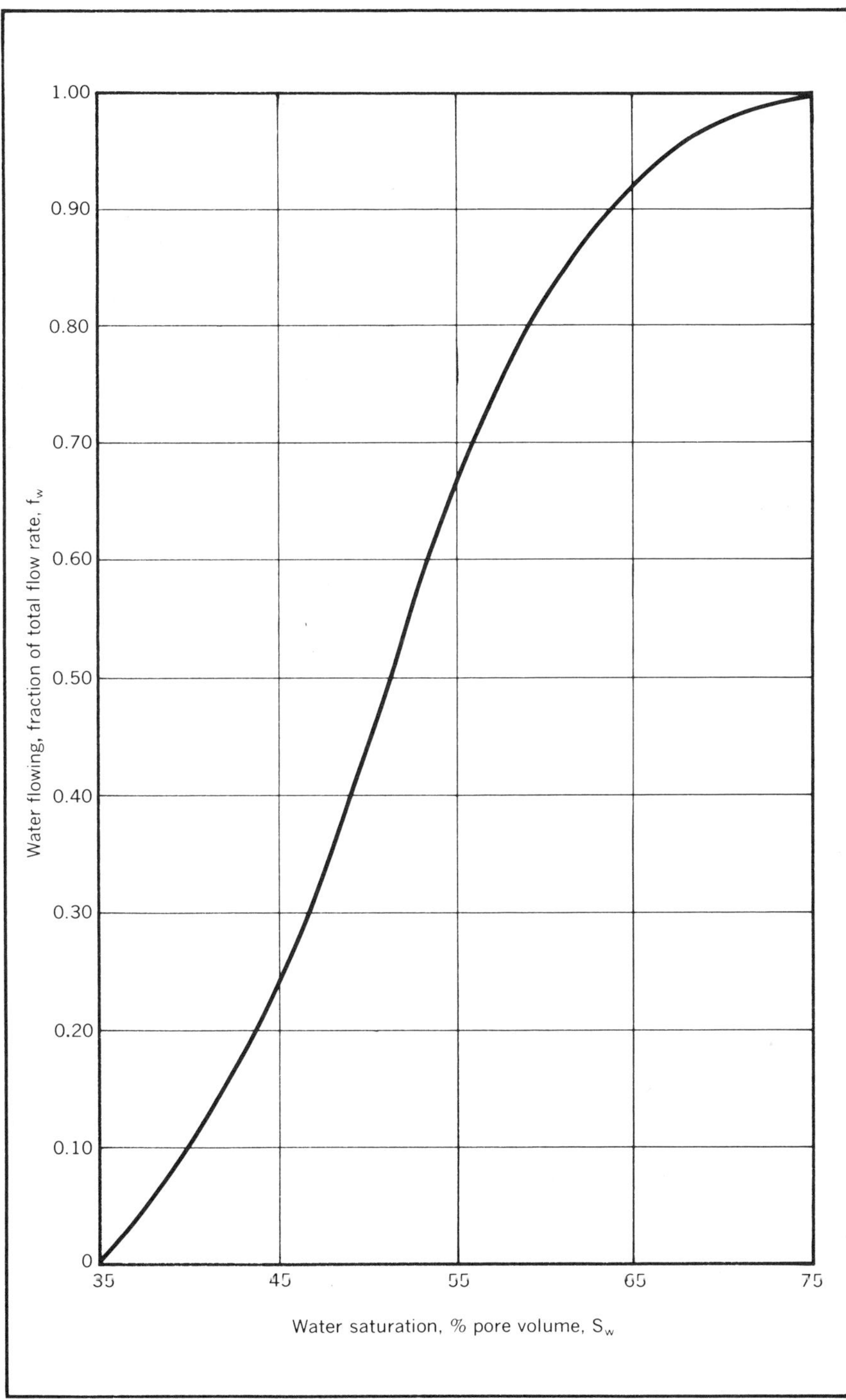

FIG. 5.21 Fractional flow curve for Problem 5.5.

ment takes place at very large radii and consequently, the behavior is closely approximated by a linear system. For example, if water encroachment took place across a 10,000-ft radius and the advance of the water eventually reached a radius of 8,000 ft then we could treat this as a linear encroachment with a width that varied from $2\pi10,000$ to $2\pi8,000$, or with an average width of $2\pi9,000$. The width is the circumference of the circle.

A radial Buckley-Leverett equation can be derived without any undue difficulty. The Fig. 5.15 plots of the displacing-fluid saturation and the displacing-fluid fractional-flow curve plotted against distance were used to derive the linear Buckley-Leverett equation. Actually the same diagrams could be used to derive the radial equation by simply substituting the square of the radius instead of the distance as the basis for the plot. The ΔX would then become $\Delta(r^2)$. Equation 5.20, the change in the number of barrels of the displacing phase in the increment during time interval, Δt, would then become $\pi\Delta(r^2)h\phi/5.615\Delta S_d$. When this expression is equated to the right-hand side of Equation 5.27 we would obtain the radial Buckley-Leverett equation.

$$\Delta(r^2)_{Sdj} = \frac{1.79q_t\Delta t}{\phi h}(\Delta f_d/\Delta S_d)_{Sdj} \tag{5.38}$$

In using this equation it is simplest to plot the saturation versus the square of the radius so that the plot will resemble the linear plots in appearance. On the surface it appears that the use of the radial Buckley-Leverett equation is about as simple as the use of the linear equation. However, for all practical purposes, the radial equation can be applied only for horizontal flow or when the gravity effect on the fractional flow is negligible. This is due to the fact that the gravity term in Equation 5.19 is a function of the velocity, (q_t/A). Since the cross-sectional area in radial flow varies directly with the radius the application becomes very difficult. To complicate matters further, the slope in a dome-type structure varies with the radius.

INTERFACE TILT DURING DISPLACEMENT

Many petroleum fields have initial water-oil or gas-oil contacts (WOC and GOC) whose subsea depth varies throughout the reservoir. This can be caused by a lateral variation in capillary pressure, but it is sometimes caused by an unbalance of the viscous forces in the reservoir. An initial tilted WOC or GOC can be caused by the flow of water beneath the stationary oil or gas. The flow of water may be due to artesian water flow or it may be caused by pressure drawdown in one of several petroleum reservoirs underlain by the same aquifer.

Even when a WOC or GOC is initially horizontal it is often discovered that the interface becomes tilted during the displacement that takes place during the producing life. In fact, this tilted condition may become so severe that the interface actually becomes unstable. The interface tilt is accentuated by radial flow around the well bore and we refer to this condition as coning. This section will attempt to provide some insight into these problems.

Initial interface tilt (hydrodynamic tilt). The simplest of the interface tilt problems is probably the initial contact tilt in the reservoir caused by the pressure gradient resulting from flow of water beneath stationary oil or gas. Under these conditions, pressure would remain constant along any horizontal line in the oil (or gas) zone because oil is not flowing and there is no pressure drop due to flow. However, along a horizontal line in the water zone there will be a pressure drop due to flow. The only way that such a situation can exist is for the interface between the two zones to tilt so that the total pressure in the water zone along a horizontal line in the direction of flow decreases because it has less water and more of the less-dense oil above it.

This phenomena was originally analyzed and published by a Shell geologist, King Hubbard, who proposed the prediction of the location of petroleum reservoirs through an understanding of this phenomena. Fig. 5.22 gives an acceptable explanation of the quantitative relationships accompanying this interface tilt. The figure shows the unlikely situation where two cable-tool wells some distance ΔH apart just miss the oil trap shown. When an effort was made to flow the wells they filled with water to the level indicated. The difference between these two levels is the pressure drop due to flow stated in terms of equivalent feet of water, Δ_{piez}. In hydraulics this is known as the piezometric pressure drop. This can be converted to the pressure drop due to flow in psi by using the pressure gradient of 0.433 psi/ft for fresh water and the water specific gravity relative to fresh water, γ_w.

$$\Delta p_{flow} = 0.433 \gamma_w \, \Delta_{piez} \tag{5.39}$$

Now consider the difference in pressures at point A and at point B. These pressures represent pressures at the interface between the oil and water. Consequently, the pressures will be the same whether we measure them in the oil or in the water. We will then write an expression for this pressure difference when measured in the water and when measured in the oil and equate these expressions to obtain the magnitude of the interface tilt.

If the pressure drop is measured in the water it is the difference between the pressure drop due to flow and the static pressure difference since there is an increase in pressure from A to B when no flow is taking place.

$$p_A - p_B = \Delta p_{flow} - \Delta p_{static} \tag{5.40}$$

$$p_A - p_B = 0.433 \gamma_w \, \Delta_{piez} - 0.433 \gamma_w \, \Delta V \tag{5.41}$$

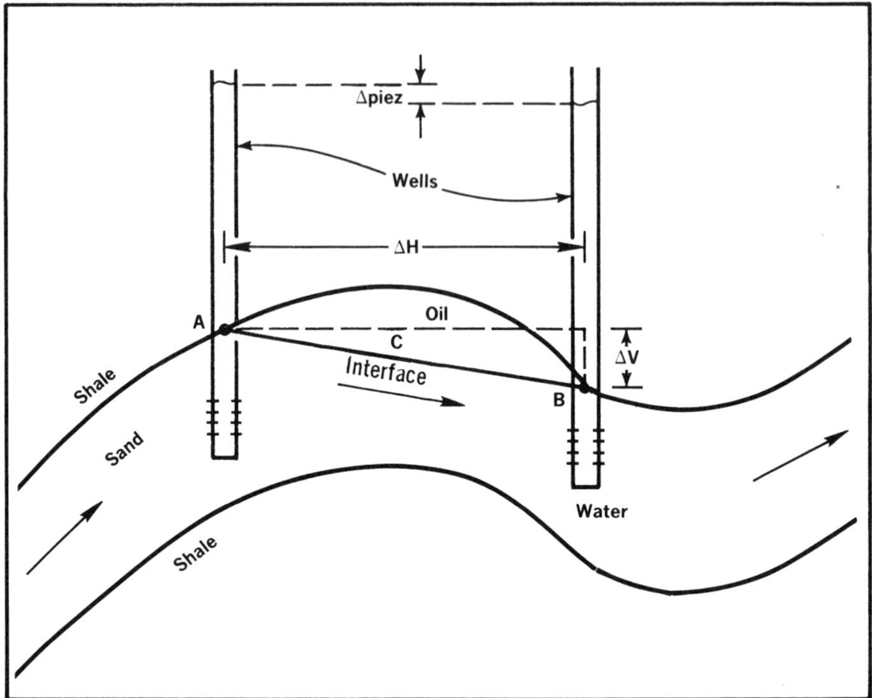

FIG. 5.22 Schematic of hydrodynamic tilt of an interface.

Now if we measure the same pressure drop on the oil side of the interface, it is due entirely to the static pressure difference in the oil because the oil is not flowing.

$$p_A - p_B = 0.433\gamma_o\,\Delta V \tag{5.42}$$

When we equate the expressions for the pressure drop, divide through by the horizontal distance, ΔH, and solve for $\Delta V/\Delta H$, we obtain the equation for the tilt or tangent of the interface dip.

$$(\Delta V/\Delta H) = [\gamma_w/(\gamma_w - \gamma_o)][\Delta_{piez}/\Delta H] \tag{5.43}$$

The initial interface tilt controls oil and gas accumulations because the formation dip must exceed the interface tilt if the oil or gas is to stay in that particular trap. Otherwise the petroleum accumulation will occur at some higher structural position.

Equation 5.43 is a convenient mathematical form of the interface tilt but it is uncommon for the change in pressure with horizontal distance to be recorded as a piezometric head. The change in the piezometric head can of course be calculated from the total pressure drop of two reservoir pressures corrected to the same datum. Correction to the same horizontal datum will assure that the pressure difference is due to flow alone and Equation 5.39 can be used to calculate Δ_{piez}.

The term interface tilt can be more explicitly called isosaturation tilt. In other words a saturation gradient still will exist vertically but the isosaturation lines will not be horizontal. Instead they will be tilted according to Equation 5.43. The saturation distribution calculated by the Buckley-Leverett equation will still be valid but it will now apply to slices of the reservoir that are parallel to the interface tilt rather than to horizontal slices.

It should be noted that on a worldwide basis it is seldom that substantial artesian flow occurs in an aquifer of such a magnitude that the interface tilt is more than a few feet per mile. A few such occurrences have been observed in the Rocky Mountain area. However, the engineer should remember that practically all reservoir flow of water follows the unsteady-state theory as opposed to the steady-state theory that would characterize artesian flow of water.

Interface tilt during linear displacement. When one fluid is displacing another in a tilted reservoir as in Fig. 5.23, where gas is displacing oil down dip, the interface will tend to be tilted because the displacing fluid is less viscous than the oil and will tend to move more readily than the oil. Consequently, the gas would override and bypass the oil if the

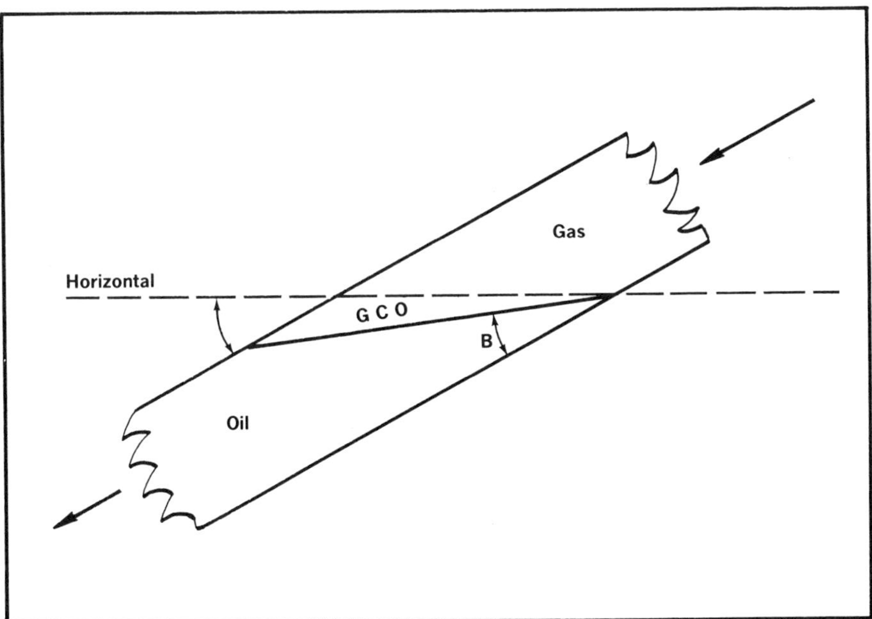

FIG. 5.23 Interface tilt during fluid displacement.

two fluids were the same density. However, the densities are different and this difference in densities tends to keep the interface horizontal. In fact the interface would remain horizontal due to the difference in the densities if the mobilities of the two fluids were equal.

The practical situation is one in which both forces are effective. The viscous forces tend to cause the interface tilt and these are balanced by gravity forces so that a stablilized interface tilt may occur during a displacement. The same phenomenon occurs when water is displacing oil updip. In this case the less-viscous water trys to flow under the oil but is opposed by the difference in gravity of water and oil.

Fig. 5.23 will be referred to in deriving the expression that relates the viscous forces, gravity forces, and stabilized interface tilt. Note the manner in which the angles α and β are measured. The angle α is simply the formation dip but β is a measure of interface tilt relative to bed dip. Relative to a horizontal position, the interface tilt will be the difference, $\alpha - \beta$. Note that when α and β are equal, the interface will be horizontal.

We will derive the interface tilt equation in much the same way Pirson derives the equation in his reservoir-engineering text.[4] This is accomplished by writing an expression for the pressure gradient at the interface in terms of the flow of the displacing fluid and writing a second expression for the pressure gradient in terms of the displaced fluid, which we will consider to be oil. The pressure gradient parallel to the interface can be obtained from Equation 1.52, which was derived in Chapter 1 and gives the flow rate for linear flow in a reservoir tilted at an angle α.

$$q = \frac{1.127 kA}{\mu} \left[(\Delta p / \Delta x)_{\text{Total}} \pm 0.433 \gamma \sin \alpha \right] \qquad (1.52)$$

In this case the direction of flow relative to horizontal is the same as the bed dip. For our purposes we need to use the direction of flow as relative to horizontal so when we apply the equation to obtain the flow velocity parallel to the interface, v_{par}, the α of Equation 1.52 will be $\alpha - \beta$. Substituting the flow velocity parallel to the interface for q/A and writing Equation 1.52 in terms of flow of the displacing phase we obtain,

$$v_{\text{par}} = \frac{1.127 k_d}{\mu_d} \left[(\Delta p / \Delta x)_{\text{Total}} \pm 0.433 \gamma_d \sin (\alpha - \beta) \right] \qquad (5.44)$$

or solving for the pressure gradient

$$(\Delta p / \Delta x)_{\text{Total}} = \left(\frac{v_{\text{par}} \mu_d}{1.127 k_d} \right) \pm 0.433 \gamma_d \sin (\alpha - \beta) \qquad (5.45)$$

Now we will write the same expression for the pressure gradient along the displaced-phase side of the interface in terms of the displaced phase, oil:

$$(\Delta p / \Delta x)_{\text{Total}} = \left(\frac{v_{\text{par}} \mu_o}{1.127 k_o} \right) \pm 0.433 \gamma_o \sin (\alpha - \beta) \qquad (5.46)$$

Note that the velocity parallel to the interface is vectorially related to the velocity parallel to the bed dip, v, so that $v_{par} = v \cos \beta$. If we equate the expressions for the pressure gradient parallel to the interface as stated in Equations 5.45 and 5.46, substitute ($v \cos \beta$) for v_{par}, and rearrange the expression we obtain

$$v \cos \beta \left[\left(\frac{\mu_o}{k_o} \right) - \left(\frac{\mu_d}{k_d} \right) \right] = \pm 0.488 (\gamma_o - \gamma_d) \sin (\alpha - \beta) \qquad (5.47)$$

To put this equation in a more convenient form it is necessary to substitute an equivalent trigonometric expression for $\sin (\alpha - \beta)$.

$$\sin (\alpha - \beta) = \sin \alpha \cos \beta - \cos \alpha \sin \beta \qquad (5.48)$$

Now substitute $\tan \alpha \cos \alpha$ for $\sin \alpha$. After making these substitutions in Equation 5.47, divide through by $\cos \beta$, substitute $\tan \beta$ for ($\sin \beta / \cos \beta$) and substitute q/A for v. When the resulting expression is solved for $\tan \beta$ we obtain

$$\tan \beta = \tan \alpha \pm \frac{q[(\mu_o/k_o) - (\mu_d/k_d)]}{0.488A(\gamma_o - \gamma_o) \cos \alpha} \qquad (5.49)$$

The minus sign, is used for gas displacing oil downdip and the plus sign is used for water displacing oil or gas updip. This is remembered by recognizing that $\tan \alpha$ must always be less than $\tan \beta$ so that the overall sign of the last term of Equation 5.43 must always be negative. For water this is accomplished by the specific gravity difference which is negative.

The engineer should carefully note that the specific gravities are relative to water in this equation even when it is applied to gas. This is contrary to the customary air base normally used for all gases.

The effect of the producing rate, q, on the interface tilt can be evaluated from Equation 5.49. Remember that the total sign of the second term which contains the rate, q, will normally be negative. Also note that the interface will be horizontal when $\tan \beta$ equals $\tan \alpha$. So the larger the rate the greater the interface tilt.

Obviously, it is possible for the rate term to become so large that $\tan \beta$ and thus β, are zero. When this occurs the interface tilt will be a maximum and the interface will be parallel to the dip of the formation. Under such conditions we would be unable to define exactly where the interface is located. Consequently, we say the interface is unstable, or "fingering" takes place. The producing rate at which an interface becomes unstable (fingers form), can be determined by setting $\tan \beta$ in Equation 5.49 equal to zero and solving for the critical rate, q_{cF}, at which this situation would occur.

$$q_{cF} = \frac{\pm .488 (\gamma_o - \gamma_d) A \sin \alpha}{[(\mu_o/k_o) - (\mu_d/k_d)]} \qquad (5.50)$$

Remember that Equations 5.49 and 5.50 represent stabilized conditions when a balance has been reached between the gravity and viscous forces. These equations give no indication as to how long it will take to reach such a stabilized condition. This time factor becomes very important when we recognize that in many cases where Equations 5.43 and 5.50 can be applied, the critical rate is too small to be economical. There is no simple method known by the author to evaluate or even estimate the time required for the reservoir to reach a stabilized condition.

If the stabilization time is critical in a particular reservoir the engineer's only recourse is to use a simulator to follow the movement of the interface with time. This can be done on a computer by assuming a sequence of steady-state flow conditions for small increments of time.

Because of the assumption of stability and the importance of the time factors involved, Equations 5.49 and 5.50 should be used in a qualitative manner or to define limits of reservoir behavior.

Interface tilt in a radial-flow system: coning.   We have considered interface tilt with only one fluid moving and interface tilt with both fluids moving at the same velocity. Both of these problems were based on linear flow systems with constant velocities and tilt angles. We will now consider the phenomena of interface tilt associated with a radial-flow system with one or both fluids flowing. This is the problem associated with the radial flow of oil into a well with gas above or water below. In radial flow the velocity varies with the radius so we will find that the interface tilt varies with the radius and thus is not constant. In this section we will talk about the concepts of coning by investigating the phenomenon for some simplified systems. However, consideration of practical solutions to the coning problem is beyond the scope of this book. This assumes that practical solutions do in fact exist. This assumption might be questioned by many experienced engineers.

Fig. 5.24 illustrates the physical aspects of water coning. This diagram shows the coning situation just before the water breaks into the bottom of the well bore. We will consider this as a stabilized situation with the cone about to reach the well but stabilized so that the water is not flowing. Under these conditions we have the same situation that existed in the previous discussion of hydrodynamic tilt except that in this case the flow is radial rather than linear and the oil is flowing rather than water. Thus, we note that the pressure at any particular horizontal position or level in the water is constant because the water is not flowing and there is an oil pressure drop due to flow. This situation can exist only if the interface is tilted so that the decrease in the pressure in the oil due to flow is offset by a change in the interface position. Thus, the change in pressure over some distance Δr will be equal to the change in the position of the interface, Δz, (over the same distance Δr) converted to an equivalent pressure change using the specific gravities of the fluids.

$$\left(\frac{\Delta p}{\Delta r}\right) = \left(\frac{\Delta z}{\Delta r}\right) 0.433(\gamma_d - \gamma_o) \tag{5.51}$$

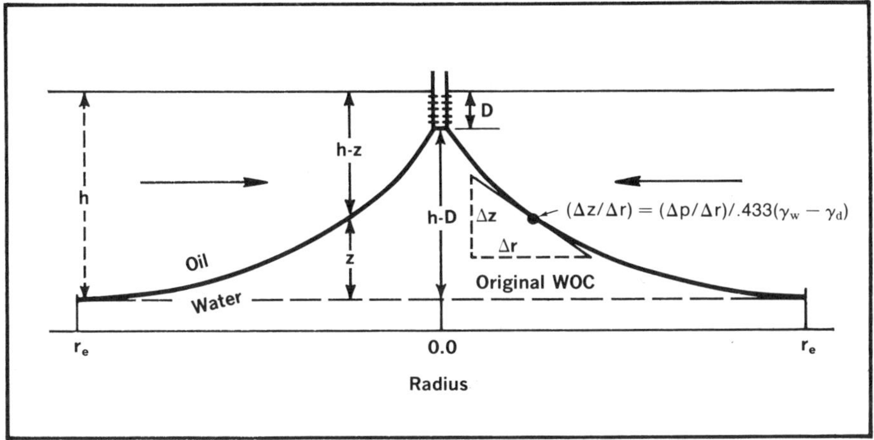

FIG. 5.24 Water coning.

We can use this expression for the pressure gradient for horizontal flow in Darcy's equation to derive the expression for the critical rate at which coning into the well will occur. To use the Darcy equation we must also evaluate the cross-sectional area, A, at a particular radius. This will be the surface of a cylinder whose radius is r and whose height is $h - z$. Thus, $A = 2\pi r(h - z)$. When these substitutions are made into the Darcy equation, we obtain:

$$q = \left(\frac{1.127kA}{\mu}\right)\left(\frac{\Delta p}{\Delta x}\right) \tag{1.3}$$

$$q = \left(\frac{1.127k}{\mu}\right)[2\pi r(h - z)]\left[\left(\frac{\Delta z}{\Delta r}\right)0.433(\gamma_w - \gamma_o)\right] \tag{5.52}$$

When this equation is applied to all the Δr increments from the well radius, r_w, out to the outer radius, r_e, we find that the sum of all the Δr/r values is $\ln(r_e/r_w)$ and the sum of all the $(h - z)\Delta z$ values from a value of $z = (h - D)$ at the well to a value of zero at the external boundary where the interface is unaffected is $(h^2 - D^2)/2$. Performing these operations on Equation 5.52 and solving the resulting expression for the critical flow rate at which the stabilized cone would reach the well bore gives the following sequence of equations:

$$q \sum_{r_w}^{r_e} (r/\Delta r) = (3.07k/\mu)(\gamma_w - \gamma_o) \sum_{h-D}^{0.0} (h - z)\Delta z \tag{5.53}$$

$$q \ln(r_e/r_w) = (3.07k/\mu)(\gamma_w - \gamma_o)[h^2 - D^2/2] \tag{5.54}$$

$$q_{cc} = \frac{1.535k(\gamma_w - \gamma_o)(h^2 - D^2)}{\mu \ln (r_e/r_w)} \tag{5.55}$$

A similar expression can be derived for gas coning if we define D as the distance from the original GOC to the top of the perforations and h is again the initial thickness of the oil zone. (h − D) is the perforated interval. The derivation is carried out in the same manner and results in the expression

$$q_{cc} = \frac{1.535k(\gamma_o - \gamma_g)(2hD - D^2)}{\mu\ln\ (r_e/r_w)} \qquad (5.56)$$

These coning equations have been derived in a manner similar to the techniques used in the Pirson text.[4] Pirson also derives an expression for the optimum placement of h_c feet of perforations in an oil zone with a gas cap above and a water zone below.

In this case, D is the distance in feet from the initial GOC to the bottom of the perforations.

$$D = h - (h - h_c)[(\gamma_o - \gamma_g)/(\gamma_w - \gamma_g)] \qquad (5.57)$$

In this equation, h is again the initial thickness of the oil zone.

Equations 5.55, 5.56, and 5.57 are not based on very realistic assumptions. One of the biggest difficulties involves the assumption that the permeability is the same in all directions. Since sedimentary formations were initially laid down in thin horizontal sheets it is natural for the formation permeability to vary from one sheet to another (vertically). This means that there is generally quite a difference between the permeability measured in a vertical direction and the permeability measured in a horizontal direction. Furthermore, the permeability in the horizontal direction is normally considerably greater than the permeability in the vertical direction. This also seems logical when we recognize that very thin, even microscopic sheets of impermeable material such as shale may have been periodically deposited. These permeability barriers would have a great affect on the vertical flow and have very little affect on the horizontal flow which would be parallel to the plane of the sheets.

The lower vertical permeability tends to minimize the coning and thus Equations 5.55 and 5.56 tend to give overly pessimistic results. In other words, the predicted critical coning rates are much too high. On the other hand, use of Equation 5.57 to predict the optimum position of perforations does not appear to be affected significantly by the difference between vertical and horizontal permeability. The error would appear to have about the same effect on gas coning as it has on water coning and the position of the oil zone perforations should not be affected.

Also note that this series of coning equations is based on the assumption of steady state. On the surface this seems to be a logical assumption for a water drive or gas-cap-drive reservoir but carefully note the nature of the steady-state system assumed. In a water drive or gas-cap-drive steady-state system we normally assume that oil is being displaced in a more or less vertical direction by gas or water at a rate equal to the production rate. But in deriving these coning equations we have assumed flow radially with gas above or water below the radial flow so that there is nothing displacing the oil radially. The inadequacy of such a model is obvious. We are assuming that all of the fluid entering the originally oil-bearing portion of the reservoir enters horizontally across the cylinder with the r_e radius and none of the entering fluid enters vertically (with a stabilized interface). In reality the exact opposite is true. None of the entering fluid enters across the r_e radius cylinder and all of the fluid enters vertically. This then means that a stabilized cone is impossible for a bottom water or vertical-gas-cap drive. Nevertheless, in the vicinity of the well bore, the flow system does approximate horizontal radial flow and the previously derived equations will give some qualitative guides for coning.

One further shortcoming of our coning equations should be noted. They give no indication of the time required for a cone to form. And, as was the case with fingering, the indicated permissible flow rates below the critical rates are generally uneconomical and the time required for the cone to form is very important.

The practical solution of coning problems is probably limited to computer modeling of the particular problem under study. As stated in the first paragraph of this section, it is beyond the scope of this book to provide a practical solution of this nature. However, almost all engineers have use of a digital computer to solve a simple coning problem. The problem should be modeled to give a realistic flow pattern based on knowledge of the reservoir. The recommended technique is to assume a sequence of steady states with new saturations calculated in each segment at the end of each time interval. The relative permeabilities can then be adjusted on the basis of the new saturations and they can be used in calculating the next steady-state flow period behavior. The general technique of generating a matrix of equations describing flow between the various segments and solving the matrix with standard computer techniques was described in Chapter 1 in connection with the calculation of the streamline and pressure distributions for irregular flow geometries. There we discussed in some detail the analysis of a five-spot flow pattern and showed in Fig. 1.17 how it could be segmented into either linear or radial segments.

The same general techniques are used in modeling a coning problem and making any single steady-state calculation except that the vertical dimension must also be introduced. The modeling will then look similar to that illustrated in Fig. 5.25. Linear flow equations would be used for vertical flow in this model and radial equations for radial flow. For proper weighting of the radial segments they should be chosen so that the ratio of outer and inner radii of each radial segment is about the same. In modeling coning problems, always take care to include radial flow segments near the well. To obtain enough detail to describe the cone using linear segments alone requires an extremely large matrix and excessive computer time. As stated previously, the matrix generated for each steady-state flow period can be solved by one of the standard math

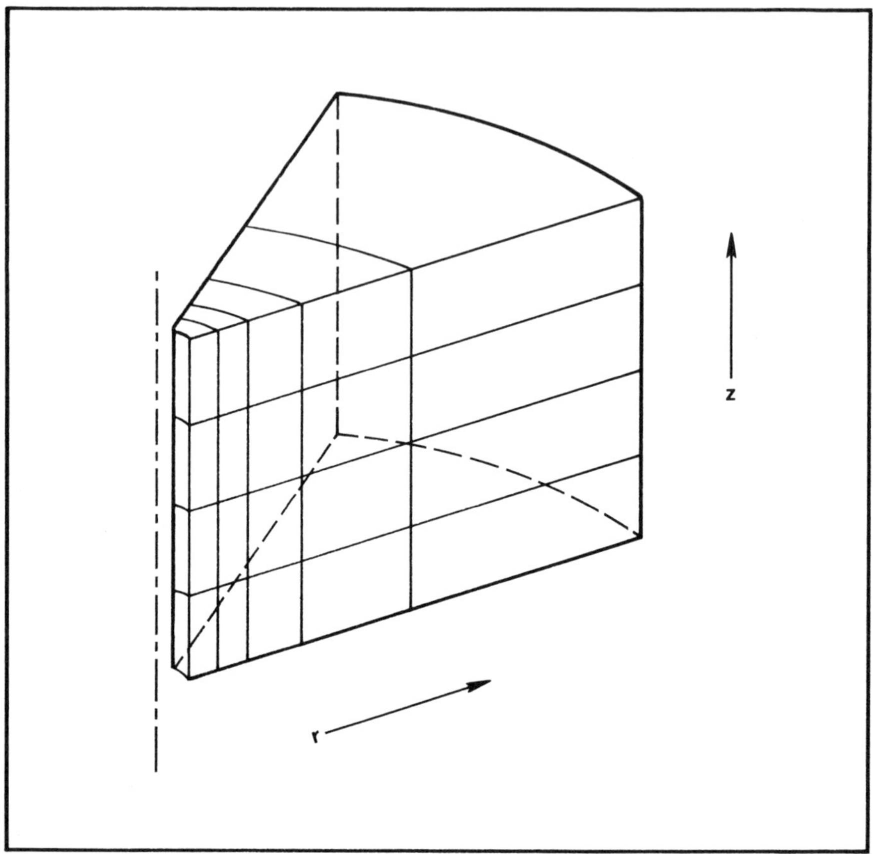

FIG. 5.25   A computer coning model with linear (vertical) and radial (horizontal) segments.

methods such as successive relaxation or the alternating direction implicit technique.[5]

Several simplified and relatively well-known coning-analysis techniques can be found in the literature[6,7] that predict critical rates for stabilized conditions. At least one method attempts to evaluate the time required for a cone to form.[8] However, all of these methods appear to have very serious limitations in their basic assumptions. Consequently, an effort should be made to obtain a computer flow model analysis for any problem that presents serious economic consequences. As before, the engineer is cautioned to only use computer programs that he fully understands and whose calculations he could duplicate by hand if given enough time.

REFERENCES

1. Katz, et. al., *Handbook of Natural Gas Engineering,* New York, N.Y.: McGraw-Hill, 1959
2. B. C. Craft, M. F. Hawkins, *Applied Petroleum Reservoir Engineering,* Englewood Cliffs, N.J.: Prentice-Hall Inc., 1959
3. J. W. Amyx, D. M. Bass, R. L. Whiting, *Petroleum Reservoir Engineering,* New York, N.Y.: McGraw-Hill, 1960
4. S. J. Pirson, *Oil Reservoir Engineering — 2nd Ed.,* New York, N.Y.: McGraw-Hill, 1958
5. M. W. Kosakowski, "An Iterative Digital Study of Water Coning in A Bottom Drive Oil Reservoir," M.S. Thesis, Ohio State Univ., 1971
6. Meyer and Gardner, "Mechanics of Two Immiscible Fluids in Porous Media," *J. Applied Physics,* 1954
7. Chaney, Noble, Henson and Rice, "How to Perforate Your Well to Prevent Oil and Gas Coning," *Oil and Gas Journal* (1956), 54
8. D. P. Sobocinski and A. J. Cornelius, "A Correlation for Predicting Water Coning Time," *Journal of Pet. Tech.* (May, 1965)
9. Carl Gatlin, *Petroleum Engineering,* Englewood Cliffs, N.J.: Prentice-Hall Inc., 1960
10. J. A. Kaplan, "A Gaussian Digital Simulation of Bottom Water Coning in A Petroleum Reservoir," an M.S. Thesis, Ohio State Univ., 1970

Material Balance

The term material balance in reservoir engineering is often criticized because the material balance equations are normally derived on a volumetric balance basis. The material balance equations could be derived strictly on a mass basis, but the derivations are somewhat more complex when carried out in this fashion. For example, we can write an expression that equates the mass of gas originally in a reservoir with the same mass of gas at some later time after a quantity of gas, G_p, in standard cubic feet, has been produced from the reservoir. At the latter time the mass of gas is represented by the pounds of gas in the reservoir plus the pounds of gas produced. However, it is simple to obtain the same equation by equating the standard cubic feet (scf) of gas originally in the reservoir to the scf of gas remaining in the reservoir plus the scf of gas produced.

Practically all reservoir engineering techniques involve some application of material balance or some implied application. The most useful applications of material balance equations require the concurrent use of flow equations. When material balance concepts are combined with flow concepts, it is possible to predict the production behavior of an oil or gas reservoir as a function of time. Without the fluid-flow concepts which tell us how rapidly fluids can be withdrawn from the reservoir at various stages of depletion, the material balance simply provides performance as a function of the average pressure in the reservoir. Thus, we might say that material balance will provide us with a cumulative production versus average reservoir pressure prediction for a reservoir and to put this on a cumulative production versus time basis it is necessary to introduce fluid-flow concepts.

318

A GENERAL MATERIAL BALANCE EQUATION

A general material balance equation will be derived that can be applied to any hydrocarbon reservoir. All of the terms of this equation are never significant at any one time. Nevertheless, by using this general equation the engineer will not need to be familiar with a multitude of special equations such as equations for material balance above the saturation pressure, material balance for solution-gas-drive reservoirs, material balance for gas-cap-drive reservoirs, material balance for water-drive reservoirs, etc. All of these problems can be handled by the one general material balance expression. The various reservoir types are considered individually as a means of developing the general material-balance equation.

Material balance in gas reservoirs. The simplest type of material balance that can be written for a reservoir is the balance written for a dry-gas reservoir. As stated previously, it would be possible to derive this equation by equating the mass of gas in the reservoir initially with the mass of gas in the reservoir at some later time plus the mass of gas produced. However, it is easier to obtain the desired expression by equating volumes. If there is no water drive and the change in pore volume with decline in pressure is negligible (which it is for a gas reservoir), we can write an expression for the volume of gas in the reservoir which remains constant, stated as a function of the reservoir pressure, p; the standard cubic feet of gas produced, G_p; the original standard cubic feet of gas in the reservoir, G; and the gas formation volume factor, B_g; evaluated at different times.

The term, gas formation volume factor, B_g, represents the ratio between the volume of gas at one particular pressure and temperature and the volume of gas at another particular pressure and temperature. The units used in this book are barrels at reservoir conditions per standard cubic foot of gas. The standard cubic foot is the volume of gas at 60° F. and 14.7 psi. Thus, from the gas equation

$$pV = znRT \tag{6.1}$$

we can write an expression for the gas formation volume factor. In the gas equation, V is the volume in standard cubic feet, z is the gas-deviation factor, n is the number of moles of gas, R is the gas constant, which is 10.73 for the units we will be using, and T is the temperature in degrees Rankine.

We desire an expression for the volume of 1 scf of gas at a reservoir pressure, p, and temperature, T, equivalent to 1/379 moles of gas since one mole of gas at standard conditions occupies 379 cu ft. When the volume of gas at reservoir conditions in standard cubic feet is converted to barrels by dividing by 5.615, we obtain an expression for B_g as follows

$$V = \frac{znRT_f}{p}$$

$$B_g = \frac{V}{5.615} = z\left(\frac{1}{379}\right)\frac{(10.73)T_f}{(p5.615)}$$

$$B_g = \frac{0.00504\ zT_f}{p} \qquad\qquad (6.2)$$

Equation 6.1 is discussed much more fully in Chapter 4 as is the evaluation of the gas-deviation factor, z. Considerable care should be taken in evaluating the gas-deviation factor when the gas formation volume factor, B_g, is calculated. Thus, engineers who are not familiar with the evaluation of the gas-deviation factor by various methods and the limitations of these methods, should obtain this knowledge from Chapter 4 before proceeding with such calculations. Others with a previous knowledge of gas-deviation factor analysis can refer to Appendix B. 7, 8, 11, and 12 to determine appropriate z factors.

With the gas formation volume factor defined, we can proceed with the development of the material balance for a dry-gas reservoir. The original barrels of gas in the reservoir can be equated to the gas in the reservoir at some future time after G_p standard cubic feet of gas has been produced, by the equation

$$GB_{gi} = (G - G_p)B_g \qquad\qquad (6.3)$$

In this equation B_{gi} is the gas formation volume factor at initial reservoir pressure and temperature. The gas formation volume factor, B_g, in the right-hand side of Equation 6.3 is based on the pressure and temperature of the reservoir after G_p standard cubic feet of gas has been produced. Note that the amount of gas remaining in the reservoir in standard cubic feet is the difference between the gas originally in the reservoir and the gas produced, $G - G_p$. Equation 6.3 is illustrated graphically in Fig. 6.1.

If this gas reservoir is under water drive, water will enter the pore space originally occupied by gas as the pressure is decreased and the original pore space occupied by gas is now occupied after G_p production, by free gas $(G - G_p)$ and encroached water, W_e. Thus, Equation 6.3 becomes

$$GB_{gi} = (G - G_p)B_g + (W_e - W_p), \qquad\qquad (6.4)$$

where W_p is the water produced, reducing the amount of original gas pore volume occupied by water. Equations 6.3 and 6.4 are used in many different ways to predict behavior of a gas reservoir in Chapter 4. Work the following problem to check your understanding of material balance for a gas reservoir. Your solution can be checked against the solution in Appendix C.

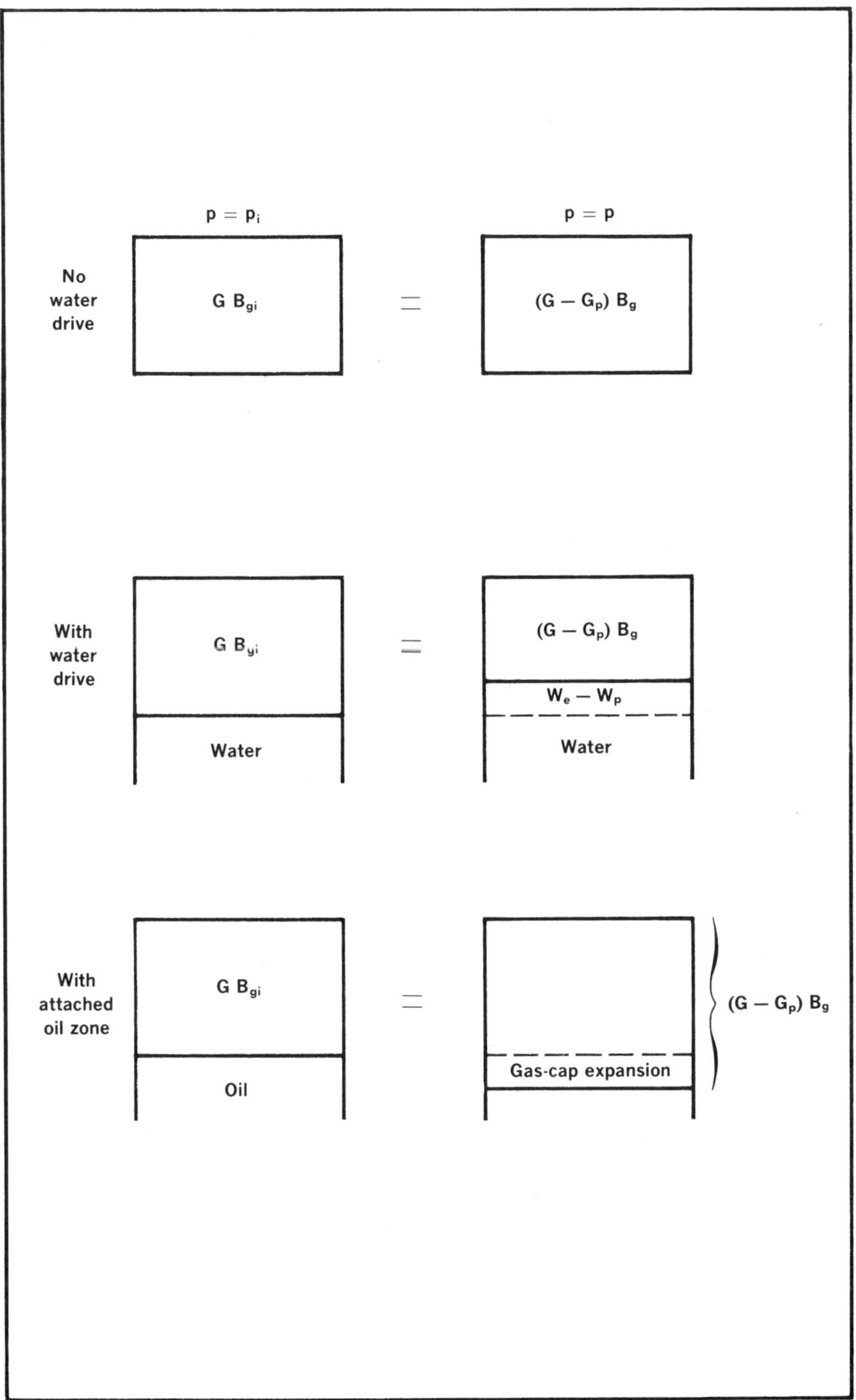

FIG. 6.1 Schematic representation of the dry gas reservoir material balance equations.

Problem 6.1A: Gas Material Balance

To illustrate the meaning of these material-balance equations, consider a reservoir that contains 400,000,000 scf of gas at an initial pressure of 3,150 psia with a temperature such that 1 scf foot of gas occupies 0.0010 bbl in the reservoir (i.e., B_{gi} is 0.0010 bbl/scf). Find the amount of production accumulated at a time when the pressure in the reservoir has declined to 2,900 psia. At this pressure 1 scf of gas occupies 0.0011 bbl of pore space in the reservoir (i.e., B_g is 0.0011 bbl/scf).

Note that if a gas reservoir is attached to an oil reservoir to form a gas-cap reservoir, an expression similar to Equations 6.3 and 6.4 can be written to describe the change in the gas-cap volume. The gas-cap volume after some production of gas-cap gas, G_{pc}, that might occur along with the oil production (by accident or poor planning, not by design) would still be expressed by the right-hand side of Equation 6.3 $(G - G_{pc})B_g$. Thus, the gas cap might expand into the original oil zone or the oil might encroach into the original gas cap. See Fig. 6.1.

$$\text{Change in Gas-Cap Volume} = GB_{gi} - (G_{pc})B_g \qquad (6.4a)$$

Work the following problem and check your solution against the solution in Appendix C.

Problem 6.1B: Gas Cap Expansion

Assume the gas reservoir of Problem 6.1A with an initial pressure of 3,150 psia and $B_{gi} = 0.0010$ reservoir bbl/scf is attached to an oil reservoir to form a gas-cap reservoir. If production from the oil-bearing portion of the reservoir causes the gas-cap pressure to decline to 2,900 psia ($B_g = 0.0011$ res. bbl/scf) with no gas-cap production, how many reservoir barrels of gas will encroach into the original oil zone? How many scf of gas will this represent?

Material balance for liquid expansion.   It is surprising the number of oil reservoirs that produce significant periods of time by expansion of liquid in the reservoir. These reservoirs are generally very large but they have limited permeability so that at the relatively small rates of production that can be obtained the average pressure in the reservoir remains above the bubble point or saturation pressure for a considerable time. During this period, production is due entirely to expansion of oil in the reservoir and the reduction in pore volume due to the decrease in reservoir pressure. For the purposes of developing the general material-balance equation for oil reservoirs let us first assume that the change in pore volume is negligible. Carefully note that this is an erroneous assumption. Much of the production for reservoirs producing above the bubble point is due to the change in pore volume accompanying the decline in reservoir pressure. However, it is convenient to add this as a separate term to the general material-balance equation.

Assuming that production is due only to liquid expansion, a material-balance equation for an oil reservoir similar to that of Equation 6.3 for a gas reservoir can be written.

$$NB_{oi} = (N - N_p)B_o \qquad (6.5)$$

In this equation, N is the volume of oil originally in the reservoir, in stock-tank barrels, N_p is the volume of oil produced (stock tank barrels) at some particular stage of depletion. The factors, B_o and B_{oi}, represent oil formation volume factors. These factors relate the volume of oil in the reservoir at a particular pressure to the volume of this oil in the stock tank. The reservoir volume of oil will shrink in traveling to the stock tank due to a change in phase — oil to gas — that accompanies the change in pressure and temperature. Note that this ratio is not the ratio of the volume of the same mass of oil or hydrocarbon at stock-tank conditions compared with the volume of the same mass at reservoir conditions as was the case with the gas formation volume factor, B_g. In the case of the oil formation volume factor the volume at stock-tank conditions contains less mass than does the related volume at reservoir conditions which includes the volume of "gas in solution."

This term "gas in solution" is used to define or describe the phenomenon of phase change that accompanies the hydrocarbon liquid's transition to the lower stock-tank pressure and temperature. The concept of gas in solution is a handy one, particularly since most of the gas liberated from the reservoir oil can be treated as having constant composition without introducing a significant error.

Fig. 6.2 illustrates the relationship between oil formation volume

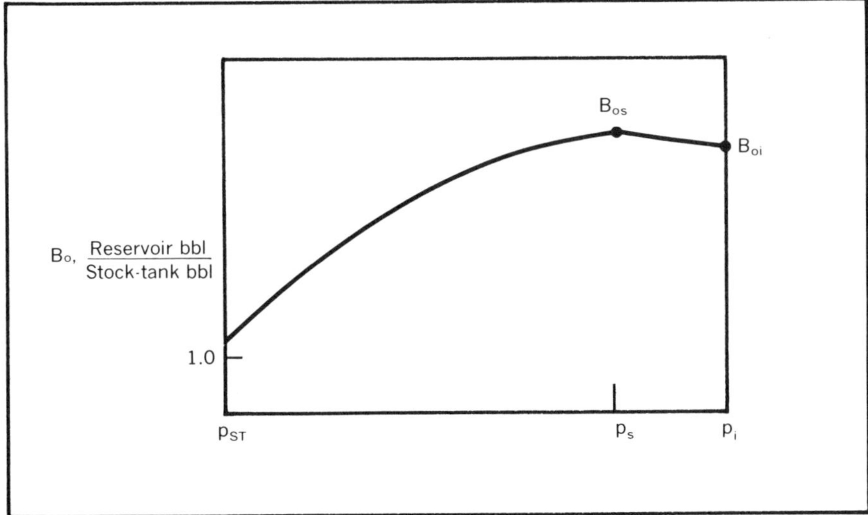

FIG. 6.2 Typical oil formation volume factor versus reservoir pressure relationship.

factor and reservoir pressure. In this diagram, p_i is the initial reservoir pressure, p_s the saturation pressure or the pressure at which the first bubble of gas forms in the reservoir, and p_{st} the stock-tank pressure.

In Equation 6.5 we are considering B_o values at or above the bubble-point pressure. Consequently, the change in B_o in this pressure range is due to the expansion of oil in the reservoir. Thus, the same mass of oil at pressure p_i will represent a lesser volume than it will at the saturation pressure, p_s. The increase in B_o is directly related to the compressibility of the liquid oil. Fig. 6.3 illustrates Equation 6.5 schematically.

Gas liberation in the reservoir. When the fluid in a reservoir reaches the saturation or bubble-point pressure, gas will be liberated. In such a case, if the reservoir existed initially at or above the bubble-point pressure, the pore volume originally occupied by oil would be occupied at some pressure less than the saturation pressure by liquid and liberated gas. This is shown pictorially in Fig. 6.3. The math expressions will be verified later. The oil volume would still be represented by the right-hand side of Equation 6.5, but the liberated gas remaining in the reservoir requires some additional analysis.

As the pressure in the reservoir declines and falls below the bubble-point pressure, gas will be liberated. As the reservoir pressure continues to decline due to production of oil and gas, additional gas will be liberated. As gas is liberated, less gas will remain in solution in the oil until the stock-tank pressure is reached and there is no longer any gas in solution relative to stock-tank conditions and all of the original solution gas has been liberated. This is illustrated in Fig. 6.4 where the gas in solution in the reservoir oil is plotted versus pressure.

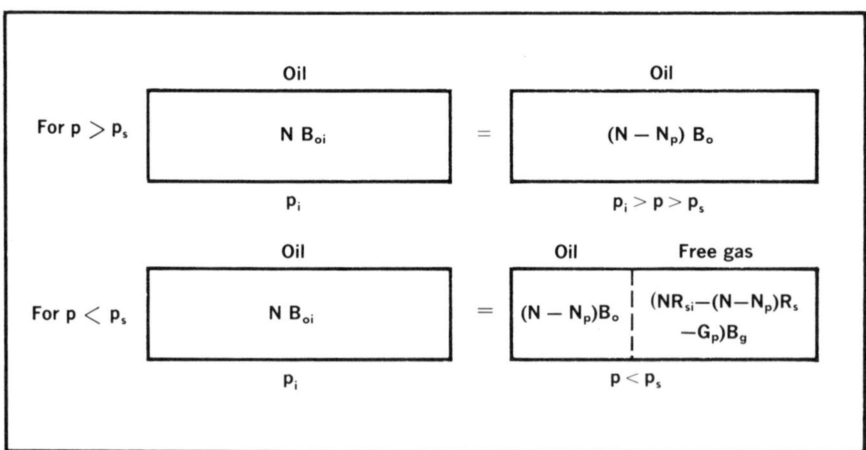

FIG. 6.3 Schematic representation of the solution gas reservoir material-balance equations.

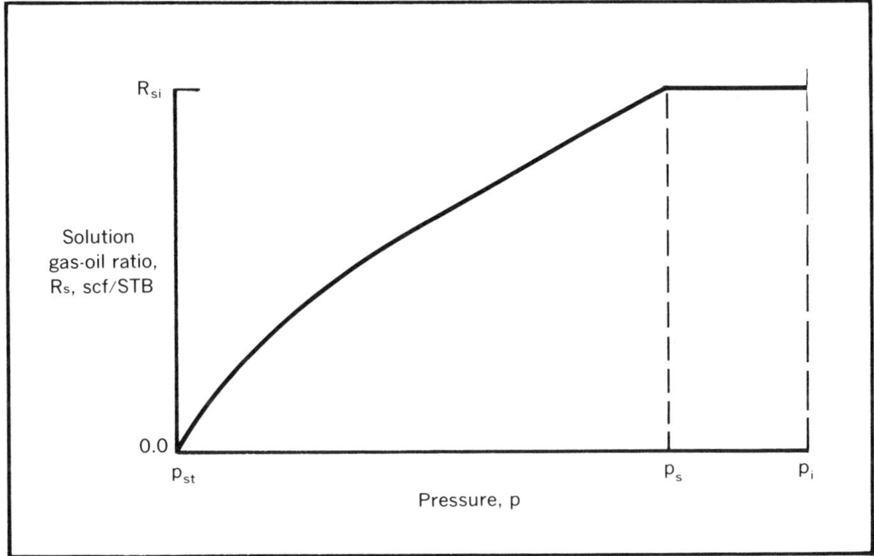

FIG. 6.4 Typical solution gas-oil ratio versus pressure relationship.

Since the oil formation volume factor is related to stock-tank barrels of oil, it is convenient to also relate the amount of gas in solution to stock-tank barrels of oil. Thus, R_s is stated in standard cubic feet per stock tank barrel (scf/STB). Note that the liberated gas will be the initial gas in solution, R_{si}, less the gas remaining in solution at a particular pressure R_s. Also note that the gas in solution above the bubble-point pressure does not change with the pressure since none of this gas is liberated until the reservoir pressure declines to the saturation pressure, p_s.

Pressure-volume-temperature (PVT) relationships. One of the best means of obtaining an understanding of the phase behavior and volume changes that take place in the reservoir during its depletion is to study the similar behavior that takes place in a PVT cell in the laboratory during the measurement of the PVT characteristics of a reservoir hydrocarbon system. Ideally, a reservoir is sampled immediately following the discovery well in a reservoir. The bottom-hole sampling of the reservoir fluid will then occur at or above the saturation pressure. This sample is taken into the laboratory and transferred to a high-pressure PVT cell.

The pressure, temperature, and volume of this cell can be controlled to permit measurement of the oil formation volume factor and the gas in solution at the various pressures to be encountered during the producing life of the reservoir. Ordinarily the pressure and temperature of the cell are first adjusted to the initial reservoir pressure and the reservoir temperature. The volume of the fluid in the cell at this time will be of some arbitrary value depending upon the size of the sample taken, the size of the cell, and the anticipated PVT analysis.

To obtain a reduction in pressure in the cell, the cell volume is increased slightly. This is accomplished physically in the laboratory by either removing mercury from the cell or moving a piston to obtain the desired volume change. If the reservoir is initially above the saturation pressure the initial change in Volume, V, with pressure will be very small as indicated in Fig. 6.5. However, once the pressure has been lowered below the saturation pressure, p_s, gas is liberated and the total volume of the cell, V_t, will begin increasing much more rapidly.

Use of a visual cell with a window will permit observation of the gas-liquid interface and evaluation of the amount of oil and gas in the cell at any particular pressure. It is observed that the oil volume is a maximum at the saturation pressure and declines as more and more of the oil changes to gas as the pressure in the cell is decreased. Once the

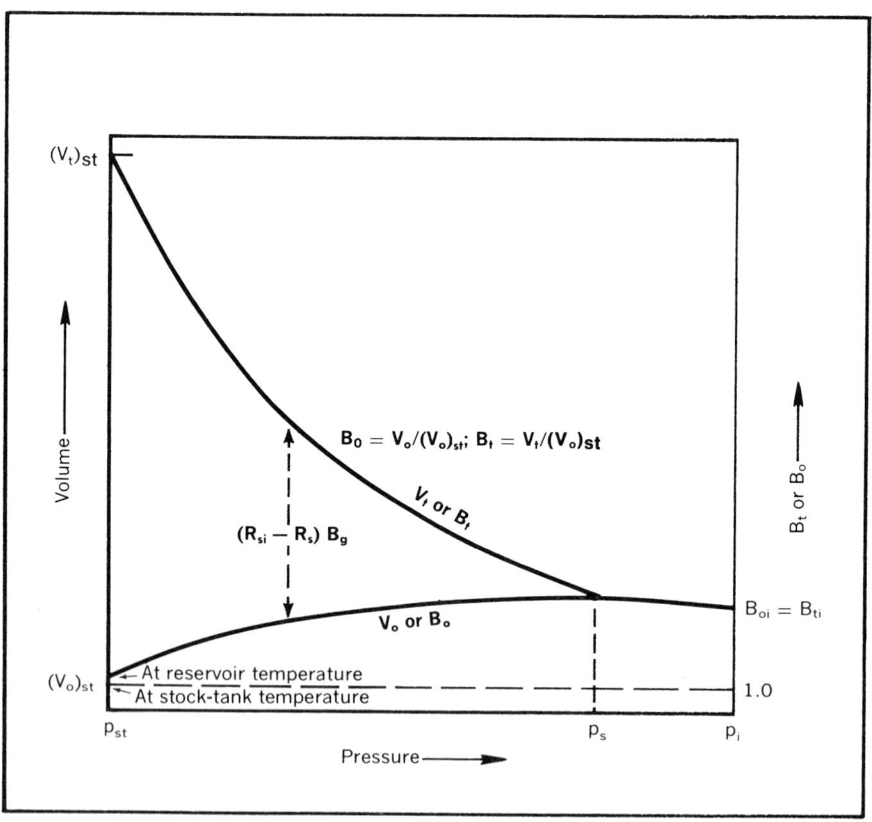

FIG. 6.5 Interrelationship of PVT data.

cell has reached the stock-tank pressure (atmospheric) it is permitted to cool to a stock-tank temperature which results in further thermal shrinkage of the oil to a volume, $(V_o)_{st}$.

Concurrently, the cell volume of gas, (indicated on Fig. 6.5 as the total volume, V_t, less the oil volume, V_o), will continue to increase with the decrease in cell pressure until at stock-tank conditions all of the gas is liberated (relative to stock-tank conditions). All of this gas was initially in solution when the cell pressure was at or above the saturation pressure, p_s. Consequently, the initial solution gas-oil ratio would be $[(V_t)_{st} - (V_o)_{st}]5.615/(V_o)_{st}$. The numerical constant is used to correct the ratio from volumes per volume to scf/STB. Note that no gas formation volume factor is necessary to correct to standard conditions since stock-tank conditions are generally taken as equivalent to standard conditions.

At other reservoir pressures the amount of free gas in the cell will be the difference between the total gas liberated to stock-tank conditions and the gas still in solution in the oil at the subject pressure. Conversely, the gas in solution at any particular cell pressure will be the difference between the total gas liberated and the free gas in the cell at any pressure corrected to a corresponding volume at standard conditions. When this difference is divided by the stock-tank oil volume, $(V_o)_{st}$, which results from the V_o volume of oil, and the result is multiplied by the number of cubic feet in a barrel to correct from a ratio of barrels/barrel to cubic feet/barrel, we obtain the gas in solution in scf/STB.

The oil formation volume factor can be determined for any pressure by calculating the ratio between the volume of oil in the cell at any pressure and the volume that results when this oil reaches stock-tank conditions, $(V_o)_{st}$. Now note that Fig. 6.5 also represents the relationship between various formation volume factors. By simply dividing all of the cell volumes by the stock-tank oil volume $(V_o)_{st}$, the stock-tank oil volume becomes 1.0 and the oil volume curve becomes the oil formation volume factor, B_o, curve.

The total volume curve becomes the total formation volume factor curve. This is a PVT parameter used by many companies. It is defined as the volume of oil and its original solution gas at reservoir conditions per stock-tank barrel of oil. The difference between the total formation volume factor curve and the oil formation volume factor curve is the illustrated volume of gas at reservoir conditions per stock-tank barrel of oil, $(R_{si} - R_s)B_g$. The difference in the solution gas-oil ratios is the volume of liberated gas and the gas formation volume factor corrects the volume to barrels at reservoir conditions. From Fig. 6.5 it is then obvious that

$$B_t = B_o + (R_{si} - R_s)B_g \qquad (6.5A)$$

The following problem should be worked and the solution checked against the solution in Appendix C to provide the engineer with some mental exercise in dealing with the PVT parameters and test his knowledge of the subject.

Problem 6.2A: PVT Lab Data Exercise

Given the following lab data from PVT analysis:

Cell pressure, psia	2,000	1,500 = p_s	1,000	500	14.7
Oil volume in cell, cc	650	669	650	615	500
Gas volume in cell, cc	0	0	150	700	44,500
Cell temperature, °F.	195	195	195	195	60

Evaluate R_s in scf/STB, B_o, and B_T at the stated pressures. Also find the compressibility factor for oil above the saturation pressure as vol/vol/psi. The gas-deviation factors at 1,000 psia and 500 psia have been evaluated as 0.91 and 0.95 respectively.

Material balance with gas liberation. With a knowledge of the gas in solution at various pressures it is possible to write an expression for the free gas in the reservoir at any particular pressure as shown in Fig. 6.3. The original gas in an oil reservoir initially found above the bubble-point pressure will be the gas in solution N R_{si}. At some later time when the reservoir pressure has declined below the bubble-point pressure due to withdrawal of oil and gas, the original gas in solution will be found in three places. The gas will either have been produced as G_p standard cubic feet; it will still be in solution in the reservoir oil remaining ($N - N_p)R_s$ standard cubic feet; or it will exist in the reservoir as free gas (as opposed to gas in solution in the reservoir). Thus, the free gas in the reservoir can be written as

$$\text{Free gas in reservoir} = \text{original gas in solution}$$
$$- \text{remaining gas in solution} - \text{produced gas}$$
$$= NR_{si} - (N - N_p)R_s - G_{ps} \tag{6.6}$$

If this volume of free gas is stated in reservoir barrels, it can be added to the remaining reservoir barrels of oil in the reservoir and equated to the original reservoir barrels of oil as illustrated in Fig. 6.3.

$$NB_{oi} = (N - N_p)B_o + [NR_{si} - (N - N_p)R_s - G_{ps}]Bg \tag{6.7}$$

One of the principle uses of material balance is to determine the original stock-tank barrels of oil in the reservoir. Consequently, material-balance equations are often solved for the initial volume of oil in the reservoir.

When Equation 6.7 is written this way we obtain

$$N = \frac{N_pB_o + B_g(G_{ps} - N_pR_s)}{B_o - B_{oi} + (R_{si} - R_s)B_g} \tag{6.8}$$

Material balance with encroaching gas and/or water. Note that the left-hand side of equation 6.7 is the original reservoir volume of the oil and the right-hand side is the same reservoir volume at some later time after a drop in pressure due to production of N_p stock-tank barrels of oil and G_{ps} standard cubic feet of gas. Consequently, if fluid encroached into the original oil-bearing portion of the reservoir from either an expanding gas cap or water encroachment, the right-hand side of Equation 6.7 would be smaller than the left-hand side by the amount of encroachment. Consequently, we could use Equation 6.7 to write an expression for the change in the gas-cap volume and water encroachment in a reservoir where such conditions exist.

$$\text{Change in gas-cap volume} = (G - G_{pc})B_g - GB_{gi}$$

$$(W_e - W_p) + \text{change in gas-cap volume} =$$
$$NB_{oi} - [(N - N_p)B_o + [NR_{si} - (N - N_p)R_s - G_p]B_g] \qquad (6.9)$$

Now work the following problems and check your solutions against the ones in Appendix C.

Problem No. 6.1C: Material Balance in a Solution Gas Drive Reservoir

An oil reservoir initially contains 4 million STB of oil at a pressure of 3,150 psia with 600 scf/STB of gas in solution. When the average reservoir pressure has dropped to 2,900 psia the gas in solution is 550 scf/STB. The initial oil formation volume factor was 1.34 and at 2,900 psia the oil formation volume factor is 1.32. If no initial gas cap or water drive is associated with this oil reservoir, how many stock-tank barrels of oil will be produced when the pressure has declined to 2,900 psia? The cumulative produced gas-oil ratio at this time is estimated as 600 scf/STB. The cumulative gas-oil ratio is defined as the total gas produced divided by the total oil produced at a particular time. Thus, G_p would be $600N_p$. The gas formation volume factor at 2,900 psia is 0.0011 res. bbl/scf.

Problem No. 6.1D: Calculation of Oil Zone Shrinkage

If our calculations are correct, we predicted that an oil reservoir would produce 101,800 STB of oil with an average gas-oil ratio of 600 scf/STB when the average reservoir pressure declined to 2,900 psia if there is no gas cap or water drive. However, assume this reservoir actually produced 130,800 STB of oil and 78,480,000 scf of gas when the average reservoir pressure reached 2,900 psia. What unexpected gas-cap gas or water invasion of the original oil zone could cause this additional production?

If the expression for the change in the gas-cap volume in Equation 6.4a is substituted for barrels of gas-cap expansion in Equation 6.9, the resulting expression can be solved for the original oil in place to obtain the general material-balance expression that includes all of the possible sources of natural reservoir energy except the change in the pore volume.

$$N = \frac{N_pB_o + B_g(G_{ps} - N_pR_s) - [(G - G_{pc})B_g - GB_{gi}] - (W_e - W_p)}{B_o - B_{oi} + (R_{si} - R_s)B_g} \qquad (6.10)$$

In Equation 6.10 the gas production is separated into gas-cap gas production, G_{pc}, and solution-gas production, G_{ps}. However, it is normally impossible to distinguish between gas-cap gas and solution gas for any given volume of produced gas. Fortunately, production of either will have exactly the same effect on the material balance and the two terms can then be added together and listed as G_p.

$$N = \frac{N_p B_o + B_g(G_p - N_p R_s) - G(B_g - B_{gi}) - (W_e - W_p)}{B_o - B_{oi} + (R_{si} - R_s)B_g} \qquad (6.11)$$

In problem 6.1B we calculated a gas-cap reservoir expansion of 40,000 res. bbl for a pressure change from 3,150 to 2,900 psia and in Problem 6.1D we calculated an oil zone shrinkage of 40,000 reservoir barrels for the same pressure change. Now we will consider this as one reservoir with a pressure change from 3,150 to 2,900 psia.

Problem 6.1E: Determining the Original Volume of Oil in the Reservoir

A gas-cap drive reservoir with no water drive has an original gas cap of 400 million scf at an initial pressure of 3,150 psia. The initial gas formation volume factor, oil formation volume factor and gas in solution are 0.001 res. bbl/scf, 1.34 res. bbl/STB and 600 scf/STB. After 78,480,000 scf of gas and 130,800 STB of oil are produced from this reservoir, the pressure is 2,900 psia and the gas formation volume factor, oil formation volume factor, and gas in solution are 0.0011 res. bbl/scf, 1.32 res. bbl/STB and 550 scf/STB. Assume that geologic control is such that the extent of this reservoir is unknown. Find the original stock-tank barrels of oil in place.

Including pore-volume changes in material balance. The compressibility of oil may be in the range of 10^{-5} volume of oil expansion per volume of oil per psia of pressure change. If the pressure is dropped 1 psi on a volume of 100,000 bbl of oil whose compressibility is 10^{-5} per psi, that volume of oil would expand 1 bbl. Although this is a very small expansion it is interesting to note that many reservoirs produce primarily by the expansion of oil for several years. The compressibility of water is even smaller, about 10^{-6} per psi of pressure change, but as it is shown in the chapters on fluid flow this expansion causes tremendous quantities of water to encroach into an oil reservoir. Thus, although the change in pore volume with a change in reservoir pressure is very small, in the range of 10^{-6} fractional volume change per psi, it nevertheless can play a primary role in the energy supplied for the production of oil, just as the compressibility of oil and water play large roles in oil and gas production.

The reduction in the pore volume that accompanies a decline in the reservoir pressure is due primarily to two factors, the reduction in the gross volume in the reservoir and the increase in volume of reservoir solids.

A microscopic look at the pore size of a sand reservoir is schematically represented in Fig. 6.6. In this diagram the pore space between the sand grains is indicated by a darkened area. A reservoir thousands of feet

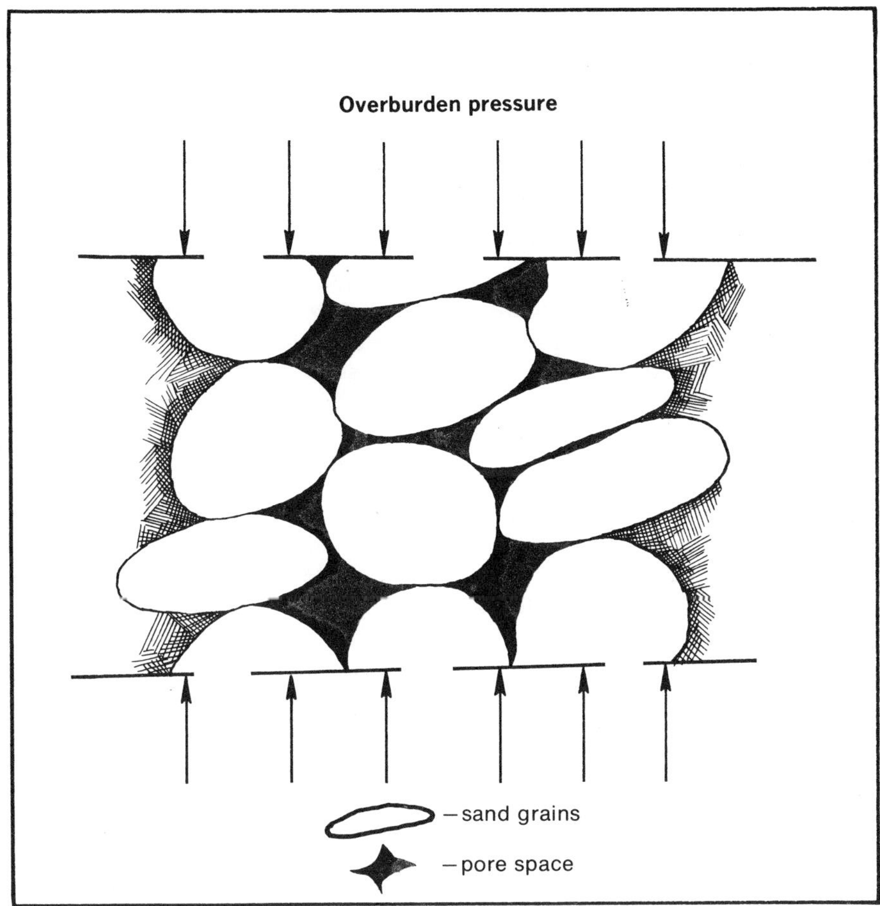

FIG. 6.6 Schematic microscopic cross-section of a sandstone reservoir (for discussion of pore volume compressibility).

underground is subjected to an overburden pressure caused by the weight of the formation above the subject reservoir. This has been shown to be equivalent to about 1 psi/ft. of depth, although the overburden pressures vary widely from area to area depending upon the nature of the structure, the consolidation of the materials and indirectly the geologic age of the materials. These overburden pressures simply apply a compressive force to the reservoir rock.

The pressure in the pore space does not normally approach that of the overburden pressure. A "normal" pressure in the pore space, which is known as the reservoir pressure, is about 0.5 psi/ft. of depth provided that the reservoir rock is sufficiently consolidated so the overburden pressure is not transmitted to the fluids in the pore space. In some areas, such as the Gulf Coast area of Texas and Louisiana, reservoir pressures may be very close to the overburden pressure due to the unconsolidated nature of the sandstones which permits the overburden pressure to be transmitted to the fluids in the pure volume.

At any rate, the overburden pressure which is trying to compress the outer dimensions of the reservoir is opposed by the strength of the reservoir rock and the pressure in the pore space. When the reservoir pressure declines due to fluid production, the force opposing the overburden pressure is reduced and the bulk volume of the reservoir is slightly reduced. At the same time note that the reservoir pressure almost completely surrounds all of the reservoir solids or grains. Consequently, any reduction in the reservoir pressure will cause an accompanying increase in the volume of the reservoir solids. Porosity is obtained by dividing the pore volume by the bulk volume of the porous media. Hence, the pore volume is equal to one minus the solids volume of the reservoir. We can also write the porosity as a function of the volume of solids.

$$\phi = \frac{V_p}{V_b} = \frac{V_b - V_s}{V_b} = 1 - \frac{V_s}{V_b} \tag{6.12}$$

In these equations ϕ is the porosity, V_p is the pore volume, and V_b and V_s are the bulk volume and solids volume, respectively. Note that a reduction in the bulk volume due to a change in the reservoir pressure and an increase in the solids volume due to a reduction in the reservoir pressure both decrease the porosity or pore volume.

The measurement of rock compressibility, c_f, has been carried out by Hall and others. Data sufficient for the empirical evaluation of the rock compressibility can be found in Appendix B.6.

Since c_f represents the fractional change in the pore volume per unit pressure change, we can write the pore volume change in barrels as

$$\Delta V_p = c_f \, \Delta p \, V_p \tag{6.13}$$

As noted previously, water exhibits a compressibility which is as large or larger than the formation compressibility. Since all known producing formations contain some water, the expansion of this water also contributes to a reduction in pore volume available for the hydrocarbons.

All formations contain some water because sedimentary formations were normally laid down in ancient seas and thus were initially filled with water. At some later time oil and gas migrated into these pore spaces and displaced the water. However, since the pore volume is made up of a wide variety of very small pores, the hydrocarbons are not capable of displacing all of the water from the pore space. Consequently, the small pores remain filled with water. This irreducible water is often called the connate water although other experts feel that any initial water saturation in an oil reservoir should be called connate water. Nevertheless, the water that is immobile to the displacing hydrocarbons will generally remain immobile while the hydrocarbons are being produced and this connate water can then expand due to the reduction in pressure in the reservoir and cause the corresponding reduction in pore volume. Water compressibilities can be determined from several sources and a wide range of water compressibilities are presented in Appendix B.3. The pore-volume change resulting from the expansion of the water can then be stated as

$$\Delta V_p = c_w \, \Delta p \, S_{wc} \, V_p \tag{6.14}$$

In this equation S_{wc} is the connate water saturation and is multiplied by the pore volume to obtain the volume of connate water. The change in pore volume resulting from the rock compressibility and the water compressibility can be conveniently combined into one term which is the sum of Equations 6.13 and 6.14,

$$\Delta V_p = (c_f + c_w S_{wc}) \, \Delta p \, V_p \tag{6.15}$$

Note that this expression for change in pore volume will have the same material-balance affect as will water encroachment or gas-cap-gas encroachment. Consequently, the term can be conveniently added to Equation 6.11 with the same sign as the gas-cap expansion and water-encroachment terms. However, the original pore volume is most conveniently stated as a function of the original oil in place

$$V_p = N \, B_{oi}/(1 - S_{wc}) \tag{6.16}$$

When we substitute this V_p expression in Equation 6.15 and add the resulting expression to Equation 6.11 we obtain

$$N = \frac{N_p B_o + B_g(G_p - N_p R_s) - G(B_g - B_{gi}) - (W_e - W_p)}{B_o - B_{oi} - (R_{si} - R_s)B_g} -$$

$$\frac{(c_f + c_w S_{wc})\Delta p \, N B_{oi}/(1 - S_{wc})}{B_o - B_{oi} - (R_{si} - R_s)B_g} \tag{6.17}$$

When this equation is rearranged to factor out the N from the ΔV_p term, we obtain the general material-balance equation

$$N = \frac{N_p B_o + B_g(G_p - N_p R_s) - G(B_g - B_{gi}) - (W_e - W_p)}{B_o - B_{oi} + (R_{si} - R_s)B_g + (c_f + c_w S_{wc}) \, \Delta p \, B_{oi}/(1 - S_{wc})} \tag{6.18}$$

Modifications of the general material balance equation. It is strongly recommended that all material-balance applications begin with Equation 6.18 so none of the significant reservoir energies are omitted in a particular application. The engineer will find that all of the terms in the general material-balance equation are never of significant size at the same time. Nevertheless, the use of this one equation seems to greatly simplify material balance since it eliminates the hodgepodge of equations necessary to describe each individual reservoir drive and its peculiar circumstances.

By using the general material-balance equation it will be unnecessary to have a special equation for expansion oil drive above the bubble point or solution-gas drive below the bubble point or water drive above the bubble point etc. Nevertheless, it should be helpful to investigate the effects of these special circumstances because most material-balance equations in the literature are limited rather than general.

For reservoirs above the bubble point note that many of the terms of the equation become zero. For example the gas-production term containing $G_p - N_p R_s$ will be zero since only solution gas will be produced above the bubble point and consequently the solution gas produced, $N_p R_s$, will be equal to the total gas produced G_p. Also the gas-expansion term containing G will be zero since there can be no free gas in the reservoir above the bubble point. In the denominator, the liberated-gas term, $R_{si} - R_s$, will be zero because R_s, the gas in solution at any particular pressure, will be equal to the gas originally in solution R_{si}.

When Equation 6.18 is applied to a reservoir that does not contain an initial free gas saturation or an initial gas cap, the term containing G will be zero as noted above. Furthermore, the gross water encroachment, W_e, will be zero if a reservoir is not affected by a water drive. However, the engineer should continue to carry the water production term, W_p, even when it is apparent that there is no water drive in the reservoir since connate water is often produced as the reservoir pressure declines even though no new water is entering the reservoir as W_e.

Once a reservoir contains any free gas the engineer will find that the pore-volume expansion term containing c_f and c_w will be insignificant in size compared with the free-gas terms. This becomes apparent when the compressibility of water and the formation is compared in magnitude with the compressibility of gas. The water and formation compressibilities, c_w and c_f, are in the range of 10^{-5} to 10^{-6}/psi while gas compressibility (not to be confused with the compressibility factor for gas) is in the range of 10^{-3} to 10^{-4} or about 100 times as great as the compressibility of the water and formation. Consequently, a gas saturation of 1% may provide as much energy as will the water and formation compressibility terms. So, once the gas saturation is substantial, the pore volume change becomes insignificant.

It should also be emphasized that the general material-balance equation does not contain any restriction as concerns application above and below the bubble-point pressure. There is no reason to break a material-balance calculation into a calculation to the bubble point and from the bubble point to values of pressure less than the bubble point as appears to be common practice in some areas. Review of the derivation procedures will indicate that there is absolutely no restriction on the initial condition applied in the derivation of these equations.

There are two reservoir properties used frequently in the general literature in connection with material balance that receive only sparing use in this book. These are the total formation volume factor, B_t, previously defined, and m, the ratio of the initial reservoir free-gas volume to the initial reservoir oil volume, (GB_{gi}/NB_{oi}). By using the total formation volume factor, B_t, all of the oil formation volume factors, B_o, can be eliminated from the equations so that only B_t is needed. Since all PVT lab procedures include the evaluation of the oil formation volume factor, the advantage of a material-balance equation that avoids the use of B_o seems to be more imaginary than real.

It is more difficult to rationalize a preference for the use of the ratio, m, in material-balance equations since it is necessary to know the original volume of free gas in the reservoir to determine m. Nevertheless, consider the conversion of Equation 6.18 to the B_t and m parameters. This can be accomplished by first substituting for the gas production, G_p, a function of the cumulative produced gas-oil ratio, N_pR_p; adding and subtracting the term $N_pR_{si}B_g$ to the numerator; substituting mNB_{oi}/B_{gi} for G; and rearranging the equation.

$$N = \frac{N_pB_o + B_gN_pR_{si} - B_gN_pR_s + B_gN_pR_p - B_gN_pR_{si} - (W_e - W_p)}{B_o + (R_{si} - R_s)R_g + (c_f + c_wS_{wc})\,\Delta pB_{oi}/(1 - S_{wc}) + mB_{oi}(B_g - B_{gi})/B_{gi} - B_{oi}}$$

$$(6.19)$$

Now substitute B_t for $B_o + B_g(R_{si} - R_s)$ and B_{ti} for B_{oi} where possible and rearrange the equation to obtain

$$N = \frac{N_p[B_t + (R_p - R_{si})B_g] - (W_e - W_p)}{B_t - B_{ti} + (c_f + c_wS_{wc})\,\Delta pB_{ti}/(1 - S_{wc}) + mB_{ti}(B_g - B_{gi})/B_{gi}} \quad (6.20)$$

GENERAL DIFFICULTIES IN APPLYING MATERIAL BALANCE EQUATIONS

There are many problems involved in applying material-balance calculations to a reservoir. One of the foremost of these is the general lack of PVT data for specific reservoirs. Ideally such PVT data should be determined for the specific reservoir in the laboratory or should be calculated from the specific reservoir hydrocarbon compositions using appropriate equilibrium constants. Under ideal conditions this requires a bottom-hole sample of the reservoir fluid before the flowing pressure at the well falls below the bubble-point pressure. Several sampling devices are available. However, once the pressure in the well is below the bubble-point pressure it becomes extremely difficult to obtain a representative sample of oil and free gas.

In some cases it is necessary to take surface samples and recombine the oil and gas in the laboratory. Here again it is difficult to know exactly what gas-oil ratio should be recombined to represent the initial reservoir fluid. One of the reservoir engineer's primary functions in a producing company should be to make certain that proper data are obtained on reservoirs. However, if samples have not been taken, much useful information and guidance can be obtained by using empirical data.

In Appendix B.20 and B.21, enough information will be found to permit determination of formation volume factors, gas in solution and the oil formation volume factors as a function of the reservoir temperature and pressure, and the oil and gas gravities. Similar data for the evaluation of gas-deviation factors are in Appendix B.7 and B.8. Since the trend of PVT data (the change in the data with pressure), is more important than the absolute values, these empirical data will often give very excellent results. Empirical PVT data do not relieve the reservoir engineer of his responsibilities in obtaining the most accurate data possible by obtaining initial bottom-hole fluid samples from reservoirs, but it does tend to minimize the results of his failures.

Even when adequate sampling procedures have been followed, problems still exist concerning the evaluation of PVT data. In the laboratory there is a choice between flash and differential types of liberation procedures to be followed. In the flash procedure all of the liberated gas remains in contact with the oil from which it is liberated as the process continues. However, in the differential process, ideally, all of the liberated gas is immediately separated from the oil. The actual reservoir situation is somewhere between these two extremes because all of the gas liberated in the reservoir is not immediately removed from the presence of the remaining oil, but as the gas is liberated in the reservoir it has a much greater mobility than the oil and consequently is produced at a much greater rate than the oil itself. Consequently, the oil in the reservoir is in contact with all of its initial solution gas for only a short period of time. Thus, the reservoir liberation process is somewhere between flash and differential.

Fortunately, the results of these two processes are not greatly different for most reservoir fluids and the reservoir fluids with great differences do not lend themselves generally to the simplified type of material-balance approach described herein. Conventional material balance assumes that the composition of the gas liberated throughout the life of the reservoir is constant. However, when the hydrocarbon system contains considerable amounts of propane, butane, and pentane (the intermediate hydrocarbons), the assumption of a constant liberated gas composition is no longer applicable and conventional material balance becomes too inaccurate for use. Such reservoirs are called volatile oil reservoirs, but they are not common. One international group of companies who desired examples of such volatile oil reservoirs searched their files and could find only three such reservoirs.

Engineers are often confused by the evaluation of PVT data empirically. Since an engineering analysis is no more accurate than the data used it is important that PVT data be determined as accurately as possible. Consequently, the reader is asked to work the following problem and check his answers against the solution in Appendix C.

Problem 6.2B: Determining PVT Data Empirically

In Problem 6.2A you evaluated the oil formation volume factors and gas in solution at reservoir pressures of 2,000, 1,500, 1,000, 500, and 14.7 psia from lab data. Now make the same evaluations empirically if the gas gravity is 0.7 and the oil stock-tank gravity is 40° API. The reservoir temperature and saturation pressure are 195° F. and 1,500 psia, respectively. Note that the evaluation of the oil formation volume factor above the saturation pressure requires the use of the oil compressibility data of Appendix B.4.

Accuracy of production data. Another type of data problem that exists in connection with material balance applications is the general inaccuracy of gas and water production. These data greatly affect the behavior of most oil reservoirs. The economic worth of these two commodities when produced from an oil reservoir is often nil since water is never sold and the gas produced with oil may not be sold. Consequently, records of the amounts of gas and water produced may be very poor. It is not at all common for field personnel to simply "boiler-house" water and gas production data with little regard for correctness.

When the gas associated with oil production is sold, there can be little question as to the amount of gas produced. However, the reservoir engineer should be fully aware of the possibility that water and gas production figures are grossly in error and should be prepared to check with field personnel on his own, in any way possible, to ascertain the accuracy of these data as well as the accuracy of the allocation of oil production to the individual wells and zones.

Since oil is sold, the total oil produced should be accurately known. However, allocation of this production to different zones and different wells may also be subject to considerable question.

Fluid-flow fundamentals. It is impossible for the reservoir engineer to make intelligent applications of material-balance without an adequate knowledge of fluid-flow fundamentals. The ratio between the gas production and oil production depends on the flow characteristics of the reservoir. Furthermore, water encroachment, W_e, also must be evaluated in these same terms. For this reason, it is virtually impossible to use material-balance concepts without applying fluid-flow fundamentals even though only a prediction based on the average reservoir pressure is desired. A much more extensive knowledge of fluid-flow fundamentals is necessary if the prediction is to be based on time rather than the average reservoir pressure.

Accuracy of reservoir-pressure data. The author believes that the biggest single difficulty in material balance is the use of inaccurate average reservoir pressures. This subject is treated thoroughly in Chapter 3 and to a lesser degree in Chapter 4 (as it applies to gas reservoirs).

It is common to leave a well shut-in for 3–4 days and assume the pressure has reached the average reservoir pressure for that well's drainage area. For a gas reservoir, the cautious engineer may require up to a 4-day shut-in period, but even this may not be enough. The pressure record must be examined to determine if the pressure is still changing after the shut-in period. It may appear to be constant but still be changing at a rate too small for detection over a short period of time. However, this small rate of change could be significant over a period of several days.

As noted in Chapter 4, some gas reservoirs require years to stabilize, making it virtually impossible to directly record the average reservoir pressure. As a guideline, it is wise to remember that you can not hope to directly record the average pressure in a drainage area until the well has been shut-in for a period of time equal to or greater than the stabilization time, t_s. This time should be calculated using the greatest distance from the well to the outer drainage boundary as r_e and the equation derived in Chapter 2.

$$t_s = \frac{0.04\phi\mu cr_e^2}{k}$$

Remember that the compressibility in this equation is based on the effective compressibility of the system. This means that it is greatly affected by the gas saturation. Consequently, the engineer should be especially careful that his average pressures are based on realistic shut-in times when substantial amounts of gas saturation exist in the reservoir.

Even when shut-in times exceed the stabilization time the recorded pressure will not exactly equal the average pressure unless this is the only well in the reservoir or all the wells in the reservoir are shut-in. However, the recorded pressure will closely approximate the average pressure except for highly permeable formations. In this case the recorded pressure after a shut-in time greater than the stabilization time will be so greatly affected by the offset producers that the pressure in the subject well will simply represent a pressure in the new (since shut-in) drainage area of an offset well.

If he does not completely understand the difficulties and recommended methods of determining the average pressure in a well's drainage area the engineer is encouraged to review this subject in Chapter 3.

PREDICTING GAS DRIVE BEHAVIOR

Whenever material balance is discussed in detail it is customary to consider its use in predicting behavior of solution-gas-drive reservoirs. We will see that the same techniques can be applied to gas-cap-drive analysis but the prediction of the produced gas-oil ratio becomes more difficult because of the trouble encountered in GOR control in actual operations.

Aside from this latter difficulty, it's still a question whether or not a theoretical prediction of reservoir behavior for a gas-drive reservoir based on material balance is more accurate than a prediction based on decline curves. This question will be better understood after the engineer has considered the problem of determining relative permeability with its large influence on the prediction. It will be noted that about the only present method of accurately determining relative permeability of a gas-drive reservoir is by calculation from reservoir behavior and extrapolation. Since past performance is necessary to obtain any accuracy from the theoretical prediction it can be argued that empirical decline-curve analysis (Chapter 7) will give just as good a prediction of the reservoir's behavior.

This situation is particularly true under full-capacity operation. It does seem clear that any prediction of gas-drive reservoir behavior should be reconciled as soon as possible with the decline-curve experience of the reservoir. A complete understanding of gas-drive reservoir prediction methods provides the engineer with an excellent understanding of the various factors that control behavior and ultimate recovery.

There appears to be an infinite variety of methods for solving this problem. Only those that appear to be most useful will be presented here.

Behavior of produced GOR. To predict the behavior of a gas-drive reservoir it is necessary for the engineer to predict the behavior of the produced GOR. The produced GOR will be constant above the saturation pressure but once gas saturation is high enough for free gas to flow in the reservoir, behavior of the GOR becomes much more complicated. The produced GOR, at any particular time is the ratio of the gas volume being produced to the volume of oil being produced. Hence, the name instantaneous gas-oil ratio is often applied to the symbol, R. Remember that the gas produced will include solution gas and free gas. Thus,

$$R = R_s + R_{flow} \tag{6.21}$$

where R_{flow} is the flowing GOR in the reservoir, q_{scf}/q_{STB} stated in SCF/ST Bbl. This expression can be expanded using the radial flow equation derived in Chapter 1.

$$q_r = (1.127k \ A_r/\mu)(\Delta p/\Delta r)_r \tag{1.6}$$

Writing q_{scf} and q_{STB} in terms of this equation applied at the well bore with the rates corrected from reservoir volumes to scf and STB, respectively, we obtain an expression for the reservoir flowing GOR and the produced GOR.

$$R_{flow} = \frac{q_{scf}}{q_{STB}} = \frac{(1.127k_gA_w/\mu_g)(\Delta p/\Delta r)_w/B_g}{(1.127k_oA_w/\mu_o)(\Delta p/\Delta r)_w/B_o} \tag{6.22}$$

$$R_{flow} = \frac{k_g}{k_o}\frac{\mu_o}{\mu_g}\frac{B_o}{B_g} \tag{6.23}$$

$$R = R_s + \frac{k_g}{k_o}\frac{\mu_o}{\mu_g}\frac{B_o}{B_g} \tag{6.24}$$

Note that the ratio of effective permeabilities in this equation is the same as the ratio of relative permeabilities, k_{rg}/k_{ro}, which is a function of the liquid saturation (See Fig. 6.7 for example data). The water saturation in a gas-drive reservoir without a water drive will be practically constant but the oil saturation is changing continually. Consequently, to determine the relative permeability ratio and the produced GOR it is necessary to evaluate the oil saturation corresponding to any cumulative oil production. This can be done by material balance.

The oil saturation is the remaining volume of oil in the reservoir, $(N - N_p)B_o$, divided by the reservoir pore volume in barrels. The pore volume can be determined from the initial oil saturation, $(1 - S_{wc})$ and the original reservoir oil volume, NB_{oi}. Thus, the material-balance expression for oil saturation can be written.

$$S_o = \frac{\text{Res. bbl oil}}{\text{bbl pore volume}} = \frac{(N - N_p)B_o}{(NB_{oi})/(1 - S_{wc})} \tag{6.25}$$

$$S_o = \frac{B_o(N - N_p)(1 - S_{wc})}{NB_{oi}} \tag{6.26}$$

A typical GOR history for a gas-drive reservoir is shown in Fig. 6.8. As noted previously, GOR will remain constant as long as reservoir pressure is above the saturation pressure. Once free gas begins to form in the reservoir, the GOR will theoretically decline slightly until gas saturation becomes high enough to permit flow. This critical saturation is referred to as the equilibrium gas saturation. Because of the difficulty of accurately measuring this value in the laboratory, it is the author's opinion that it is generally thought to be much greater than it actually is. As a matter of fact, it is seldom that this declining GOR portion of the history can be distinguished in practice. However, part of this difficulty is undoubtedly due to the fact that the entire reservoir does not reach the saturation pressure at the same time, tending to mask this GOR decline.

Once gas begins to flow, the GOR increases rapidly. This rapid increase is due to the rapid increase in relative permeability ratio with a change in saturation. It is also caused by the extreme difference in the viscosities of gas and oil. As noted previously, the ratio of gas viscosity to oil viscosity at low pressure may be 100, and even at high pressure the viscosity ratio will be high.

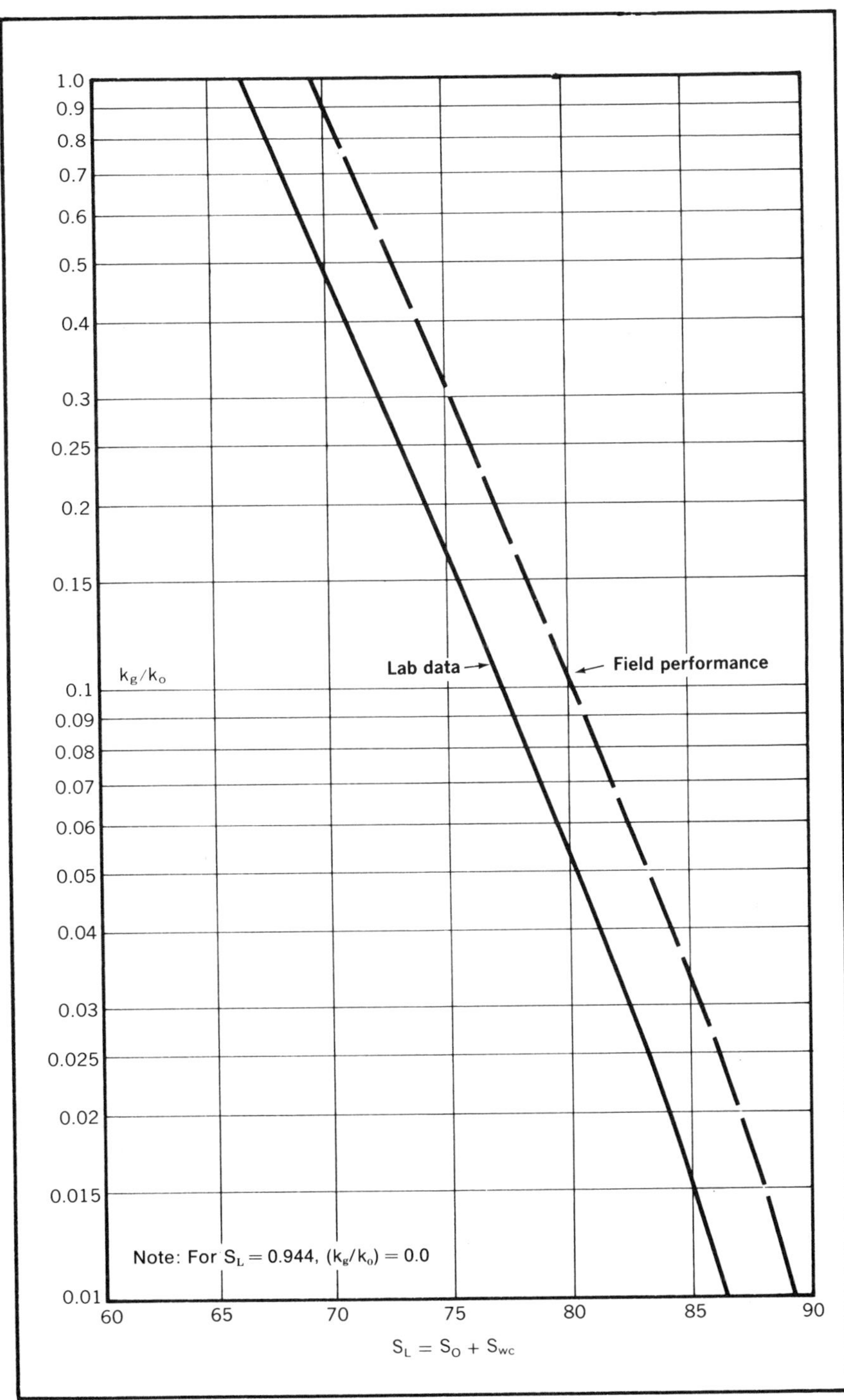

FIG. 6.7 Comparison of laboratory-measured and field-calculated relative permeability.

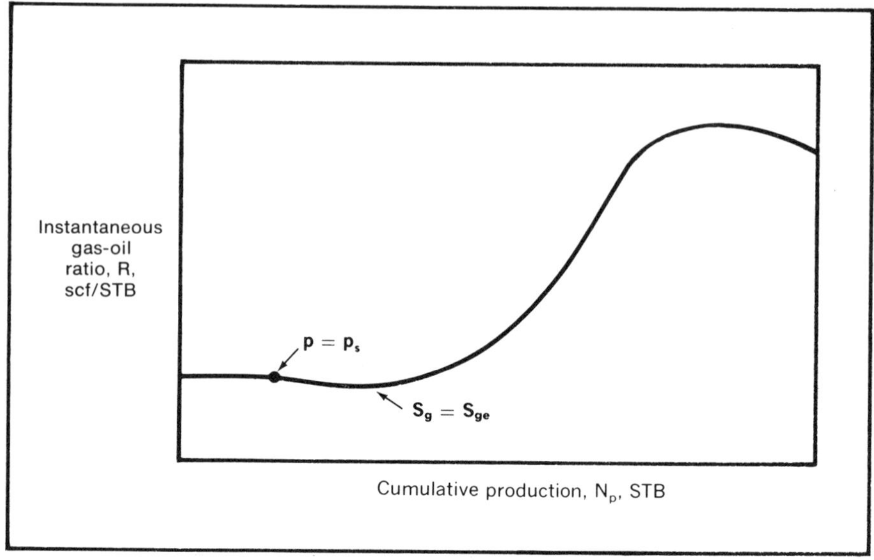

FIG. 6.8 Gas-oil ratio history for a solution-gas-drive reservoir.

During the late life of the gas-drive reservoir the GOR may decline as indicated in Fig. 6.8. This decline is sometimes difficult to understand since the relative permeability and the viscosity ratios continue to increase. This decline in GOR during late life is due to the change that takes place in the gas formation volume factor, B_g. Remember that the gas formation volume factor is inversely proportional to the reservoir pressure. Consequently, B_g continues to increase in size as the reservoir depletes. The change in B_g at high reservoir pressures is small, but at low pressures it changes more rapidly than the relative permeability and viscosity ratios, resulting in a decline in the produced GOR.

The Schilthuis method.[1] Since material-balance concepts were originated by Schilthuis, it seems that a method relying basically on the form of his original equation should be so named.

Starting with the general material balance, Equation 6.18, application to a reservoir below the saturation pressure with no water drive eliminates the water-encroachment term, W_e, and the pore volume expansion term which, as noted previously, is insignificant when free gas is present in the reservoir. Consequently, for a solution-gas-drive or gas-cap drive reservoir, Equation 6.18 becomes

$$N = \frac{N_p B_o + B_g(G_p - N_p R_s) - G(B_g - B_{gi})}{B_o - B_{oi} + (R_{si} - R_s)B_g} \qquad (6.27)$$

Solving this equation for N_p we obtain

$$N_p = \frac{N[B_o - B_{oi} + (R_{si} - R_s)B_g] + G(B_g - B_{gi}) - B_g G_p}{B_o - B_g R_s}$$ (6.28)

Thus, to predict the cumulative oil production at a particular stage of depletion, it is necessary to know the original oil and gas in the reservoir and the original reservoir pressure which fixes N, G, B_{oi}, R_{si}, and B_{gi}. The stage of depletion fixes the reservoir pressure at that time which in turn fixes the other PVT data in the equation. However, we still must evaluate the gas production, G_p, for this stage of depletion before we can proceed with the calculation of N_p.

The cumulative gas production, G_p, is determined from the produced GOR and the cumulative oil production. However, as discussed previously, the produced GOR varies with oil saturation in the reservoir as it continues to decline with continued oil production. Consequently, it is necessary to multiply an increment of oil production, ΔN_p, by the average produced GOR that existed during the production of that increment to obtain the increment of gas production. If all of these products are summed we obtain the total G_p.

$$G_p = \sum_{j=1}^{j=n} R_j \, \Delta N_{pj}$$ (6.29)

However, we know from Equation 6.24 that the GOR is a function of the relative permeability ratio, which is in turn a function of the oil saturation in the reservoir. Furthermore, Equation 6.26 shows that the saturation is a function of the cumulative oil production, N_p, which is what we are to calculate by Equation 6.28. Consequently, Equation 6.28 requires a trial and error solution.

The most direct procedure for obtaining this solution is to (1) set the reservoir pressure, (2) assume G_p, (3) calculate N_p by Equation 6.28, (4) calculate S_o by Equation 6.26, (5) determine the relative permeability ratio based on the liquid saturation from data such as that of Fig. 6.7, (6) calculate G_p using Equations 6.29 and 6.24, (7) compare the calculated G_p with the assumed G_p from Step 2, and (8) repeat the calculations from Step 2 through 8 if the assumed and calculated values do not agree satisfactorily, or (9) go back to Step 1 and set a new pressure if the assumed and calculated G_p values do agree.

Although this appears to be the most direct approach to the problem it is not necessarily the simplest one. If a computer is used to solve prediction problems of this type it will make little difference to the engineer which procedure is followed except for some small difference in computer time that might result. However, if it is necessary for the engineer to use a desk calculator he can select easier methods. The problem is which item should be used for the trial and error iteration in satisfying Equations 6.29, 6.28, 6.26, and 6.24.

The Tarner method.[2] Tarner appears to be the first person to suggest iteration on the produced GOR at the state of depletion to be calculated or at the time when N_p barrels of oil have been produced (which we are calculating). The iteration can be carried out by extrapolating a plot of instantaneous GOR versus reservoir pressure to the next average reservoir pressure at which cumulative production of oil and gas is desired. Data for the plot can be previously calculated data or a plot of actual data. In either case the GOR determined by extrapolation is used as the assumed GOR, R_n, that exists after N_{pn} barrels of oil have been produced. With the GOR plot completed the cumulative gas production, G_{pn}, can be calculated using Equation 6.29. Actually the engineer will already have calculated or know the cumulative gas production for some recent time $G_{p(n-1)}$ and the new cumulative gas production, G_{pn}, will be

$$G_{pn} = G_{p(n-1)} + [(R_n + R_{(n-1)})/2](N_{pn} - N_{p(n-1)}) \tag{6.30}$$

When this expression is substituted into Equation 6.28 for G_p, the result is an expression containing only one unknown, N_{pn}.

$$N_{pn} = \frac{N[B_o - B_{oi} + (R_{si} - R_s)B_g] + G(B_g - B_{gi})}{B_o - B_g R_s} -$$

$$\frac{B_g\left[G_{p(n-1)} + \left(\dfrac{R_n + R_{n-1}}{2}\right)(N_{pn} - N_{p(n-1)})\right]}{B_o - B_g R_s} \tag{6.31}$$

This expression can be solved for N_{pn}.

$$N_{pn} = \frac{N[B_o - B_{oi} + (R_{si} - R_s)B_g] + G(B_g - B_{gi})}{B_o - B_g R_s + (R_n + R_{n-1})B_g/2} -$$

$$\frac{B_g[G_{p(n-1)} - (R_n + R_{n-1})N_{p(n-1)}/2]}{B_o - B_g R_s + (R_n + R_{n-1})B_g/2} \tag{6.32}$$

Once N_{pn} has been calculated, based on an R_n assumed from an extrapolated plot of the produced GOR versus reservoir pressure, it is possible to determine the oil saturation in the reservoir at this time, S_{on}, using Equation 6.26 and N_{pn} calculated from Equation 6.32. Based on this saturation, the permeability ratio can be determined from given data and R_n can be calculated from Equation 6.24. If the assumed and calculated R_n GOR are in satisfactory agreement, the engineer can proceed with the calculation for the next lowest pressure of interest. If the R_n values do not agree sufficiently, it is necessary to adjust the extrapolated GOR, and repeat the calculations until R_n by extrapolation and R_n calculated do agree.

The procedure for performing a Tarner-type prediction for a gas-drive reservoir can be summarized as follows:

(1) Extrapolate a plot of previously observed or calculated GOR versus reservoir pressure to the next reservoir pressure at which the cumulative production, N_{pn}, is desired. This will provide a gas-oil ratio, R_n, corresponding to the cumulative production desired, N_{pn}.

(2) Use Equation 6.32 to calculate N_{pn}, using previously known cumulative oil and gas production and produced GOR where indicated by the subscript $(n - 1)$.

(3) Use Equation 6.26 to calculate the oil saturation, S_{on}, corresponding to the cumulative production prediction, N_{pn}.

(4) Calculate the produced GOR corresponding to N_{pn} and S_{on} using Equation 6.24. This will give a calculated R_n and should be based on a k_g/k_o ratio corresponding to S_{on}.

(5) If the produced GOR's determined by exprapolation in (1) and calculated in (4) agree to an acceptable accuracy, assume a new lower reservoir pressure and repeat (1) through (4). Note that the calculated R_n will now be a "previously calculated GOR and will be R_{n-1} for the next sequence of calculations.

(6) If the produced GOR's determined by extrapolation and calculation do not agree to an acceptable accuracy, it is necessary to return to step (1) and adjust the produced GOR extrapolation until the calculated and extrapolated R_n agree satisfactorily.

It is seldom that more than two iterations on R_n are necessary to obtain acceptable accuracy. Also note that subsequent calculations automatically adjust for errors tolerated in the produced GOR in previous calculations. We will see that with some other popular gas-drive prediction techniques the errors tend to accumulate.

A recommended procedure. One of the biggest difficulties in performing an accurate prediction of the performance of a gas-drive reservoir is the difficulty of obtaining relative permeabilities characteristic of the producing formation. Several problems exist in this connection. As noted in Chapter 1, the relative-permeability data obtained differs widely from one sample to another and the problem of determining an average or effective single relationship to represent a particular reservoir has not been satisfactorily solved. Also, the actual procedures followed in the lab leave much to be desired since the size of the sample used is limited by the core diameter and the wettability of the core has been subjected to change by its contact with drilling fluids and the fluids necessary to core preparation in the laboratory.

So laboratory-measured relative-permeabilities are seldom characteristic of the producing reservoir. Consequently, it is recommended that a prediction of a gas-drive reservoir behavior based solely upon lab-measured data be used only as a last resort and that such a prediction be corrected as soon as meaningful reservoir performance data can be obtained. Use past reservoir behavior to calculate the effective relative permeability ratios corresponding to various average reservoir saturations. Then extrapolate these calculated data to represent the data for future reservoir saturations.

To be more specific, the past oil production at various average reservoir pressures is used to calculate the average oil saturations, S_o, using Equation 6.26. The corresponding relative-permeability ratios are calculated from Equation 6.24 based on the average produced GOR at that time. When the oil saturations are corrected to equivalent liquid saturations (by adding the connate water saturation) and a plot of the log of the relative permeability versus liquid saturation is prepared, it will be found that the slope of this plot generally is the same as the slope of the similar lab data. However, the field-calculated data are generally considerably displaced from the lab data. By drawing a line parallel to the lab data and positioning the plot on the basis of the field-calculated points, a reasonable estimate can be obtained for the relative permeability characteristics of the reservoir.

The recommended procedure for predicting the behavior of a gas-drive reservoir is:

(1) Use as much reservoir production and pressure history as possible to calculate k_g/k_o versus S_o data. Use Equation 6.24 and observed average producing GOR to determine k_g/k_o. Reservoir oil saturations can be calculated from Equation 6.26.

(2) Correct the reservoir oil saturations to reservoir liquid saturations by adding the connate water saturation and plot the liquid saturations versus the permeability ratio data on semi-log paper.

(3) Plot the lab relative-permeability data on the same graph prepared in (2) and extend the field-calculated permeability data parallel to the lab data.

(4) Assume the extrapolated field data of (3) is representative of the reservoir and proceed with the Tarner method described in the last section.

To check your understanding of this procedure work the following problem and check your solution against the solution in Appendix C.

Problem No. 6.3: Predicting the Behavior of a Gas Drive Reservoir

The following data apply to a solution-gas-drive reservoir:

$$N = 10,025,000 \text{ STB}$$
$$S_{wc} = 22\%$$
$$p_i = 3,013 \text{ psia}$$
$$p_s = 2,496 \text{ psia}$$

p	B_o	R_s	B_g	μ_o/μ_g	R
3,013	1.315	650	0.000726	53.9	650
2,496	1.325	650	0.000796	56.6	650
1,302	1.233	450	0.001616	102.61	2,080
1,200	1.224	431	0.001807	108.96	?
1,100	1.215	412	0.001998	115.20	?

A. Calculate S_o and k_g/k_o when the reservoir pressure is 1,302 psia, cumulative production is 1,179,000 STB and 1,123 MMscf of gas. Note that this point falls on the field plot of k_g/k_o vs S_L in Fig. 6.7.
B. Calculate the cumulative oil and gas production when the reservoir pressure reaches 1,200 psia. Base your $R_{1,200}$ guess on extrapolation of the observed GOR in Fig. 6.9 and your k_g/k_o values on the trend established by the field data in Fig. 6.7.
C. Repeat the calculations of Part B for a reservoir pressure of 1,100 psia.

This analysis lends itself to a simple but very valuable computer program due to the trial-and-error iterations on R_n involved in the Tarner portion of the calculations. The Schilthuis method is also greatly enhanced by programming it for computer solution due to the trial-and-error iterations on the cumulative gas-oil ratio, R_p.

Note that the Schilthuis method was based on a finite equation while

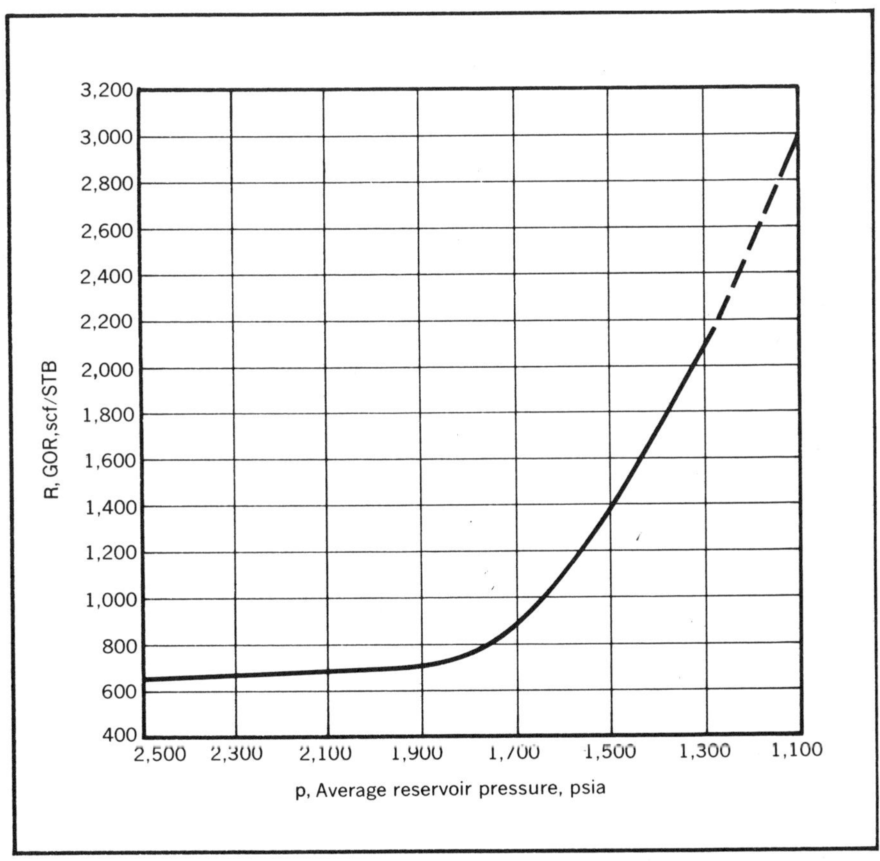

FIG. 6.9 Production history for Problem 6.3

the basic equation for the Tarner method might be considered to be a finite difference form of the material balance.

The Muskat method. We will now consider a prediction technique that is based on a differential form of the material balance equation.

$$\frac{\Delta S_o}{\Delta p} = \frac{\dfrac{B_g S_o}{B_o}\dfrac{\Delta R_s}{\Delta p} + \dfrac{S_o k_g \mu_o}{B_o k_o \mu_g}\dfrac{\Delta B_o}{\Delta p} + (1 - S_o - S_w)B_g \dfrac{\Delta(1/B_g)}{\Delta p}}{1 + \dfrac{k_g \mu_o}{k_o \mu_g}} \qquad (6.33)$$

The theoretical basis for this equation can be seen by recognizing that the equation for the reservoir oil saturation, Equation 6.26, is stated as a function of the produced oil, N_p. Now note that N_p is defined in Equation 6.28 as a function of the produced gas, G_p, and the produced gas is a function of the mobility ratio of the gas and oil, $k_g \mu_o / k_o \mu_g$, through Equations 6.24 and 6.29. Consequently, these relationships can all be combined into one equation and differentiated with respect to the pressure to obtain Equation 6.33. Craft and Hawkins[4] present a much simpler derivation by equating the producing GOR by material balance to the producing GOR by fluid flow.

On the surface it appears that Equation 6.33 avoids the trial-and-error-type solution necessary in the Schilthuis and Tarner methods because we normally know everything in the right-hand side of the equation. Note that $\Delta R_s / \Delta p$, $\Delta B_o / \Delta p$, and $\Delta B_g / \Delta p$ represent the slope of a curve of these PVT data versus pressure. Consequently, if we were only interested in the slope of reservoir oil saturation versus pressure, at a particular pressure, we could apply Equation 6.33 without resorting to trial and error. However, what is needed is a plot of oil and gas production versus the reservoir pressure. This requires that we determine the finite change in reservoir oil saturation for some finite change in pressure. This means we must apply the equation on a finite-difference basis as far as $\Delta S_o / \Delta p$ is concerned, but on a differential basis as far as the other Δ terms are concerned.

In other words, we use Equation 6.33 to calculate the finite change in oil saturation, ΔS_o, accompanying some finite change in pressure, Δp. The slopes $\Delta R_s / \Delta p$, $\Delta B_o / \Delta p$ and $\Delta B_g / \Delta p$ are evaluated at the average pressure but the parameters in the equation that are a function of the saturation, S_o and k_g / k_o, are not theoretically fixed until we know the change in saturation, ΔS_o, for which we are solving. Thus, the Muskat prediction is also theoretically a trial-and-error solution in which S_o average and k_g / k_o are guessed until it is found that they are satisfied by the calculated change in oil saturation, ΔS_o.

In spite of the noted theoretical limitation on the Muskat prediction, the technique can be used without trial and error when a digital computer is used. Such an analysis can use such small pressure increments that the change in S_o is small and no significant error is encountered in the calculations due to this inaccuracy. The accuracy is further helped by basing S_o and k_g/k_o on a saturation equal to that at the end of the last calculated pressure change plus half of the change in saturation for the last pressure change, $S_{o(n-1)} + 0.5\ \Delta S_{o(n-1)}$. Even when these precautions are taken, the engineer should make certain that periodically during the program the results of a Muskat prediction are checked against a finite material-balance equation such as Equation 6.28. Without such a check errors can accumulate from each step in the prediction because the technique is not self correcting as is the case with the Schilthuis and Tarner prediction techniques.

The Muskat method represents one of the original applications of the digital computer to reservoir engineering and consequently, most companies have a program that provides a solution-gas-drive prediction based on the Muskat technique. The comments on the lack of self correction should not be taken to mean that a Muskat computer program should be set aside in favor of a program for the Tarner method. It does mean that the author recommends you check your program. Furthermore, if you do not presently have a computer program for a tank-type prediction of a solution-gas-drive reservoir, the Tarner method described herein is recommended for use due to its more direct approach to the problem and its self-correcting characteristics.

Material balance in partially saturated reservoirs. When a reservoir is thick or has large areal extent, there will be a period of the productive life when a portion of the reservoir is above the bubble-point pressure while another portion is below the bubble-point pressure. When possible, a material-balance application should avoid these periods of time, because periods of analysis or prediction should be for times above or below the period when the reservoir is only partially saturated. However, in large, thick reservoirs it may be impossible to avoid analysis during these periods. It then becomes necessary to use a special material balance because a discontinuity exists in the PVT data at the bubble-point pressure and the average PVT data for the reservoir will not exist at the average pressure.

This situation can be best illustrated by considering a very thick reservoir such as the one whose pressure history is shown in Fig. 6.10A, with PVT data in Fig. 6.10B. This example is concerned with a solution-gas-drive reservoir whose initial pressure varies from 2,700 psia at 5,000 ft to 3,100 psia at 6,000 ft. The reservoir is 200 psi undersaturated at 5,000 ft and 600 psi undersaturated at 6,000 ft. Such a situation is common due to the large pressure difference that exists in reservoirs with considerable thickness or closure.

Note that at the initial pressures in the reservoir there is no difficulty in determining an initial average B_o. The pressure range from 2,700 to 3,100 covers a virtual straight-line portion of the B_o and R_s curves. Thus, the average pressure of 2,900 psia can simply be used as a basis for obtaining the average B_{oi} and R_{si} values. The same is true when the pressure in the reservoir varies from 2,600 to 3,000 psia and from 2,500 to 2,900 psia. Also note that once the pressures in all parts of the reservoir have fallen below the bubble-point pressure of 2,500 psia, there is little difficulty in using the B_o and R_s at the average reservoir pressure as the average PVT data. When the reservoir pressure varies from 2,000 to 2,400 psia note from Fig. 6.10B that this range encompasses a virtual straight line on B_o and R_s curves. The same situation exists when the reservoir pressure varies from 2,100 to 2,500 psia.

Working with the 400-psi pressure differential from top to bottom of the reservoir, and looking only at increments of 100-psi decline, there are three pressure distributions where the average pressure could not be used to obtain the average B_o and R_s values. These pressure ranges (2,200–2,600; 2,300–2,700; and 2,400–2,800 psia) do not encompass a straight-line portion of the B_o and R_s curves but lie in the discontinuity of data that exists at the bubble-point pressure. For example, consider the reservoir-pressure range from 2,300 to 2,700 psia. If you attempted to evaluate B_o and R_s at the average pressure of 2,500, a B_o would be obtained that would be the maximum for any portion of the reservoir and R_s would still be R_{si}, although half of the reservoir is below the bubble point.

Some companies use rather complicated material balances based on the effective compressibility of all fluids in the reservoir and the rock compressibility for analysis of a reservoir while it is partly saturated. However, it is clear that if production is uniformly distributed throughout the reservoir it is only necessary to determine weighted averages for PVT data to use the material-balance equations we have been employing throughout this chapter. The validity of this statement can be understood when the derivations of these equations are reviewed. It is seen that the expressions are based on differences in the volumes of reservoir fluids at two different pressures, with the PVT data used to express the gas and oil volumes at reservoir pressures and temperatures. Originally we assumed the same average reservoir pressure existed throughout all of the pore volume but the same equations can be used if we use appropriately weighted average PVT data. The gas in solution, R_s, and the oil formation volume factor, B_o, should be weighted by the reservoir oil volumes. The gas formation volume factor should similarly be weighted on the basis of the reservoir gas volume.

For example, assume that the data of Figs. 6.10A and B represent a depletion-type reservoir (no water drive). Further, assume that the volume of the reservoir in each depth increment from 5,000 to 6,000 ft is the same and that we want to predict the cumulative oil production when the pressure in the reservoir is 2,400 psia at 5,000 ft and 2,800 psia at 6,000 ft.

Since the initial reservoir pressures vary from 2,700 to 3,100 psia, well above the saturation pressure of 2,500 psia, the initial PVT data should be based on the average pressure of 2,900 psia as noted previously. Thus, B_{oi} and R_{si} would be 1.296 res. bbl/STB and 800 scf/STB,

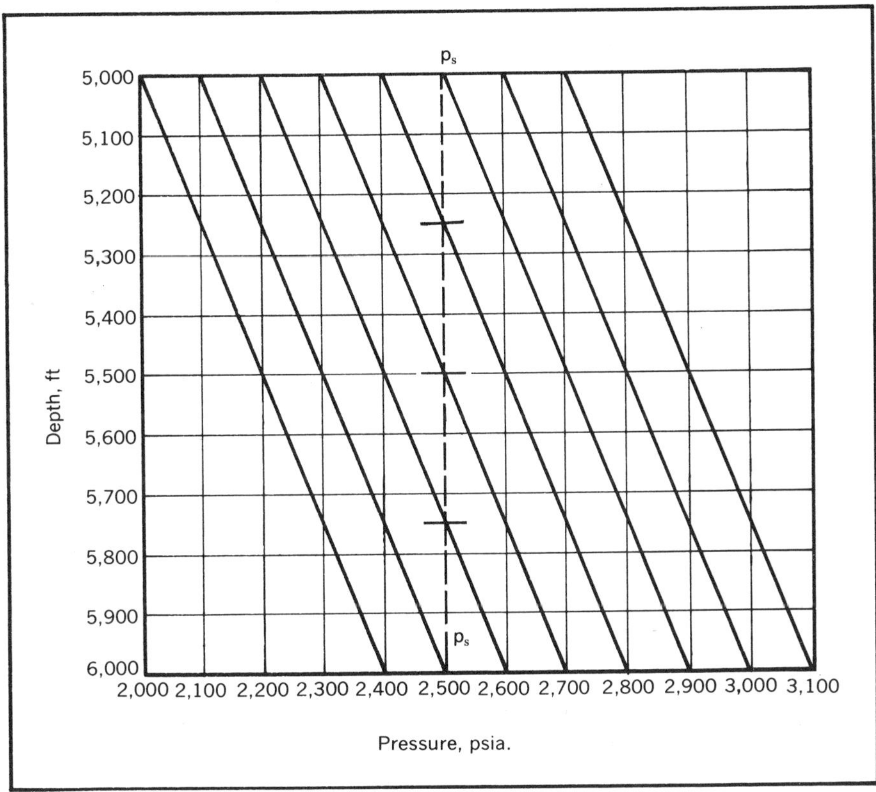

FIG. 6.10A   Pressure versus depth depletion of a thick reservoir

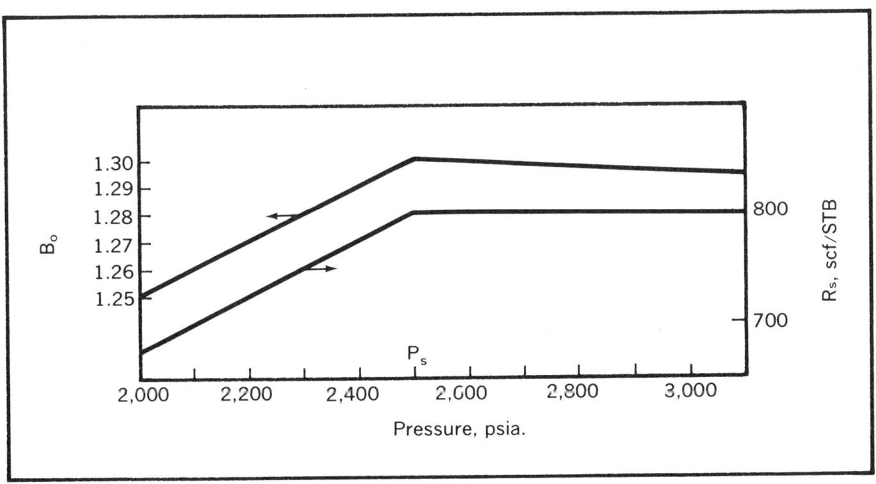

FIG. 6.10B   PVT data for the reservoir of Fig. 6.10A.

respectively. The initial gas volume factor, B_{gi}, would be calculated on the basis of the average pressure of 2,900 psia.

When the reservoir pressure ranges from 2,400 to 2,800 psia note that the top 250 ft, or $\frac{1}{4}$ of the reservoir, is below the saturation pressure of 2,500 psia and the bottom 750 ft, or $\frac{3}{4}$ of the reservoir, is still above the saturation pressure. Since the free gas volumes in the reservoir will all be in the undersaturated portion of the reservoir, the gas formation volume factor, B_g, should be calculated on the basis of the average pressure in the undersaturated zone, or $(2,400 + 2,500)/2 = 2,450$ psia.

However, the oil formation volume factor and gas in solution representing this state of depletion should be an average weighted on the basis of reservoir oil. The B_o and R_s average for the undersaturated portion of the reservoir can be based on the average pressure in this zone, 2,650 psia, or 1.2985 and 800, respectively. The average B_o and R_s in the portion of the reservoir below the saturation pressure would be based on the same average pressure of 2,450 psia cited previously as the basis for calculating the B_g. Then B_o and R_s averages for the top 250 ft at this time would be 1.295 and 787, respectively. The appropriately weighted average oil formation volume factor, B_o, would then be $0.75(1.2985) + 0.25(1.2950)$ or 1.2976. Similarly, the weighted average gas in solution, R_s would be $0.75(800) + 0.25(787)$ or 797 scf/STB. Use of these PVT data and the other appropriate reservoir parameters in a regular material-balance equation would permit the calculation of the cumulative oil and gas production, N_p and G_p.

Calculations for a partially saturated reservoir are normally simplified somewhat because the gas saturation is less than the equilibrium gas saturation. Thus, no free gas flow takes place in the reservoir. If free gas does flow, the calculations become complicated because the gas saturation varies with depth.

It may appear that the corrections are unimportant. However, many examples can be found where the use of PVT data weighted as above is extremely important to the accuracy of the prediction.

Gas-cap complications. The discussion of gas-drive reservoir predictions has been based, for the most part, on the assumption that none of the encroaching gas from the gas cap will be produced. As discussed later this is certainly the most desirable operating condition but it is generally difficult to obtain due to the current demand for oil. When an engineer can logically foresee that a reservoir will be producing gas-cap gas his prediction of the reservoir behavior becomes considerably more difficult.

In states where governmental agencies control the maximum gas-oil ratio under which production can take place without a penalty, the reservoir prediction can be based on the assumption that this maximum gas-oil ratio will prevail beginning at some particular stage of depletion. Without such an artificial control the assumed gas-oil ratio for various stages of depletion will be difficult.

PREDICTION OF WATER DRIVE RESERVOIR BEHAVIOR

When oil or gas is a natural water drive the prediction of reservoir behavior becomes even more difficult than for gas-drive reservoirs. We have added to the effects of the solution and gas-cap drives the additional productive force of water encroachment. When there is only one hydrocarbon phase flowing in the reservoir (a gas reservoir or an oil reservoir above the saturation pressure) the analysis is fairly simple. Many discussions of reservoir predictions are based on the assumption that this is the situation. However, with most wells in the U.S. producing at their peak capacity, it is seldom that a domestic water-drive reservoir is found, that is producing above the saturation pressure. Consequently, the prediction of the behavior of a water drive reservoir is difficult.

To predict performance of a water-drive reservoir, it is necessary to predict water encroachment independent of material balance. We will see that water encroachment is not only a function of the aquifer flow characteristics but is also a function of pressure history and time. Up to this point our predictions have been independent of time but a water drive introduces time as an important parameter. Since water-drive reservoirs producing below the bubble point will also be producing by solution gas drive or gas-cap drive we must also include the ideas covered under the gas-drive prediction section in a water-drive prediction.

The logical extension of the techniques of predicting the behavior of a gas drive reservoir to a water-drive reservoir would be to choose a pressure at which the production is desired, iterate in some way on the produced gas-oil ratio as before and iterate a second time (a double iteration) on the time at which the chosen pressure will be reached. However, we will find that generally it is simpler to set the time and iterate on the pressure. In other words we will choose a time at which we desire to know the cumulative oil and gas production, iterate on the pressure to be obtained in the reservoir at that time and iterate on the produced gas-oil ratio.

Calculating water encroachment. Many methods will be found in the literature for predicting water encroachment.[5] However, in practically all cases it is the author's opinion that water encroachment has taken place as predicted by unsteady-state compressible fluid flow theory. Consequently, only this approach will be presented herein.

Water encroachment follows the constant-pressure solution to the radial diffusivity equation described in Chapter 2. This solution appears to have been originally developed by Hurst[6] for predicting behavior of an aquifer. This is the only type of reservoir application that has any practical use for the dimensionless size, r_e/r_w, reservoirs for which the original data were developed. These data[6] cover dimensionless reservoir sizes of 10 or less which has no physical meaning unless the well radius, r_w, is the internal radius of an aquifer which is furnishing water encroachment for a hydrocarbon reservoir.

In Chapter 2 we learned that a general solution could be obtained for a reservoir producing by unsteady-state with a constant pressure maintained in the reservoir at some particular radius, r. This general solution takes the form of a relationship between a dimensionless group of terms that includes the cumulative flow across the radius, r, and another dimensionless group of terms that includes the time corresponding to the cumulative flow. Furthermore, we learned in Chapter 2 that this general solution can be applied to any reservoir problem by using the appropriate reservoir constants for the subject reservoir.

To be more specific, the constant-pressure solution relates the dimensionless cumulative flow function, Q_{tD}, to the dimensionless time function, t_D, as presented in Fig. 2.11 and Tables 2.2 and 2.3 with the functions defined as (from Chapter 2)

$$Q_{tD} = \frac{Q}{1.12\phi hcr^2\theta\Delta p} \tag{2.23A}$$

and

$$t_D = \frac{6.33kt}{\phi\mu cr^2}. \tag{2.10A}$$

This last equation is Equation 2.10 with the diffusivity constant, η, expanded according to Equation 2.8. Equation 2.23A is Equation 2.23 with the parameter, θ, introduced to account for the fraction of a circle represented by the flow system.

Physically, these functions represent the cumulative flow, Q, that will take place across a radius, r, if the pressure at this radius is suddenly dropped an amount, Δp, and this pressure is maintained constant at this radius for a time, t. However, constant pressures are not generally maintained at the original oil-water or gas-water contacts because the pressure drop at this radius is due to the production of oil and gas from the hydrocarbon-bearing portion of the reservoir. If the wells in the reservoir are being produced at their peak capacity it is not likely that the water encroachment will be able to match the hydrocarbon withdrawal rates and the pressure at the original hydrocarbon-water contact will continue to decline. In such a case it is necessary to calculate the water encroachment by superposition.

The application of Equation 2.23A to the calculation of water encroachment by using superposition is discussed in some detail and illustrated with a problem in Chapter 4. Some of the more important ideas in this type application will be repeated here for convenience.

When Equation 2.23A is solved for the cumulative flow, Q, we obtain

$$Q = 1.12\phi hcr^2\ \theta\Delta pQt_D$$

When this equation is applied to a sequence of pressure drops, $(\Delta p_j$'s), by using the dimensionless cumulative flow function, Q_{tDj}, that is appropriate for the effective dimensionless time, t_{Dj}, of each Δp_j, we obtain

$$Q = 1.12\phi hcr^2\theta \sum_{j=1}^{j=n} (\Delta p_j \, Q_{tDj}) \tag{2.36}$$

Care should be exercised in determining the pressure drops that occur at the original hydrocarbon-water contact especially if the reservoir has a low permeability. In such a case it is unwise to assume that the drop in average pressure in the hydrocarbon-bearing portion of the reservoir is the same as the drop in pressure at the original hydrocarbon-water contact.

The engineer should also evaluate the group of constants in Equation 2.36 ($1.12\phi hcr^2\theta$) from past reservoir performance whenever possible because of the difficulty in determining these reservoir parameters for the aquifer. When this group is treated as one constant, B, Equation 2.36 becomes

$$Q = B \sum_{j=1}^{j=n} (\Delta p_j \, Q_{tDj}) \tag{2.36A}$$

Using past reservoir performance the cumulative water influx can be calculated by material balance. Then by plotting the cumulative water influx at various times versus the summation term for those times, a straight-line plot should be obtained whose slope is the constant, B.

Once the aquifer constant has been evaluated it is then possible to predict the cumulative water influx for any pressure history by using Equation 2.36A.

The prediction procedure. The preferred prediction technique is to estimate the pressure that will exist in the reservoir at a particular time and calculate the water influx, Q, by using the Q_{tD} functions (Equation 2.36A) and W_e using material balance. When a pressure is found that will make $Q = W_e$, we can assume this to be the correct pressure and cumulative water encroachment, then predict pressure for the next desired time.

When stated this way the procedure sounds simple. However, let's consider the details of the calculations. As soon as a pressure is guessed we can calculate Q by fluid flow using Equation 2.36A. As noted previously, the aquifer constant should be evaluated from past reservoir performance whenever possible. The engineer should also be careful to determine the pressure at the original hydrocarbon-water contact and not simply assume that the change in average pressure in the hydrocarbon-bearing portion is the same as the change in pressure at the original hydrocarbon-water contact.

Now consider the material-balance calculation of the water encroachment, W_e. The Equation 6.18 form of the general material-balance equation can be solved for the cumulative water encroachment, W_e, to provide a means for making this calculation.

$$W_e = N_p B_o + B_g(G_p - N_p R_s) + W_p - G(B_g - B_{gi})$$

$$- N[B_o - B_{oi} + (R_{si} - R_s)B_g + (c_f + c_w S_{wc}) \Delta p\, B_{oi}/(1 - S_{wc})] \quad (6.34)$$

We must first estimate how many barrels of oil will be produced during the next time period in order to determine N_{pn}. This means we must estimate how many wells will be invaded by water during the next time period based on our tentative calculation of Q by fluid flow. Then using productivity indices (see Chapter 3), a calculated average oil saturation at the producing wells, and relative permeability data, we could determine how many barrels of oil would be produced during the period. However, to determine the oil saturation it is necessary to know the production so that this represents a trial-and-error solution.

To use Equation 6.34 we must also know the cumulative gas production at the end of the prediction period, G_{pn}. This entails calculating the instantaneous gas-oil ratio at the end of the prediction period, R_n, and using ΔN_p for the period along with the average gas-oil ratio to calculate G_{pn}.

Calculation of the oil saturation in the producing portion of the reservoir presents a real problem. A technique that will work in all cases is virtually impossible to derive. Consequently, we will look at one method and hope that the introduction to the problems involved will give the engineer sufficient background to permit him to make similar calculations under other conditions.

First, we will assume that wells that have been invaded by water will not be produced. In other words, we are assuming that as soon as a well starts cutting water it is shut-in. When demand is very high this may not be a realistic assumption. However, it should be clear after consideration of the simplified assumptions treated here that the technique can be modified to include the effect of wells producing from the invaded zone.

In addition to assuming that all production takes place from the uninvaded zone we will assume that all bypassed hydrocarbons in the water zone can be represented by oil and that the fractional pore volume of oil bypassed is a constant, S_{oBY}. Note that this is inconsistent with our knowledge of fluid displacement as taught in Chapter 5. Nevertheless, the relatively high pressure normally associated with the water encroachment and relatively constant average water saturation in the advancing water bank tend to make the inaccuracy of this latter assumption relatively insignificant. Based on this assumption note that the pore volume in the invaded zone can be stated as $(W_e - W_p)/(1 - S_{oBY} - S_{wc})$. Recognizing this relationship and that the expression for the total original oil zone pore volume is, $NB_{oi}/(1 - S_{wc})$, the equations for the reservoir oil, pore volume, and oil saturation in the uninvaded zone should be clear. Further explanation follows the listing of the equations.

$$\text{Res. oil in uninvaded zone} = B_o(N - N_p) - S_{oBY}(W_e - W_p)/(1 - S_{oBY} - S_{wc}) \quad (6.35)$$

$$\text{Pore volume in uninvaded zone} = [NB_{oi}/(1 - S_{wc})] - (W_e - W_p)/(1 - S_{oBY} - S_{wc}) \quad (6.36)$$

$$S_{oUN} = \frac{(N - N_p)B_o - S_{oBY}(W_e - W_p)/(1 - S_{oBY} - S_{wc})}{[NB_{oi}/(1 - S_{wc})] - (W_e - W_p)/(1 - S_{oBY} - S_{wc})} \quad (6.37)$$

In Equation 6.35 the reservoir oil in the uninvaded zone is equated to the difference between the oil originally in the reservoir and the total of the oil produced and that remaining in the water-invaded zone. In Equation 6.36 the pore volume in the uninvaded zone is equated to the difference between the original pore volume and the pore volume of the invaded zone. The average oil saturation in the uninvaded zone is then the ratio of Equations 6.35 and 6.36 as indicated in Equation 6.37.

Once the oil saturation in the uninvaded zone has been determined, the engineer can estimate the average productivity index, J, for the prediction period. This is done by using Equation 3.4 from Chapter 3.

$$J_1/J_2 = (k_o/B_o\mu_o)_1/(k_o/B_o\mu_o)_2 \quad (3.4)$$

With an average productivity index, J_1, known for a particular set of reservoir conditions, the productivity index, J_2, can be calculated for the conditions at the end of the prediction period. Then an average productivity index for the period can be calculated and applied to the estimated number of producing wells using the average reservoir pressure during the period to check the originally estimated value of N_{pn}. If the assumed and calculated N_{pn} do not check sufficiently, this portion of the calculation must be repeated until an acceptable check is obtained.

Only after an acceptable check has been obtained between the calculated and assumed N_{pn} based on an assumed average reservoir pressure at the end of the prediction period is it wise to move on to the calculation of the water influx by material balance. Equation 6.34 can be used for this calculation only after the cumulative gas production at the end of the prediction period has been determined.

If the water encroachment by fluid flow, Q, matches the water encroachment by material balance, W_e, we can then conclude that the assumed pressure at the end of the prediction period is correct.

If Q does not check W_e we must go back and adjust the assumed pressure and repeat all of the calculations until the two values are the same. If the water encroachment by material balance is greater than the water encroachment by fluid flow, a smaller average hydrocarbon reservoir pressure needs to be assumed. If Q is greater than W_e it is necessary to assume a larger pressure.

Since this method of calculating behavior of a water-drive reservoir involves the simultaneous satisfaction of several iterations it will probably be helpful to list stepwise the procedure that can be followed in the calculations.

A Procedure for Predicting the Behavior of a Water Drive Reservoir

Assume: No free gas exists in the water bank; Piston-like displacement; A uniform saturation throughout the unaffected zone; Producing wells are shut-in whenever the water front reaches them.

(1) Estimate the average reservoir pressure in the uninvaded portion of the reservoir at the end of the next prediction period. Base this on a plot of average reservoir pressure versus time extrapolated to the end of the prediction period. Determine the corresponding pressure at the original water-oil contact at this time.

(2) Calculate Q by Equations 2.36A and 2.10A. The aquifer constant B should have been previously evaluated from past performance.

(3) Assume which wells will be producing during the next prediction period. Base this on an inspection of the water encroached area at the start of the period and its advance as a result of the water encroachment Q calculated in (2) and the bypassed oil in the invaded zone.

(4) Estimate N_{pn}. Initially base this on the history (or predicted history) of the field production rate versus time and an extrapolation of this relationship.

(5) Evaluate S_{oUN} at the beginning and end of the period. If the limiting assumptions are acceptable use Equation 6.37 for this purpose.

(6) Determine the average productivity index during the prediction period for the wells that will be producing during this period. Use Equation 3.4.

(7) Based on the results of (6) and the number of wells producing, calculate N_{pn}. If N_{pn} calculated does not agree sufficiently with N_{pn} estimated in (4) repeat steps (4) thru (7) until an acceptable agreement is obtained.

(8) Calculate G_{pn} based on S_{oUN} at the end of the period. Use Equations 6.24 and 6.30.

(9) Calculate W_e by material balance. Base on Equation 6.34 if the inherent assumptions of this equation are acceptable.

(10)A. If $Q = W_e$, proceed to the prediction for the next time period.

(10)B. If $Q > W_e$ repeat (1) through (10) assuming a larger pressure in (1).

(10)C. If $Q < W_e$ repeat (1) through (10) assuming a smaller pressure in (1).

This procedure can be modified to account for about any assumptions that are necessary. If gas saturations exist in the water bank the calculation of the water bank size must be modified. If the saturation distribution is to be considered along a streamline (other than piston-like distribution), it will be necessary to construct a saturation distribution plot in step (3) and check it after the amount of water encroachment has been determined. This will also mean that the estimation of the producing rates during the time period will have to be based on a different saturation for each well. If wells will be produced to some water cut greater than zero, this must also be taken into account in determining the prediction of production during the prediction period.

Modeling the reservoir prediction problem. As indicated by the previous section, prediction of reservoir behavior is not a simple problem. Many different approaches have been taken but one of the methods most used by major oil companies is the modeling of the reservoir in some way. The prediction of the behavior of the model is then applied to the actual reservoir.

The first modeling was concerned mainly with the prediction of water encroachment for complex systems where the permeability and other reservoir characteristics varied from one portion of the reservoir and aquifer to the other, the water drive was stronger in some sectors of the aquifer than in others, the placement of wells made it imperative that a detailed knowledge of the water encroachment in the various portions of the reservoir be known, etc. Since compressible fluid flow is governed by the same basic equations as the unsteady-state flow of electricity it was found possible to model the reservoir with a series of resistors and condensors and use results of the electric analog to predict water-drive behavior. The procedure was a trial-and-error one in most cases. An electric analog was designed by trial and error that would fit the past performance of the reservoir and this model was then used to predict the future reservoir behavior.

With the advent of the digital computer it was found that virtually any electrical circuit could be solved so that it was unnecessary to actually build an electric analog model. However, at the same time it was found that the reservoir could be segmented and a material balance carried on each segment with flow between segments calculated using the digital computer. It was unnecessary to go the route of the intermediate electric analog model. The number of segments considered depends upon the detail of the analysis desired and on the user's ability to pay for the computer time. What the computer does basically is to run a material balance on each segment similar to those detailed in the previous sections and also calculate the movement of fluids between segments.

There is no doubt that the computer reservoir model is a powerful tool for solving reservoir prediction problems. However, like any other cookbook, it is extremely dangerous when it is not used intelligently. Unfortunately, the author does not believe that there has yet been designed a computer reservoir model program with all of the precautions and ramifications necessary for it to handle any reservoir prediction problem. The same rule of thumb applies to computer reservoir models that applies to any computer program. If the engineer can not perform the same calculations that the computer program is making, given enough time, he does not sufficiently understand the program to use it intelligently. The engineer is again cautioned about simply furnishing data to a computer expert or computer section and taking the results returned by them as being the fact.

This general computer problem is considerably increased for the computer reservoir model because these programs are so extremely complex that virtually no one using the program understands it completely. This is particularly true of purchased software. The same situation also often prevails where a change in personnel has taken place. In the author's opinion an engineer should make every effort to solve his reservoir prediction problems by using tank-type analysis (with or without a computer) before he turns to a two-or-three-dimension reservoir. With the knowledge he has gained in working on the tank-type analysis, he will be in a much better position to determine whether or not the data generated by the computer reservoir model is garbage or useful data. If the computer results are unexpected he can then determine why they are different from his expectations. In many cases more than one set of conditions will fit the past performance of a reservoir.

INCREASING PRIMARY RECOVERY

It should be obvious by now that there are many steps that can be taken to increase the ultimate primary recovery from a reservoir. Some of these steps can be surmised from the previous discussions in this book and others have been specifically noted when various subjects have been discussed. Nevertheless, it seems wise to list here some of the steps that have been mentioned while pointing out some of the methods that have not.

At this point we get involved with the problem of semantics when we attempt to define primary recovery. Strictly speaking we could define secondary recovery as any production obtained by using artificial energy in the reservoir. This would automatically place pressure maintenance through gas or water injection in the secondary-recovery category. However, the author finds that most engineers prefer to think of pressure maintenance as an aid to primary recovery and it will be so treated in this section.

It appears that we can logically classify the measures that can be taken to improve oil recovery during primary production as well-control procedures and reservoir-control procedures. Under reservoir control we will discuss mainly pressure maintenance while many different ideas will be discussed under well control.

Well control. It should be stated that any steps taken to increase producing rate from an oil or gas reservoir, will generally increase the ultimate recovery from that reservoir by placing the economic limit further along the cumulative-production scale. We recognize that there is a particular rate of production at which the producing costs exactly equal the operating expenses. Producing a well below this rate results in a net loss. If onset of this rate can be forestalled by increasing the productive capacity of a well it is clear that additional oil will be produced before the economic rate is reached. Consequently, any steps taken to improve a well's capacity through fracturing, acidizing, paraffin control, and any other means actually increases ultimate production from that well.

It is or will be clear that production of gas and water robs an oil reservoir of energy. If the energy loss can be minimized, a larger ultimate production will result. The same can be said for the production of water from a gas reservoir.

Proper control of the individual well rate is a big factor in the control of gas and water coning or fingering. The engineer is referred to Chapter 5 for a quantitative and qualitative description of this problem. This general problem is not restricted to just water-drive and gas-cap drive reservoirs. In a solution-gas-drive reservoir it may be possible to produce a well at too high a rate from an ultimate recovery standpoint because excessive drawdown of the producing well pressure results in an excessive gas-oil ratio and a corresponding waste of solution gas. The engineer should be aware of this possibility and test wells in a solution-gas-drive reservoir to see if the gas-oil ratio is sensitive.

The engineer should also be aware of the fact that excessive drawdown in a solution-gas-drive reservoir (through high producing rates) often causes paraffin deposition in the tubing and occasionally in the reservoir itself. Keeping gas in solution in the oil by keeping the well pressure as high as possible will result in minimizing the paraffin deposition. Deposition of paraffin in the tubing is not serious when compared with the deposition of paraffin in the reservoir. Given enough time and money, the paraffin can be cleaned from the tubing and flow lines. However, it is questionable whether or not paraffin deposited in the pores of the formation around the well bore can be removed. Consequently, the operator should be very careful to avoid such deposition in the formation.

The proper positioning of wells in a reservoir also plays a big part in the control of gas and water production. It is obvious that wells should be positioned as far as possible from the original gas-oil, water-oil, and gas-water contacts to minimize the production of unwanted gas and water. The positioning of the producing wells must also be consistent with the needs for reservoir drainage, the total reservoir producing capacity, and the cost of development.

In determining the proper well spacing to use in a particular reservoir, the engineer should make certain that he is giving full recognition to the pressure distribution that will prevail in the drainage area of a well at the time the economic limit is reached. There is no question that in a continuous reservoir there is no limit on the amount of reservoir that can be affected by one well. However, the engineer should be concerned with the additional oil that will be recovered prior to reaching the economic limit by adding additional drainage volume (increasing the drainage radius) of a well. In very tight reservoirs we may be able to accomplish only a small reduction in the reservoir pressure in the additional reservoir volume. This effect may be nearly offset by the reduction of the well rate caused by the increase in the drainage radius. Thus, care should be exercised to make certain that the greatest well spacing possible is also the most economical well spacing.

Total reservoir control. The effect of water and gas production on recovery in an oil reservoir can be shown by solving Equation 6.18 for the produced oil.

$$N_p = \frac{N[B_o - B_{oi} + (R_{si} - R_s)B_g + (c_f + c_w s_{wc})\Delta_p B_{oi}/(1 - S_{wc})] - B_g G_p + G(B_g - B_{gi}) + W_e - W_p}{B_o - R_s B_g} \quad (6.38)$$

Here we see that the oil production obtainable at a particular reservoir pressure is almost directly reduced by the reservoir volume of gas and water produced. Furthermore, study of the derivation of the material-balance equation will show that the cumulative gas production, G_p, is the net produced gas—produced gas less injected gas. Similarly, if the water encroachment, W_e, is defined as the natural water encroachment, then the produced water, W_p, must represent the net water produced—water produced less water injected. So, if water or gas can be injected without adversely affecting the amount of water or gas produced, the amount of oil produced at a particular reservoir pressure will be increased by about the same reservoir volume.

Injection of gas or water into the reservoir without adversely affecting the amount of produced gas or water is generally most readily accomplished by injecting the gas into the original reservoir gas cap and injecting the water into the original water-bearing portion of the reservoir. Such procedures are referred to as pressure maintenance since the decline in reservoir pressure with time is reduced or eliminated.

The cost of injecting produced water into the original water-bearing portion of the reservoir is relatively small considering that the operator must dispose of the water in some way. Due to the cost of gas compression and the price of natural gas, a much more careful look must be taken at the economic affects of injecting produced gas back into the gas cap.

Whenever a gas cap or water drive does not exist in a reservoir, pressure maintenance as such is generally difficult or impossible to accomplish unless a considerable amount of closure exists in the reservoir. The closure of a reservoir is the difference between the minimum subsea depth in the reservoir body (the top of the formation) and the maximum subsea depth in the reservoir. When closure is sufficient it may be possible to use the difference in densities of water and oil or gas and oil to keep the phases separated so that high gas-oil or water-oil ratios do not accompany the injection of gas or water into the reservoir. When closure is small, the engineer will find that attempts at pressure maintenance results in conventional secondary recovery by frontal displacement with the accompanying inefficiency of pattern sweep and more adverse effective mobility ratios. Mobility-ratio affects in conventional pressure maintenance are minimized by advantageous gravity (density difference) affects.

The engineer should give careful consideration to the use of a miscible material in any reservoir where a strong water drive or gas-cap drive exists. In many instances it has been possible and profitable to inject a material that is miscible with both the oil and the formation water between the immiscible oil and water and thus greatly improve the displacement efficiency of the system. The same thing has been done in gas drive reservoirs where a fluid miscible to both the oil and gas has been injected to separate the immiscible oil and gas phases and thus increase the displacement efficiency.

In both cases it is necessary to choose a miscible fluid that has a gravity intermediate to the two reservoir fluids being separated (water and oil, or oil and gas). Thus, in very permeable reservoirs where gravity forces are large compared to viscous-flow forces the injected miscible material flows by gravity to a position between the two immiscible reservoir fluids and results in the miscible displacement of the oil by the miscible injected fluid and the miscible displacement of the miscible injected fluid by the water or gas. The efficiency of a miscible displacement is very high. However, if the formation has a low permeability the injected miscible material will not form a layer between the two immiscible phases quickly enough and the system will be much less effective. The engineer should also take care in determining the amount of miscible fluid necessary to maintain the miscibility of the system. Determination of this amount is still largely controversial but the problem is discussed in Chapter 8.

The engineer may encounter the term "Maximum Efficient Rate" in his work. Theoretically, this refers to the maximum rate at which a reservoir can be produced without adversely affecting the ultimate recovery of the reservoir. Many reasons might be cited for the existence of a maximum efficient rate for a particular reservoir. There may be critical rates at which coning or fingering takes place or at which the produced gas-oil ratio in a solution-gas-drive reservoir is adversely affected or at which paraffin deposition occurs in the reservoir. However, the last two examples are actually maximum efficient well rates rather than reservoir rates.

The term maximum efficient rate (MER) seems to have originated from a different reservoir phenomenon. It can be shown in the laboratory that the residual oil saturation that exists after displacement of the hydrocarbons in a reservoir by water, is less if the reservoir contains a free gas saturation prior to the displacement. The theory is that the free gas takes the place of some of the residual oil and thus the displacement results in a larger overall oil recovery.

For this phenomenon to work in the reservoir as it does in the laboratory would mean that the gas saturation at the moving water-oil contact could be maintained at or above the equilibrium gas saturation. It would be very difficult to maintain a gas saturation at the advancing water-oil contact that is significant.

When the gas saturation at the advancing water-oil contact exceeds the equilibrium gas saturation, an oil bank will form and reduce the gas saturation in the front of the advancing oil bank to the equilibrium gas saturation. However as the oil bank moves past this immobile gas, the reservoir pressure at this point will increase due to the water bank coming closer and the mobile phase now being entirely liquid instead of part mobile gas. As the pressure increases, the equilibrium gas will tend to go back into solution in the oil so that by the time the water bank reaches this point the equilibrium gas will probably have disappeared.

This is conjecture on the part of the author but it nevertheless does appear unlikely that it is possible from a practical point of view to reduce the residual oil in a water bank by increasing the gas saturation in the water bank.

REFERENCES

1. Schilthuis, Ralph J., "Active Oil and Reservoir Energy," *Trans. AIME* (1936), pg. 33.
2. J. Tarner, "How Different Size Gas Caps and Pressure Maintenance Programs Affect Amount of Recoverable Oil," Oil Weekly, June 12, 1944, Vol. 144.
3. M. Muskat, "The Production Histories of Oil Producing Gas-Drive Reservoirs," Journal of Applied Physics (1945).
4. B. Craft and M. Hawkins, *Applied Petroleum Reservoir Engineering*, Prentice-Hall Inc., Englewood Cliffs, N.J. 1959.
5. Stewart, Callaway, and Gladfelter, "Comparison of Methods for Analyzing a Waterdrive Field, Torchlight Tensleep Reservoir, Wyoming," Trans. AIME (1955), 204, 197.
6. A. F. Van Everdingen and W. Hurst, "The Application of the Laplace Transformation to Flow Problems in Reservoirs," Trans. AIME (1949).

7 | Decline-Curve Analysis

Many years ago it was discovered that a plot of oil-production rate versus time could be extrapolated to provide an estimate of the future rates of production for a well.[1] With the future rates known it was possible to determine the future total production or reserves of the well. This represented the beginning of the art that has since become more of a science known as decline-curve analysis.

Decline-curve analysis may be one of the most misused, and at the same time, one of the most neglected reservoir engineering techniques.[2] Decline-curve analysis can only be used as long as the mechanical conditions and reservoir drainage stay constant in a well and the well is produced at capacity. These limitations lead to much misuse.

On the other hand, the more theoretically inclined petroleum engineer may not appreciate decline-curve analysis and fail to use it to augment or back up his theoretical prediction.

Typical decline-curve analysis performed by a reservoir engineer consists of plotting well production versus time on semilog paper and attempting to fit these data with a straight line which is then extrapolated. The reserves are calculated by reading some average production rate per year for the extrapolated production rates and calculating the reserves on this basis. The typical engineer probably realizes that a hyperbolic decline-curve analysis would give a more accurate prediction. However, the complexity of the hyperbolic techniques with which he is familiar, and the excuse that late-life differences between a constant percentage (straight line on semilog paper) and hyperbolic decline affects his present worth value very little, leads him to use the easiest method, the constant percentage decline. In this chapter we will see that equations (rather than reading points) can be used more easily and more accurately in determining reserves and that hyperbolic decline curve analysis is little, more difficult than constant percentage decline once the engineer is familiar with the methods.

No effort will be made to theoretically justify the hyperbolic or constant-percent decline of a well's productivity. Those readers who find this subject of interest are referred to Folkert Brons[2] work referenced in this Chapter.

DECLINE RATE DEFINITION

The various methods of decline-curve analysis[3,4,5] are based on the manner in which the rate of decline varies with time, rate, etc. Consequently, it is important that we carefully define the decline rate. When production rate is plotted versus time we observe that the rate declines with time as indicated in Fig. 7.1A. The decline rate is the fractional change in rate with time.

$$a = -\frac{(\Delta q/q)}{\Delta t}$$ (7.1)

The graphical interpretation of this definition is illustrated in Fig. 7.1A. Consequently, the decline rate at any particular time can be determined graphically by noting the slope of the rate-versus-time curve at the time of interest and dividing the slope by the rate at that particular

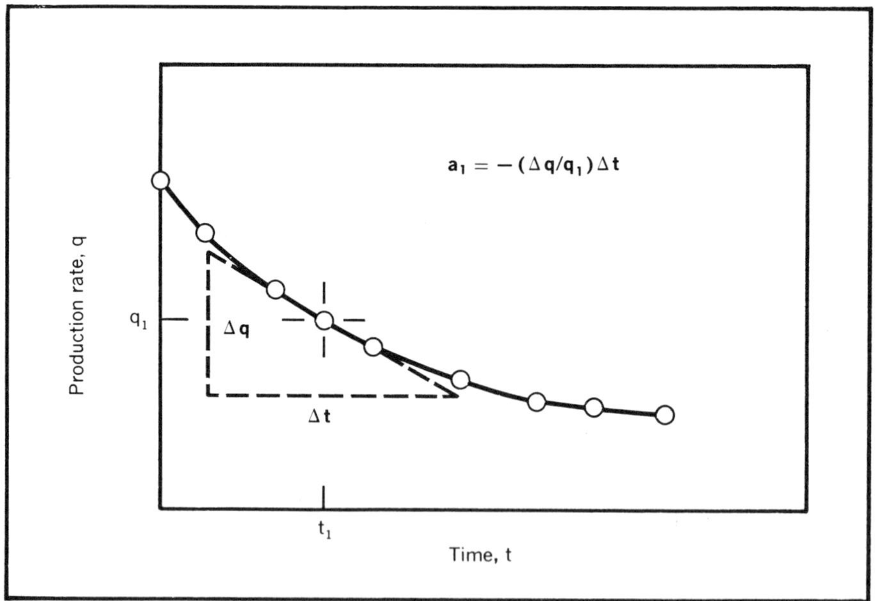

FIG. 7.1A Rate of decline definition — linear plot.

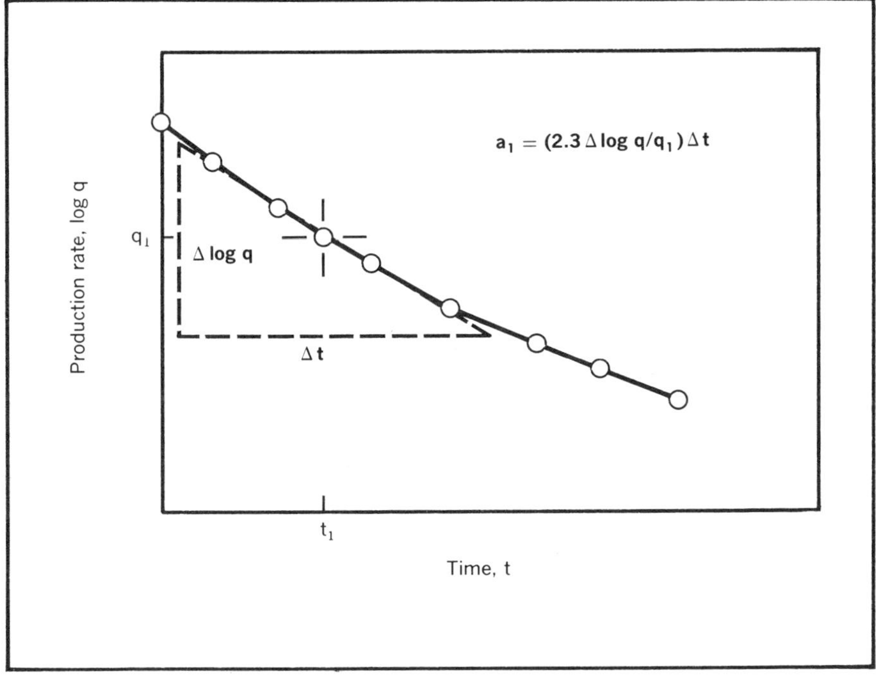

FIG. 7.1B Rate of decline definition — semilog plot.

time. Note that for the decline rate, a, to be constant the slope must decline at the same rate that the rate, q, declines.

The mathematician will immediately recognize that the expression of Equation 7.1 is equal to the change in the natural log of q ($\Delta \ln q$), with respect to the time.

$$a = \frac{-\Delta \ln q}{\Delta t} \tag{7.2}$$

Thus, the slope from a plot of the natural log of the rate versus time would be the decline rate, a. However, as noted previously, natural log paper is not in common use so that it is convenient to use logarithms to base 10. Since the natural log is equal to 2.3 times the log to base 10, Equation 7.2 becomes,

$$a = \frac{-2.3\, \Delta \log q}{\Delta t} \tag{7.3}$$

The graphical interpretation of this equation is illustrated in Fig. 7.1B. It is also convenient to recognize that the slope of a semilog plot can be obtained by determining the change in the linear scale that occurs over a one-cycle change in the log scale (e.g. 1,000 to 100, 0.1 to 0.01, etc.). This is true because the change in the log that occurs over a one-cycle change in the rate is exactly equal to 1.0. Consequently, the slope of a plot of the natural log of rate versus time (Equation 7.2) can be obtained by determining the change in time per cycle from a log to base 10 plot and dividing it into 2.3.

$$a = \frac{2.3}{(\Delta t/\text{cycle})} \qquad (7.4)$$

Note that in all of these equations for the decline rate the units are the reciprocal of time. Physically it may be helpful to consider the units to be a fractional change in the rate per unit of time. Also note that the rate units used are stock tank barrels per day. This is the only Chapter in this book in which the symbol q is used in stock tank barrels without some appropriate subscript.

CONSTANT-PERCENTAGE DECLINE

Constant-percentage decline is also known as exponential decline since the mathematical expression that defines this type decline is an exponential equation. Constant-percentage decline[6,7] is much more widely used than the other mathematical forms even though engineers generally agree that hyperbolic decline more nearly describes the decline characteristics of most wells. The simplicity of the constant-percentage decline technology makes it attractive to the engineer and when decline rates are small the additional accuracy added by using the hyperbolic decline may not be significant for many purposes.

The constant-percentage decline-rate equation. As the name implies, the constant-percentage decline is based on the assumption that the decline rate, a, does not change with time. We can then use the basic definition of decline rate, as in Equation 7.1, to derive the mathematical expression for constant-percentage decline. When we rearrange Equation 7.1 as,

$$a \, \Delta t = -\left(\frac{\Delta q}{q}\right), \qquad (7.5)$$

it can be applied to any very small increment of time. If it is then applied to all of the time increments from one rate, q_i, when the time is taken as zero to another rate, q, corresponding to a time, t, and all of these equations are added together, we would obtain,

$$a \sum_{0}^{t} \Delta t = - \sum_{q_i}^{q} (\Delta q/q) \qquad (7.6)$$

$$a \, t = \ln q_i - \ln q \qquad (7.7)$$

This equation shows that a plot of the natural log of the rate, ln q, versus the time, t, will give a straight line. Differentiation of this equation will yield Equation 7.2 and show that the slope of the ln q versus t plot will be the decline rate, a. Equation 7.7 can also be stated in log to base 10 values.

$$a\,t = 2.3 \log q_i - 2.3 \log q \tag{7.8}$$

When this expression is differentiated with respect to time, Equation 7.3 is obtained. Thus, a plot of log q versus t will give a straight line whose slope is $-a/2.3$.

Conversely, we recognize that a well whose plot of the log of rate versus time yields a straight line is experiencing constant-percentage decline and rates at any future time can be calculated from Equation 7.7 or 7.8 or a time could be calculated that would correspond to the time necessary for the rate to decline to some particular value, for example the economic limit rate for the well.

Many engineers prefer to use the exponential form of Equation 7.7 for their calculations.

$$a\,t = \ln\left(\frac{q_i}{q}\right) \tag{7.9}$$

$$\left(\frac{q_i}{q}\right) = e^{a\,t} \tag{7.10}$$

$$q = q_i\, e^{-at} \tag{7.11}$$

This equation, or Equation 7.7 or 7.8 may be used to calculate future production rates or times. A decline curve generally does not immediately follow a constant-percentage decline. A plot of rate versus time on semilog paper does not generally approach a straight line immediately but continues to curve at a lesser rate as the production life continues until it is possible to approximate the plot with a straight line. An example of such a decline curve after the data have been smoothed is shown in Fig. 7.2.

Decline curves are generally maintained on a month-to-month basis by calculating a daily average production rate for the month and plotting this rate or by simply plotting the month's production versus time so that the rate scale will be in barrels per month. Generally there is no effort made to adjust the time scale for the difference in the lengths of the months. Neither is the monthly volume generally adjusted for the difference in the number of days in each month.

Regardless of the method used to plot the raw data, this plot normally exhibits a wide fluctuation around the trend of the data. This fluctuation may be due to a variation in periods of downtime, weather difficulties, pipeline runs, etc. To make the data easier to interpret it can generally be smoothed by calculating averages for periods of time and plotting the averages at the middle of the time increment used for averaging. It may

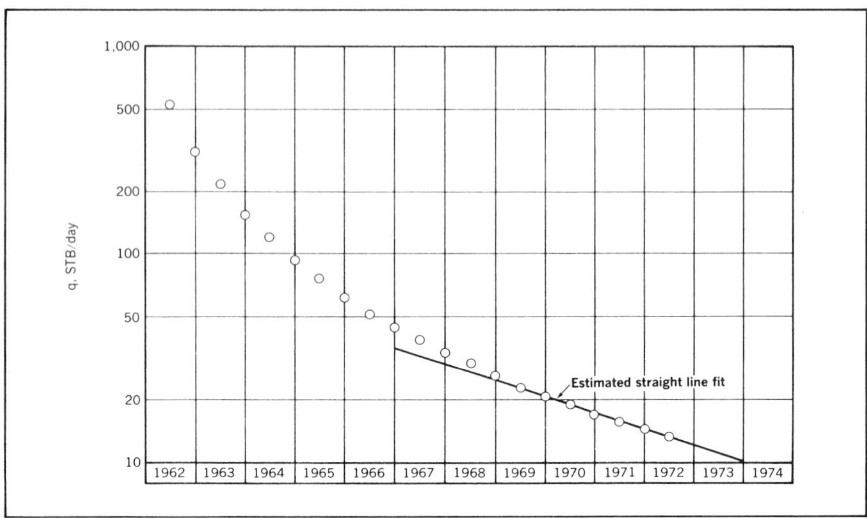

FIG. 7.2 Smoothed decline curve data.

be necessary to determine averages on the basis of 3-month, 6-month, or even 1-year periods to obtain a smooth production decline trend.

To make certain you understand the principles of constant-percentage decline, work the following problem and compare your solution with the solution in Appendix C.

Problem No. 7.1A: Using Constant-Percentage Decline to Calculate the Future Life and Rates of a Well

Fig. 7.2 represents smoothed data for a well. It is assumed that since June 1970 the well has followed a straight-line decline on this semilog paper. What will be the producing rate in 5 years? What is the life of the well as of June 1972 if the economic limit is 1 bopd?

Determining reserves during constant-percentage decline. Most engineers desiring a total production figure for some period representing constant-percentage decline will read a series of averages for short time periods and calculate the total on that basis. For example, assume that we needed to determine the amount that would be produced from the well represented by Fig. 7.2 for the period June 1972 through December 1973. The simple approach is to read the average for the three 6-month periods. Thus, averages of 12.6, 11.5, and 10 b/d might be read at September '72, March '73, and September '73, respectively. From these values we calculate a total production for the period of 6,315 bbl (12.6 + 11.5 + 10.5) 182.5. The 182.5 figure represents the number of days in ½ year.

This technique is useful and simple for short periods but is cumbersome when applied to longer periods of time which require long extrapolations and the reading of many points to obtain satisfactory accuracy. Consequently, it is generally easier to calculate total production accumulated while constant-percentage decline occurs from one rate to another at a constant-decline rate by equation.

$$\Delta N_p = \frac{(q_1 - q_2)}{a} \qquad (7.12)$$

To apply this equation and calculate the increment of production, ΔN_p, it is only necessary to know the rate at the start of decline, q_1, the rate at the end of the period, q_2, and the decline rate, a. Since we normally use our rates in barrels per day it is necessary to use the decline rate in a fraction of rate change per day. The units of days^{-1} gives very small numbers and the engineer may find it convenient to use a time unit of months or years in Equation 7.12. Care should be taken to make certain that the time base of the rates matches the time units of the decline rate.

Equation 7.12 is derived by recognizing that the cumulative production during a time from t_1 to t_2 is,

$$\Delta N_p = \sum_{t_1}^{t_2} (q \ \Delta t) \qquad (7.13)$$

By rearranging Equation 7.1 we can obtain an expression for $q\Delta t$ which can be substituted into Equation 7.13 to obtain,

$$\Delta N_p = - \sum_{q_1}^{q_2} (\Delta q/a) \qquad (7.14)$$

When the Δq's are summed from q_1 to q_2 we obtain Equation 7.12.

If we consider the initial rate, q_1, in Equation 7.12 as a constant, note that the ΔN_p value is inversely proportional to the rate, q_2.

$$\Delta N_p = \left(\frac{q_1}{a}\right) - \left(\frac{q_2}{a}\right) \qquad (7.15)$$

Consequently, a plot of N_p versus the rate, q_2, would give a straight line whose slope would be $-(1/a)$, as long as the decline follows the constant-percentage decline. This type of plot can be very useful, especially in evaluating or understanding remedial workovers.

In Fig. 7.3 is shown a plot of the rate versus cumulative production for a well that is assumed to have entered constant-percentage decline before remedial work and followed constant-percentage decline after this work. In this case the reservoir drainage was unaffected by the workover because the primary mobile oil appears to have remained unchanged. Here the primary mobile oil is assumed to be all of the oil that would be produced if it was possible to produce the well to a rate of zero. For con-

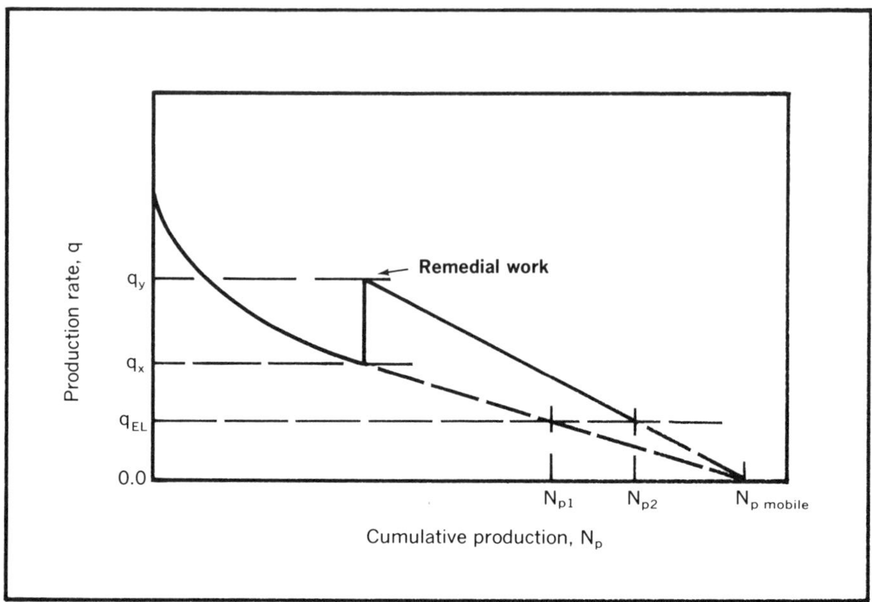

FIG. 7.3 Use of a rate-cumulative decline curve.

stant-percentage decline, the mobile oil remaining in the reservoir at any time could be calculated using Equation 7.12 with q_1 as the current rate and q_2 equal to zero.

$$\text{Remaining prim. mobile oil} = \left(\frac{q_1}{a}\right) \tag{7.16}$$

If the remaining primary mobile oil in a reservoir is unchanged by a workover that increases the producing rate from q_x to q_y, note that the decline rate will also be increased in the same ratio for the remaining primary mobile oil to remain the same according to Equation 7.16. Conversely, if a well in constant-percentage decline before and after a workover has a plot of rate versus time that extrapolates to the same cumulative production for a rate of zero before and after the workover, we can conclude that the workover has not increased the drainage of the well. This generally seems to be the case for fracturing, acidizing, and similar methods of stimulating a well-production rate. In such a case the increase in the ultimate production will be the difference in the cumulative production that would be achieved at the economic limit rate as indicated by the difference in N_{p1} and N_{p2} in Fig. 7.3.

This relatively small increase in ultimate primary recovery does not mean that the workover economics are unattractive. When corresponding lifes are calculated using Equation 7.11 and the appropriate rates and decline rates, it will be seen that the remaining life of the field is drastically shortened which means that all of that operating expense difference (due to the difference in the life) is saved, in addition to the savings in the present-worth value of the production.

Work the following problem and compare your solution with the solution in Appendix C.

Problem No. 7.1B: Use of a Rate Versus Cumulative Plot During Constant Percentage Decline

In Problem 7.1A a straight-line extrapolation of Fig. 7.2 was used to calculate the remaining life as of June 1972. Now assume that in June 1972 this well was fractured and the producing rate increased to 53 b/d without increasing the drainage volume of the well. Assume that the well again entered constant-percentage decline. What is the remaining primary mobile oil in June 1972? What will be the increase in the ultimate primary production and the change in the well life if the economic limit is 1.0 b/d? Note that it is not necessary to prepare a rate-versus-cumulative plot to solve this problem.

HYPERBOLIC DECLINE

Engineers who have been faced year after year with the task of determining reserves for a company in an area where solution-gas reservoirs exist will be aware of the shortcomings of the constant-percentage decline type of analysis. Such an engineer has undoubtedly experienced the situation in which year after year he tried to use a straight-line extrapolation of a semilog rate—time plot and year after year he had to increase the reserves on these leases because the decline continued to flatten. If he turned to hyperbolic methods to predict reserves from his decline curves he probably found that the best known hyperbolic methods were too time consuming or too insensitive for his use. The engineer probably reached a situation in which he simply extrapolated his decline curves using a favorite French curve in a manner that is more of an art than a science. It is hoped that this discussion of hyperbolic decline-curve analysis will provide methods that are relatively simple to use and much less arbitrary than the French-curve method.

The nature of the loss ratio and log log plot methods of analyzing hyperbolic decline will be covered along with some of the difficulties associated with the practical application of these methods.

The hyperbolic decline equations. In working with constant percentage, we found the decline rate, a, to be constant, but the decline rate in hyperbolic declines varies according to the equation,

$$\left(\frac{a}{a_i}\right) = \left(\frac{q^n}{q_i^{\,n}}\right), \tag{7.17}$$

where the constant, n, represents a number between (but not including) zero and 1.0. Note that when n is zero q^n and $q_i^{\,n}$ would be 1.0 and the decline rate, a, would be equal to a_i which would indicate that the decline rate is constant percentage decline.

We will later see that when n is 1.0 a special type of decline known as harmonic decline prevails. In this case we can see that the decline rate is proportional to the rate. In the more general case of hyperbolic decline we see by Equation 7.17 that the decline rate is proportional to some fractional power of the rate.

To obtain the rate-time hyperbolic decline equation we can first substitute for the decline rate in Equation 7.17 according to the definition of decline rate in Equation 7.1.

$$\frac{-(\Delta q/q\,\Delta t)}{a_i} = \left(\frac{q^n}{q_i^{\,n}}\right) \tag{7.18}$$

Now a_i and q_i can be treated as constants and the variables can be separated and summed between limits.

$$a_i \sum_0^t \Delta t = -q_i^{\,n} \sum_{q_i}^q q^{-(n+1)}\,\Delta q \tag{7.19}$$

In this expression note that q_i is taken as the rate at the time when t is zero. This would mean that a_i is the decline rate at the time when t is zero. When the expression is integrated between limits we obtain

$$a_i\,t = -q_i^{\,n}\left[\frac{q^{-n}}{-n} - \frac{q_i^{\,-n}}{-n}\right] \tag{7.20}$$

This can be rearranged to obtain

$$na_i t = \left(\frac{q_i^{\,n}}{q^n}\right) - 1 \tag{7.21}$$

When this equation is solved for rate we obtain the hyperbolic rate-time decline-curve equation.

$$q = \frac{q_i}{(1 + na_i t)^{1/n}} \tag{7.22}$$

In using this equation remember that a_i is the decline rate when the rate, q_i, prevails which is the time equivalent of zero. Also note that the time is that required for the rate to decline from q_i to q. The constant n, is defined in the definition of hyperbolic decline in Equation 7.17. Thus, in applying Equation 7.22 to calculate the rate at some particular time or the time required to reach a particular rate it is necessary to know one more parameter than was necessary in applying the similar constant-percentage rate equation. The difficulty of determining the hyperbolic-decline constant, n, from past rate-time data is the major difficulty of hyperbolic-decline analysis. Once this constant is determined it is relatively simple to graphically determine the decline rate corresponding to a q_i and calculate the rate, q, corresponding to any time, t. These same parameters can also be used to calculate the production accumulating during the time, t, when the production rate was declining from q_i to q.

The hyperbolic-decline cumulative-production equation is derived in the same manner that the constant-percentage decline cumulative-production Equation, 7.12, was derived. Equations 7.13 and 7.14 apply to any type of decline curve.

$$\Delta N_p = \sum_{t_1}^{t_2} q \Delta t \qquad (7.13)$$

$$\Delta N_p = \sum_{q_1}^{q_2} (\Delta q/a) \qquad (7.14)$$

To apply Equation 7.14 to hyperbolic decline we will substitute for the decline, a, according to Equation 7.17.

$$\Delta N_p = \sum_{q_1}^{q_2} (-q_i{}^n/a_i) \, q^{-n} \, \Delta q \qquad (7.23)$$

When this expression is integrated between q_1 and q_2 we obtain the hyperbolic decline cumulative production equation.

$$\Delta N_p = [q_i{}^n/a_i(1-n)][q_1{}^{(1-n)} - q_2{}^{(1-n)}] \qquad (7.24)$$

Thus, once the hyperbolic constant, n, has been determined, along with a decline rate, a_i, corresponding to any rate, q_i, the cumulative production between any two rates, q_1 and q_2, can be calculated.

The curve-fitting method.[8] It is believed that the simplest means of extrapolating hyperbolic decline performance and evaluating the future life and reserves is by comparing the actual decline-curve data with a series of hyperbolic-type curves that represent various combinations of n and a_i. Once the engineer has determined which of the hyperbolic-type curves most closely fit his data he will have determined the n, a_i, and q_i which can be used to calculate future rates, reserves, etc. using Equations 7.22 and 7.24. Alternatively he may be able to directly read the future rates and cumulative production values from some type curves.

Examples of the hyperbolic-type curves are shown in Figs. 7.4, 7.5, and 7.6. These curves have some peculiarities that should be noted. They are basically rate versus time and cumulative production versus time curves for various combinations of n and a_i. In all cases q_i, (the rate when the time, t, is zero), is 1.0. Thus, the Y axis for the rate-time curves can be considered to be the rate if q_i is 1.0 or it can be taken as the ratio of the rates, q/q_i. Since the rate is plotted on a log scale the units

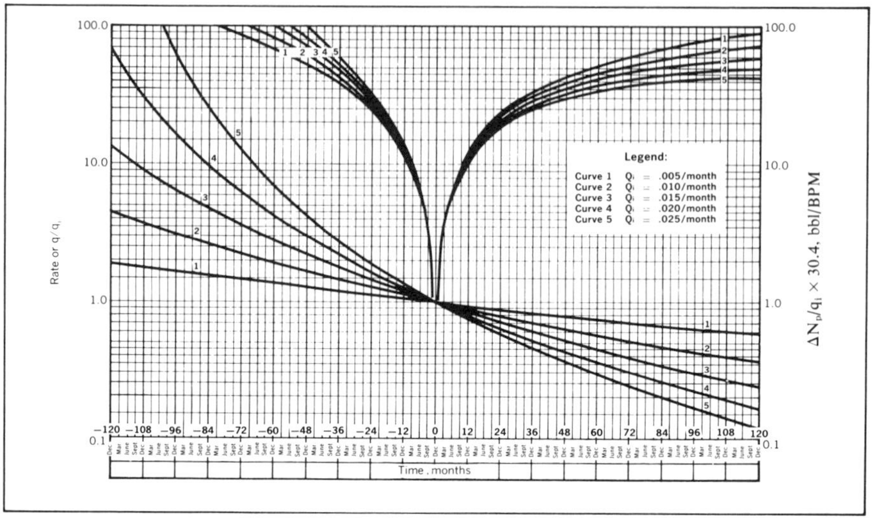

FIG. 7.4   Hyperbolic decline curves for n = 0.3

do not affect the shape of the curve. Consequently, we could consider the hyperbolic-type rate curves to be plots of the rate stated in q_i units. Examination of Equation 7.22 will show that the slope of the log of q plotted versus the time, t, will be the same as the slope of log q/q_i plotted versus time.

Due to the extremely wide range of values associated with the

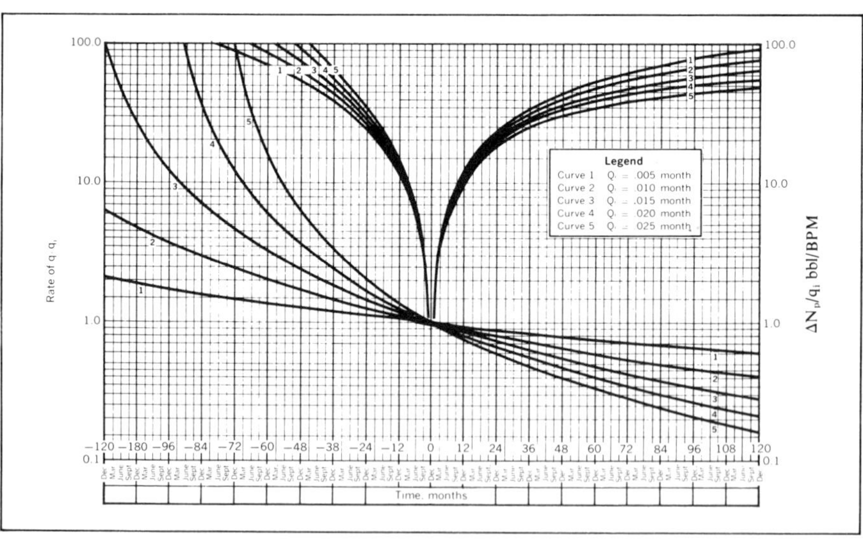

FIG. 7.5   Hyperbolic decline curves for n = 0.5

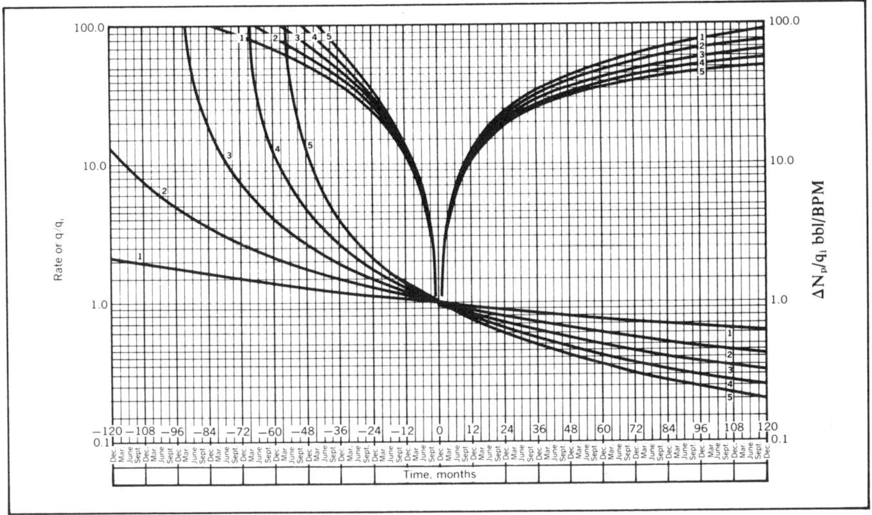

FIG. 7.6 Hyperbolic decline curves for n = 0.7

various n and a_i combinations it simplifies the presentation of the type curves to place the zero time in the center of the plot rather than at the left-hand side of the plot as is customary. The use of negative times does not mathematically interfere with the application of Equations 7.22 and 7.24.

The cumulative curves are presented also in units of q_i (i.e., as $\Delta N_p/q_{iBPM}$). The same curves could then represent a ΔN_p for a q_{iBPM} of 1.0. Notice also that each rate-time curve is represented by two cumulative production curves. One curve gives $\Delta N_p/q_{iBPM}$ for the production cumulative from some negative time to the graph zero time and the second curve gives the cumulative production from the chart zero time to some later time.

It would be awkward to attempt to use Figs. 7.4, 7.5, and 7.6 in the scale presented for predicting hyperbolic-decline behavior since normal graph paper is not this size. Also, a more extensive suite of curves should be used to obtain the greatest accuracy. Consequently, tabular data for plotting curves for various values of n are presented in Appendix B. It is suggested that for the greatest utility the hyperbolic-type curves be plotted on the particular size semilog graph paper the engineer customarily uses for his decline curve plots. The data plots on Codex No. 4272, 20 years by months by three 3-in. cycles (11 × 17-in. paper), can be obtained from the author.

To use these hyperbolic data plots it is necessary to make a plot of the actual (or smoothed) decline curve data on onion skin weight semilog paper with the same linear log cycle size and time scale as that of the type curves. Any rate units (e.g. barrels per day, month, or year) can be used for the plot of the actual data but the time scale must be the same as that of the hyperbolic-type curves. Once the plot of the actual data is completed its shape should be compared with the shape of the family of hyperbolic-type curves to determine which one best fits the actual data. In comparing the actual data with the type curves it is only necessary to keep the vertical (or horizontal) axes of the actual and type plots parallel. Once the best fit is obtained the actual chronological time and producing rate (q_i), corresponding to the chart time of zero should be noted as well as the a_i and n for the type curve that gave the best fit. With q_i, a_i, and n fixed it is then possible to apply Equations 7.22 or 7.24 to calculate rates and cumulatives as desired.

The mathematical application of these equations may be difficult for some engineers since they involve the use of numbers to fractional or negative powers. For this reason, and in the interest of saving time, a wide range of hyperbolic-type curve data should be plotted so that as many answers as possible can be directly read from the type curves. An example calculation will probably be most useful in determining the exact method of applying the type curves.

In Problem 7.1 the smoothed decline-curve data of Fig. 7.2 was extrapolated with a straight line as a constant-percentage decline. However, if we analyze the data using the hyperbolic-type curves we will obtain somewhat different and more accurate answers. The procedure would be to plot the data on a semitransparent graph paper with the same cycle size and time scale as that of the type curves of Figs. 7.4, 7.5, and 7.6. The resulting curve of actual data would then be compared with the type curves to determine which curve fits best. When this is done the best curve fit is obtained as indicated in Fig. 7.7. From this curve fit we should observe that the actual rate corresponding to the chart time of zero is 15.6 b/d, or 474 bbl/month, (15.6 × 30.4) and the chronological time corresponding to the chart time of zero is July 1971. The curve fit has an n of 0.5 and an a_i corresponding to the zero time of 0.015/month.

In Problem 7.1 we evaluated the rate 5 years after June '72 based on a constant-percentage decline extrapolation. Note that we can read a rate for this time directly from the hyperbolic curve fit. June '72 represents a +11 time on the curve fit so that 5 years or 60 months later should represent a chart time of 71. At this chart time q/q_i is about 0.42 and the rate would be 0.42 × 15.6 or 6.5 b/d. This compares with a constant-percentage extrapolation of 5.6 b/d, a difference of nearly 20%. Note that the total production accumulating during the 5-year period following June '72 can be obtained by reading values from the type curves. Reading the $\Delta N_p / q_{iBPM}$ figure for curve number 3, corresponding to the number 3 rate-time curve which we fit, we obtain at a chart time of 71 a figure of 46.5. Thus, for the 71 months, from a chart time of zero, this well produced 46.5 × 474 or 22,040 bbl of oil. However, we want the production for the last 60 months. Thus, reading the cumulative curve at a chart time of 11 months corresponding to June 1972 we obtain 10.0. Thus, for the 11 months preceding June '72 or from a chart time of zero, a total of 10.0 × 474 or 4,740 bbl of oil will have been produced. Subtracting this from the 22,040 bbl, we obtain the production for the next 5 years following June '72 as 22,040 − 4,740 or 17,300 bbl. We should obtain this figure more readily by the calculation 474 (46.5 − 10.0).

When the period for which cumulative production is required brackets the chart zero it is necessary to read both the cumulative curves for positive and negative chart times. To get the cumulative production from December '68 through June '72, for example, the negative cumulative production curve reads 41.0 for December '68 or a chart time of −31 months. Thus the 31 months of chart time preceding the chart time of zero would give total production of 41.0 × 474 or 19,434 bbl and the cumulative production from December '68 through June '72 would be 19,434 + 4740 or 24,174 bbl. Again note that the calculation would be more readily performed as 474 (41.0 + 10.0).

When the times of interest fall off the 20-year span of the type curves it is necessary to use Equations 7.22 and 7.24. For example in Problem 7.1A we were asked to calculate the remaining life as of June 1972 from a constant-percentage decline interpretation of Fig. 7.2 and in Problem 7.1B we had to calculate the reserves for the same date. Now assuming the hyperbolic curve fit of Fig. 7.7 let us calculate the life and reserves for this well. We will first solve Equation 7.22 for the time.

$$t = [(q_i/q)^n - 1.0]/n\,a_i \qquad (7.25)$$

When we substitute the previously determined parameters into this equation and use q equal to the economic limit of 1.0 b/d we obtain

$$t = [(15.6/1.0)^{0.5} - 1.0]/(0.5)(0.015) = 395 \text{ months}$$

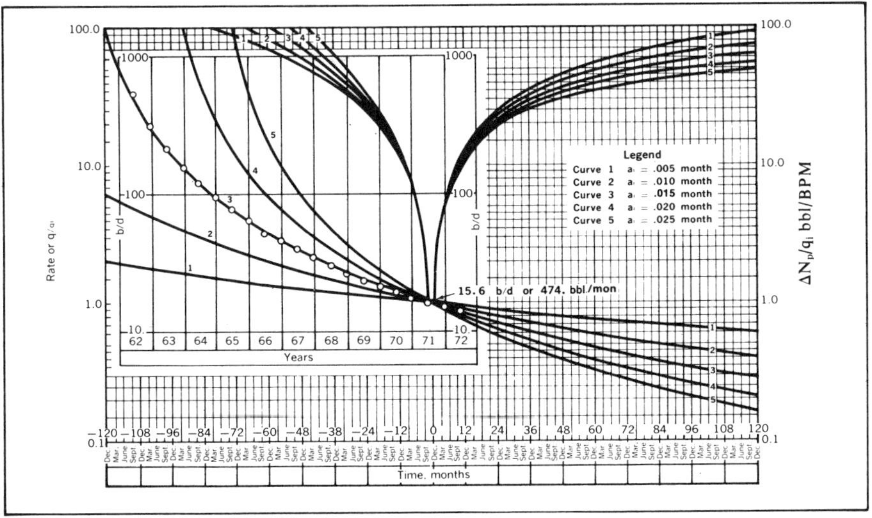

FIG. 7.7 Hyperbolic curve fitting technique

Note that this is the time required for the rate to decline from q_i or 15.6 b/d to 1.0 b/d and we desire the time for the rate to decline from the rate on June '72 or 13.5 to 1.0 b/d. Since it required 11 months for the rate to decline from 15.6 to 13.5 b/d the desired life will be $395 - 11$ or 384 months.

Note from this calculation that it is necessary for the q_i of Equation 7.25 to match the a_i. Since we only know the decline rate at the chart time of zero, we must make our time calculations from this rate or use Equation 7.17 to calculate the decline rate, a, corresponding to the desired rate and use this as a_i along with the desired rate as q_i in Equation 7.25. Then the calculated time would not have to be corrected.

To calculate the reserves as of June '72, use Equation 7.24 to calculate the cumulative production between the rate on June '72 of 13.5×30.4 or 410 bbl/month and the economic limit of 30.4 bbl/month.

$$\Delta N_p = [q_i^n/a_i(1 - n)][q_i^{(1-n)} - q_2^{(1-n)}] \tag{7.24}$$

$$= [474^{0.5}/0.015(1 - 0.5)][410^{(1-0.5)} - 30.4^{(1-0.5)}]$$

$$= 42{,}785 \text{ bbl}$$

To make certain you understand the use of the hyperbolic-type curves work the following problem and compare your solution with the solution in Appendix C.

Problem No. 7.2: Application of the Hyperbolic Type Decline Curves

Given the following production history of a well:

Month	Production, STB
1	2,580
2	2,100
3	2,090
4	1,780
5	1,860
6	1,470
7	1,510
8	1,250
9	1,330
10	1,220
11	1,090
12	1,150
13	982
14	940
15	883
16	850
17	713
18	700
19	743

Overlay Figs. 7.3, 7.4, or 7.5 with a sheet of transparent paper and plot a decline curve using these production data. It may be necessary to replot averages to smooth the data sufficiently for a good analysis. Find the remaining life and reserves if the economic limit is 100 bbl/month. Find the remaining life and reserves if the economic limit is 10 bbl/month.

Some additional guidance in the use of the hyperbolic-type curves may prove helpful. For a particular hyperbolic constant, n, you will generally be able to fit more than one curve. This is true because an illustration representing a family of curves for one n is actually a plot of various segments of one mathematical curve. Consequently, for most plots the data curves overlap. In some cases you may be able to fit three different curves on the same illustration. In such cases the life and reserves calculated would still be the same. When there is a choice, always fit the curve that will place your last data point nearest the chart time of zero. By doing this you will be more apt to have cumulative production curves available so that you need not make calculations with Equations 7.22 and 7.24.

When you have hyperbolic data plots prepared from Appendix F data remember that any reproduction of these plots must be by a method that will give exact size copies. Otherwise, the basic graph size will be different from the graph size used for your production data plot and curve fitting will be inaccurate.

Other methods. Undoubtedly, the best known and most widely used method of decline-curve analysis involves a log-log trial-and-error plot of the rate-versus-time data until a straight line is obtained. The procedure involves plotting the log of the rate versus the log of the time plus some constant. The constant is varied until a straight line is obtained. The technique is illustrated in Fig. 7.8.

The validity of the log-log data plot may not be readily apparent but Equation 7.22 can be rearranged to show that the method is theoretically correct. For several years the author did not realize that the log-log plot was theoretically sound but Mr. Leo Shrider of the Bureau of Mines and some of his colleagues showed me that Equation 7.22 can be rearranged and the log of the equation taken to give,

$$\log q = \log \left[q_i (n \, a_i)^{-1/n} \right] + (-1/n) \log (n \, a_i + t) \qquad (7.27)$$

This transformation is accomplished by multiplying the numerator and denominator of Equation 7.22 by the expression $(n a_i)^{-1/n}$. When the equation is rearranged and the log is taken of both sides, Equation 7.27 results. Thus, the log of the rate, q, plotted versus the log $(n a_i + t)$ will give a straight line whose slope is $-1/n$. This is true because the first term of Equation 7.27 contains only constants. Consequently, when plots of log q versus log $(C + t)$ are made using various values for C, a straight-line plot will be obtained when C is equal to $n a_i$, and the slope of this straight line will be $-1/n$.

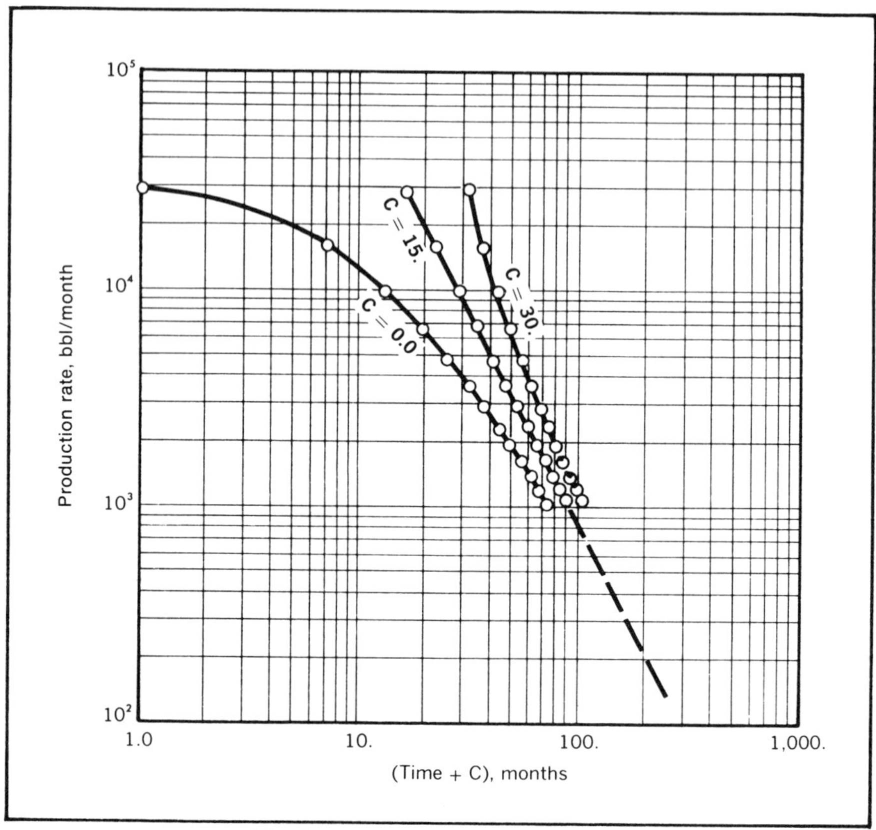

FIG. 7.8   Log-log extrapolation of hyperbolic decline

Normally the engineer who uses the log-log hyperbolic plot simply extrapolates the straight line to obtain the rates, times, and cumulative production figures he desires. He could evaluate n from the slope and once n is evaluated a_i could be determined from the C (or na_i), that resulted in the straight line. Then Equations 7.22 and 7.24 could be used to calculate the future rates and cumulatives. Generally engineers who have attempted to use the log-log analysis method have found it to be time consuming and insensitive. By using a large enough C it is generally possible to force a straight line from the production data, even when it does not fit a hyperbolic equation.

Another common method for analyzing hyperbolic data is the loss-ratio method. This technique uses the relationship between the decline rate and the hyperbolic constant.

$$n = \frac{\Delta\,(1/a)}{\Delta t} \qquad\qquad (7.28)$$

This equation can be derived from Equations 7.17 and 7.22. First raise both sides of 7.22 to a power n to obtain a value for q^n. Now substitute this expression for q^n in Equation 7.17 and solve for $1/a$. When the resulting expression is differentiated with respect to time Equation 7.28 will result.

As proposed for use in the Campbell evaluation text production rates at the end of equal time intervals are tabulated. The reciprocal of the decline rate for each period is calculated using the production rate at the end of the period as q in Equation 7.1. The differences in successive reciprocals are also calculated. The n can then be calculated by Equation 7.28. Constant reciprocals, $1/a$, imply exponential decline, while constant differences in reciprocals for the same length of time, indicate hyperbolic decline and will give a constant calculated n for each time period.

The use of the rate at the end of a period of time, Δt, in calculating the decline rate, a, by Equation 7.1, will provide an accurate value of a only if Δq is very small compared to the rate. Consequently, a constant n can be obtained by this method only when the decline rates are small. At high decline rates hyperbolic data which exactly fits Equation 7.22 will not yield a constant hyperbolic constant, n, by the loss-ratio method. The calculation of n by this method is also very sensitive to small variations of the actual data from hyperbolic decline and may be too involved when many hyperbolic declines must be evaluated as in the case of an annual company reserve report. One big advantage of the loss-ratio method is that a hyperbolic decline can be forced by averaging the calculated n values and then using this average n for the rate-time extrapolation.

Several computer methods for obtaining hyperbolic extrapolations of rate-time data are available in the literature. However, it does not appear that any of these methods can consistently obtain the best fit desired by most engineers.

Mobile oil by hyperbolic decline. As noted in the discussion of constant-percentage decline methods we define primary mobile oil for decline-curve purposes as the oil that would be produced if it was possible to continue producing a well to a rate of zero. In the case of hyperbolic decline this would mean that q_2 of Equation 7.24 would be zero. Then if we let q_1 equal q_i, ΔN_p of Equation 7.24 would be the remaining primary mobile oil.

$$\text{Remaining primary mobile oil} = \frac{q_i}{a_i(1-n)} \qquad (7.29)$$

This equation then indicates a result similar to that noted for constant-percentage decline. There it was noted that if remedial work did not increase the drainage volume of a well, the remaining primary mobile oil volume would remain unchanged and any increase in the current rate of production would have to be matched by a corresponding increase in the decline rate. Note that the same conclusion can be reached based on hyperbolic decline if we further specify that the hyperbolic decline constant, n, would be the same before and after the remedial work.

HARMONIC DECLINE

A third type of decline is generally recognized. This is the decline that occurs if the n of Equation 7.17 is 1.0. In this case the decline rate, a, is proportional to the rate, q. This is sometimes encountered when production is controlled predominantly by gravity drainage, and is known as harmonic decline. Examination of Equation 7.22 shows that the rate-time equation for this type decline is,

$$q = \frac{q_i}{(1 + na_i t)} \tag{7.30}$$

The cumulative-production equation cannot, however, be obtained from Equation 7.24, because the value of the second bracket in the equation would be zero when n is 1.0. Thus, we must return to Equations 7.13 and 7.14 and derive a specific equation for harmonic decline. Note that when n is 1.0 and the decline rate, a, is proportional to the rate, q, the decline rate can be stated as a function of the rates and the initial decline rate, a_i, as $(q/q_i)a_i$. This expression can be substituted for the decline rate in Equation 7.14 and the equation can be solved to obtain the cumulative-production equation for harmonic decline.

$$\Delta N_p = -\sum_{q_1}^{q_2} \left(\frac{\Delta q}{a}\right) \tag{7.14}$$

$$\Delta N_p = -\sum_{q_1}^{q_2} \frac{\Delta q}{(q/q_i)a_i} \tag{7.31}$$

$$\Delta N_p = \left(\frac{q_i}{a_i}\right) \ln\left(\frac{q_1}{q_2}\right) \tag{7.32}$$

The harmonic analysis can be performed in much the same way that the hyperbolic analyses are performed although no type curves are available in this book for this purpose since harmonic decline is seldom encountered.

If the engineer works with harmonic decline often it would be a simple matter to prepare data plots (similar to the hyperbolic-type curves) by using Equations 7.31 and 7.32. These plots would have the same general shape as the hyperbolic-type curves. However, note from Equation 7.32 that a plot of the cumulative production versus the rate on semilog paper would give a straight line. This should simplify the application of the rate-cumulative curves analysis as compared to a similar hyperbolic analysis.

REFERENCES

1. Cutler, W. W., Jr., "Estimation of Underground Oil Reserves by Well Production Curves," *Bull.*, USBM (1924) 228.
2. Brons, Folkert, "On the Use and Misuse of Production Decline Curves," API Paper No. 801–39E (1963).
3. Arps, J. J., "Analysis of Decline Curves," *Trans.*, AIME (1940) *160*, 228–247.
4. Arps, J. J., "Estimation of Primary Oil Reserves," *Trans.*, AIME (1956) 182–191.
5. Chatas, A. T. and Yanhee, W. W., "Applications of Statistics to the Analysis of Decline Curves," *Trans.*, AIME (1958) 399–401.
6. Gray, K. E., "Constant Percentage Decline Curve," *Oil and Gas J.* (Aug. 20, 1960).
7. Gray, K. E., "How to Analyze Yearly Production Data for Constant Percent Decline," *Oil and Gas J.* (Jan. 1962).
8. Slider, H. C., "A Simplified Method of Hyperbolic Decline Curve Analysis" JPT Forum, *Petroleum Technology* (March, 1968).
9. Campbell, John M., *Oil Property Evaluation*, Prentice-Hall, (1959) 177–208.

Waterflooding and Its Variations

This chapter may well be the most valuable one in this book. There appears to be more opportunity for the engineer to conserve petroleum energy and make profits for his company through effective engineering of a waterflood than in any other single method. The author was tempted to treat the broader subject of secondary recovery in this chapter and even so advised the publishers originally. However, his personal knowledge and experience is in the field of waterflooding and an attempt to treat in situ combustion or some other exotic recovery technique in this book would result in the recitation of a literature search. Furthermore, with the exception of steam flooding, it is the author's opinion that there is currently very little possibility for economic application of any secondary recovery technique outside the area of waterflooding and its variations. This is not meant to imply that there have been no economic successes using miscible displacement, in situ combustion, or other exotic recovery techniques. However, it is meant to suggest that such opportunities occur so infrequently that they are presently more in the area of research and development than in the realm of engineering.

Because steaming of reservoirs to increase recovery does appear to have considerable worth as an engineering tool, the author wishes for some practical experience in this area that would justify his discussion of this subject. However, since he does not possess such practical experience he will reluctantly decline the opportunity to discuss and expound upon well steaming and go on to matters of water flooding. Engineers overseeing reservoirs with viscous oils, especially if they are very permeable, should consider steaming to improve recovery. Many useful articles can be found listed in the *General Index to Publications of the Society of Petroleum Engineers of AIME, 1953–1966* under "thermal recovery of oil."

Most of the fundamentals of waterflooding are useful in all types of frontal secondary recovery methods and in fact in many cases in natural water drive or gas-cap drive applications. Many of the concepts covered in this chapter are also fundamental to all secondary-recovery operations. For example, the effect of permeability distribution is important in all secondary-recovery methods including such nonfrontal techniques as steaming. Consequently, the concepts covered in this chapter are important to the engineer regardless of whether or not he is faced with reservoir problems in waterflooding.

As has been the case through out this book the author does not claim that the methods presented herein are the only useful and practical methods of waterflooding analysis. However, the objective is to present methods that can be completely understood by the engineer and that have proven reasonably accurate over the years. By using methods that he understands, the engineer can intelligently apply these methods and to a large extent avoid misapplication of the calculating techniques. Consequently, we will not discuss in detail such well known but highly complicated techniques as the Leighton and Higgins (Bureau of Mines) method. Only two prediction techniques will be presented in sufficient detail to permit their practical use. One of the methods will be useful for theoretical predictions and the other has proven very useful and sur prisingly accurate for the prediction of waterflood behavior by analogy.

However, before concerning ourselves with the rate-time behavior prediction we will consider the problem of predicting the total flood recovery which is where every waterflood begins. Discussion of the total waterflood recovery prediction will require our consideration of many of the fundamental ideas involved in flooding such as pattern sweep, vertical or stratification sweep, evaluation of the residual oil, and determining the bulk swept volume.

This discussion of predicting total flood recovery will be followed by a discussion of recommended methods of predicting oil and water production rate versus time. A comparison of recommended methods with some of the more common published methods will be presented. Then the very important subject of flooding variations using mobility-ratio-control additives and residual-oil-reducing additives will be considered so the engineer can determine which procedure or combination of procedures will result in the most profit for his company.

Before we consider detailed flood calculations we will first take a close look at the flood displacement mechanism to provide a good physical picture of the displacement taking place in the reservoir.

THE WATER FLOOD DISPLACEMENT MECHANISM

A waterflood displacement is governed by the same fractional flow and Buckley-Leverett equations that describe all immiscible displacement processes. If there is no free gas present in the reservoir at the time the flood is initiated, the saturation distribution along any streamline (flow line) in the reservoir can be calculated by using the equations of Chapter 5. However, if the reservoir contains a free gas saturation at the time the flood is initiated the analysis presents some additional problems.

The engineer's first conclusion concerning the displacement of oil and free gas concurrently in the reservoir is that it would require three-phase relative permeability data. Fortunately, this is not generally the case. What does generally occur is depicted in Fig. 8.1. The lower portion of this figure shows a plot of water and total liquid (oil + water) saturation along a streamline before the displaced oil reaches the producing face. The basis for the lower diagram may be better understood by considering the upper diagram. This considers various classifications of reservoir pores grouped together. At the bottom of the diagram is shown the pore space originally occupied by immobile water, here called connate water. The next horizontal strip of the diagram (not a horizontal strip of reservoir) represents the pores originally occupied by immobile

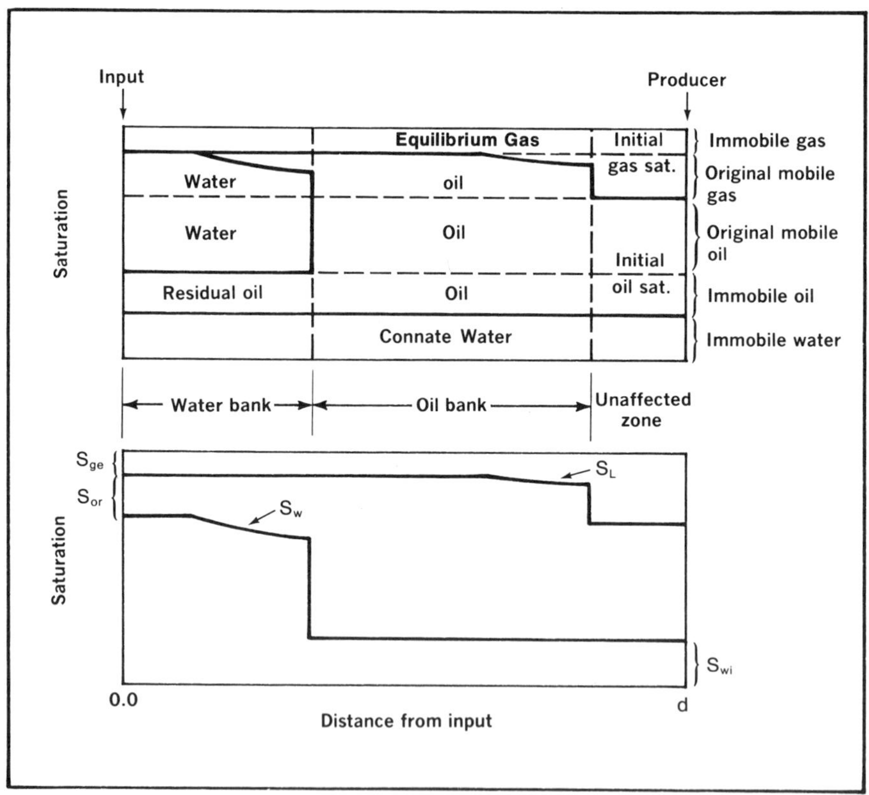

FIG. 8.1 Saturation distribution in a waterflood

oil here termed residual oil. The top strip represents the comparable gas pore volume, that is, the pores originally occupied by immobile gas (shown as being unrealistically high here). For all practical purposes these three groups of pores can be considered as inactive or dead pores. If their fluids cannot be moved they can play no important part in the displacement process.

However, the pores initially containing the mobile oil and mobile gas do play an active part in the displacement. The right-hand side of the diagram indicates the "unaffected" portion of the reservoir; thus, this represents the original saturations in the reservoir. On the left is the water bank which indicates that the water entering the formation displaces the original mobile oil and oil in the original mobile gas pores. In between the "unaffected" zone and the water bank we find a zone where the pore space originally occupied by mobile gas is almost completely filled with oil so that there is very little free gas in the oil bank except for that which is immobile. The question is how does this occur?

This question can be answered physically by considering what happens to the oil displaced by water. This displaced oil must choose a path to follow. It is obvious that it cannot enter pores occupied by immobile fluids. Consequently, it has a choice of traveling through pores occupied by mobile oil or pores occupied by mobile gas. This means that oil displaced from the water bank can either displace oil from the pore space originally occupied by mobile oil or it can displace gas from pores originally occupied by mobile gas. The reservoir oil is normally about 100 times as viscous as gas and occupies pores that are smaller in size than the gas-occupied pores. What happens is then obvious. The displaced oil follows the path of least resistance and displaces the highly mobile gas rather than the much less mobile oil and an oil bank is formed.

Since free gas displaced from the oil and water banks will be traveling through the "unaffected" zone of Fig. 8.1, it is not strictly accurate to say that it is unaffected. However, the free gas is so highly mobile compared to the reservoir liquids that it will generally flow through to the producing wells with a negligible increase in the reservoir pressure that existed at the start of the flood. This point will be further emphasized when we discuss bypassing of outside producers by the oil bank (later in this chapter). It should be specifically noted that it is seldom that any substantial quantity of equilibrium gas remains in the reservoir after the flood front passes because the increase in pressure causes free gas to go back into solution. Also, equilibrium gas is difficult to measure in the lab and is seldom as high in the reservoir as indicated by lab data. If appropriate relative-permeability data such as that in Fig. 8.2 are available, it is possible to determine the three-phase saturation distribution along a linear or radial displacement by using the linear or radial Buckley-Leverett equations derived in Chapter 5.

$$\Delta X_{Sdj} = \left(\frac{5.615 q_t \Delta t}{\phi A}\right)\left(\frac{\Delta f_d}{\Delta S_d}\right)_{Sdj} \tag{5.15}$$

$$\Delta(\mathrm{r}^2)_{\mathrm{Sdj}} = \left(\frac{1.79 q_t \Delta t}{\phi h}\right)\left(\frac{\Delta f_d}{\Delta S_d}\right)_{\mathrm{Sdj}} \qquad (5.38)$$

It would first be necessary to calculate the fractional-flow curves for water displacing oil and for oil displacing gas and plot the data as in Fig. 8.3 so that the slopes could be used to determine the position of the water saturation and liquid saturation lines as in the lower part of Fig. 8.1.

It is recommended that the reader solve the following problem and compare his solution with the solution in Appendix C.

Problem No. 8.1: Saturation Distribution in a Waterflood

A large peripheral flood contains a ring of producers 1,000 ft within the surrounding injectors. If we treat this section of the reservoir as a linear displacement, what will be the saturation distribution in the most permeable zone when the first oil production increase occurs? The most permeable section's cross-sectional area is 150,000 sq ft and it has an effective injection rate of 3,000 b/d. Assume no free gas goes back into solution. Note that this assumption is only for the purposes of this problem. It is, of course, possible to calculate the amount of gas going back into solution due to the reservoir pressure increase.

Porosity = 0.2
Initial oil saturation = 40%
Initial water saturation = 40%
Initial gas saturation = 20%

Use the Fig. 8.3 fractional-flow curves calculated from relative permeabilities.

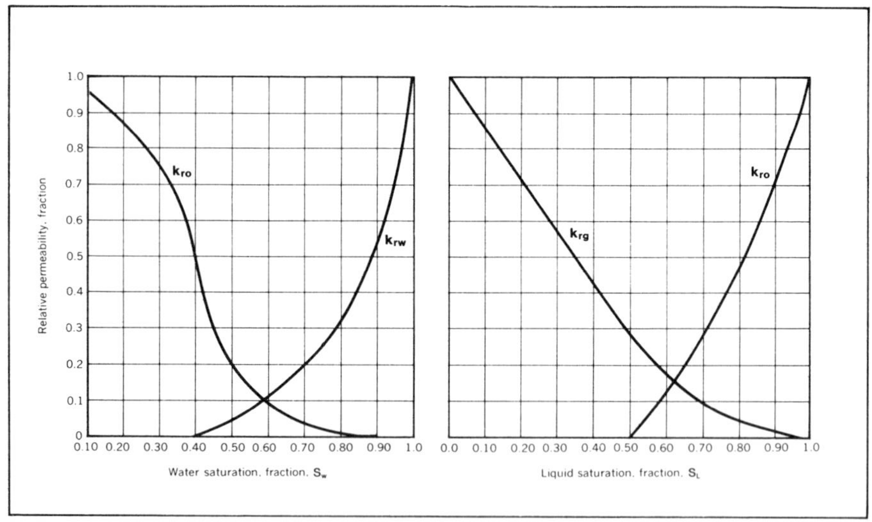

FIG. 8.2 Relative permeability data

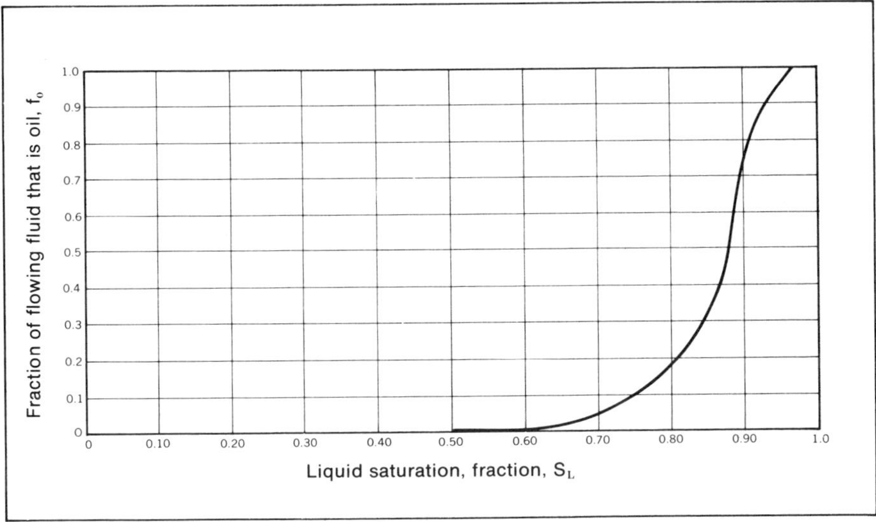

FIG. 8.3A Fractional flow curve for oil displacing gas

The engineer should be very careful to make certain that the formation of the oil bank and its affect on the performance of a frontal displacement is always taken into account in his engineering analysis. Many, and possibly even most, of the studies of frontal displacements in general, and waterflooding in particular, assume that there is no free gas in the reservoir at the time the secondary-recovery process is initiated.

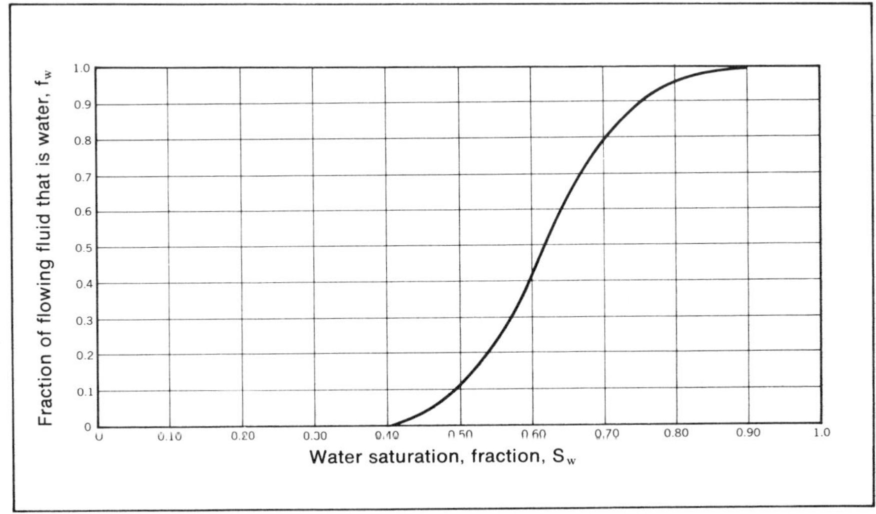

FIG. 8.3B Fractional flow curve for water displacing oil

This can lead to erroneous conclusions. It may be justified by the investigator, engineer, or researcher by stating that secondary recovery should be initiated before a free gas forms in the reservoir. This is theoretically indisputable, but the fact remains that it is seldom that a reservoir with many different producers and royalty owners can agree upon a formula for participation in a unit, before the reservoir begins to approach the economic production limit for primary production. Consequently, it is seldom that secondary recovery is initiated before a substantial free-gas saturation forms in the reservoir.

PREDICTING TOTAL FLOOD RECOVERY

Much money has been spent needlessly on detailed engineering studies and actual field flooding operations due to an engineer's inability to accurately predict the total flood recovery of a reservoir. Conversely, many flooding opportunities have been missed for the same reason. The basic calculation need not be elaborate to obtain a reasonably accurate estimation of recovery. The following simple equation should yield accurate results if the reservoir parameters in the equation are evaluated with care.

$$N_{pf} = 7{,}758 \phi E_t V_{sw} \left[\left(\frac{S_{oP}}{B_{oP}} \right) - \left(\frac{S_{or}}{B_{or}} \right) \right] \qquad (8.1)$$

This equation is based on the difference in the oil in the gross swept volume, V_{sw}, at the start of the flood and the oil in the gross swept volume at the abandonment of the flood. In the equation, the residual oil, S_{or}, represents the oil saturation that would remain at abandonment in a part of the reservoir that has been flushed by water. The saturation at the start of the flood or after primary is denoted by the symbol, S_{oP}. We will see that much of the reservoir volume included in the gross swept volume, V_{sw}, is not flushed by water, so that an overall efficiency, E_t, a fraction, is introduced to account for this difference.

Thus the expression, $7{,}758 \phi E_t V_{sw}$, represents the volume of the reservoir flushed by water. The constant 7,758 converts the acre-foot units of the gross swept volume, V_{sw}, to barrels. Thus, this group of terms, when multiplied by (S_{oP}/B_{oP}), gives the stock-tank barrels of oil in the reservoir at the start of the flood. The same group of terms when multiplied by (S_{or}/B_{or}) gives the stock-tank barrels of oil in the reservoir after flooding and the difference is Equation 8.1, the flood recovery.

All of the factors in Equation 8.1 are interrelated. The efficiency, E_t, used must take into account the method of evaluating the gross swept volume, V_{sw}, and the residual oil saturation, S_{or}, which in turn must be correlated with the basis for the sweep efficiency, the gross swept volume, and the porosity. Consequently, Equation 8.1 appears to be a simple volumetric-balance equation but it is useless or even dangerously misleading unless the engineer considers the basis and interrelationship of its reservoir parameters very carefully. We will consider the elements of the equation individually.

It does not appear that the evaluation of porosity or formation volume factors, should present many new problems. However, engineers do not always limit the basis for their statistical evaluation of porosity to the area included in the gross swept volume. Many times the gross swept volume considered eliminates the edge areas of a reservoir that cannot be swept and thus eliminates some of the areas of marginal porosity. Also, in evaluating the oil formation volume factor at abandonment, remember that the oil probably does not have much more gas in solution than it contained at the start of the flood. The trapped free gas may go back into solution as the reservoir pressure is increased by water injection but this amount is generally negligible. In any event the residual oil will be far from saturated with gas after flooding has increased the reservoir pressure. It is seldom that any sizable error is introduced by assuming the oil formation volume factor at the start of the flood to be equal to the formation volume factor at abandonment.

Determining oil saturation at the start, S_{oP}. The oil saturation at the start of the flood, S_{oP}, is generally best evaluated by a material-balance equation based on the entire reservoir instead of just the gross swept volume of the flood; we assume a uniform saturation throughout the reservoir at the start of the flood. A material-balance equation subscripted specifically for our purposes is convenient for this calculation.

$$S_{oP} = \frac{(N - N_{pP})B_{oP}(1 - S_w)}{NB_{oi}} \qquad (8.2)$$

The symbol N_{pP} represents the cumulative stock-tank barrels of oil production at the start of the flood or after primary, denoted by the subscript P. The equation represents the remaining reservoir oil in the reservoir, $(N - N_{pP})B_{oP}$, divided by the pore volume of the reservoir determined from the initial oil saturation and volume of oil in place, $NB_{oi}/(1 - S_w)$.

The original stock tank volume of oil in the reservoir (before any primary production), N, is calculated by material balance or volumetrically in the same way it would be evaluated for any reservoir. The following equation is very similar to the one derived in Chapter 6 on material balance.

$$N = \frac{7758\phi(1 - S_{wc})V_{bP}}{B_{oi}} \qquad (8.3)$$

The only thing new about this equation is use of the symbol, V_{bP}, for the bulk volume. The subscript P is used to make certain the engineer recognizes the difference between the reservoir volume drained under primary production, V_{bP}, and the reservoir volume drained under secondary, V_{sw}. We will now consider the specific difference between these two volumes.

Determining gross swept volume, V_{sw}.   One of the common errors made by engineers in calculating total flood recovery is to assume that the swept volume is equal to the total reservoir volume without adjusting overall sweep efficiency or residual oil. Since all of these parameters are interrelated there can be no right or wrong value for one without a discussion of the others. Consequently the engineer should recognize that there can be no correct value for any one of these parameters without first fixing the others. However, the methods discussed here will give reasonably accurate results if applied conscientiously and consistently.

The necessity for differentiating between primary drainage volume and swept volume is shown in Fig. 8.4. This idealized reservoir is bounded by a shale out at the zero thickness isopach as indicated. Originally it was developed with five wells and produced to near depletion with natural solution-gas-drive energy. During this phase of production it is clear that the entire reservoir would be drained by the five pressure sinks represented by the five producing wells. However, when the four

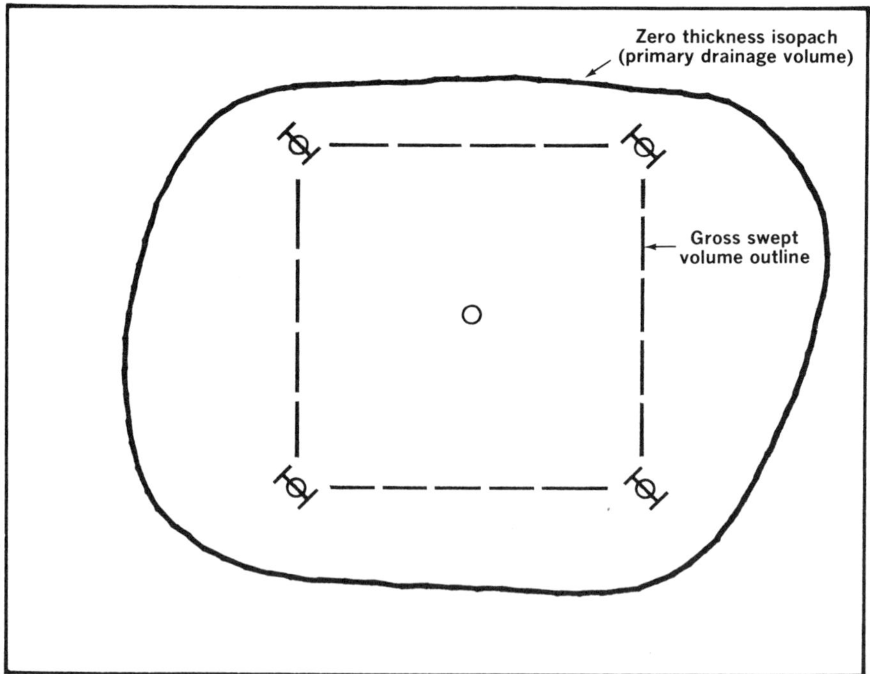

FIG. 8.4   Comparison of primary drainage volume and gross swept volume

outside wells are converted to injection wells and the fifth well continues to produce, it is clear that during the economic life of this flood (before the water cut becomes uneconomical) little of the oil between the gross swept volume outline and the zero thickness isopach will be produced. Initially the water injection into the four wells will move out radially forming an oil bank ahead of the water bank until the adjacent oil banks meet. The oil will be squeezed from between the two water banks with about half of it going toward the producing well and being produced and the other half moving toward the reservoir boundary to be trapped there. Once all of the gas has been displaced from the gross swept volume, either by being produced or by being moved toward the reservoir boundary, most of the injected fluid will move along streamlines in the gross swept volume. This will occur because a pressure sink will exist only at the producing well.

Late in the flood life, after the reservoir has substantially reached steady state from a pressure standpoint, some flow will take place from outside the gross swept volume toward the producing well. It can be shown that steady-state streamlines will cover the entire reservoir. However, the velocity along the streamlines outside the gross swept area is so low compared to the velocities in the flood area that they have little significance, especially when we recognize that they occur so late in the flood life.

We have discussed Fig. 8.4 based on the assumption that there was a substantial gas saturation in the reservoir at the time the flood was initiated. However, note that essentially the same situation results when there is no free gas in the reservoir when the flood is initiated. In this case the reservoir would quickly approximate steady-state conditions but substantial flow would still take place only within the flood area. Consequently, the gross swept volume would remain the same.

It should be noted that there are some published articles[1] on flood sweep that conclude that flow does take place outside the gross swept area. However, these studies do not attach quantitative values to the streamline flow outside the flood area so for all practical purposes we can consider recovery to occur only in the area marked in Fig. 8.4 as the gross swept volume.

Consideration of Fig. 8.4 was only for the purposes of demonstrating why there must be a difference between the recovery volume considered in primary and in secondary recovery. In most situations the delineation of gross swept volume is much more complicated than the simple case described in Fig. 8.4. Consider the more complex flood geometry depicted in Fig. 8.5. The solid line indicates the recommended gross swept volume if the gas saturation at the start of the flood, S_{gs}, is less than the equilibrium gas saturation, S_{ge}. This would mean that gas would not flow and an oil bank would not form. In this case the recommended method for delineating the gross swept volume is simply to consider parts or elements of the flood geometry. In this five-spot pattern we have

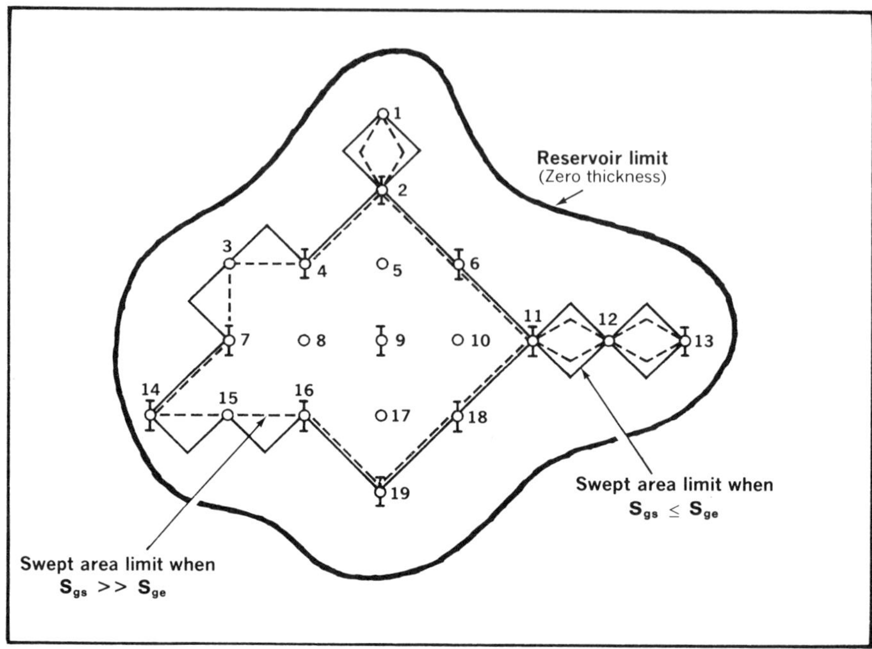

FIG. 8.5   Estimation of gross swept volume in a flood

included one-fourth of a five-spot between wells 1 and 2, one-half of a five-spot between wells 3, 4, and 7, three-fourths of a five-spot between wells 7, 14, 15, and 16, etc.

When a substantial gas saturation exists in the reservoir at the start of the flood, delineation of the gross swept area is much more difficult — the edge producers cause the problem. Since the oil bank is pushed past the outside producers before the reservoir pressure in that vicinity is increased, the oil-producing rate is increased very little prior to the time the water bank reaches the outside producer. A look at the pressure distribution associated with outside producers during the early flood life will help provide an understanding of this phenomenon.

Consider Fig. 8.6 which is a pressure cross-section of a portion of Fig. 8.5 through wells 9, 5, 2, and 1, and on to the reservoir limit. This could represent the pressure distribution in one strata (uniform permeability) of a reservoir that has a substantial gas saturation at the start of the flood such that an oil bank would form. Now consider the time when the oil bank front reaches the producers and the first oil-rate increase is realized in wells 5 and 1. Note that at this time there is a considerable increase in pressure around well 5 and a negligible increase around well 1 because the pore volume between 1 and the reservoir limit contains a large saturation of gas that has been compressed only a small amount

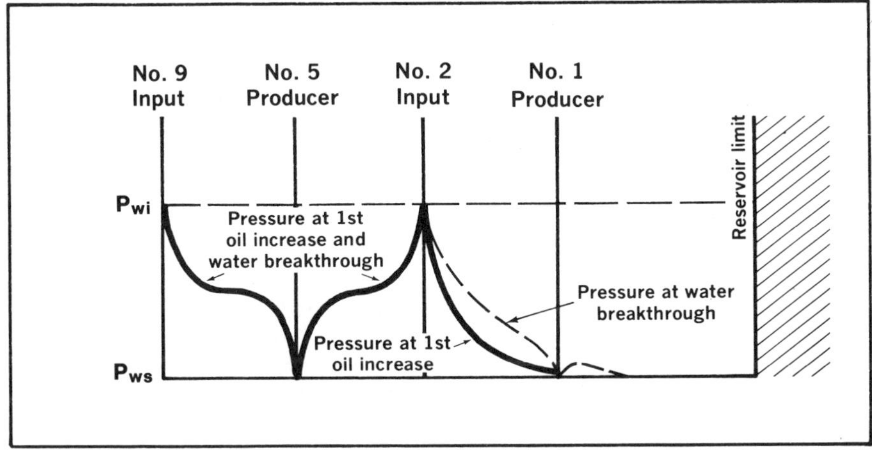

FIG. 8.6 Pressure distribution around edge wells with a substantial gas saturation at the flood start (See Fig. 8.5).

while all of the gas around 5 has been produced. The radial pressure gradient in the gas zone will be negligible compared with the pressure gradient in the oil and water banks because the gas viscosity is only about $\frac{1}{100}$ of the liquid viscosities at low pressures.

Since the oil-production rate is a function of oil saturation and pressure drop in the drainage area, the rate increase in well 5 at this time will be quite large while the rate increase in well 1 will be relatively small. Both wells will have the same oil saturation but well 1 will have a very small pressure drop available for production.

At water breakthrough the pressure in the well 1 drainage area has still increased only a small amount so most of the oil is simply pushed past well 1 and is trapped in the original gas-occupied pore volume between the flood area and the reservoir limit. Remember that the pressure shown is the result of radial flow from the injection well with the pressure drops due to the production from well 1 superimposed upon it. This results in the pressure hump between the producer and the reservoir limit. To the left of the hump, fluid is flowing toward the producing well, while to the right of the hump, oil and gas continue to migrate toward the reservoir boundary. Eventually the hump will reach the reservoir boundary and all flow will be toward the producer.

Consequently, there will be a considerable difference in the gross swept volume of a producing well that is surrounded by injection wells, as is well 5 in Fig. 8.5; a producing well affected directly be three injection wells, for example well 15; a producing well affected by two injection wells, 3 and 12; and a producing well affected by only one injection well, well 1. A reasonable gross swept volume is obtained by (1) delineating the gross swept volume with a line constructed by joining adjacent edge producing and injection wells and adjoining edge injection wells but not adjoining edge producing wells, and (2) adding one-half quadrant for isolated producers such as wells 1 or 12.

This technique is illustrated in Fig. 8.5. Adjoining producers and injection wells are connected (with the line delineating the gross swept area) between wells 4 and 3, 3 and 7, 14 and 15, and 15 and 16. Adjoining edge injection wells are joined by the delineating line between 2 and 4, 7 and 14, 16 and 19, 19 and 18, 18 and 11, 11 and 6, and 6 and 2. There are no adjoining edge producing wells in this flood. However, suppose a producing well, 1A, existed midway between wells 1 and 3. Then the delineation would go from 2 to 1A to 4 to 3. It would not go from 1 to 1A to 3. In the pattern that exists in Fig. 8.5, note that one-half quadrant has been included in the gross swept volume between wells 1 and 2, 11 and 12, and 12 and 13. These rules of thumb may be varied with the initial gas saturation and the area between the outside wells and the zero thickness isopach. They are simply cited as an example of a procedure that gives reasonable values most of the time.

This discussion emphasizes the difficulty involved in evaluating a pilot flood in a reservoir. In such a case only one or two elements of a pattern are normally used; and if the outside pattern wells are producing wells, it can be seen that much of the oil can be pushed past the wells without a corresponding increase in pressure and oil production. Furthermore, note that if the injection wells are placed on the outside of the pattern, the oil within the pattern is contained but most of the injected water will tend to go into the unpressured reservoir outside of the pattern. (This effect will be discussed at greater length later in this chapter). Both of these effects can make a pilot operation result in very little additional oil recovery while a full-scale flood of the reservoir could be extremely successful.

To test your knowledge of this section work the following problem and check your solution against the solution in Appendix C.

Problem No. 8.2A: Evaluation of Gross Swept Volume

In subsequent parts of this problem we will predict the total flood recovery for the reservoir in Fig. 8.7. However, in this part the engineer is asked to determine the original (before primary) stock tank volume of oil in place, the reservoir saturations at the start of the flood, and the acre-feet of gross swept volume based on the best 20-acre five-spot flood pattern. Use only the wells that presently exist as either injection or producing wells. Note that a 20-acre five-spot is based on 10-acre well spacing. Use the 10-acre grid on Fig. 8.7 to estimate the reservoir volumes by estimating areas based on the grid and average thicknesses for the individual areas.

Reservoir Data:

Porosity, 20%; Connate water, 20%
Original B_o, 1.2
B_o at the flood start, 1.1; Cumulative Production at the flood start, 364,500 STB

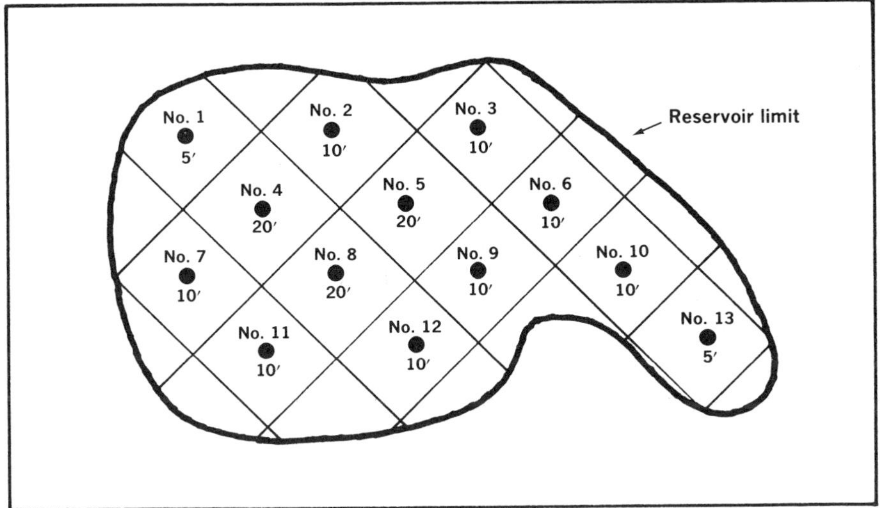

FIG. 8.7 Reservoir for Problem 8.2 with superimposed 10-acre grid.

Determining residual oil saturation. The residual oil saturation should represent the oil that cannot be moved regardless of the amount of displacing phase moved through the formation. It does not include the inability to make the injected fluid contact all of the formation. In this respect it might be thought of as a laboratory residual oil saturation. In water-oil relative-permeability data it would represent the oil saturation below which oil will not flow.

Ideally the residual oil saturation, S_{or}, should be based on relative-permeability tests or other laboratory tests on actual reservoir cores designed to determine the residual oil saturation. Theoretically, such tests can be run on old cores that have been in storage for years, by using a restored-state method. The core is restored to its initial reservoir saturation by first cleaning all of the old oil and water from the cores, restoring the wettability (generally destroyed in the cleaning process), and running displacements with first water and then oil. The core is then again displaced with water to reach a residual oil saturation. However, if good laboratory data are not available, the residual oil saturation can be determined by analogy or by using the reported oil saturation from conventional core analysis.

In cutting a conventional core, the core is subjected to mud-filtrate invasion that may compare with flooding. (See Chapter 5, Determining Original WOC and GOC). In addition, the core is subjected to a solution-gas drive when it is removed from the well and the pressure is reduced to atmospheric, causing the liberation of the solution gas and the corresponding displacement of oil from the core. Thus, core-analysis oil saturations often closely approximate the residual oil obtained after solution-gas drive and water displacement.

When available, core-analysis oil saturations should always be compared with laboratory-measured residual oil values. So much difficulty is involved in determining residual oil saturations in the laboratory (especially control of the core wettability) that it is not unusual for an engineer to use the core-analysis oil saturation in preference to the saturations resulting from laboratory analysis.

The total displacement efficiency combines several different efficiencies including the horizontal or sweep efficiency, the vertical or stratigraphic efficiency, and the conformance efficiency. The total efficiency is then the product of these three or

$$E_t = (E_H)(E_V)(E_C) \qquad (8.4)$$

Determining ultimate horizontal sweep efficiency. Most waterfloods are developed on a pattern such as those illustrated in Fig. 8.8. By far the most widely used pattern is the five-spot so most of our discussions will be concerned with this although the same general methods can be applied to most patterns. Results from peripheral flooding, where the injection wells are along the perimeter of the reservoir, are more difficult to predict.

Even if piston-like displacement efficiency could be obtained in the reservoir, it would be impossible to displace all of the mobile oil from a patterned element before any of the displacing fluid is produced because it is necessary to inject into a point (well) in the reservoir and produce from a point (well) in the reservoir. This is demonstrated in Fig. 8.9 which depicts the water-bank front at various stages of development in a five-spot element. This is one five-spot in a theoretically infinite pattern with wells 1, 2, 3, and 4 producing and well 5 being a water-injection well.

The water bank tends to cusp into the producing wells due to the greater mobility of the water (k_w/μ_w) as compared with the mobility of the oil in the oil bank (k_o/μ_o) and the fact that the most direct flow path in the five-spot is much shorter than the least direct flow path along the edges of the five-spot quadrant (quarter of a five-spot). Thus, water tends to bypass oil because it will move more readily, it has a shorter distance to travel along the most direct streamline, and the pressure gradient is greater along the most direct streamline. Fig. 1.12 in Chapter 1 more fully indicates the conditions in a five-spot quadrant by showing the streamlines and pressure distribution.

After the water front breaks into the producing well, there will be a continued advance of the water front throughout the five-spot, but the advance will be retarded considerably because of the mobility ratio. Many studies have been made to relate the percent of a five-spot swept by the water bank at different stages of depletion for different mobilities.

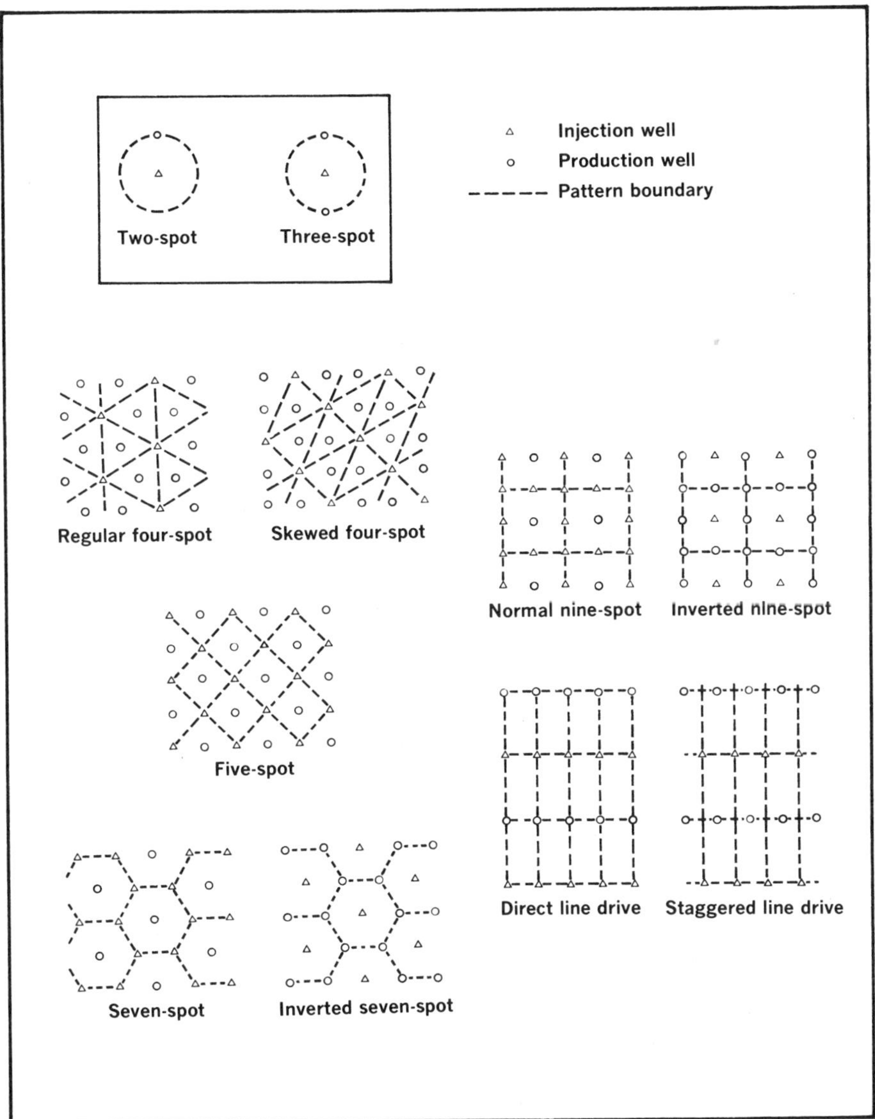

FIG. 8.8 Flooding patterns (after Craig[10], SPE Monograph No. 3)

The results of one such investigation are shown in Figs. 8.10A and 8.10B. The appendix of SPE Monograph Volume 3 contains much more data of this type covering many other flood patterns. These data are for homogeneous sands, and we will see that most sands perform like a series of parallel homogeneous sands stacked on top of each other, each with its

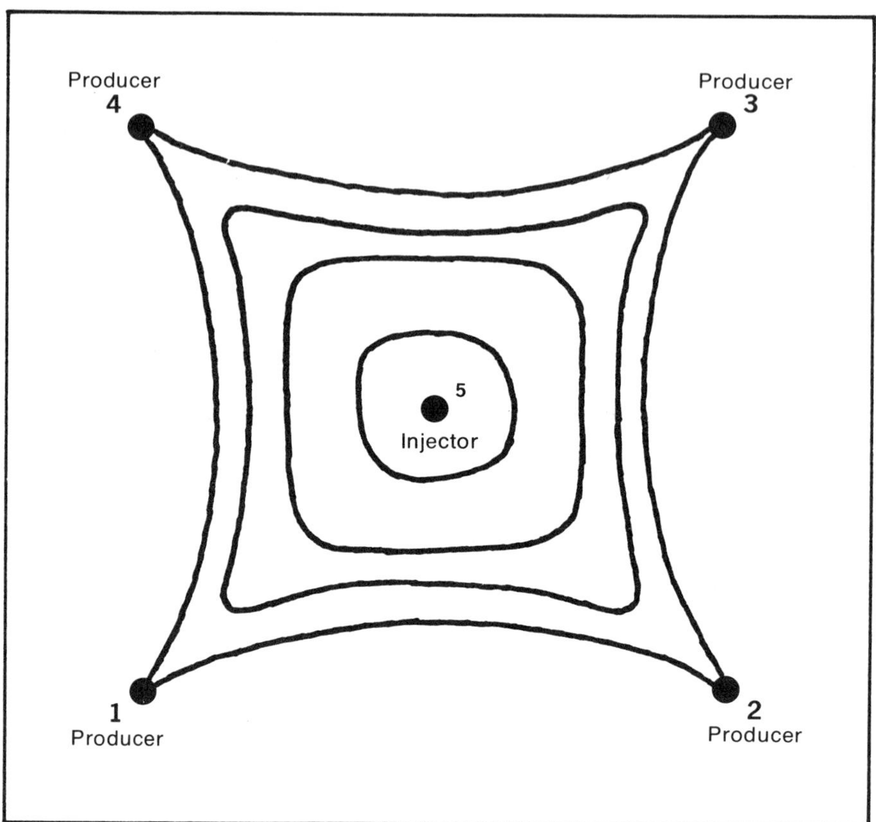

FIG. 8.9   Water-front cusping in a five-spot.

own permeability and porosity. Consequently, we will see that by the time the economic produced water-oil ratio is reached in a flood, most of the strata either have had sufficient water throughput to give them a 100% horizontal sweep or the water bank has not yet reached the producing well, and no sweep efficiency correction is necessary. Consequently, in estimating the ultimate sweep efficiency we can generally consider the horizontal sweep as being equal to 1.0. The incomplete sweep of the strata before water breakthrough at the producing well is included in the vertical sweep efficiency. This will be clearer after we consider the vertical sweep efficiency. However, we will see later that horizontal sweep data such as that of Figs. 8.10A and 8.10B are very valuable and necessary when predicting the rate-versus-time behavior of a waterflood rather than just the ultimate recovery.

The conformance efficiency is used by some engineers to account for lateral variations in permeability due to shale or low-permeability

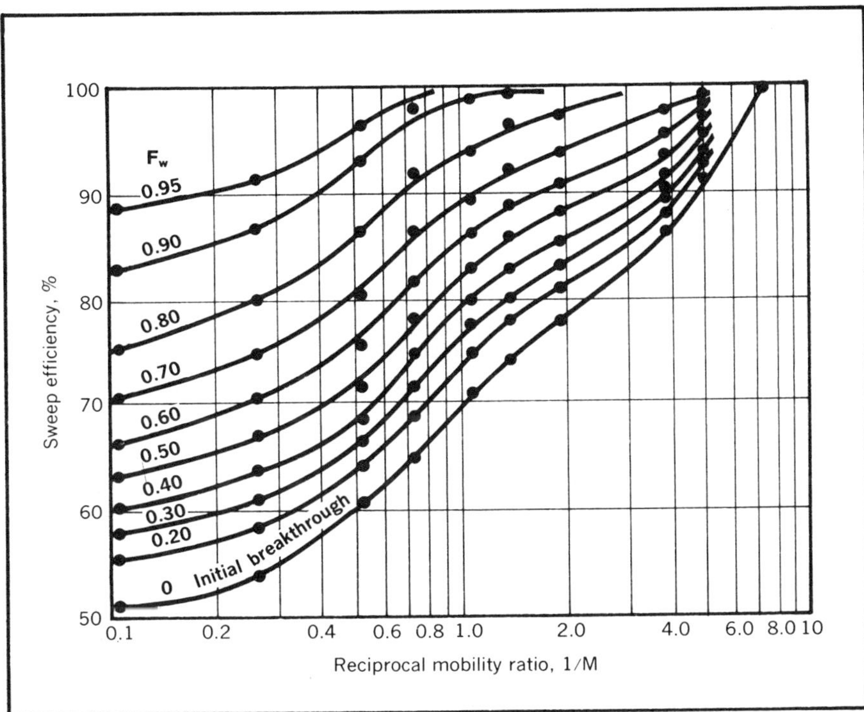

FIG. 8.10A Effect of mobility ratio on sweep efficiencies for the five-spot pattern; f_w is the reservoir cut and $M = \lambda_w/\lambda_o$. (After Dyes, Caudle, and Erickson, *Trans. AIME*.)

lenses. Other engineers use it to cover any displacement inefficiencies that are not included in the horizontal or vertical sweep. In this book we will consider the conformance efficiency to be 1.0 at all times. It is included in Equation 8.4 for completeness.

Determining ultimate vertical sweep efficiency. As noted above, most reservoirs appear to perform as though they were comprised of a series of parallel reservoirs of different permeabilities. This seems to be true even when there is little evidence of actual vertical stratification. In the latter case, it is probably due more to a microscopic permeability distribution or effective pore-size distribution than it is to actual stratification. Nevertheless, most of the early methods proposed for predicting the behavior of a waterflood concerned themselves almost exclusively with the effect of permeability stratification. Probably the best-known method of waterflood prediction is the Stiles method that accounts *only* for permeability stratification in predicting the behavior history of a waterflood; but, by the time the flood reaches its economic limit the inaccuracies inherent in the Stiles method have become unimportant,

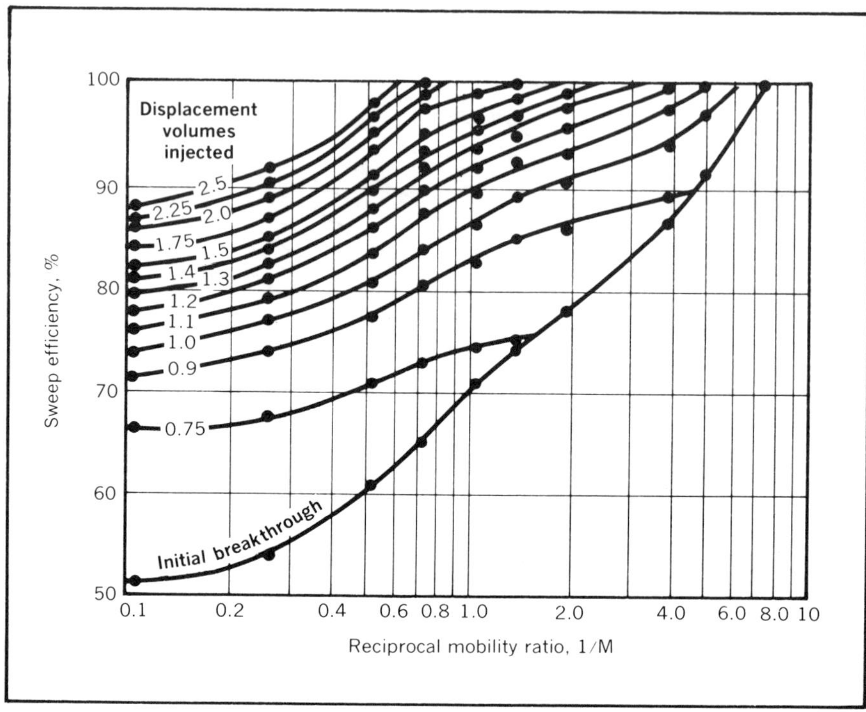

FIG. 8.10B   Effect of mobility ratio on the displaceable volumes injected for the five-spot pattern. $M = \lambda_w/\lambda_o$. (After Dyes, Caudle, and Erickson, *Trans. AIME.*)

and the fractional recovery, R, is an excellent estimate of the ultimate vertical sweep efficiency. Consequently, it is recommended that the Stiles method, with one modification, be used to predict the ultimate vertical sweep efficiency.

The forms of the Stiles equations presented in the Craft and Hawkins text[2] are probably the easiest to understand and the simplest to derive. Basically, the results of a Stiles analysis provides a relationship between the fractional water cut, f_w, of the total (oil plus water) producing rate and the fraction of the mobile oil that has been produced, R. It will be shown that these two performance parameters are governed by the following equations.

$$f_w = \frac{Ac_j}{[Ac_j + (c_t - c_j)]} \tag{8.5}$$

$$R = \frac{[h_j k_j + (c_t - c_j)]}{h_t k_j} \tag{8.6}$$

where

$$A = \frac{B_o(k_r/\mu)_w}{(k_r/\mu)_o} \tag{8.7}$$

The meaning of the capacity symbols, c, and thickness symbols, h, can best be explained by example. Examine Fig. 8.11. The permeability-distribution curve is a plot of permeability versus the thickness of the formation that has a permeability equal to or greater than the subject permeability. For example, 15 ft (h_j), of the total reservoir thickness of 29 ft, (h_t), has a permeability equal to or greater than 127 md, (k_j). The capacity curve is obtained from the permeability-distribution curve by

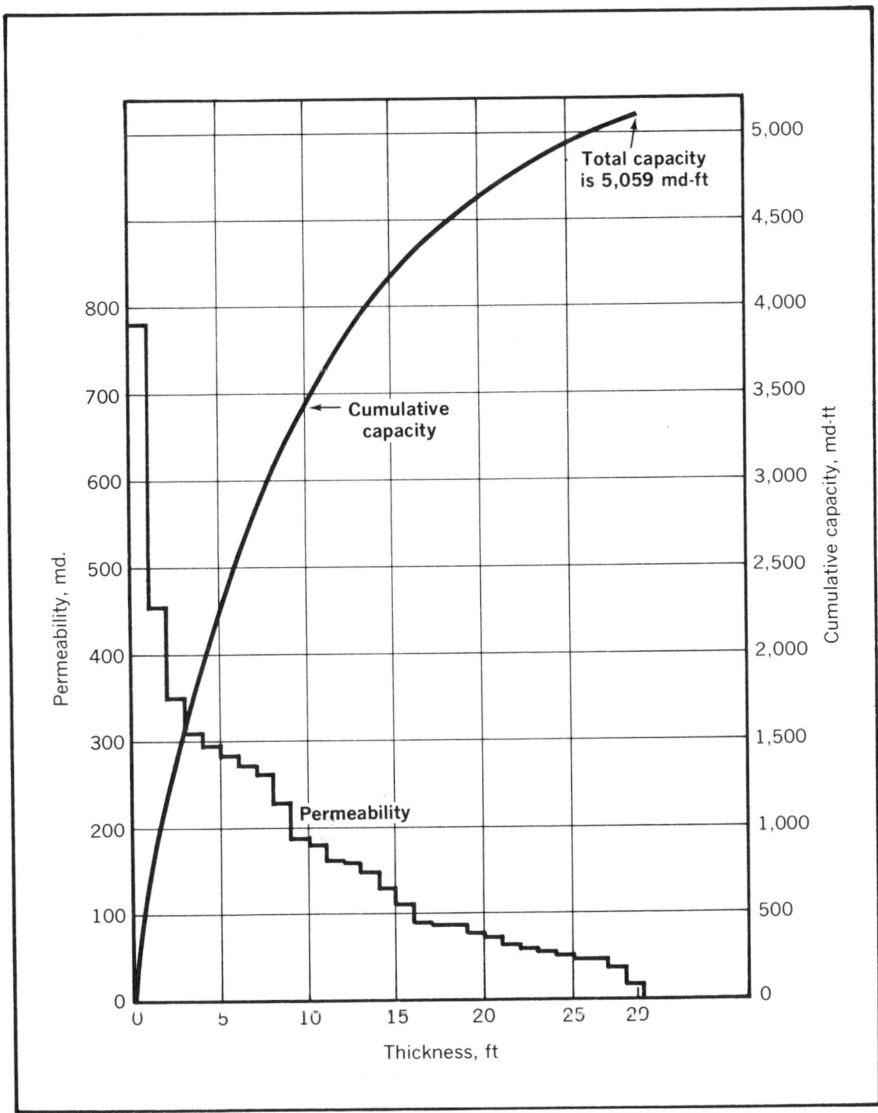

FIG. 8.11 Permeability and capacity distribution

summing the products of the individual permeabilities and the thickness they represent from a zero thickness to the subject thickness. Mathematically, the capacity is the integral of the permeability distribution curve from a thickness of zero to the thickness of interest. Thus, the capacity curve represents the permeability-thickness of the most permeable part of the reservoir being considered. For example, the most permeable 15 ft of the reservoir thickness represented in Fig. 8.11 has a capacity of 4,187 md-ft compared with the total reservoir capacity of 5059 md-ft.

Equation 8.5 is based on the assumption that at any particular time the thickness producing oil is the least permeable portion of the reservoir and the thickness producing the water is the most permeable portion of the reservoir. Thus, if the mobility of water and oil were the same, the ratio of oil-to-water production would be the ratio of the capacity of the reservoir thickness producing oil to the capacity of that producing water. However, since water normally flows more readily than oil, the difference in mobilities of the two fluids must be taken into account by weighting the capacities with the mobility ratio.

An example may help in understanding Equation 8.5. Assume that 15 ft of the formation represented by Fig. 8.11 has been flooded out so that 15 ft of the formation produces water with the remaining 14 ft producing oil. (There was no free gas saturation in the reservoir at the start of the flood.) If mobilities of oil and water are the same, the rate of production of each would be proportional to the capacity of the thicknesses from which each is producing and the water cut would be $4,187/[4,187 + (5,059 - 4,187)]$. However, suppose the fluid in the water-producing thickness flows two times as easily as the fluid in the oil-producing thickness. Then the water cut would be $(2)(4,187)/[(2)(4,187) + (5,059 - 4,187)]$. These figures represent the water cut at reservoir conditions. To correct to stock-tank conditions the oil-production rate must be corrected to stock-tank conditions by dividing by the oil formation volume factor, B_o. In Equation 8.5 this is done by including the oil formation volume factor in the parameter, A.

At the instant the production in the 15th most permeable foot of the formation goes from oil to water, notice that all of the oil will have been produced from the 15 most permeable feet, (h_j), which will be 15/29ths, (h_j/h_t). Thus, we can conclude that the fraction of the total mobile oil that has been recovered from the h_j thickness after it begins producing water is

$$(R)_{hj} = \left(\frac{h_j}{h_t}\right) \tag{8.8}$$

Also note that some oil has been produced from all of the thickness with lesser permeabilities, those still producing oil, $h_t - h_j$. For example, if we assume that fluid velocity in the individual zones is proportional to the permeability, then at the instant the 15th ft with a permeability of

127 md, (k_j), is flooded out, (the water front just reaches the producer in this zone) the 16th ft with a permeability of 109 md will have had (109/127), of the oil in that foot displaced. This fraction (k_i/k_j) will be filled with water. Thus, we could say that the section of the reservoir that is less permeable than the zone that has just been flooded out, the thickness producing oil, $(h_t - h_j)$, will have a fractional oil recovery of the total mobile oil of

$$(R)_{h_t-h_j} = \sum_{i=j}^{i=t} [(k_i/k_j)h_i/h_t]$$ 8.9)

Note that k_j is a constant at any particular time and thus k_j and h_t can be removed from the summation. Then the sums of the $k_i h_i$ products will represent the capacity of the $(h_t - h_j)$ thickness. This capacity is the difference between the capacity of the total thickness and the capacity of the h_j thickness or $(c_t - c_j)$. Equation 8.9 then becomes

$$(R)_{h_t-h_j} = \frac{(c_t - c_j)}{k_j h_t}$$ (8.10)

If we add this expression to the fraction of total mobile oil recovered from the completely flooded-out portion of the reservoir at this time, as shown in Equation 8.8, we obtain an expression for the total fraction of recovery, R, when a thickness h_j has been flooded out.

$$R = \left(\frac{h_j}{h_t}\right) + \frac{(c_t - c_j)}{k_j h_t}$$ (8.11)

When the first term of this equation is multiplied by k_j/k_j it is possible to rewrite the equation as Equation 8.6.

Stiles assumed that Equations 8.5 and 8.6 could be used to obtain a curve of the fractional recovery, R, versus the water cut, f_w, and that this curve could then be interpreted in terms of oil and water producing rates versus time. The Stiles method has no provision for an initial gas saturation except to assume that the first water injected displaces all of the free gas from all of the permeability zones before any oil is produced from any zone. This is of course inconsistent with the basic permeability distribution concept. Also, the Stiles method does not permit the variation of pattern or horizontal sweep with cumulative injection into a zone. Neither of these inaccuracies should introduce much prediction inaccuracy late in the flood life. All of the free gas will have been displaced from practically all zones that will ultimately affect performance, and practically all individual permeability zones that do not have a 100% pattern sweep late in the flood life will be so tight they will have little influence on the ultimate recovery.

The engineer should note that the method also omits consideration of many other phenomena that influence the behavior of a flood such as the variation in saturation along stream-line (the Stiles method assumes piston-like displacement), the variation in average mobility along a streamline with time, the variation in saturations, residuals, and porosities from zone to zone, etc. However, for most floods these errors are relatively unimportant.

In using the Stiles method to determine the flood efficiency at flood abandonment it is not necessary to generate the entire recovery-versus-water-cut curve or even all of the permeability and capacity-distribution curves. With the water-cut fraction at abandonment assumed, the corresponding c_j can be calculated from Equation 8.5. The total capacity, c_t, could be evaluated from the average permeability and total thickness if it is not more accurately known. The thickness corresponding to c_j can then be evaluated by summing the capacities from the least permeable bed in an increasing permeability direction until a capacity of $c_t - c_j$ is reached. The thickness corresponding to this capacity will be $h_t - h_j$ from which the thickness, h_j, can be evaluated. Once h_j and the corresponding k_j are known, the corresponding fractional recovery, R, can be determined from Equation 8.6.

In using Equations 8.5 and 8.7 care should be taken in evaluating mobility for the zone producing oil. Mobility for the zone producing water can be taken as the mobility of water in the water bank without introducing any significant error. However, note that the zone producing oil is flowing oil at the producing well but is flowing water at the injection well. This should be taken into account in evaluating fluid mobility in the oil-producing zone. For example, for a five-spot pattern, mobility of the zone producing oil is effected by the mobility of oil in the oil bank which exists at the producing well, and the mobility of water in the water bank that exists at the injection well.

We will see later that the effective mobility in a zone can generally be closely estimated by considering only the resistance to flow at the injection and producing wells. The resistance to flow is the reciprocal of the mobility, λ, and since resistances in series are additive, (like pressure drops in series), we can write an expression for the resistance to flow in the oil producing zone.

$$\left(\frac{1}{\lambda_{oz}}\right) = \left(\frac{1}{\lambda_w}\right) + \left(\frac{1}{\lambda_o}\right).$$

Then the mobility in the oil zone is the reciprocal of this or,

$$\lambda_{oz} = \frac{1}{[(1/\lambda_w) + (1/\lambda_o)]}.$$

If we develop a similar expression for the resistance to flow and the corresponding mobility in the water producing zone we obtain,

$$\left(\frac{1}{\lambda_{wz}}\right) = \left(\frac{1}{\lambda_w}\right) + \left(\frac{1}{\lambda_w}\right)$$

$$\lambda_{wz} = \frac{\lambda_w}{2}$$

Then the expression for the mobility ratio is obtained by dividing the expression for λ_{wz} by the expression for λ_{oz} to obtain,

$$\frac{\lambda_{wz}}{\lambda_{oz}} = \frac{[1 + (\lambda_w/\lambda_o)]}{2}.$$

Since Stiles A is $(\lambda_{wz}/\lambda_{oz})B_o$, we can write an expression similar to Equation 8.7 for the effective A as,

$$A_{eff} = \frac{B_o[1 + ((k_r/\mu)_w/(k_r/\mu)_o)]}{2} \qquad (8.7A)$$

Evaluating in this way amounts to a minor modification of the Stiles method. The example analysis provided by Stiles simply based "A" on mobility of water in the water bank and mobility of oil in the oil bank.

Work the following problem and check your solution against the solution in Appendix C to test your knowledge of this section and provide a background for the discussions that follow.

Problem No. 8.2B: Calculation of the Total Displacement Efficiency and Recovery From a Waterflood

In Problem 8.2A the saturations at the start of the flood and the gross swept volume were evaluated for a proposed flood. Now complete this problem by calculating the ultimate flood recovery and stating it as (1) stock-tank barrels, (2) percent of oil originally in the reservoir and (3) percent of oil originally in the swept area. Assume the economic water cut as 95%. $S_{wc} = 0.2$.

In performing this calculation it will be necessary to first calculate the total displacement efficiency based on the Stiles method. Assume that the oil formation volume factor does not change significantly during the flood, that the average conventional core-analysis oil saturation is 19%, and that the permeability distribution of the reservoir (listed in descending order of magnitude) is as follows:

Permeability, md	Thickness, ft
780	1
455	1
348	1
309	1
295	1
282	1
271	1
260	1
227	1
185	1
179	1
160	1
158	1
147	1
127	1

Permeability, md	Thickness, ft
109	1
87	1
85	2
76	1
70	1
61	1
57	1
54	1
50	1
46	2
35	1
15	1
Average 174	Total 29

Use Fig. 8.11A for relative-permeability data. Note that the residual oil and connate water indicated on the relative-permeability curve does not agree with the data of this problem. The oil formation volume factor is 1.1, oil viscosity 6.38 cp and water viscosity 0.7 cp. In 8.2A we determined that the oil saturation after primary is 0.561, the original oil in place before primary was 155.2×10^4 STB and the gross swept volume under flood is 900 acre-ft.

This problem emphasizes that the fractional recovery from a reservoir is dependent upon the basis for the fraction. The problem illustrates the difference due to using the swept reservoir volume or the total reservoir volume. Obviously, there will also be a difference depending upon whether the basis is mobile oil or total oil. If the swept volume is used the method of delineating the swept volume will be very important. The point is that when you are comparing recovery efficiencies of floods, be careful that all of the bases are the same.

Permeability-distribution data used in the above example implies that some previous study of the reservoir has led to the conclusion that a particular strata in the reservoir is the most permeable strata and has a particular permeability. For example the 1-ft interval between 5 and 6 ft from the top of the reservoir might have the peak permeability in each well and an average permeability of 780 md. Similarly, the strata 3 ft from the bottom of the pay zone may be the least permeable and have an average permeability of 15 md. A strict interpretation of this type would then mean that reservoir thickness in each well would have to be 29 ft. However, we learned in the problem that this permeability distribution could be used as a general distribution that applies to all of the reservoir regardless of the thickness variation throughout the reservoir.

Generally reservoirs are not sufficiently uniform from well to well to permit assumption that the 5th ft from the top is the most permeable or the 3rd ft from the bottom is the least permeable. Whenever a reservoir can be so characterized, permeability distribution should be based on such an analysis. However, in most cases we find that it is impossible to

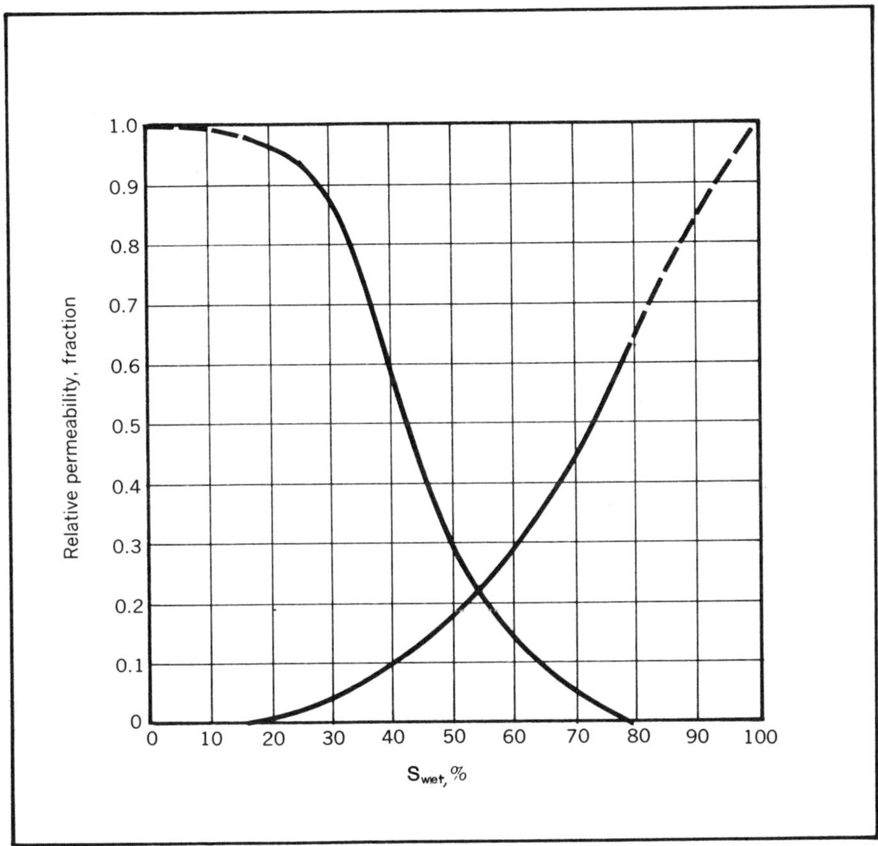

FIG. 8.11A　Example relative permeability data

so characterize a reservoir. Nevertheless, flooded reservoirs behave like stratified reservoirs. Thus, we must approach the stratification problem from a statistical rather than a geological standpoint to obtain realistic predictions.

Permeability distribution can be put on a statistical basis by dividing thickness values by total thickness and permeability values by average permeability. Then the thickness would be the fraction of total thickness and the permeability would be a multiple of average permeability. Since the average permeability on this scale would be 1.0, and the total thickness would be 1.0, the total capacity would be 1.0 × 1.0 or 1.0, and the individual values of capacity would be a fraction of total capacity. This dimensionless capacity curve can be obtained mathematically by dividing individual capacity values by total capacity and, as before, dividing individual thickness values by total thickness.

In most reservoirs it is found that the dimensionless form of permeability and capacity distribution gives a prediction that best fits actual flood behavior. The thickness scale is used as an actual percent of samples or percent of core-analysis measurements. The dimensionless permeability distribution curve then represents the percent of samples with a permeability greater than a particular dimensionless permeability multiplied by the average permeability. In a reservoir where

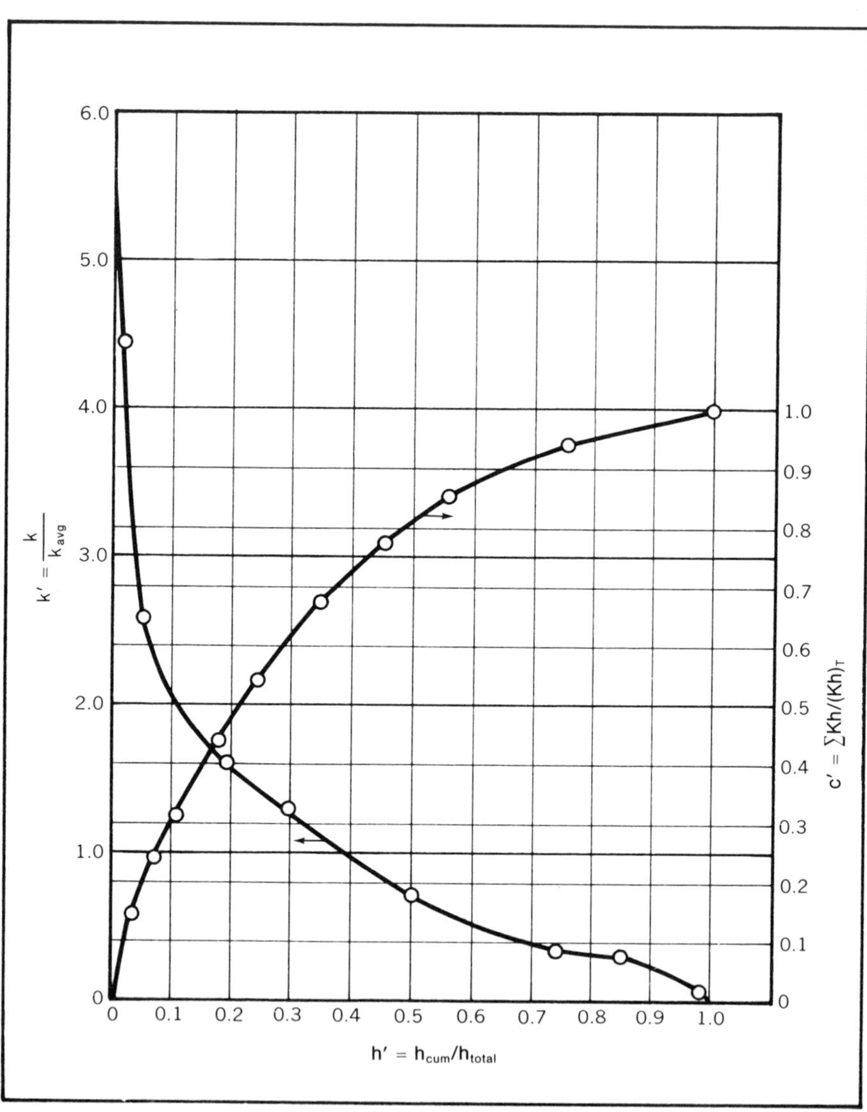

FIG. 8.12 Dimensionless permeability and capacity distribution

several thousand core-analysis measurements have been made, the data for the permeability-distribution curve is obtained by counting the number of samples with a permeability greater than a particular value and converting this observation to one point on the curve. For example, suppose a total of 5,000 core-analysis measurements have been made in a reservoir which has an average permeability of 100 md. Then if we find that 500 of the samples have a permeability greater than 300 md, one point on the permeability-distribution curve would be 3.0, (300/100), versus 0.1, (500/5,000).

When a dimensionless type of permeability distribution is used the two Stiles equations become

$$f_w = \frac{(A\ c_j')}{[A\ c_j' + (1 - c_j')]} \qquad (8.12)$$

$$R = \frac{[h_j' k_j' + (1 - c_j')]}{k_j'} \qquad (8.13)$$

In these equations the prime(') denotes the dimensionless thickness, permeability, and capacity. These dimensionless forms of the Stiles equations are derived from Equations 8.5 and 8.6 by dividing the numerator and denominator of both equations by $h_t k_{avg}$ where k_{avg} is the average permeability. When Fig. 8.11 is converted to a dimensionless permeability and capacity distribution we obtain Fig. 8.12.

PREDICTING RATE VERSUS TIME PERFORMANCE

Most practicing engineers can do a creditable job of predicting total flood recovery. However, the prediction of the rate at which oil will be recovered is a much more difficult task. From a standpoint of profitability it is almost as important as predicting total ultimate recovery. Due to the time value of money the rate at which production is obtained can be the difference between a profitable flood and an unprofitable flood, both of which reach the same ultimate recovery. Much has been written and is available in the literature on the prediction of waterflood behavior. No effort will be made here to review all of the various contributions of the engineers and researchers. Generally each method seems to concentrate on one particular difficulty in predicting the behavior of a flood. Stiles,[3] as noted previously, emphasized the need to account for the permeability distribution; Suder-Calhoun[4] accounted for the oil bank formation; Prats et al.[5] worked with the variation of the injectivity into a particular strata with time; Higgins-Leighton[6] accounted for the variation of saturation along a particular streamline; Dyes-Caudle-Erickson[7] were chiefly concerned with sweep efficiency; Dykstra-Parsons[8] predicted residual saturations empirically etc. These authors and researchers would certainly, (and rightly) point out that they also took into account many other factors. The point is that these were the items that dominated these methods.

The recommended method.   The basic bookkeeping method probably comes close to the method presented by F. F. Craig in the SPE Monograph on Waterflooding.[9] Others may believe it is more similar to the method used by Michael Prats[5] and his associates. At any rate the idea is not original. It may be of interest that for many years in teaching this recommended method the author referred to it as the Modified Prats method. The bookkeeping method assumes a knowledge of injectivity of each permeability zone as a function of cumulative injection into that zone and a knowledge of recovery of each zone as a function of cumulative injection into that zone. Many engineers will feel that once they have the data necessary to perform the bookkeeping calculations, they have practically solved the problem. There is something to be said for this conclusion. The problem of generating the basic data for the bookkeeping system will be discussed later in the chapter.

The bookkeeping method is necessary because of the variation in the injectivity, resistance to flow, or conductance of a particular permeability strata during its flood life. The strata may start its flood life with essentially only gas flowing which would mean that the fluid flowing would have high mobility. Later on all of the gas may have been displaced from the strata and much less mobile oil and water will be flowing in the strata. Still later all of the mobile oil may have been displaced from the strata and the less mobile oil will have been displaced by the normally more mobile water. We learned previously that various strata do not take water at the same rate due to permeability differences. Consequently, no two strata of different permeabilities will be at the same state of depletion at the same time. What this means then is that the relation between the injectivity or resistance to flow in the various strata is continually changing as the flood developes.

For example, a particular strata in a flood may originally take 10% of the total injected water but later on it may be taking only 4% of the water, but still later the take may increase again to 8 or 9%. We cannot determine recovery from each zone unless we know how much water is in each zone at a particular time.

We will see later that the injectivity rate into a particular strata may be approximated by a history such as that shown in Fig. 8.13, if minor variations are omitted and the reservoir contains a substantial gas saturation at the start of the flood. These data will be more useful to us in a dimensionless form where the injection rate is stated as the ratio of initial rate divided by the existing rate into this strata and cumulative injection into the strata is stated as a function of strata pore volume. Replotting the data of Fig. 8.13 gives us a Fig. such as 8.14.

As noted above, before our bookkeeping procedure can be initiated we must also have a recovery curve for each of the permeability strata.

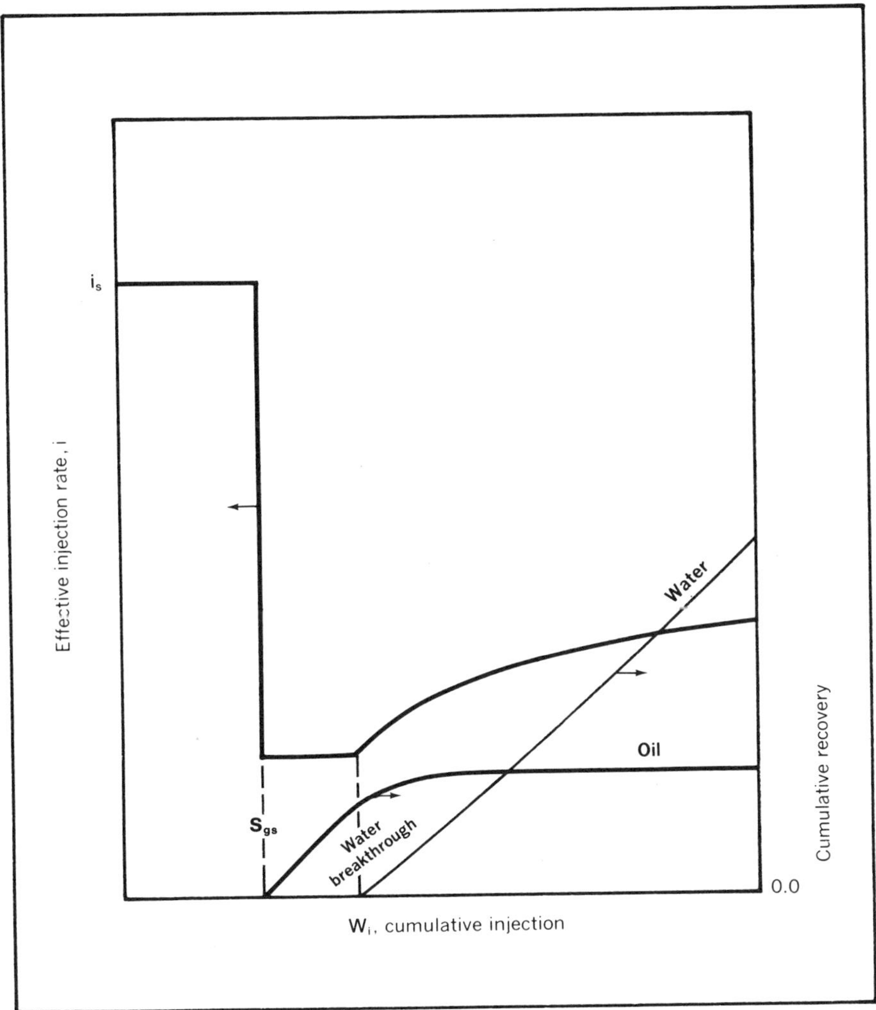

FIG. 8.13 Approximate variation of the injection rate in a strata of a five-spot ($S_{gs} > S_{ge}$)

The recovery data should be in the form presented in Fig. 8.15 with cumulative oil production from the strata stated as a function of pore volume and cumulative injection again stated as a function of pore volume. The oil formation volume factor is included in the recovery term so that it will represent a fractional recovery in the reservoir with recovery in barrels at stock-tank conditions being, N_{pf}.

It will be shown that by using the injectivity data (such as Fig. 8.14) the permeability distribution, and other reservoir parameters, it is possible to construct a plot of (W_i/V_p) versus a function of time, t_r, for each of

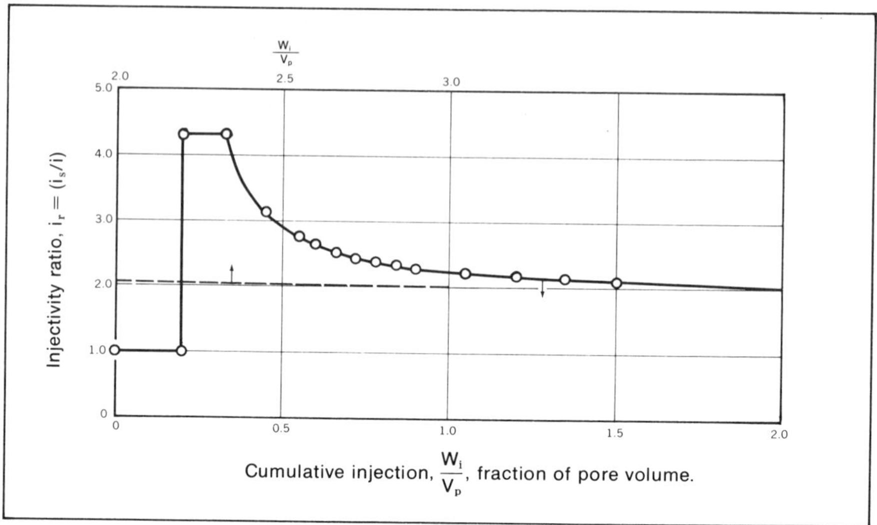

FIG. 8.14 Example injectivity ratio data

the permeability zones of the reservoir. Such a set of data is illustrated in Fig. 8.16 for a reservoir where only three zones are considered. The reduced time, t_r, is the injection time divided by a group of reservoir constants that does not vary from zone to zone. Consequently, the relationship between the cumulative injection into each zone at a particu-

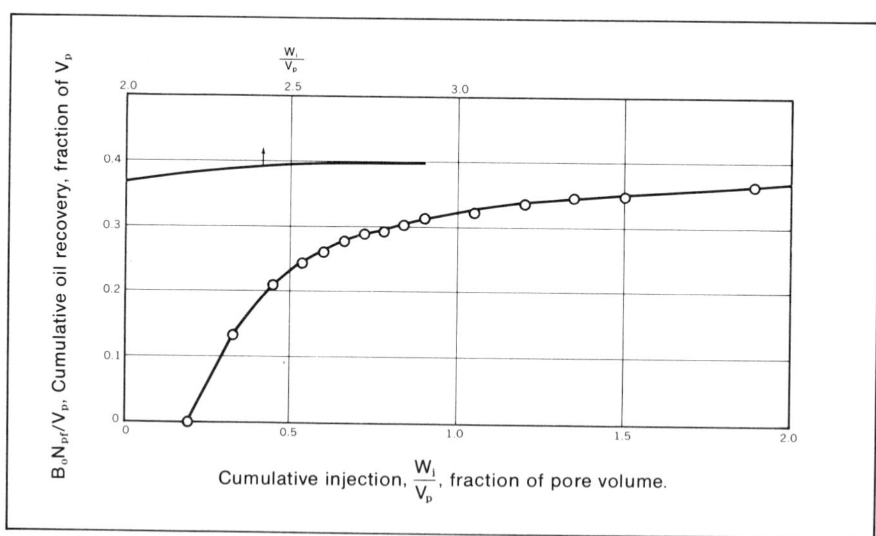

FIG. 8.15 Example recovery data

lar time can be determined from a plot such as Fig. 8.16 for when zone 1 has a cumulative injection of 1.0 bbl water/bbl PV zone 2 has a cumulative injection of 0.45 and zone 3 has a cumulative injection of 0.2.

Now since our basic data include recovery versus cumulative injection for each zone we can determine oil production for each zone at this particular time represented by this particular t_r value. Since we know the pore volume for each zone we can convert recovery and cumulative injection data to barrels. The cumulative injection and injection rate can then be used to calculate injection time in days.

To determine the basis for the plot of cumulative injection versus t_r for a particular zone, note that the increment of time, Δt, necessary for the injection of an increment of water, ΔW_{in}, into a particular strata at an average rate, i_n, is $\Delta W_{in}/i_n$ and the total time required for the injection of W_{in} cumulative barrels of water into a particular strata will be the summation of the time increments.

$$t = \sum \left(\frac{\Delta W_i}{i}\right)_n \tag{8.14}$$

Since our given data are not in barrels, W_{in}, and barrels per day, i_n but are rather stated as W_i/V_p and i_s/i, we must substitute functions of these values for W_i and i in Equation 8.14.

$$i_n = \frac{i_{sn}}{(i_s/i)_n} \tag{8.15}$$

and

$$(W_i)_n = \left(\frac{W_i}{V_p}\right)_n \left(\frac{\phi Ah}{5.615}\right)_n \tag{8.16}$$

The second set of parentheses in Equation 8.16 is the pore volume stated as a function of the horizontal area, A, affected by each injection well and the thickness, h, both stated in feet.

When Equations 8.15 and 8.16 are substituted into equation 8.14 we obtain

$$t = \sum \frac{\Delta(W_i/V_p)_n (\phi Ah/5.615)_n (i_s/i)_n}{i_{sn}} \tag{8.17}$$

This expression can be further simplified when we substitute a function of the capacity, kh, for the initial injection rate into the nth strata, i_{sn}. Any flow rate can be equated to the transmissibility, kh/μ, multiplied by the pressure drop, a geometric constant, and a numerical constant. If we group all of the parameters together that are the same for all of the strata we can show that the injection rate into each strata initially is proportional to the capacity of the strata, $(kh)_n$.

$$i_{sn} = (\text{numerical constant})(\text{geometric constant})\Delta p(kh/\mu)_n \tag{8.18}$$

and

$$i_{sn} = (\text{constant})(kh)_n \tag{8.19}$$

Note that the effective viscosity of the strata, μ_n, is included in the group of constants. This is done because the effective viscosity in the reservoir initially will be based on essentially the same saturation throughout the reservoir and thus μ_{sn} will be the same for all strata. At any other time in the flood history the saturations will differ from strata to strata; but at the start of the flood they will differ only to the extent that the primary drainage differs from strata to strata. This will have a negligible effect on the difference in the initial injection rates from strata to strata.

When Equation 8.19 is substituted for the initial injection rate of a strata, i_{sn}, in Equation 8.17 we note that the thickness, h_n, cancels and the equation can be arranged to obtain

$$t = \left[\frac{A}{(\text{constant})(5.615)}\right] (\phi/k)_n \sum \Delta \left[\frac{(W_i/V_p)_n}{(i_s/i)_n}\right] \tag{8.20}$$

Remember that the constant includes only reservoir parameters that are the same for all strata. Note that the drainage area, A, and the numerical constant 5.615 are also the same for all strata. Consequently, all of the "constants" can be grouped with the real time, t, which is also the same for all strata, to obtain a correlating reduced time, t_r, which is also the same for all strata.

$$t_r = \frac{t}{[A/5.615(\text{constant})]} = (\phi/k)_n \sum (i_s/i)_n \Delta(W_i/V_p)_n \tag{8.21}$$

This expression can be used together with the injectivity data, such as that shown in Fig. 8.14, to construct curves as in Fig. 8.16. Equation 8.21 is used to calculate the t_r for any strata at any time. The equation is solved graphically. Note that the summation term is graphically represented by the area under the curve of Fig. 8.14 from $(W_i/V_p) = 0$ to (W_i/V_p) equal to the value of interest.

The procedure to follow in constructing t_r curves such as those of Fig. 8.16 is to first determine the ratio of the porosity and absolute permeability for a particular zone. Any consistent units can be used since we are only interested in the relationship of the (W_i/V_p) values for different strata at a particular t_r. Then to calculate the t_r values assume some value for (W_i/V_p) and determine the area under the injectivity ratio curve from zero to the assumed value of W_i/V_p. When this area is multiplied by the ϕ/k ratio for that strata the reduced time, t_r, corresponding to that particular assumed (W_i/V_p) is determined and one point on the reduced time curve for that strata has been defined. The procedure is repeated until all the curves have been evaluated for all of the strata for the range of water injection desired. Now work the following problem and check your solution against the solution in Appendix C.

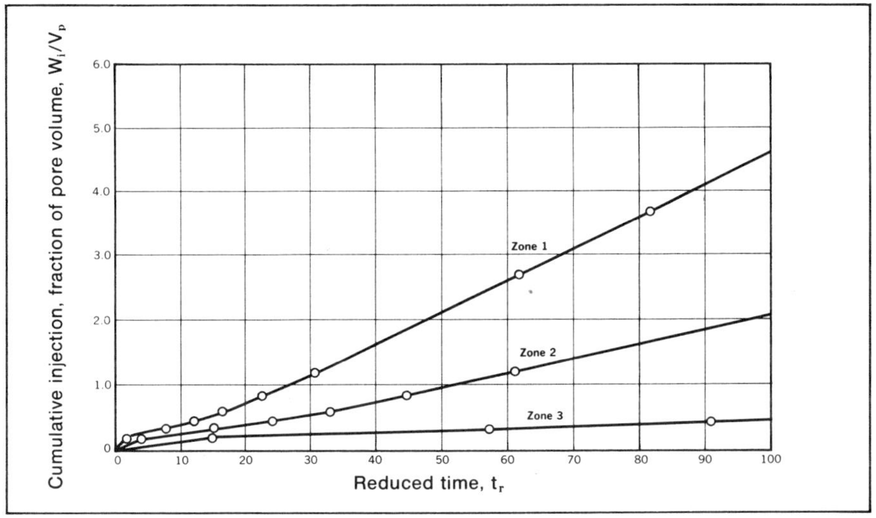

FIG. 8.16 Reduced time, t_r, curves

Problem 8.3A: Calculation of Reduced Time Curves for Use in a Waterflood Prediction

To demonstrate the recommended method assume a 10-acre five-spot in a reservoir can be treated as having three zones with properties as shown.

Zone no.	h, ft	k, md	ϕ %	V_p bbl
1	5	25	25	97,000.
2	10	10	20	155,000.
3	15	2	15	175,000.

For all zones: $S_{gi} = 20\%$, $S_{os} = 65\%$, $S_{wc} = 15\%$, $S_{or} = 25\%$, $\mu_w = 0.7$ cp, $\mu_o = 4.15$ cp.

Assume i_r and recovery curves as shown in Figs. 8.14 and 8.15 respectively, and relative permeabilities as in Fig. 8.11A. Note that the connate water and residual oil values differ from those indicated by the relative-permeability data.

In later parts of Problem 8.3 we will determine strata injectivity curves, strata recovery curves, and the five-spot recovery in barrels of oil and water versus time. At this time the engineer is asked to verify the reduced time curves of Fig. 8.16.

Converting reduced-time data to production versus injection. Once reduced-time curves have been generated they can be interpreted in terms of cumulative total injection and production of both oil and water. The reduced-time curves show the relationship between cumulative injection into each zone at any particular reduced time. Consequently, even though we do not know directly the calendar time represented by any particular time, we can conclude that at that unknown real time the cumulative total injection will be some particular value. This becomes meaningful when we recognize that a particular cumulative injection into a particular zone represents a particular cumulative production from that zone. Thus, for some unknown time we can determine the cumulative total injection and corresponding total oil and water recovery. When such numbers are calculated for enough reduced times we have data representing cumulative oil and water production versus cumulative injection. By assuming some value for the injection rate these data can in turn be used to calculate the cumulative oil and water production versus time.

To determine the cumulative injection representing a particular reduced time we plot reduced-time data as in Fig. 8.16 and read the pore volumes of cumulative injection into each zone for that reduced time. Each of the pore volumes of cumulative injection can then be converted to barrels by multiplying by the pore volume of that zone. The total injection will then be the sum of injections into each zone.

$$(W_i)_{Total} = \sum_{n=1}^{n=y} (W_i/V_p)_n \, V_{p_n} \tag{8.22}$$

The cumulative oil recovery corresponding to this cumulative injection can be determined by finding the pore volumes of oil that correspond to the pore volumes of cumulative injection for each zone. This is accomplished by using the basic recovery data such as that of Fig. 8.15. Once the pore volumes of reservoir oil recovered from each zone has been determined, these figures can be converted to stock-tank barrels by multiplying by the appropriate pore volume and converting to stock-tank barrels by dividing by the oil formation volume factor.

$$(N_{pf})_{Total} = (1/B_o) \sum_{n=1}^{n=y} (B_o N_{pf}/V_p)_n \, V_{pn} \tag{8.23}$$

In this equation the oil formation volume factor, B_o, is taken outside the summation sign because it is the same for all zones in most cases. In some cases where reservoirs are very thick, and the pressure and oil formation volume factor vary from zone to zone, B_o for each zone should be left inside the summation sign as B_{on}.

If we assume that the free-gas saturation in an oil-producing zone is negligible, we can determine the amount of water production from each zone by material balance based on our knowledge of oil recovery, cumulative injection, and the saturations at the start of the flood. All of the water injected into a particular zone will either be in that zone or will have been produced. Furthermore, all of the injected water that remains in the zone has either displaced free gas or oil. We have shown that we can evaluate the cumulative oil production at any particular time. We also know how much free gas was originally in the zone and if we assume that the gas saturation in an oil-producing zone is zero, we then know that all of the original free gas has been produced.

When we put these observations together we find that the pore volumes of injection, less the pore volumes of gas produced, less the pore volumes of oil production, gives the pore volumes of water produced. When this is multiplied by the appropriate zonal pore volume we have the barrels of water that have been produced from a particular zone. The summation of water produced from each zone will result in the total water recovery from the flood.

$$(W_p)_{total} = \sum_{n=1}^{n=y} [(W_i/V_p)_n - S_{gin} - (B_oN_{pf}/V_p)_n] \, V_{pn} \qquad (8.24)$$

In applying this equation the engineer should exercise care that no negative water-production figures are included for any zone. The nature of the terms is such that if the oil production is zero the gas saturation in the zone would not be zero so that the volume of free gas produced would not be the initial gas saturation. This would mean that the injected water volume would be greater than the initial gas saturation and a negative number would be obtained for the volume of water produced. Thus, Equation 8.24 should only be applied to those zones producing oil.

Now work the following problem and check your solution against the solution in Appendix C to check your understanding of this section.

Problem No. 8.3B: Calculation of Oil and Water Production Rates from Reduced Time Curves

In problem 8.3A the reduced time curves were calculated for a flood. Now using these reduced time curves (Fig. 8.16) and the recovery curves of Fig. 8.15, calculate and plot the oil production and water production versus cumulative injection to a cumulative injection value of about 320,000 barrels or a reduced time of 40. Assume $B_o = 1.0$

Recovery curves for individual strata. There is much data in the literature that is usable for determining recovery curves for individual strata. We previously referred to one such set of data for a five-spot flood, Figs. 8.10A and 8.10B. Dyes, Caudle, and Erickson[7] published similar data for other patterns as did Kimbler, Caudle, and Cooper.[9] All of these data can be found in Appendix D of the SPE Monograph on waterflooding.[10] Since the problem lends itself to an interesting and not too difficult analysis either by use of flow models or computer models, there is much other data in the literature that could be used for the same purpose.

Remember that what is required are data that will provide cumulative recovery versus time for a single layer of a homogeneous flood pattern element. Most of the data will assume piston-like displacement or a minimal drag zone (zone containing mobile oil behind the flood front) characteristic of a very permeable sand. However, the error introduced by using the piston-like concept seems minimal compared to the effect of sweep efficiency. The Higgins and Leighton method[6] could be used to obtain the strata recovery curve and include the variation in saturation along each streamline if sufficient reservoir data and computer time are available.

Fig. 8.10B will be used to illustrate the method of determining a strata recovery curve from published data. Note that the recovery must be obtained from the percent of the five-spot area swept and the corresponding cumulative injection.

The percent of the five-spot area occupied by water must be equal to the percent of mobile hydrocarbons (oil and gas) displaced from the reservoir by the injected water. Thus, the percent of the five-spot area occupied by injected water will be equal to the oil and gas production caused by water injection. Now note from Fig. 8.10B that if the reciprocal of the mobility ratio is greater than about 7.5, 100% of the reservoir will be swept before breakthrough. This means that the mobility ratio needs to be less than (1/7.5) or 0.133. Since the viscosity ratio of oil to gas approaches 0.01, it appears safe to assume that the mobility ratio for oil displacing free gas will generally be less than the critical 0.133. Consequently, we can assume that mobile free gas in a strata is produced before any displaced oil is produced. Thus, for a particular strata we can assume that before gas fillup, the injected water volume is equal to the reservoir volume of free gas produced. After gas fillup the oil produced will be equal to the injected water in the reservoir less the reservoir volume of free gas in the strata at the start of the flood. When the swept area is 100%, all of the mobile oil and gas will have been displaced from the reservoir. Thus, when the swept area is less than 100% we can state the oil and gas production as

$$\text{Oil and gas produced} = E_H(S_{os} - S_{or} + S_{gi})V_p \qquad (8.25)$$

Now if the initial gas volume is subtracted from the oil and gas production and the resulting expression solved for the reservoir barrels of oil production, stated as a fraction of pore volume, we obtain,

$$B_o N_{pf}/V_p = E_H(S_{os} - S_{or} + S_{gi}) - S_{gi} \qquad (8.26)$$

This expression assumes that all of the gas is mobile but it is easily modified to include immobile gas.

To obtain the corresponding cumulative water injected from the displacement volumes injected it is only necessary to recognize that one displacement volume is equal to the mobile oil saturation at the start of the flood plus the mobile gas saturation at the start multiplied by the pore volume. Then the cumulative water injection as a fraction of the pore volume is

$$\left(\frac{W_i}{V_p}\right) = (DVI)(S_{os} - S_{or} + S_{gi}) \qquad (8.27)$$

Using Fig. 8.10B it is then possible to read corresponding values of the percent of the five-spot swept and displacement volumes injected for a particular reciprocal of the mobility ratio. These corresponding values can then be converted to reservoir barrels of oil production, as a fraction of pore volume, and water injection, as a fraction of the pore volume, by using Equations 8.26 and 8.27.

Work the following problem to make certain you understand this conversion procedure.

Problem No. 8.3C: Calculation of Individual Strata Recovery Curves

In Problem 8.3A and 8.3B calculations were made predicting oil recovery and water production versus cumulative effective injection based on the individual strata injectivity and recovery curves of Fig. 8.14 and 8.15, respectively. Verify Fig. 8.15 by calculating the points representing gas fillup, water breakthrough, and displacement volumes injected of 0.75 and 1.5. Fig. 8.11A represents the relative permeability for this reservoir and the oil and water viscosities are 4.15 and 0.7 cp, respectively. The initial gas saturation, connate water, and residual oil are 20%, 15%, and 25% respectively.

Injectivity curves for individual strata. As pointed out, there is much data in the literature that can be used to generate the desired recovery data for a particular strata. On the other hand there is little or no data available for determining the injectivity curves for an individual strata. Fortunately, we can approximate an injectivity curve of acceptable accuracy in most cases by observing a few fundamental concepts.

To approximate the injectivity curve we must first recognize that practically all of the pressure drop in a pattern occurs at the wells, whether they are injection or producing wells. A study of the pressure distribution in a five-spot pattern quadrant will emphasize this point. Fig. 1.12 in Chapter 1 verifies that about 80% of the total pressure drop between the injection and producing wells occurs in about 4% of the pattern around the injection well and 4% of the pattern surrounding the producing well. Thus, accounting for the pressure drop at the injection and producing wells will provide a reasonably accurate estimate of the individual strata injectivity variations in the flood history.

Looking at the same phenomenon physically, the cross-sectional area between the injection and producing wells is very large compared with the very small cross-section at the wells, so the wells provide practically all of the resistance to flow. This is true not only in a five-spot but in virtually any waterflood pattern. Thus, we will model the flood pattern by assuming it represents two radial systems "back to back" as shown in Fig. 8.17. In illustrating the use of simple geometry to approximate the behavior of a more complex geometry, we showed in Chapter 1 that the model of Fig. 8.17 gives a flow equation almost the same as the exact analytical equation. We will further expand this idea by assuming that saturations on the injection side are the same as satura-

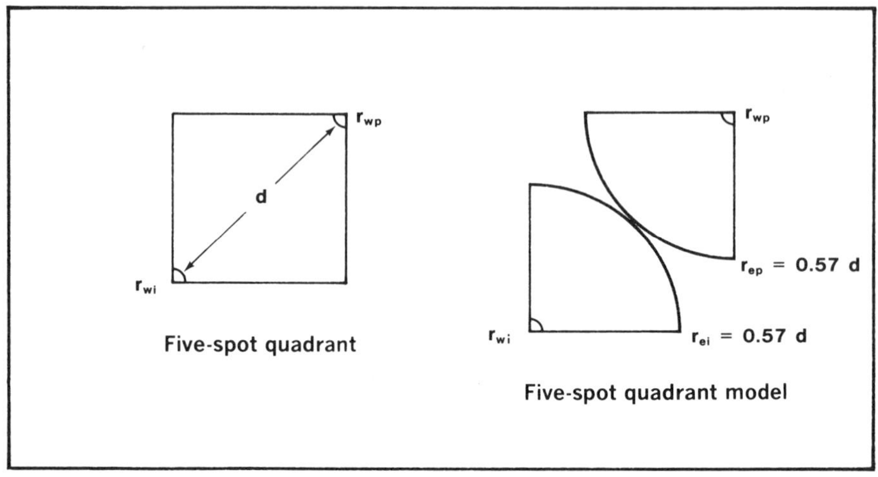

FIG. 8.17 Modeling a five-spot

tions at the injection well, and that saturations on the producing side are the same as saturations at the producing well.

Using the model of Fig. 8.17 we can say that the total pressure drop between the injection and producing wells is the sum of the pressure drops in the injection and producing sides of the pattern.

$$p_{wi} - p_{wp} = \Delta x_i + \Delta x_p \tag{8.28}$$

We can determine an expression for the two pressure drops from the steady-state radial-flow equation as a function of the injection rate, i.

$$p_{wi} - p_{wp} = 0.141i \ln(r_{ei}/r_{wi})(\mu_i/k_i)/h + 0.141i \ln(r_{ep}/r_{wp}) \\ (\mu_p/k_p)/h \tag{8.29}$$

When we solve this equation for injection rate, and assume the radius ratio on the injection and production sides are equal, we obtain,

$$i = \frac{(p_{wi} - p_{wp})/0.141 \ln(r_e/r_w)}{(\mu_i/k_i) + (\mu_p/k_p)} \tag{8.30}$$

Now note that the numerator of this equation will be the same for all zones at any particular time. Thus, the ratio of injection rates into two zones at a particular time will be the ratio of the denominators of Equation 8.30. By writing the numerator of Equation 8.30 as a constant we obtain,

$$i = \frac{\text{constant}}{[(\mu_i/k_i) + (\mu_p/k_p)]} \tag{8.31}$$

To obtain the initial injection rate into each zone, i_s, we will consider only water to be flowing at the injection well and free gas to be flowing at the producing well. This would represent a typical situation when the reservoir had been substantially depleted by solution-gas drive prior to the start of the flood. However, if the gas saturation is very low at the start of the flood the initial injection rate should be based on the assumption that oil is being produced.

If we assume that production is substantially all free gas, note that the initial injection based on Equation 8.31 would be,

$$i_s = \frac{\text{constant}}{[(\mu_w/k_w) + (\mu_g/k_g)]} \qquad (8.32)$$

When this equation is applied to a substantially depleted solution-gas-drive reservoir, note that the reciprocal of gas mobility will be negligible compared to the reciprocal of water mobility in the water bank. Thus, we can consider the reciprocal of gas mobility to be zero and Equation 8.32 becomes,

$$i_s = \frac{\text{constant}}{(\mu_w/k_w)} \qquad (8.33)$$

Now note that based on our assumption that injectivity will not change until a substantial change in saturations occur at the injection or producing wells, and that all of the free gas will be produced from a particular strata before any oil-rate increase occurs, the injection rate will remain constant until gas fillup has occurred in the strata. Thus, the injectivity ratio, i_s/i, would be 1.0 until W_i/V_p equals S_{gi}. At that time the oil bank will reach the producing well and oil will start flowing. Then during the time between gas fillup and water breakthrough (BT) into the producing well, the injectivity ratio according to Equation 8.31 would be,

$$i \text{ for fillup to BT} = \text{constant}/[(\mu_w/k_w) + (\mu_o/k_o)] \qquad (8.34)$$

In this equation water mobility should still be evaluated at the saturations existing in the water bank and oil mobility should be evaluated at the saturations existing in the oil bank. A ratio of Equations 8.33 and 8.34 will provide the injectivity ratio for the period from gas fillup to first water breakthrough.

$$(i_s/i) \text{ for fillup to BT} = 1 + (\mu_o/k_o)(k_w/\mu_w) \qquad (8.35)$$

Recognizing the last term as the mobility ratio, M, we obtain,

$$(i_s/i) \text{ for fillup to BT} = 1 + M \qquad (8.36)$$

During the period when water and oil are both being produced from a particular strata the evaluation of the injectivity ratio is more difficult. It is necessary to modify the five-spot model, Fig. 8.17, so that both oil and water are flowing in parallel streamlines in the producing side of the model as shown in Fig. 8.18. Use of this model means that the pressure drop on the production side of the model has to be based on oil or water flow rate but the flow-rate base would not be the same as the flow or

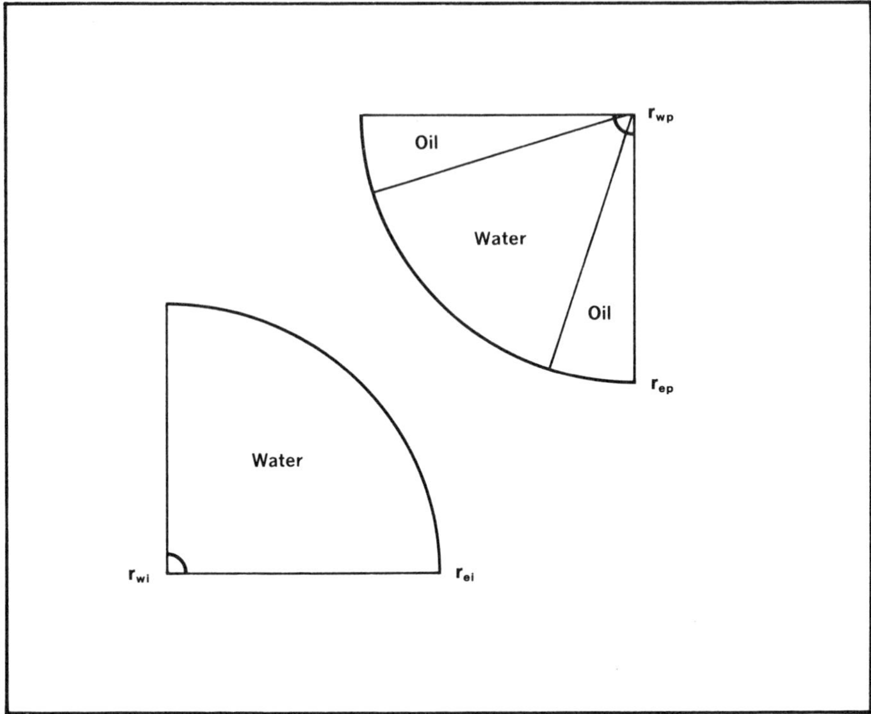

FIG. 8.18   Five-spot quadrant model after water breakthrough

injection-rate base used to calculate pressure drop on the injection side. Thus, Equation 8.31 can not be used to determine injectivity ratio when both oil and water are being produced from the strata.

It is convenient to state the pressure drop on the producing side of the five-spot as a function of water flow rate, q_w, and what we will define as the "apparent" thickness flowing water, h_w.

$$p_{wi} - p_{wp} = \Delta p_p + \Delta p_i \qquad (8.28)$$

$$p_{wi} - p_{wp} = 0.141i \ln(r_e/r_w)(\mu_w/k_w)/h + 0.141q_w \ln(r_e/r_w)(\mu_w/k_w)/h_w \qquad (8.37)$$

Since there is not 100% sweep of the quadrant at breakthrough the cross-sectional area flowing water is only a fraction of the total cross-sectional area in the producing well. If the mobility ratio is 1.0, the fraction of the total cross-sectional area in the producing well that is flowing water would be the same as the water cut. For example, if the water cut was 0.667 with a mobility ratio of 1.0, then 66.7% of the total cross-sectional area would be flowing water. We can adjust the total thickness flowing water more readily than we can adjust the total cross-sectional area with the same result. Thus, if the mobility ratio was 1.0, we could say that the thickness flowing water would be the product of water cut and total thickness, $f_w h$. However, the mobility ratio is seldom 1.0, so that we must adjust the actual water cut on the basis of the mobility ratio to obtain the water cut that would exist if the mobility ratio was 1.0. This adjusted f_w can then be applied to the total thickness to obtain the "thickness" that is flowing water.

Modifying our previous example, suppose that at a particular stage of depletion a strata has the same water cut of 0.667 but the mobility ratio is 2.0. Now to get the water cut that would exist at this time if the mobility ratio was 1.0, the actual oil cut of 0.333 has to be weighed by 2.0 so that the water cut with mobility of 1.0 would be,

$$0.667/(0.667 + (0.333)(2.0)), \text{ or } 0.5$$

so that in this example, one-half of the total thickness could be considered as flowing water. In equation form we could then write the apparent thickness flowing water as

$$h_w = f_w h / [f_w + (1 - f_w)M] \tag{8.38}$$

Substituting for h_w in Equation 8.37 according to Equation 8.38, substituting $f_w i$ for q_w, and solving the equation for the injection rate we obtain,

$$i \text{ after fillup} = \frac{(p_{wi} - p_{wp})h/0.141 \ln(r_e/r_w)}{(\mu_w/k_w)[1 + f_w + M(1 - f_w)]} \tag{8.39}$$

Note that the numerator is the same as the numerator of Equation 8.30 which is the constant of Equation 8.33. Thus, the injectivity ratio is obtained by dividing Equation 8.33 by Equation 8.39.

$$(i_s/i) \text{ after fillup} = (1 + M) - (M - 1) f_w \tag{8.40}$$

This equation actually applies to any time after gas fillup (as indicated) which includes the period of time during which the water cut is zero. We discussed the period from fillup to breakthrough previously and showed that the injectivity ratio would be as shown in Equation 8.36. Note that Equation 8.40 gives the same expression when f_w is zero. Thus, Equation 8.40 and the realization that the injectivity ratio will be 1.0 prior to the time of gas fillup provides the entire history of the injectivity ratio for a particular strata.

In using Figs. 8.10A and B to evaluate the injectivity-ratio history for a strata, it is necessary to determine the displacement volumes injected (DVI) at a particular water cut to calculate the injectivity ratio at that particular stage of injection. Since the water cut and DVI's injected do not appear on the same graph it is necessary to relate them on the basis of the swept area which appears on both graphs. For example, for a reciprocal mobility ratio of 0.3 note from Fig. 8.10B the sweep efficiency is 0.8 when 1.1 DVI's have been injected. Now turn to Fig.

8.10A and note that when the sweep efficiency is 0.8 with a mobility ratio of 0.3, f_w is about 0.78. Thus, when 1.1 DVI's have been injected f_w would be 0.78. When these values are converted to W_i/V_p and i_s/i using Equations 8.27 and 8.40, respectively, one point for the injectivity-ratio plot is determined.

To make certain you understand this procedure work the following problem and compare your solution with the solution in Appendix C.

Problem No. 8.3D: Calculation of Individual Strata Injectivity Curves

In Problems 8.3A and B, calculations were made predicting the oil recovery and water production versus cumulative effective injection based on the individual strata injectivity and recovery curves of Fig. 8.14 and 8.15, respectively. Verify Fig. 8.14 by calculating the points representing gas fillup, water breakthrough, and displacement volumes injected of 0.75 and 1.5. Fig. 8.11A represents the oil and water relative permeability for this reservoir and the oil and water viscosities are 4.15 and 0.7 cp, respectively.

PREDICTING WATER FLOOD PERFORMANCE BY ANALOGY
(OR FROM PILOT FLOOD RESULTS)

It is not always obvious just how the results of a previous flood project can be used to predict the results in a proposed project. Ideally the model and a proposed flood should be exactly the same; then results would be exactly as predicted. However, this is never the case. When the results of a previous flood are used to predict the results of a proposed flood, we are invariably faced with many differences in the two projects. These include differences in size, percent of injection lost outside the swept area, injection pressure, reservoir capacity, the state of depletion, mobility ratio, permeability profile, etc.

It is impossible to account for variations in all of these factors. However, if the permeability distribution and the mobility ratios are roughly the same, and the states of depletion are similar, prediction by analogy will probably result in a more accurate prediction than will prediction by straight theoretical considerations. The most accurate predictions will be obtained by using both methods. The methods presented in this section are especially valuable in expanding the results of a pilot flood to a full-scale flood prediction. Many profitable floods have been missed due to the misinterpretation of pilot-flood results.

The engineer should carefully note that the methods presented in this section are not in any way meant to replace the theoretical methods of predicting flood performance. The method is proposed as a supplement to theoretical methods and as a means of obtaining a prediction when sufficient reservoir data are not available for a theoretical prediction or when time does not permit a theoretical prediction. Even under these circumstances prediction by analogy may be impossible if the reservoirs are not similar. The proposed method supplements theoretical techniques the same way that decline-curve analysis supplements theoretical primary production predictions.

As noted previously, flood recovery predictions fall into two general categories—prediction of total ultimate flood recovery and prediction of rate versus time. Prediction of total ultimate recovery is not generally as difficult as the prediction of rate versus time since it is, to a large extent, independent of injection rates, percent of injection that is ineffective, and many other parameters that affect displacement efficiency but do not significantly alter ultimate recovery.

This section will be concerned with analogy methods of predicting total flood recovery, effective injection rate, producing rate versus time, and with the application of these methods to a pilot flood.

Predicting total flood recovery. The most logical basis for predicting total flood recovery is the total secondary plus primary recovery from the swept area as a fraction of total pore volume.

$$\text{Fractional recovery in swept area} = \frac{(\text{Total recovery, STB/acre-ft})B_{os}}{7,758\phi}$$

$$(8.41)$$

The total recovery to be used in Equation 8.41 should be determined very carefully.

$$\text{Total recovery, STB/acre-ft} = \frac{\text{STB primary production}}{\text{acre-ft primary drainage}}$$

$$+ \frac{\text{STB waterflood production}}{\text{acre-ft gross swept volume}} \quad (8.42)$$

In using this equation, primary production is defined as the cumulative production up to the start of the flood. Also note that the primary recovery and secondary recovery per acre-foot are based on different drainage volumes. Care should be exercised in making certain that all of the primary drainage volume is included in this term. A good rule of thumb is to assume that original oil in the reservoir will be produced by the nearest producing well if it is produced during the solution-gas-drive phase of production. Unequal production rates, completion dates, and other factors should be used to adjust this rule of thumb.

The acre feet in the gross swept volume can be determined by any reasonable technique since the same technique will be applied to both the model and proposed flood. Nevertheless, care should be exercised that the gross swept volume is not excessive. The techniques recommended and discussed in the section, "Determining Gross Swept Volume," can be used for this purpose.

Once the necessary parameters have been determined for the application of Equations 8.41 and 8.42 to the model and proposed flood, the prediction of total recovery is a matter of calculating the fractional total recovery in the swept area for the model and then reversing this procedure by using the fractional total recovery in the swept area for the model as the basis for calculating the volume of waterflood production for the proposed flood. To check your knowledge of this section, work the following problem and check your solution against the solution in Appendix C.

Problem No. 8.4A: Predicting Total Flood Recovery by Analogy

In subsequent parts of this problem, effective injection rates and the performance in terms of rate versus time will be determined. At this time it is desired to calculate the total flood recovery for the proposed flood using the data obtained for the "Old" flood.

Data on old flood

Formation thickness,	20 ft
Average porosity,	18%
Estimated connate water	25%
Original B_o	1.25
B_o at flood start	1.01
Primary drainage volume	10,000 acre-ft
Gross swept flood volume	8,000 acre-ft
Oil production to flood start	2,410,000 STB
Waterflood recovery	1,570,000 STB

Data on proposed flood

Average porosity	20%
Estimated connate water	20%
Original B_o	1.20
B_o at flood start	1.00
Primary drainage volume	15,000 acre-ft
Swept flood volume	14,000 acre-ft
Cumulative oil production at flood start	3,645,000 STB

Predicting effective injection rates. Prediction of effective injection rates for a waterflood involves two important considerations. First, it is necessary to account for the difference in the total injection rates that result from a difference in the transmissibility kh/μ, the pressure drop $(p_{wi} - p_{wp})$, and geometry. Just as important is the necessity for estimating how much of the total injection rate enters the gross swept volume of the reservoir and how much is lost outside the gross swept volume. For example, the total injection rate in a single five-spot consisting of four injection wells and one producing well as in Fig. 8.4, might be 400 b/d, but the effective injection rate might initially be only one-fourth of this total amount. In other words, with radial flow into all four injection wells, only one-fourth would enter the swept volume.

If there is an insignificant gas saturation at the start of the flood, radial flow would not prevail and virtually 100% of the injected fluid would enter the gross swept volume since there are no other paths available. However, when a substantial free-gas saturation exists at the start of the flood, the estimation of the percent of total injection that enters the gross swept volume is much more difficult. Referring to Fig. 8.5, note that if the initial gas saturation is substantial, initially radial flow will exist around all injection wells. This would mean that Well 19 would have about 25% of its water entering the gross swept volume while Well 18 would have 50% entering the swept volume and Well 9 would have 100% entering the gross swept volume.

It can be argued that whenever the gross swept volume is pressured to a point higher than that of the reservoir outside the gross swept volume, a larger percentage of the injection from the edge wells will enter the region outside the flood due to the differences in reservoir pressure. However, if the zero thickness contour is near the injection wells, it seems likely that eventually the reservoir outside the gross swept volume will be filled and attain a pressure greater than the average pressure in the gross swept volume. Then virtually all of the injected fluid will enter the gross swept volume since pressure sinks exist at the producing wells.

Thus, as a first estimate, it is suggested that edge wells in a flood with a substantial initial gas saturation be treated as though radial flow was taking place throughout the flood life so that the percentage of each well's injection that is entering the gross swept volume would be calculated as illustrated above. This assumption can be checked for the old flood during its late life by comparing the total injection rate with the total fluid production rate. By late life virtually all of the gas will have been displaced from all of the zones with significant permeability so that if 100% of the injected water is entering the gross swept volume there would be a one-to-one ratio between injection rate and the total (oil plus water) producing rates.

Study of the steady-state injection rate representing a particular flood geometry will result in methods for calculating the injection rate for the proposed flood based on experience gained from the old flood. For example, the steady-state flow equation for a five-spot flowing one fluid of viscosity μ was shown in Chapter 1 to be

$$i = \frac{3.54kh(p_{wi} - p_{wp})}{\mu\left(\ln \dfrac{d}{r_w} - 0.619\right)} \tag{1.34}$$

From this equation we note that the injection rate is proportional to the transmissibility, kh/μ, the pressure drop between the injection and producing wells $(p_{wi} - p_{wp})$, and the reciprocal of the geometry factor $(\ln (d/r_w) - 0.619)$. Consequently, whenever these terms are different for the old and proposed floods the total injection rate for the old flood can be corrected to the estimated total injection rate for the proposed flood by ratio.

$$i_{pro} = i_{old} \frac{(kh/\mu)_{pro}}{(kh/\mu)_{old}} \frac{(p_{wi} - p_{wp})_{pro}}{(p_{wi} - p_{wp})_{old}} \frac{\left(\ln \dfrac{d}{r_w} - 0.619\right)_{old}}{\left(\ln \dfrac{d}{r_w} - 0.619\right)_{pro}} \tag{8.43}$$

The viscosity values can probably be accurately enough depicted by using an arithmetic average of the water and oil viscosity. The pressure at the water-injection well, p_{wi}, can be calculated from the well depth, wellhead pressure, and water density.

$$p_{wi} = 0.433\gamma D + p_{wh} \tag{8.44}$$

Another analogy technique, often overlooked in predicting total injectivity for a new flood, is the use of initial oil-production test data for the well at time of completion (peak primary production). Except for highly depleted reservoirs it should always be safe to assume that an injection pressure can be used that is at least as high as the initial reservoir pressure. Under these conditions the injection rate for most five-spot pattern floods would be one-half or less of the initial productivity of the well, if k/μ for injected water is equal to the initial k/μ for oil. This is true because the injection pressure must force water into an injection well and force the fluids out of the producing well in a five-spot whereas the initial reservoir pressure in the reservoir must only force oil out of the producing well. Another way of saying this is that the pressure drop during flooding results in flow into the radial injection-well system and out of the radial producing-well system which requires twice as much energy as the initial production test where the pressure drop from the reservoir to the producing well causes flow only out of the radial producing-well system.

Note that it was stated above that the injection rate would be one-half or less of the initial productivity of the well. The "or less" is added to account for differences that may exist in the relative permeability of water in the water bank and oil initially produced under primary conditions. Relative permeability characteristics are such that this ratio may be as much as 3 or 4 in some cases. A comparison of the relative permeability to the wetting phase at an irreducible nonwetting-phase saturation and the relative permeability of the nonwetting phase at the irreducible wetting-phase saturation will illustrate this phenomenon.

Thus, we can simply correct the initial primary oil-production rate for the differences in viscosity and relative permeability and show that a well converted to injection with an injection pressure equal to the initial reservoir pressure will have an injection rate related to the well's initial oil-producing rate as

$$i = q_{oi}B_{oi}(\mu_{oi}k_{rw}/\mu_w k_{roi})/2 \tag{8.45}$$

Predicting oil production rate versus time. To predict production rate versus time we must account for several major differences between the model and prototype that affect the rate-versus-time curve. These include effective injection rate, cumulative effective injection at a particular time, difference in size of the floods, difference in the patterns, and difference in oil saturation at the start of the flood. The rate-versus-time curve can be normalized for the effects of these differences by plotting production rate as a fraction of current effective injection rate versus cumulative effective injection plotted as a multiple of total ultimate flood recovery. The assumption then is that such a plot will be the same for similar floods. This plot is illustrated in Fig. 8.19 which shows such a plot for three different floods.

The rate is stated as a ratio of effective injection rate since most floods, after they have had an initial increase in oil-producing rate, are very sensitive to injection rate. A change in injection rate will cause a proportional change in oil-producing rate in a few hours or at most a few days. Using this ratio for the calculation of oil-producing rate in the proposed project also gives an adjustment for the differences in injection rates between the two floods.

The cumulative effective injection must also be adjusted for the differences in flood size. If the same saturations existed at the start of the flood in both reservoirs and the same residual oil was to be experienced, use of the pore volume to normalize the cumulative effective water injection would be logical. However, under these conditions normalizing with the total recoverable oil would provide exactly the same degree of normalization. Use of ultimate total flood recovery for this purpose accounts for mobile oil differences between the two reservoirs. Since the correction for the difference in mobile oil is so inexact, it is suggested that this analogy technique of predicting rate versus time for a waterflood be limited to those reservoirs that are in about the same state of depletion at the time of flood initiation. A detailed modification of the rate-versus-time curve could be accomplished where mobile oil differences are considerable, but such a detailed correction is beyond the scope of this chapter.

To indicate the validity of the normalization approach, a normalized plot was prepared for three floods with substantial production history. Since floods normally experience injection difficulty, and are thus difficult to keep in balance, it is difficult to find a number of similar old floods. The writer had data available on only three "similar" floods in sufficient detail to prepare the subject plots. Consequently, the data fit as indicated in Fig. 1.19 is not as good as would normally be expected. However, a study of the reservoir data for the three floods as indicated in Table 8.1 will reveal that the similarity of the plots in Fig. 1.19 is certainly very good under the circumstances. Note first that while two of the reservoirs are in the Benoist formation the third is in an entirely different formation 1,000 ft deeper. Consequently, we might expect the best fit with the two

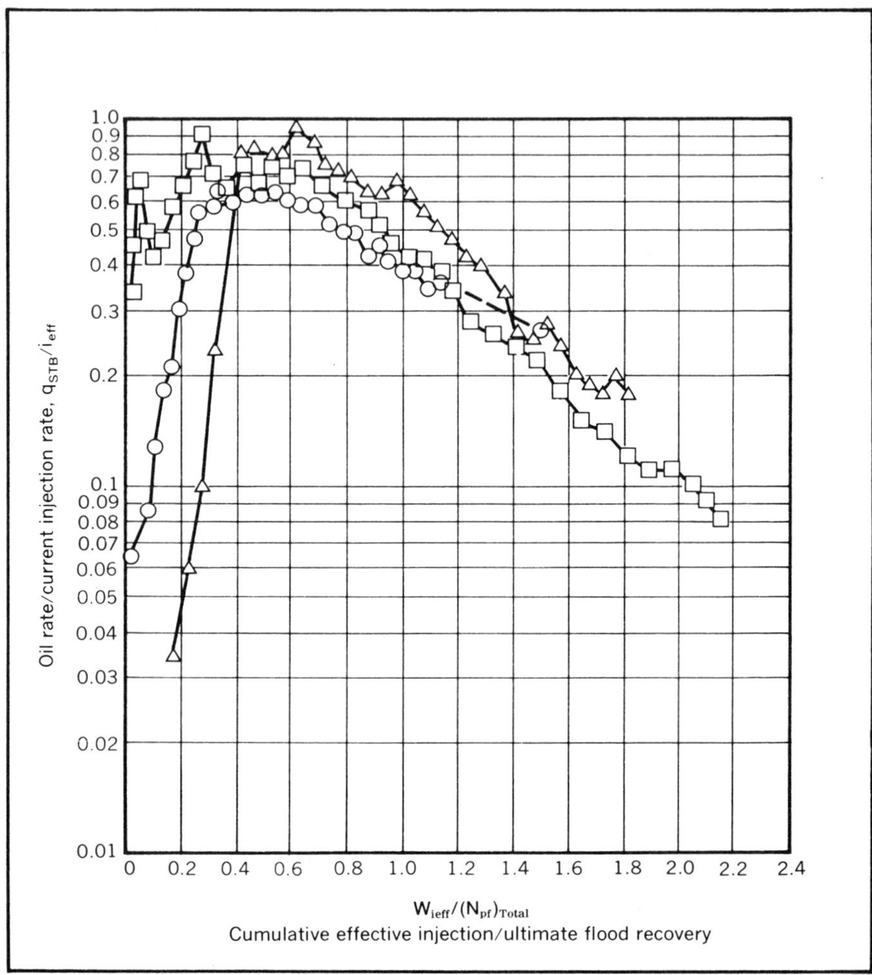

FIG. 8.19   Normalized performance data for three Illinois waterfloods (See Table 8.1 for reservoir characteristics)

Benoist floods. The parameters of the two Benoist reservoirs are represented by squares and circles in Fig. 8.19.

Note that these two reservoirs represent vastly different permeabilities, flood areas, thicknesses, injection rates, and five-spot spacings. Yet we would expect the normalization to overcome these differences and provide a good curve fit except that the cumulative primaries at the start of the flood are so different. This results in greatly different gas saturations in the reservoir at the start of the floods and affects the time (effective cumulative injection/ultimate flood recovery) of the first production increase. The plots reflect this situation in much the same way we would

Table 8.1

COMPARISON OF RESERVOIR PARAMETERS FOR ILLINOIS FLOODS
SHOWN IN FIG. 8.19

	Benoist Sand	Benoist Sand	Tar Springs Sand
Depth, ft	1,400	1,300	2,300
Porosity %	19.0	20.5	19.0
Permeability, md	110	250	450
Area, acres	500	300	230
Thickness, ft	27	14	?
Flood pattern	10-acre 5-spot	20-acre 5-spot	20-acre 5-spot
Avg. inj. rate, b/d	8,500	1,800	1,500
Cumulative oil Production at flood start, STB/acre-ft	184	284	?

anticipate with a good fit between the two normalized plots during most of the life but with the one Benoist flood—with the lower gas saturation—responding to injection somewhat earlier than the other which had higher gas saturation. This comparison might suggest a method of adjusting an analogy prediction for differences in initial saturations.

Actually the fit of the two Benoist plots would be much better in spite of the differences in gas saturations at the start except that one flood was initiated on a staggered-time basis, resulting in the early oil-rate peaks.

The plot for the Tar Springs flood (triangles) is included to indicate that reservoirs from different geologic formations are sometimes suitable for predicting flood behavior by analogy. The plot fit with the Benoist reservoirs is not too inaccurate and the availability of detailed information on flood balance, spacing irregularities, and other performance could conceivably result in an excellent fit. Reliable reservoir thickness data are not available on this reservoir since the wells were drilled with cable tools. Consequently, gas saturation at the start of the flood is not known.

Thus, the proposed method for predicting rate versus time by analogy for reservoirs with similar reservoir characteristics and saturations at the start of the flood is to use the model data to determine the relationship between (q_o/i_{eff}) versus (W_{ieff}/N_{pf}). This relationship is then assumed to hold equally for the proposed flood and the data are converted to q_o versus W_{ieff} for the proposed project by using the i_{eff} and N_{pf}, respectively, to convert the data determined for the old flood. Note that it is not necessary to make a plot like Fig. 8.19 for either the old or proposed flood. The old data can be used to calculate corresponding values of (q_o/i_{eff}) and (W_{ieff}/N_{pf}). These values can then be converted to corresponding values of q_o and W_{ieff} or time by applying the values of i_{eff} and N_{pf} for the proposed flood. To check your knowledge of this section work the following problem and check your solution against the solution in Appendix C.

Problem No. 8.4A: Prediction of the Rate vs. Time Performance by Analogy

In Problem 8.4B the engineer was asked to predict the total flood recovery by analogy. For this same reservoir he is asked to predict the oil production rate versus time if Fig. 8.20 represents the rate-versus-time history of the old flood. The following reservoir data are also necessary for this prediction.

Data on old flood

Formation depth	3,000 ft
Average permeability	100 md
Total injection-well thickness	401 ft
Average injection rate at 600 psi wellhead pressure (fresh water)	1,700 b/d
Estimated effective injection rate	90%
See Fig. 8.20 for oil rate, injection rate, and cumulative injection versus time	

Data on proposed flood

Formation depth	2,500 ft
Average permeability	75 md
Total injection-well thickness	780 ft
Estimated wellhead pressure	800 psi
Estimated effective injection rate	95%

WATERFLOODING VARIATIONS

Much effort and money has been expended to find more efficient and economical means of recovering oil from subsurface reservoirs. Research and development continue on insitu combustion and various forms of miscible displacement. However, the most useful ideas that have come out of modern research on recovery methods have been variations of waterflooding.

A long list of waterflooding variations could be compiled but for the purposes of evaluating the possible application of these variations to specific reservoirs they will be treated here as falling into two categories. The first of these reduces residual oil in that portion of the reservoir contacted by the water. The second category includes those techniques that control mobility ratio and thus increase the amount of formation contacted by water. This latter effect is accomplished by increasing sweep efficiency, stratification efficiency, and frontal water saturation. We will consider means of including both of these effects in our predictions.

Adjusting mobility ratios. Probably the most widely used variation of waterflooding currently in use is addition of polymer to the injected water to increase its viscosity. The mechanism by which the polymer

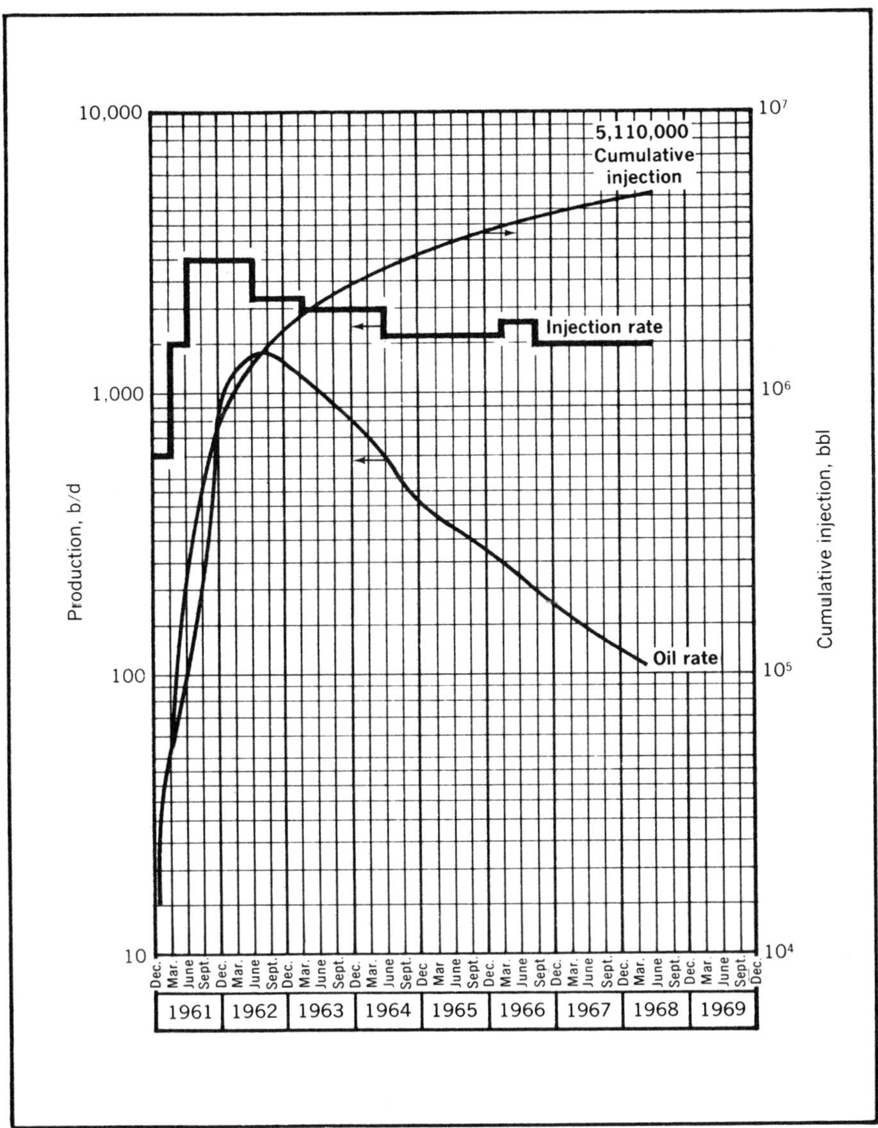

FIG. 8.20 Production and injection data for a flood; data for Problem 8.4B

reduces water mobility is not clearly understood at the present time. Decreasing the mobility of the displacing water has several desirable effects and at least one highly undesirable effect.

A decrease in mobility ratio will increase the sweep efficiency as will be seen in the Dyes-Caudle-Erickson data of Figs. 8.10A and 8.10B.

For example, using an M of 4 or reciprocal of 0.25, we determine that at breakthrough the strata sweep is 53% and at f_w of 0.95, the strata sweep efficiency is 92% after injection of 2.5 injection volumes. Now if we quadrupled the water viscosity, M would be 1.0 and the reciprocal would be 1.0. The sweep efficiency at breakthrough would be 70%, and when f_w is 0.95, the sweep efficiency would be about 100% with about 1.8 displacement volumes injected.

A decrease in mobility ratio will also increase vertical efficiency. To put this effect in some perspective, consider the ultimate sweep efficiency at the economic limit as indicated by the Stiles method. As shown previously, the strata sweep efficiencies play a minor role in the ultimate overall efficiency which is affected mainly by the vertical or stratification sweep.

In Problem 8.2B we found that using an effective mobility of about 4.0 for A_{eff}, we obtained an efficiency of 0.76. If we quadrupled the water viscosity (from .7 cp to 2.8 cp) the comparable efficiency would be 0.93. The engineer is again reminded that these efficiencies are too high due to the high lower limit on the permeability distribution considered. Nevertheless, improving (reducing) the mobility ratio greatly improves the vertical sweep efficiency.

The decrease in mobility ratio also affects the microscopic displacement efficiency as seen in the Buckley-Leverett-type calculations. Decreasing the mobility ratio by simply changing the viscosity ratio changes the f_d versus S_d relationship and thus increases the frontal S_d saturation. This effect is illustrated in Fig. 8.21 and generally shifts the displacing phase saturation behind the front by decreasing all S_d values greater than the S_d equivalent of residual oil saturation. However, if the change in mobility ratio is due also to a change in relative permeability characteristics (as presently seems likely), the change in saturation distribution would be more complex.

The obvious disadvantage of decreasing water mobility is the adverse effect on injection rate obtained for a particular injection pressure. This effect seems to limit economic applications to high-permeability reservoirs and high-viscosity oils. The high-viscosity oil is necessary because the percent decrease in the injection rate (compared with a normal flood) resulting from the use of viscous water to displace a high-viscosity oil is much less than the percent decrease associated with the use of viscous water to displace low-viscosity oil. For example, consider the situation that exists for a five-spot flood. We noted previously that for all practical purposes we can consider all of the resistance to flow to exist at the wells. Thus, over most of the life of a flood only water is flowing at the injection wells while mostly oil is flowing at the producing wells. In this situation we may be justified in considering the arithmetic average of the oil and water as being characteristic of the effective viscosity in the five-spot flood. If this is the case, note that a water viscosity of 1.0 combined with a low-viscosity reservoir oil of say 0.7 would give an "effective" viscosity

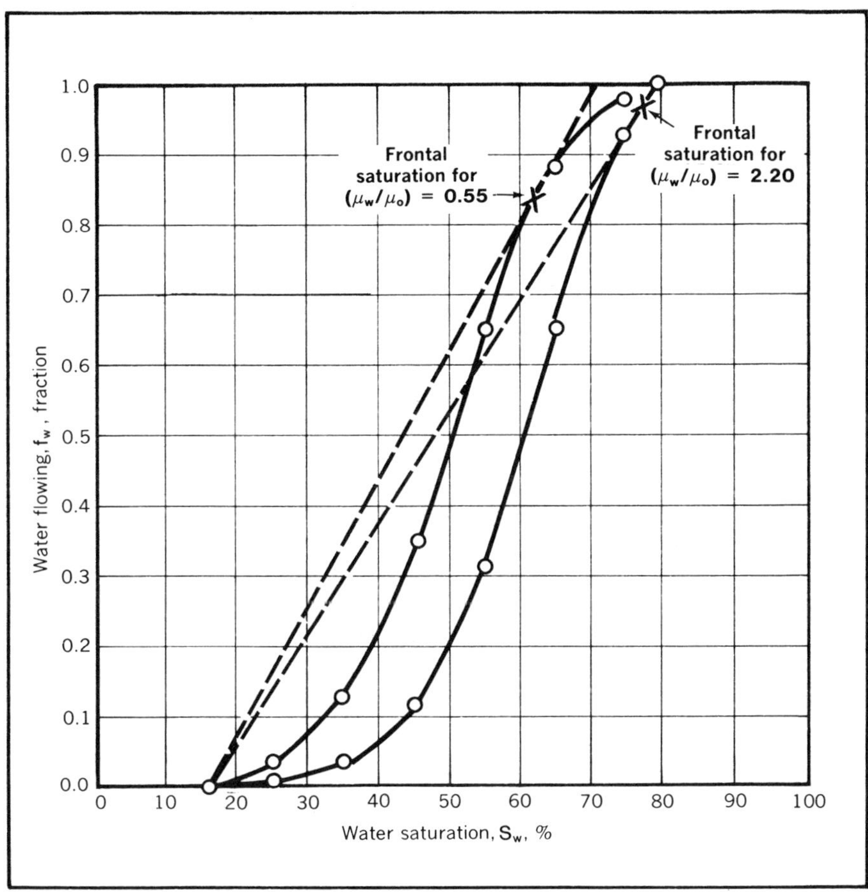

FIG. 8.21 Comparison of f_w curves for different viscosity ratios

of 0.85. If a water thickener that increases the water viscosity to 4.0 is employed the effective viscosity would be increased to 2.35 and the average injection rate might be reduced by a factor of about 2.75 (2.35/0.85).

Now compare this with the effect of the same thickened water on the injection rate where the reservoir oil viscosity is 7.0. Without the thickener, average viscosity would be 4.0 and with the thickener the average would be 5.5, and the injection rate would be reduced by a factor of only 1.38. Thus, it is clear that the thickening of the water has a much less pronounced effect on the injection rate in reservoirs containing high-viscosity oils.

At the same time, to move very viscous oil at economic rates it is necessary to have relatively high permeability reservoirs. Hence, the statement that economic applications of thickened water seem to be limited to reservoirs of relatively high permeability and high-viscosity oils. Claims have been made that the polymer solution flowing in the porous media exhibits shear thickening characteristics which would further increase the sweep and stratification efficiencies by slowing down the flow rate along the high-velocity flow paths. However, this effect has not been sufficiently established to warrant its inclusion in any practical evaluation of a proposed application. In fact some data have indicated shear thinning characteristics.

The same recommended method of predicting flood behavior can be used to predict the results of a polymer flood. The easiest way to accomplish such an application is simply to assume that the thickened water will be used throughout the flood so that the application is a matter of making a prediction with an adjusted mobility ratio. However, such a prediction will represent the maximum result obtainable from a thickened waterflood of that mobility ratio. This would indicate whether further study is advisable but it certainly would not give a realistic prediction of the most economic application of polymer flooding. Since it is generally uneconomic to thicken all the water in a flooding operation, the normal procedure is to use a slug of thickened water amounting to somewhere between 10 and 40% of the pore volume and then displace this with water. Lab and field tests show that the thickened water affects the porous media in such a way that following injection of the slug of thickened water the formation exhibits a greatly reduced permeability to water.

Consequently, it is necessary to assume that displacement of the slug is by a water with an abnormally high viscosity which gives the same effect on the performance as would a reduced effective permeability. To make an accurate prediction of the performance of a polymer flood it is thus necessary to know the cost of thickening water to various effective mobility ratios and to know the residual effect of the polymer on the permeability to normal injection water. Care should be exercised to make certain that the supplier of the polymer also provides you with the polymer absorption data for your formation. The absorption is generally substantial.

When slug-type displacement is used, the generation of injectivity and recovery curves for use in the "recommended" method is somewhat more difficult than the process previously described for a regular waterflood. An acceptable recovery curve can be obtained by simply using the Dyes-Caudle-Erickson data together with the mobility ratio existing between the mobile oil and the fluid directly displacing it, i.e. the slug. This approach would not be sufficiently accurate if a very small slug size, say less than 15% of the pore volume, is used but generally such a small slug is impractical due to the dispersion that takes place between the slug and the water displacing it.

Determining the slug distribution in the various zones. To determine the injectivity curves to be employed in the recommended method it is necessary to first determine the portion of the slug that will enter each permeability zone. This is most readily obtained by applying the prediction method to the reservoir as though the thickened water would be continuously applied.

In so doing, reduced-time curves such as those of Fig. 8.16 are obtained. With these curves the cumulative injection corresponding to any particular reduced time can be determined (assuming continued injection of the thickened water). This technique was used in solving Problem 8.3B. The solution is shown in Appendix C.

Once we have determined the amount of slug we wish to inject into the formation we can determine the reduced time that corresponds to this particular cumulative effective injection. Then a figure such as 8.16 can be used to determine the pore volumes of injection into each zone that corresponds to this reduced time. These values can be converted to cumulative injection of slug into each zone by applying the appropriate pore volume to each of the values.

For example, suppose in Problem 8.3B that we were planning to inject a slug of fluid until the injected fluid broke through into the producing wells. The solution to 8.3B in Appendix C shows that this would occur when 86,700 bbl of slug had been injected. The solution also shows that this occurs at a reduced time of 7.63 and that at this particular time the pore volumes of injection into zones I, II, and III are 0.33, 0.24, and 0.1, respectively.

Since the pore volumes of the three zones are given as 97,000, 155,000, and 175,000 bbl, respectively, we can calculate the water injected into each zone at this time as $0.33 \times 97,000$, $0.24 \times 155,000$, and $0.1 \times 175,000$ barrels, respectively.

After the amount of slug going into each zone has been determined, the previous principles discussed for the determination of an injectivity curve for each zone can be applied and the prediction calculations can be repeated using these new injectivity curves.

The recovery data of Dyes, Caudle, and Erickson, found in Figs. 8.10A and B apply theoretically to only two phase flow; however, the error introduced by the introduction of the third phase generally has a minimal effect on the recovery curve whenever a slug of sufficient size is used to keep the water displacing the slug from the oil being displaced by the slug.

Determining injectivity curves for a slug type displacement. Once the amount of slug that will enter each zone has been determined (as outlined above) it is possible to calculate an injectivity ratio versus cumulative injection in pore volumes curve for each zone. This, of course, is required for the recommended method of performance prediction. Each zone will now have an injectivity curve that is unique since all of the zones will have different amounts of slug injected into them.

The determination of injectivity curves for a slug type operation is necessary to the prediction of many different secondary recovery mechanisms and thus is not just peculiar to a thickened water type flood. It is difficult to generalize the procedure for determining the injectivity curves for a slug type process but here again it is hoped that the development of a series of equations for one or two sets of circumstances will permit the engineer to formulate equations for other circumstances.

If the saturations in the reservoir at the start of the flood include a substantial gas saturation then the expression for the initial injection rate, Equation 8.33, can be adapted to our slug injection.

$$i_s = \text{constant}/(\mu_{sl}/K_{sl}) \qquad (8.44)$$

Thus, until the injected slug is equal to the initial mobile gas saturation or water injection starts the injectivity ratio, i_s/i, will be 1.0. The other equations for the injectivity ratio will also apply until the slug injection is discontinued and water injection or injection of another displacing phase is initiated.

Using the same principles employed in deriving Equation 8.40, we can derive the following injectivity ratio equations:

When water is being injected and oil and slug are being produced,

$$i_r = \frac{1}{M_{w/sl}} + f_{sl} + M_{sl/o}(1 - f_{sl}) \qquad (8.45)$$

When water is being injected and water and slug are being produced,

$$i_r = \frac{1}{M_{w/sl}}(1 + f_w) + (1 - f_w) \qquad (8.46)$$

There will be a period when water, slug, and oil are being produced after the injected water breaks through the slug. This situation is not reflected in Equation 8.46 which is based on only water and slug production. However, this will not introduce much of an error in most cases for two reasons. First, most of the oil will have been produced from a particular zone before water breakthrough and secondly, every effort will be made to have the slug-oil mobility ratio near one for recovery and economic reasons.

If the mobility ratio of the slug to the oil was 1.0, Equation 8.46 would not need to be modified to account for the concurrent production of water, slug and oil. Consequently, it is suggested that Equation 8.45 be used until water breakthrough and then Equation 8.46 be used to obtain the injection ratio. The point of water breakthrough can be determined from Figs. 8.10A and B based on water injection alone and using the water-slug mobility ratio.

As noted before, a thickened water flood is most often applied to a reservoir containing a high viscosity oil that has experienced very little primary production so that little or no free gas is present in the reservoir at the time the flood is initiated. In such cases initial injection would take place with oil flowing at the producing well and the pusher slug flowing at the injection well. Applying Equation 8.31 to obtain the initial injection rate we obtain

$$i_s = (\text{constant})(k_{slug}/\mu_{slug})/(1 + M_{slug/o}) \tag{8.47}$$

Where $M_{slug/o}$ is the mobility ratio between the slug and the oil.

We can then use Equation 8.31 repeatedly to obtain the injection rate at various stages of displacement stated as a function of the geometrical constant and the slug mobility in the denominator. This then permits the formulation of i_r for these various stages of displacement. For the slug flowing at the injection well and oil at the producer, $i_r = 1.0$. For the slug flowing at the injection well and oil and slug flow at the producer,

$$i_r = \frac{1 + f_{slug} + M_{slug/o}\,(1 - f_{slug})}{1 + M_{slug/o}} \tag{8.48}$$

For water flowing at the injection well and oil and slug flow at the producer,

$$i_r = \frac{M_{slug/w} + f_{slug}\,(1 - M_{slug/o}) + M_{slug/o}}{1 + M_{slug/o}} \tag{8.49}$$

For water flowing at the injection well and water and slug at the producer,

$$i_r = \frac{M_{slug/w}\,(1 + f_w) + (1 - f_w)}{1 + M_{slug/o}} \tag{8.50}$$

Equations 8.48, 8.49, and 8.50 can be used for flowing fractions of 1.0 or 0.0 and consequently they describe the entire range of i_r values based on the assumption that all of the resistance to flow exists only at the wells. Once the strata injectivity and recovery curves have been generated the same recommended procedure can be employed in predicting the behavior of a thickened water flood.

Use of micellar solutions. Oil companies may eventually perfect a micellar solution to economically displace virtually all oil from a reservoir contacted by the solution. Note that all oil can be displaced only from the reservoir contacted. There is still no magic way to overcome the vertical and sweep efficiencies. A carefully designed micellar solution will have an advantageous viscosity but the mobility ratio advantage obtainable with a polymer can not generally be achieved. To approach any possibility of economic use it is necessary to use a thickened water behind the slug of micellar solution. Consequently, a prediction utilizing a micellar solution can be achieved in much the same way described for predicting the thickened waterflood. The equations derived for predicting the injectivity curves when a slug is used can be applied here but the "slug" will now be the micellar solution. The recovery curves used will simply reflect a piston-like displacement with zero residual oil.

REFERENCES

1. R. L. Slobod and B. H. Caudle, "X-Ray Shadowgraph Studies of Areal Sweepout Efficiencies," *Trans. AIME* (1954) 201, 81.
2. B. C. Craft and M. F. Hawkins, *Applied Petroleum Reservoir Engineering,* Prentice-Hall Inc., Englewood Cliffs, N. Y., 1959.
3. W. E. Stiles, "Use of Permeability Distribution in Water Flood Calculations," *Trans.,* AIME (1949) *186,* 9–13.
4. E. Suder and J. C. Calhoun, Jr. "Water-Flood Calculations," *Drilling and Production Practice,* API (1949), p. 260.
5. M. Prats, C. S. Mathews, R. L. Jewett, and J. D. Baker, "Prediction of Injection Rate and Production History for Multifluid Five-Spot Floods," *Trans.,* AIME (1959) *216,* 98–105.
6. R. V. Higgins, and A. J. Leighton, "Computer Method to Calculate Two Phase Flow in Any Irregularly Bounded Porous Medium," *J. Pet. Tech.* (June, 1962) 679–683.
   Sweepout Efficiencies," *Trans. Aime* (1952) *195,* 265.
7. A. B. Dyes, B. H. Caudle and R. A. Erickson, "Oil Production After Breakthrough—As Influenced by Mobility Ratio," Trans. *AIME* (1954) *201,* 81.
8. H. Dykstra, and H. L. Parsons, "The Prediction of Oil Recovery by Waterflooding," *Secondary Recovery of Oil in The United States,* 2nd ed., API, New York (1950) 160–174.
9. H. C. Slider, "New Method Simplifies Prediction of Water Flood Behavior," *Petroleum Engineer,* Feb. 1961.
10. F. Craig, *Reservoir Engineering Aspects of Waterflooding,* SPE Monograph No. 3, 1971.
11. H. C. Slider, "Predicting Waterflood Performance by Analogy," unpublished SPE Paper 3442 (1971), Presented at Regional SPE Meetings in Los Angeles and in Evansville, Ind.

Reservoir Engineering Symbols in Approximate Alphabetical Order
(With dimensional units)

		Dimensions
a	Fractional change in the rate per day (decline rate)	$(day)^{-1}$
a_i	Decline-rate at q_i	$(day)^{-1}$
A'	Area	sq cm
A	Stiles mobility ratio	dimensionless
A	Area	sq ft
B	Angle between interface and bed	degrees
B_o	Oil formation volume factor	dimensionless
B_{os} or B_{op}	B_o at flood start	dimensionless
B_g	Gas formation volume factor	reservoir bbl/scf
B_{gi}	Gas formation volume factor at p_i	reservoir bbl/scf
B_{oi}	Oil formation volume factor at p_i	dimensionless
B_{oabd}	B_o at abandonment	dimensionless
B_t	Total formation volume factor	dimensionless
B_w	Water formation volume factor	dimensionless
c	Capacity; also a constant	md-ft
c'	Capacity as fraction of total, c_T	fraction
c	Compressibility	volume/volume/psi
c_f	Formation (rock) compressibility	volume/volume/psi
c_g	Gas compressibility	volume/volume/psi
c_R	Reduced gas compressibility, $c_g p_c$	dimensionless
c_o	Oil compressibility	volume/volume/psi
c_w	Water compressibility	volume/volume/psi
c_e	Effective compressibility	volume/volume/psi
c	Capillary length/core length	dimensionless
d	Distance between input and producer in five-spot	ft
d	Distance to boundary	ft
DR	Damage ratio	dimensionless
D	Depth or distance	ft
DVI	Displaceable volumes injected	dimensionless
E	Efficiency	dimensionless

		Dimensions
E_V	Vertical efficiency: hydrocarbon pore space invaded (affected, contacted) by the injected fluid divided by the hydrocarbon pore space enclosed in all layers behind the injected fluid when $E_H = 1.0$	dimensionless
E_H	Pattern sweep efficiency: hydrocarbon pore space enclosed behind the injected fluid front divided by total hydrocarbon pore space when $E_V = 1.0$	dimensionless
E_C	Conformance efficiency: accounts for by-passed volumes due to lateral permeability variations, lenses and other difficulties.	dimensionless
E_T	Volumetric efficiency: product of pattern, conformance and vertical efficiencies	dimensionless
Ei	Exponential integral	dimensionless
f_d	Fraction of a phase in the total flow rate, q_t	dimensionless
f_{dfr}	f_d at the front of the displacing phase	dimensionless
f_{dPf}	f_d at the producing face (well)	dimensionless
f_{dno}	f_d with no gravity	dimensionless
f_w or F_w	Fraction of water in q_t	dimensionless
f_g	Fraction of gas in q_t	dimensionless
G	Total initial gas in place in reservoir	scf
G_p	Cumulative gas produced	Mscf
h'	Thickness as fraction of total, h_T	dimensionless
h	Net thickness (general and individual bed)	ft
h_g or h_o	Net thickness of gas or oil zone	ft
h_t	Total thickness	ft
i	Injection rate	res. b/d
i_{eff}	Effective injection rate	res. b/d
i_s	Initial injection rate	res. b/d
i_r	Injection rate ratio i_s/i	dimensionless
j	Mathematical counter	
J	J function	dimensionless
J	Productivity index	STB/d/psi
J_s	Specific productivity index	STB/d/psi/ft
k	Absolute permeability (fluid flow)	darcies
k'	Permeability as fraction of average permeability	dimensionless
k_e	Effective permeability	darcies
k_g	Effective permeability to gas	darcies
k_r	Relative permeability	dimensionless
k_o	Effective permeability to oil	darcies
k_{rg}	Relative permeability to gas	dimensionless
k_{ro}	Relative permeability to oil	dimensionless
k_{rw}	Relative permeability to water	dimensionless
k_w	Effective permeability to water	darcies
k_i	Effective permeability at input	darcies
k_p	Effective permeability at producer	darcies
ln	Natural logarithm, base e.	

Dimensions

log	Common logarithm, base 10	—
L	Length of system	ft
m	Ratio of initial free-gas volume to oil volume	dimensionless
m	Slope	various
M	Mobility ratio (λ displacing/λ displaced)	dimensionless
n	Exponent of back-pressure curve, gas well	—
n	Number of moles of gas	moles
n	Math counter	—
n	Exponent of hyperbolic decline	—
N	Initial oil in place in reservoir	STB
N_p	Cumulative oil produced	STB
N_{pP}	Cumulative oil produced during primary	STB
N_{ps}	N_p at flood start	STB
N_{pf}	Oil produced during flood	STB
$(N_{pf})_{Total}$	Oil produced from all zones during flood (performance prediction)	STB
p	Pressure	psia
P_c	Capillary pressure	dynes
p_b	Bubble-point (saturation) pressure	psia
p_e	External boundary pressure	psia
p_i	Initial pressure	psia
p^*	Initial pressure necessary for a well to be infinite acting	psia
p_{wi}	Injection well bottom-hole pressure, flowing	psia
p_{wp}	Producing well bottom-hole pressure	psia
p_r	Pressure at r	psia
p_{sc}	Pressure, standard conditions	psia
p_{tD}	Dimensionless pressure function at dimenless time t_D	dimensionless
p_w	Bottom-hole pressure, general	psia
p_{wf}	Bottom-hole pressure, flowing	psia
p_R	Reduced pressure, p/p_c	dimensionless
p_s	Static reservoir pressure	psia
p_{skin}	Additional pressure drop in damaged zone	psia
q'	Flow rate	cc/sec
q	Production rate or flow rate	res. b/d (STB/d in Chap. 7)
q_{cC}	Minimum rate at which a cone reaches a well	res. b/d
q_{cF}	Minimum rate at which a finger forms	res. b/d
q_i	Reference rate in decline curve analysis	STB/d
q_g	Gas production rate at well	Mscfd
q_{gr}	Gas production rate at r,	Mscfd
q_o	Oil production rate	res. b/d
q_{oi}	Initial oil producing rate	res. b/d
q_r	Flow rate at r	res. b/d
q_w	Water production rate	res. b/d
q_t	Total flow rate of all fluids	res. b/d
q_{EL}	Rate at the economic limit	res. b/d or STB/d
Q	Cumulative fluid flow	res. bbl

		Dimensions
Q_i	Cumulative influx of the displacing phase	res. bbl
Q_{tD}	Dimensionless fluid influx function at dimensionless time t_D	—
r	Radial distance	ft
r_D	Dimensionless reservoir size	—
r_e	External boundary radius	ft
r_w	Well radius	ft
R	Gas constant	varies
R	Producing gas-oil ratio	scf/STB
R	Stile's per cent recovery from a flood pattern	dimensionless
R_p	Cumulative gas-oil ratio	scf/STB
R_s	Solution gas-oil ratio (gas solubility in oil)	scf/STB
R_{si}	Initial solution gas-oil ratio	scf/STB
s	Skin factor	dimensionless
S	Saturation	dimensionless
S_g	Gas saturation	dimensionless
S_d	Displacing phase saturation	dimensionless
S_{di}	Initial displacing phase saturation	dimensionless
$\bar{S}_d$	Average S_d	dimensionless
$(S_d)_{pf}$	S_d at producing face	dimensionless
S_{df}	Displacing phase saturation at front	dimensionless
S_{ge}	Equilibrium gas saturation	dimensionless
S_{gi}	Gas saturation at flood start	dimensionless
S_L	Total (combined) liquid saturation	dimensionless
S_o	Oil saturation	dimensionless
S_{oP}	Oil saturation after primary	dimensionless
S_{os}	Oil saturation at flood start	dimensionless
S_{or}	Residual oil saturation	dimensionless
S_w	Water saturation	dimensionless
S_{wc}	Connate water or irreducible water saturation	dimensionless
t	Time	days
Δt	Change in time or shut-in time	days
t_D	Dimensionless time	dimensionless
t_{DS}	Dimensionless stabilization time	dimensionless
t_r	Reduced time (for flood prediction method)	—
t_s	Time to reach pseudosteady state (stabilize)	days
T	Temperature	varies
T_f	Formation temperature	°R or °F + 460
T_{sc}	Temperature, standard conditions	°R
v'	Velocity	cm/sec
V	Volume	various
V_b	Bulk volume	acre-ft; cu ft
V_p	Pore volume	cu ft; bbl
V_{pT}	Total formation pore volume	cu ft; bbl
V_{BP}	Bulk volume under primary drainage	acre-ft
V_{sw}	Gross swept volume	acre-ft
V_o	Volume of oil	varies
$(V_o)_{st}$	Volume of oil at stock-tank conditions	varies

		Dimensions
V_t	Total volume	varies
$(V_t)_{st}$	Total volume at stock-tank conditions	varies
$(W_i)_{Total}$	W_i sum for all zones (waterflood prediction method)	res. bbl
W_e	Cumulative water influx (encroachment)	res. bbl
W_i	Cumulative water injected	res. bbl
W_p	Cumulative water produced	res. bbl
ΔW_i	Water injected during Δt	res. bbl
$(W_p)_{Total}$	Water produced during flood from all zones (Prediction method)	res. bbl
X_f	Distance to front	ft
x'	Distance	cm
S or x	Distance	ft
Y	Math counter	—
z	Gas deviation factor (compressibility factor)	dimensionless
z_r	Gas deviation factor, reservoir	dimensionless
z_{avg}	gas deviation factor at average pressure	dimensionless
W_{ieff}	Effective cumulative water injected	res. bbl
α (alpha)	Angle of formation dip	degrees
β (beta)	Acute angle between interface and bed	degrees
σ (sigma)	Interfacial tension	dynes/cm
γ (gamma)	Specific gravity	dimensionless
γ_g	Gas specific gravity	dimensionless
γ_o	Oil specific gravity	dimensionless
γ_w	Water specific gravity	dimensionless
η (eta)	Hydraulic diffusivity ($6.33k/\phi\mu c$)	sq ft/day
λ (lambda)	Mobility (k/μ)	darcies/cp
λ_g	Gas mobility	darcies/cp
λ_o	Oil mobility	darcies/cp
λ_w	Water mobility	darcies/cp
μ (mu)	Viscosity	cp
μ_d	Viscosity displacing phase	cp
μ_g	Gas viscosity	cp
μ_i	Fluid viscosity at input	cp
μ_o	Oil viscosity	cp
μ_p	Fluid viscosity at producer	cp
μ_w	Water viscosity	cp
Φ (phi)	Porosity	fraction
θ (theta)	Fraction of circle	fraction
θ (theta)	Wetting angle	degrees

Appendix 

Table B.1

WYLIE'S EMPIRICAL RELATIVE PERMEABILITY EQUATIONS
FROM PETROLUEM PRODUCTION HANDBOOK

By Frick, McGraw-Hill, 1962

I. *Oil-gas relative permeabilities* (for drainage cycle relative to oil)

$$S^* = \frac{S_o}{(1 - S_{wi})}$$

Where S_{wi} is the irreducible water saturation.

	K_{ro}	K_{rg}
A. Unconsolidated Sand — Well Sorted	$(S^*)^{3.0}$	$(1 - S^*)^3$
B. Unconsolidated Sand — Poorly Sorted	$(S^*)^{3.5}$	$(1 - S^*)^2 (1 - S^{*1.5})$
C. Cemented Sand, Oolitic Lime, and Vugular Lime	$(S^*)^{4.0}$	$(1 - S^*)^{2.0} (1 - S^{*2.0})$

II. *Water-oil relative permeabilities* (for drainage cycle relative to water)

$$S^* = \left[\frac{S_w - S_{wi}}{1 - S_{wi}} \right]$$

Where S_{wi} is the irreducible water saturation.

	K_{ro}	K_{rw}
A. Unconsolidated Sand — Well Sorted	$(1 - S^*)^{3.0}$	$(S^*)^{3.0}$
B. Unconsolidated Sand — Poorly Sorted	$(1 - S^*)^2(1 - S^{*1.5})$	$(S^*)^{3.5}$
C. Cemented Sand, Oolitic Lime, and Vugular Lime	$(1 - S^*)^2(1 - S^{*2.0})$	$(S^*)^{4.0}$

450

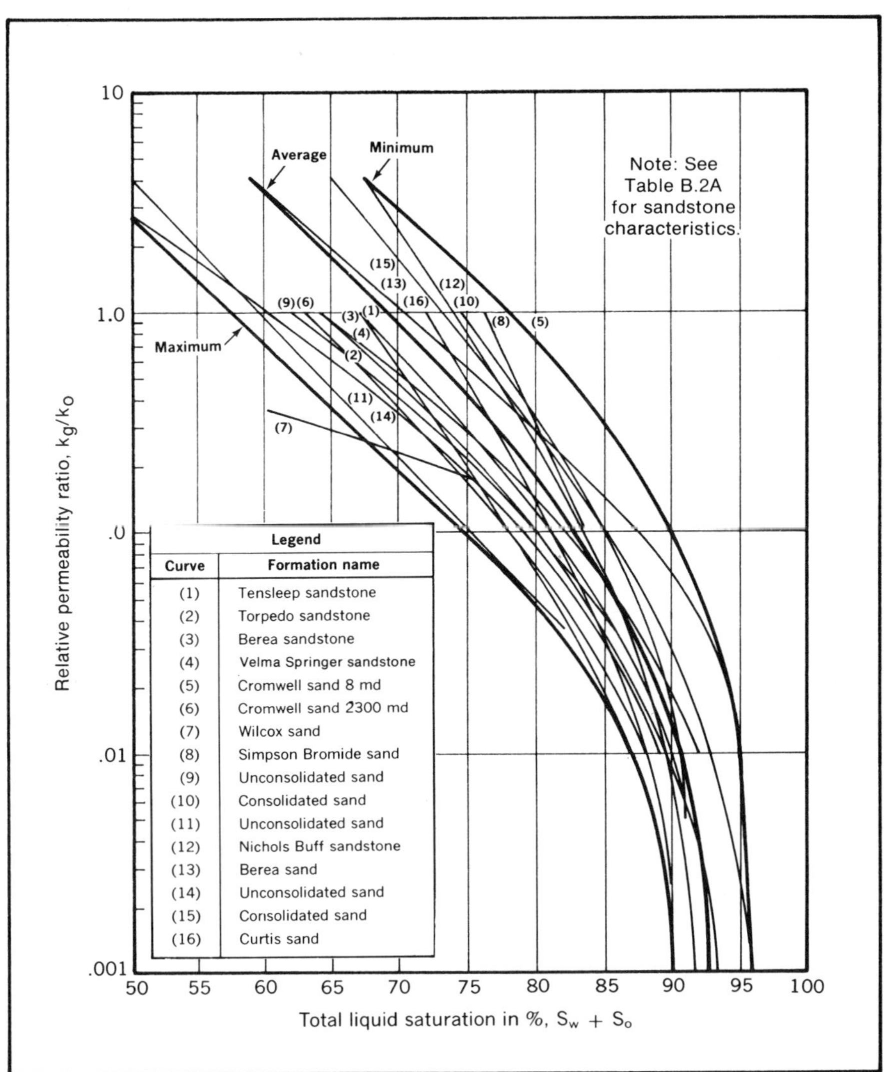

FIG. B.2A Relative permeability data for typical reservoirs (From Arps and Roberts, "The Effect of the Relative Permeability Ratio, the Oil Gravity, and the Solution Gas-Oil Ratio on the Primary Recovery from a Depletion Type Reservoir," *Trans. AIME,* 1955)

Table B.2A

SANDSTONE RELATIVE PERMEABILITY RATIO CURVES (SEE FIG. B.2A)

Curve no.	Formation name	(Md.) perm	(%) Porosity	Comments	Reference
1	Tensleep Ss.				e
2	Torpedo Sandstone				e
3	Berea Sandstone				e
4	Velma Springer Ss.				e
5	Cromwell Sand.	8			e
6	Cromwell Sand.	2300			e
7	Wilcox Sand	Oklahoma		Largely unconsolidated	a
8	Simpson Bromide Ss.	Oklahoma		Consolidated	a
9	Unconsolidated Ss.			Unconsolidated	a
10	Consolidated Sands.	500	21.8	Consolidated	a
11	Unconsolidated Sands			Unconsolidated	b
12	Nichols Buff Sand	500	21.8	Consolidated	c

13	Berea Sand	115	?		d
	Maximum			Typical of unconsolidated	
	Minimum			Highly cemented sandstones	
	Average			Average consolidated sand or sandstone	

Reference

a. Elkins, L. E.: "The Importance of Injected Gas as a Driving Medium in Limestone Reservoirs as Indicated by Recent Gas-Injection Experiments and Reservoir-Performance History," *Drilling and Production Practices* API (1946), 160.

b. Leveret, M. C., and Lewis, W. B.: "Steady Flow of Gas-Oil-Water Mixtures through Unconsolidated Sands," Trans. AIME (1941) 142, 107.

c. Botset, H. G., "Flow of Gas-Liquid Mixtures through Consolidated Sand," Trans. AIME (1940) 136, 91.

d. Richardson, J. G., Kerver, J. K., Haffold, J. A., and Osoba, J. S.: "Laboratory Determination of Relative Permeability," Trans. AIME (1952) 195, 187.

e. Arps, J. and Roberts, T. G., "The Effect of the Relative Permeability Ratio, the Oil Gravity, and the Solution Gas-Oil Ratio on the Primary Recovery from a Depletion Type Reservoir," *Trans.* AIME (1955), 204, 120.

Table B.2B

LIMESTONE AND DOLOMITE RELATIVE PERMEABILITY RATIO CURVES (SEE FIG. B2.B)

Curve no.	Formation name	(Md.) perm.	(%) Porosity	Comments	Ref.
1	San Andreas Lime				d
2	Pettit Lime				d
3	Oolithic Lime				d
4	Devonian Chert A				d
5	Devonian Chert B				d
6	Penn Reef A				d
7	Penn Reef B				d
8	Straun Reef				d
9	Palo Minto Reef				d
10	Intergranular La. A	78	18.3	Basically intergranular porosity	a
11	Intergranular La. B	3.0	17.6	Basically intergranular porosity	a
12	Reef limestone C	99.0	17.0	Very vuggy pores – some calcite crystals in them	a
13	Reef limestone D-1	36.3	28.4	Large vugs, fossils, very fine pores	a
14	Reef limestone D-2	47.2	35.6	Large vugs, fossils, very fine pores	a
15	Fractured La. E	55	12.8	Large cavities connected with hairline fracs	a

16	Fractured La. F	2.4	30.3	Large fractures connected together	a
17	Fractured La. F	2.4	30.3	Large fractures connected together	a
18	San Andreas Dol. A			West Texas pool	b
19	San Andreas Dol. B			Limestone pool West Texas	b
20	Pan Handle Dolomite			Pan Handle Region Texas	b
21	Pettit Lime			Louisiana	b
22	Hunton Lime			Oklahoma	b
23	Wasson D01	78	23.6		b
24	Permian Dolomite			An average of 26 samples	c
	Avg. Six Ls. Fields				c
	Average			Vugular Types Limestone	
	Minimum			Fractured Chert	
	Maximum			West Texas Permian dolomite	

References

a. Stewart, C. R., Craig, F. F., Jr., and Morse, R. A.: "Determination of Limestone Performance Characteristics by Model Flow Tests," Trans. AIME (1953) 198, 93.

b. Elkins, L. E.: "The Importance of Injected Gas as a Driving Medium in Limestone Reservoirs as Indicated by Recent Gas-Injection Experiments and Reservoir-Performance History," *Drilling and Production Practices*. API (1946), 160.

c. Balnes, A. C. and Fitting, R. U., Jr.: "An Introductory Discussion of the Reservoir Performance of Limestone Formations," Trans. AIME (1945) 160, 179.

d. Arps, J and Roberts, T. G., "The Effect of the Relative Permeability Ratio, the Oil Gravity, and the Solution Gas-Oil Ratio on the Primary Recovery from a Depletion Type Reservoir," *Trans.* AIME (1955), 204, 120.

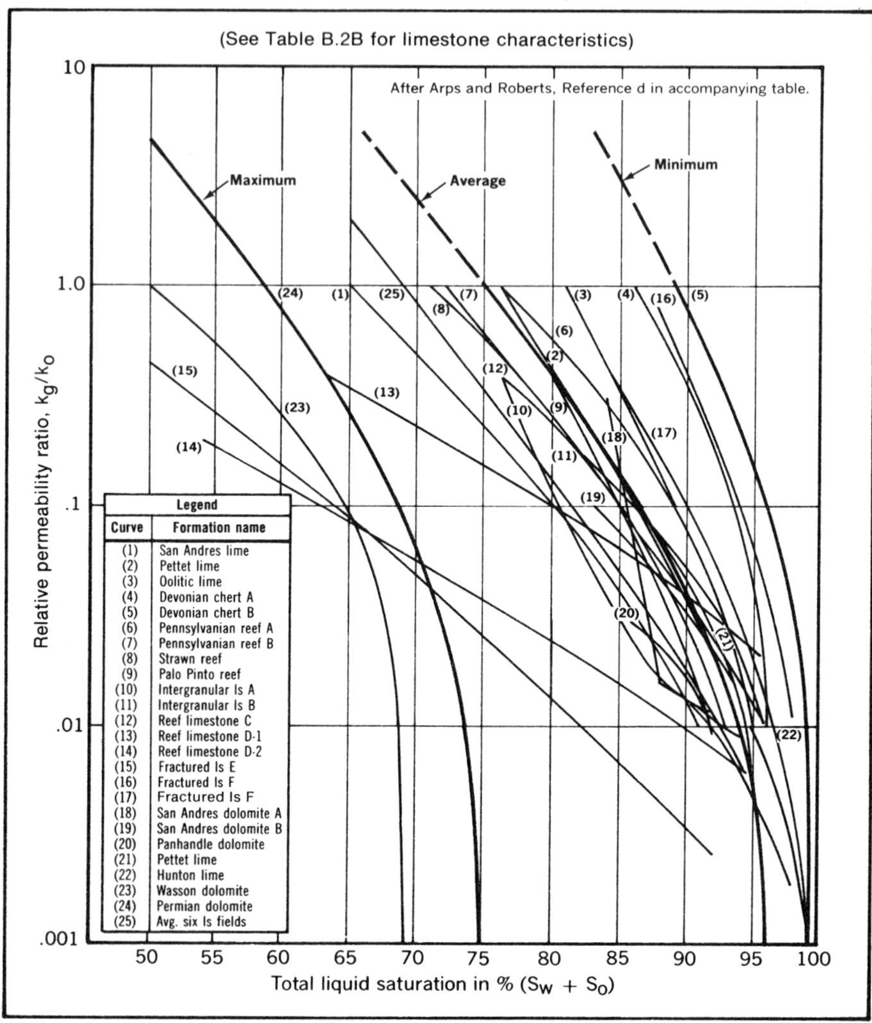

FIG. B.2B Relative permeability data for typical reservoirs (From Arps and Roberts, "The Effect of the Relative Permeability Ratio, the Oil Gravity, and the Solution Gas-Oil Ratio on the Primary Recovery from a Depletion Type Reservoir," *Trans. AIME*, 1955)

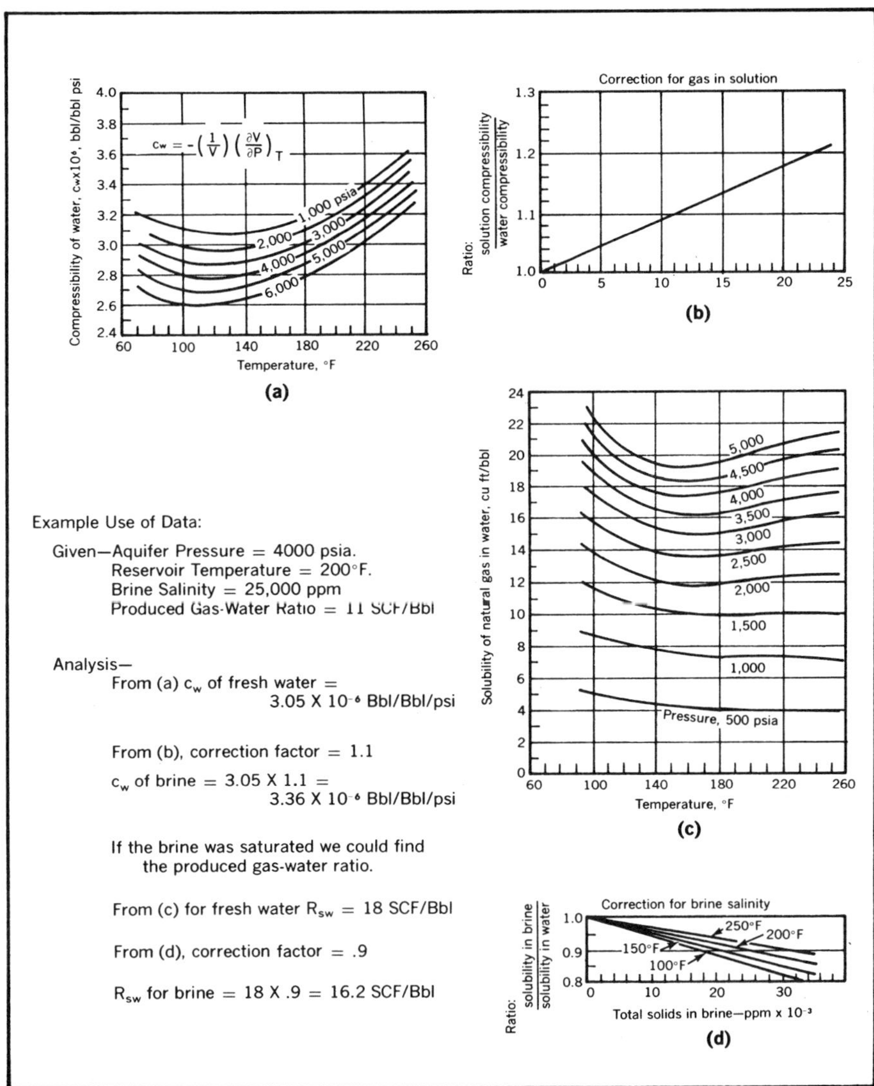

Example Use of Data:

Given—Aquifer Pressure = 4000 psia.
Reservoir Temperature = 200°F.
Brine Salinity = 25,000 ppm
Produced Gas-Water Ratio = 11 SCF/Bbl

Analysis—
From (a) c_w of fresh water =
3.05 X 10^{-6} Bbl/Bbl/psi

From (b), correction factor = 1.1

c_w of brine = 3.05 X 1.1 =
3.36 X 10^{-6} Bbl/Bbl/psi

If the brine was saturated we could find the produced gas-water ratio.

From (c) for fresh water R_{sw} = 18 SCF/Bbl

From (d), correction factor = .9

R_{sw} for brine = 18 X .9 = 16.2 SCF/Bbl

FIG. B.3 Compressibility and gas in solution for water. (From Dobson & Standing, "Pressure-volume-Temperature and Solubility Relations for Natural Gas-Water Mixtures," *Drilling and Production Practices*, API, 1944

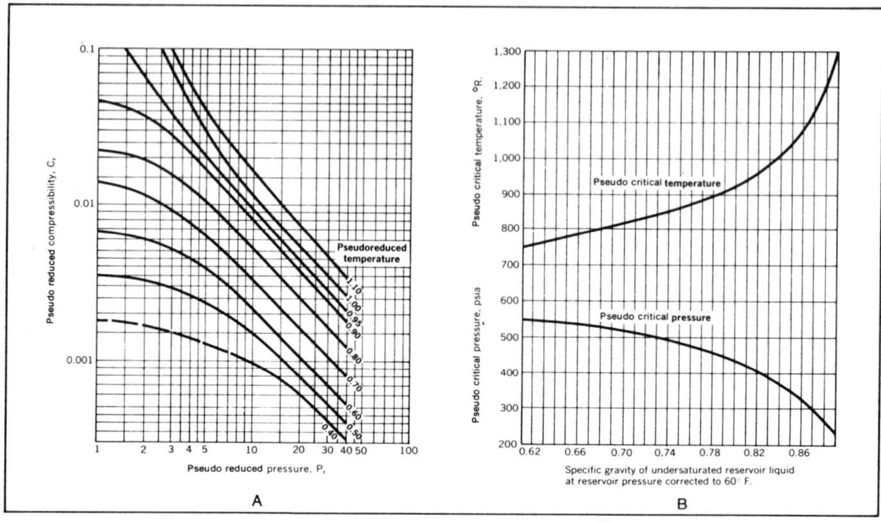

FIG. B.4 Empirical determination of oil compressibility (From Trube, "Compressibility of Undersaturated Hydrocarbon Reservoir Fluids," *Trans. AIME,* 1957)

Table B.4
EMPIRICAL DETERMINATION OF OIL COMPRESSIBILITY

Example determination:

A reservoir oil has a solution gas-oil ratio of 600 scf/STB. The gas has a specific gravity of 0.7 and the stock-tank oil is 40° API. Reservoir pressure is 3,500 psia, reservoir temperature 200° F., oil formation volume factor 1.3, and the original reservoir bubble-point pressure was 4,000 psia. Find the oil compressibility.

Solution: First calculate the density of the reservoir oil at reservoir conditions.

$$\text{Res. oil density} = (\text{STO wt} + \text{solution gas Wt})/B_o$$
$$= [(0.7 \times 350) + (29/379)(600)]/1.3$$
$$= 224 \text{ lb/bbl}$$

Res. oil specific gravity $= 224/350 = 0.64$

To determine the reservoir oil specific gravity at 60° F., enter Fig. B.4C at the reservoir temperature of 200° F. and specific gravity of 0.64 and follow the correlation lines to 60° F. and read a specific gravity at 60° F. as 0.705.

From Fig. B.4B read the pseudocritical pressure and temperature as 520 psia and 820° R.

Reduced temperatures and pressures can then be calculated.

$$T_r = (200 + 460)/820 = 0.805$$
$$p_r = 3,500/520 = 6.731$$

Now the reduced oil compressibility can be read from Fig. B.4A as 0.008 and the oil compressibility can be calculated as

$$c_o = c_r/p_c$$
$$= 0.008/520 = 1.54 \times 10^{-5}/\text{psi}$$

Note that any oil that does not contain free gas can be considered undersaturated.

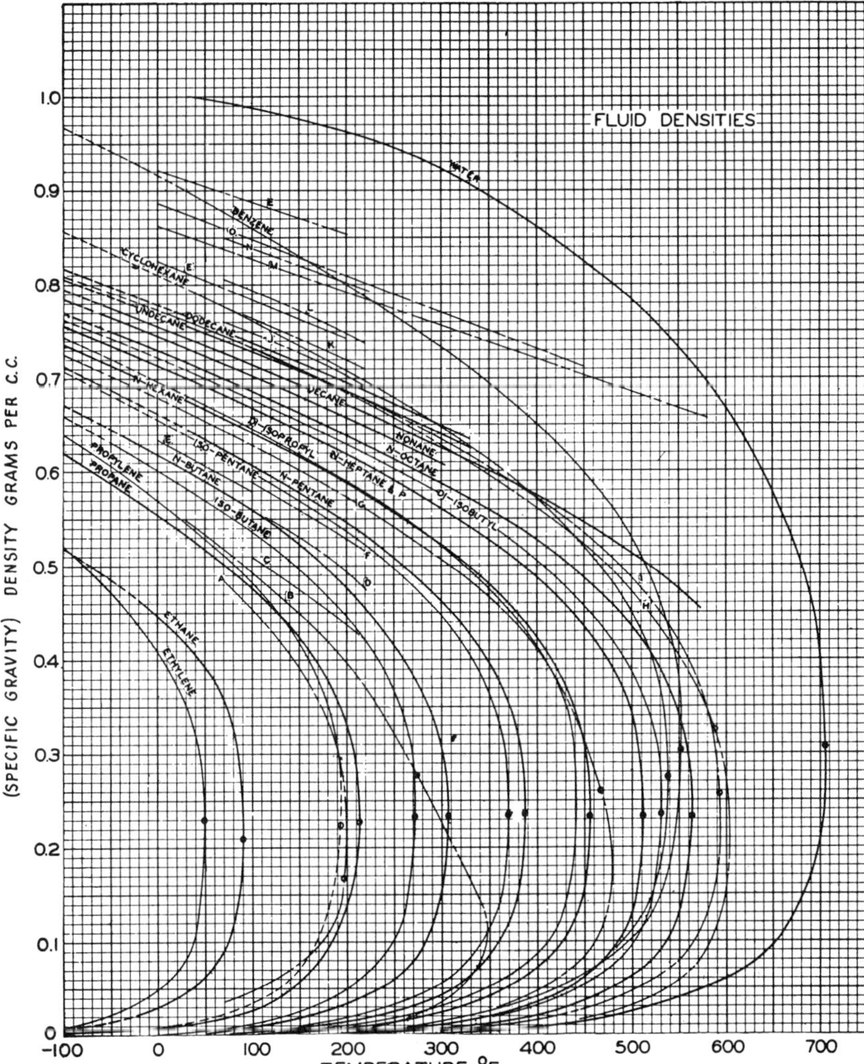

FIG. B.4C FLUID DENSITIES for hydrocarbons (From Brown et al., Natural Gasoline and the Volatile Hydrocarbons, 1948, Courtesy Gas Processors Association)

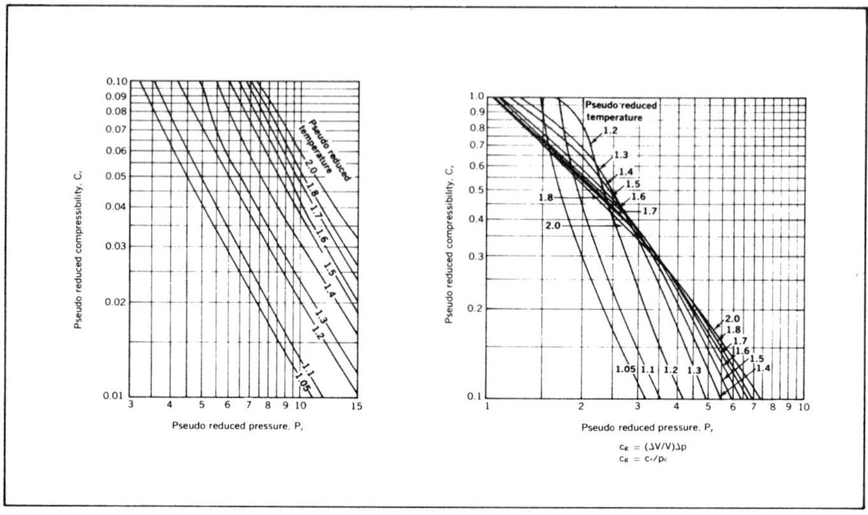

FIG. B.5 Compressibility of natural gases (From Trube, "Compressibility of Natural Gases," *Journal of Petroleum Technology,* 9(1):69(1957))

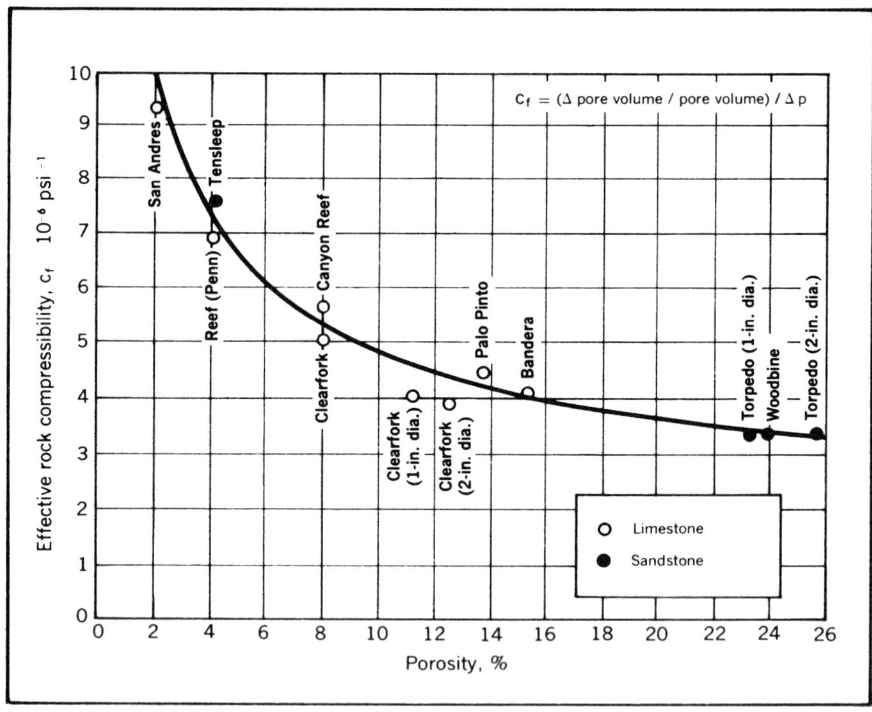

FIG. B.6 Formation compressibility (From Hall, "Compressibility of Reservoir Rocks," *Trans. AIME,* 1953)

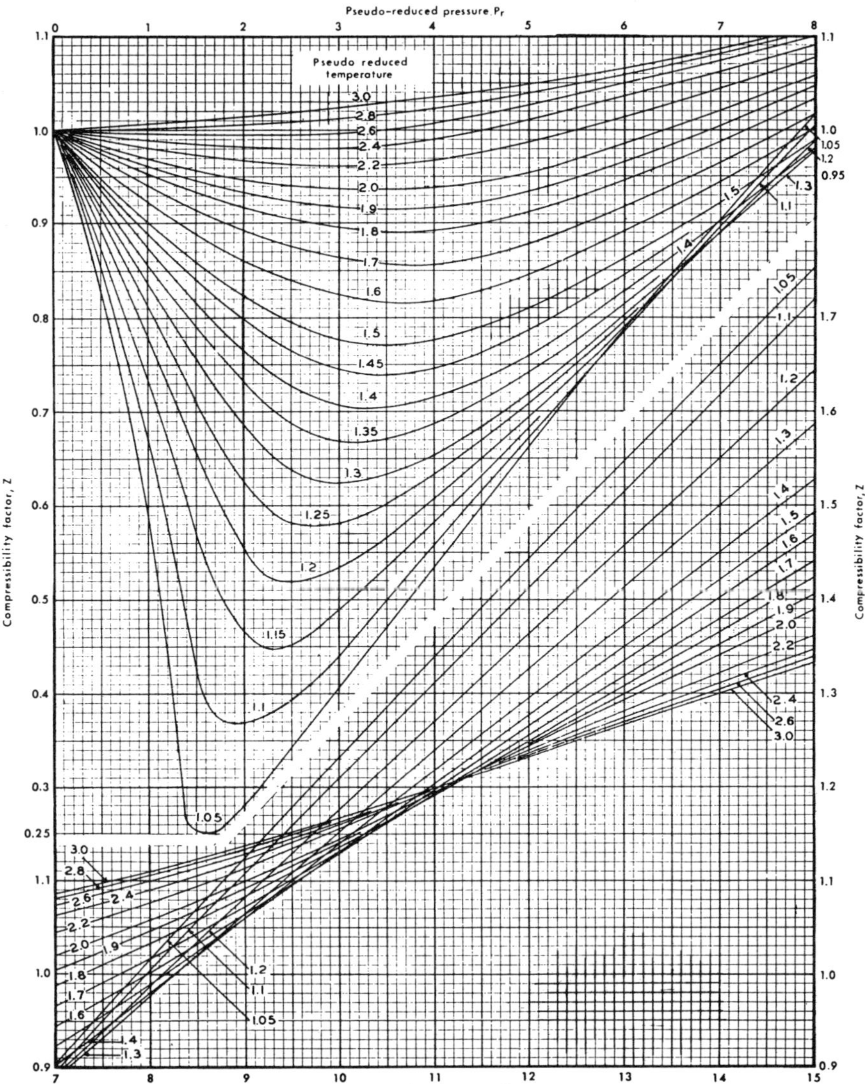

FIG. B.7 Gas deviation (compressibility) factors (From Standing and Katz, "Density of Natural Gases, *Trans. AIME,* 1942)

Note: T_c and p_c should be calculated from the gas composition and the critical values of Appendix B.11 or alternatively obtained from the empirical data of Appendix B.8.

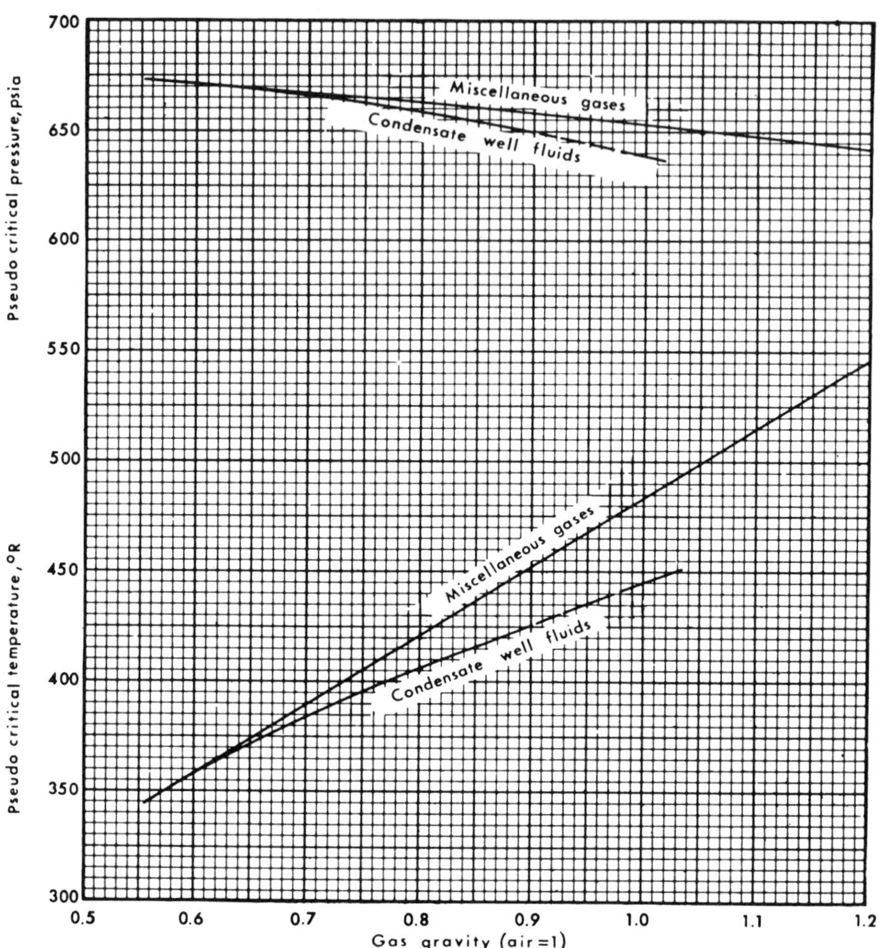

FIG. B.8   Pseudo critical properties of natural gas (From Brown et al., *Natural Gasoline and the Volatile Hydrocarbons.* Courtesy Gas Processors Association)

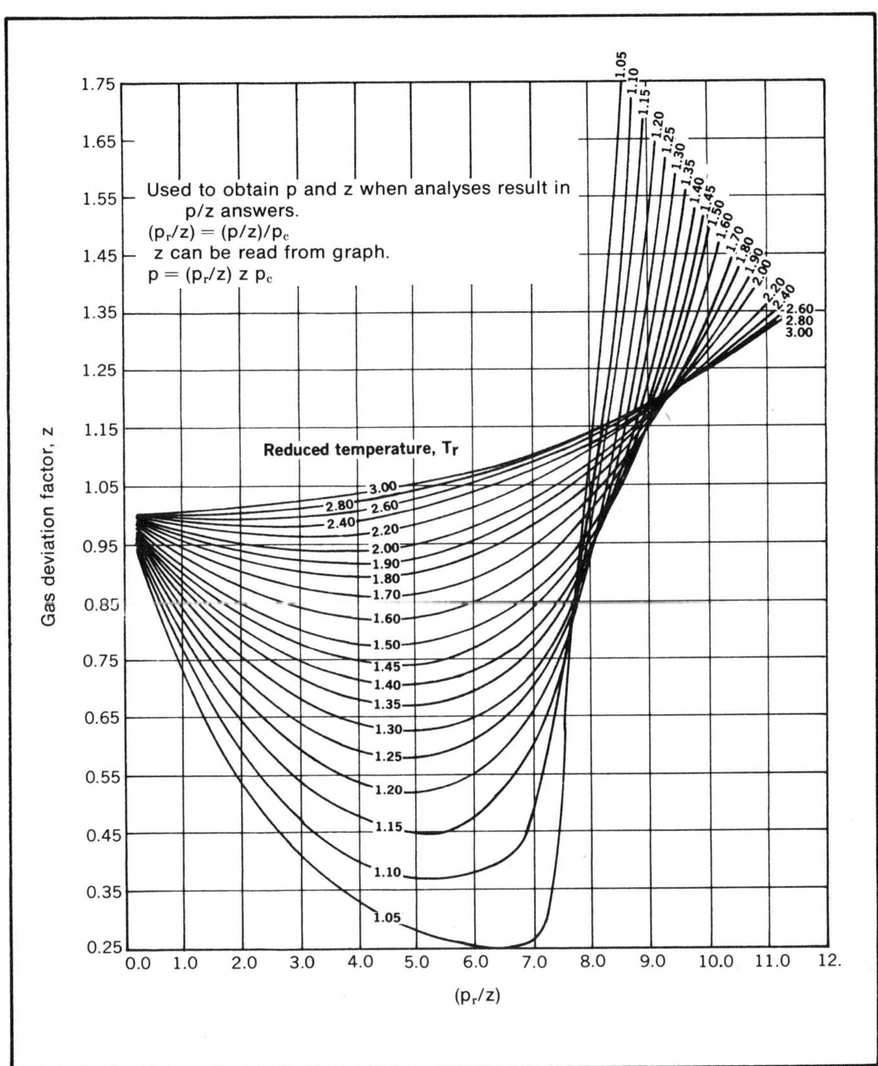

FIG. B.9 Gas deviation factor, z, vs. (p_r/z)
(unpublished data from Slider's Gas Reservoir Engineering Short Course)

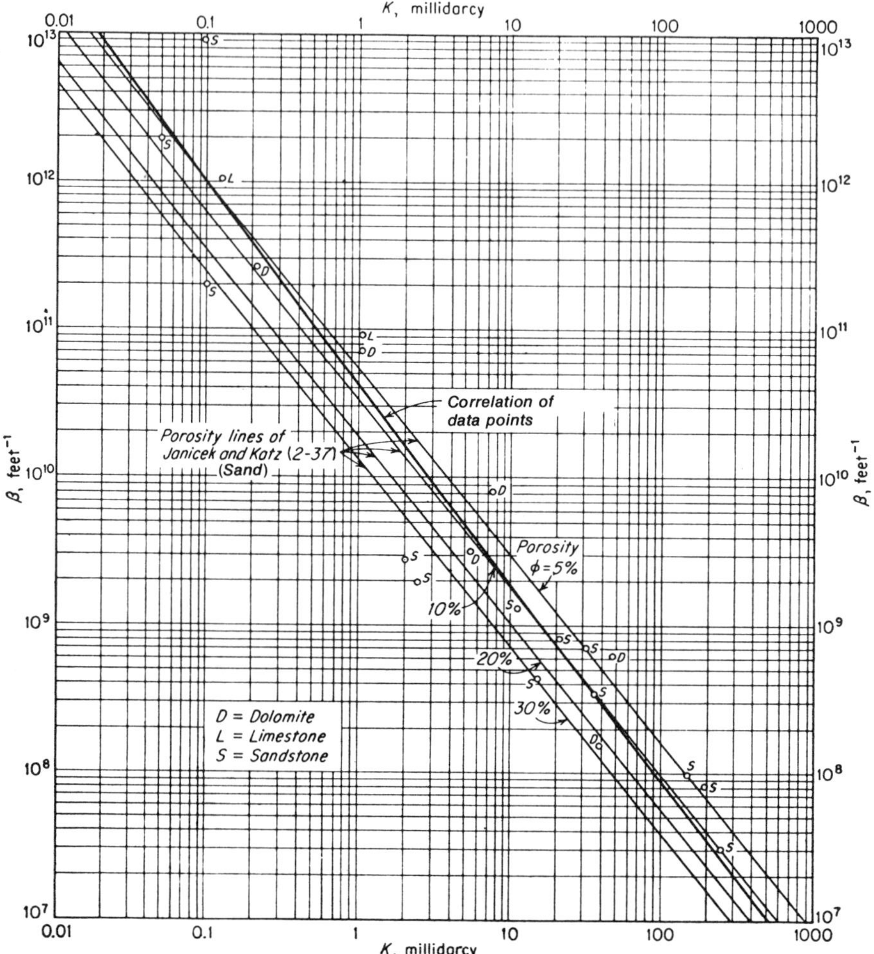

FIG. B.10 Gas turbulence factor (From *Handbook of Natural Gas Engineering*, McGraw-Hill Book Co., 1959)

Porosity Lines of Janicek and Katz—*not a correlation of plotted data points*—from Janicek and Katz, "Application of Unsteady State Calculations, preprint, Univ. of Mich. Publishing Services, Ann Arbor, 1955.

Correlation of Data Points—from Katz and Correli, "Flow of Natural Gas from Reservoirs, course notes, Univ. of Mich. Publishing Services, Ann Arbor, 1955.

Table B.11

PHYSICAL PROPERTIES OF PETROLEUM COMPONENTS

(After Eilerts and Others, Phase Relations of Gas-Condensate Fluids, U.S. Bureau of Mines Monograph 10, Vol. I (New York: American Gas Association, 1957), 427–434.)

Compound	Molecular weight	Boiling point at 14.7 psia, $°F.$	Critical constants	
			Pressure, p_c, psia	Temperature, $T_c°R.$
Methane	16.04	−258.7	673.1	343.2
Ethane	30.07	−127.5	708.3	549.9
Propane	44.09	−43.7	617.4	666.0
Isobutane	58.12	10.9	529.1	734.6
n-butane	58.12	31.1	550.1	765.7
Isopentane	72.15	82.1	483.5	829.6
n-pentane	72.15	96.9	489.8	846.2
n-hexane	86.77	155.7	440.1	914.2
n-heptane	100.2	209.2	395.9	972.4
n-octane	114.2	258.2	362.2	1024.9
n-nonane	128.3	303.4	334	1073
n-decane	142.3	345.4	312	1115
Air	28.97	−317.7	547	239
Carbon dioxide	44.01	−109.3	1070.2	547.5
Helium	4.003	−452.1	33.2	9.5
Hydrogen	2.016	−423.0	189.0	59.8
Hydrogen sulfide	34.08	−76.6	1306.5	672.4
Nitrogen	28.02	−320.4	492.2	227.0
Oxygen	32.00	−297.4	736.9	278.6
Water	18.02	212.0	2109.5	1165.2

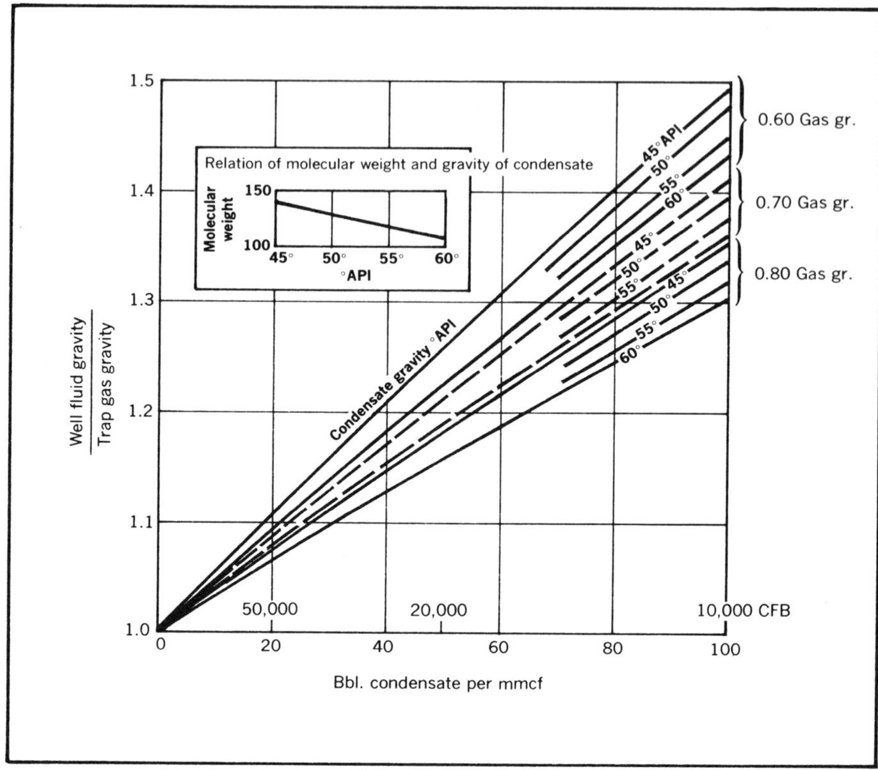

FIG. B.12 Effect of condensate volume on the ratio of surface gas gravity to well-fluid gravity (After M. B. Standing, Volumetric and Phase Behavior of Oil Field Hydrocarbon Systems. Chevron Research Co., 1952)

Table B.13

HYPERBOLIC DECLINE CURVE FUNCTIONS FOR $n = 0.1$

Decline rates listed, (a), are at time = 0 and in months^{-1}.

Time, months	$a = 0.005$		$a = 0.010$		$a = 0.015$		$a = 0.020$		$a = 0.025$	
	q/q_i	N_p	q/q_i	N_p	q/q_i	N_p	q/q_i	N_p	q/q_i	N_p
−120.000	1.857	−165.604	3.591	−239.974	7.275	−367.843	15.555	−601.200	35.401	−1056.928
−117.000	1.827	−160.078	3.470	−229.383	6.888	−346.603	14.378	−556.326	31.823	−956.197
−114.000	1.798	−154.639	3.355	−219.147	6.523	−326.492	13.299	−514.832	28.638	−865.598
−111.000	1.770	−149.287	3.243	−209.250	6.179	−307.444	12.308	−476.442	25.800	−784.022
−108.000	1.742	−144.019	3.136	−199.883	5.855	−289.396	11.398	−440.902	23.269	−710.493
−105.000	1.715	−138.834	3.032	−190.432	5.550	−272.292	10.562	−407.979	21.008	−644.142
−102.000	1.688	−133.730	2.932	−181.486	5.262	−256.078	9.792	−377.466	18.986	−584.207
−99.000	1.661	−128.706	2.836	−172.833	4.991	−240.704	9.084	−349.167	17.176	−530.012
−96.000	1.835	−123.760	2.744	−164.465	4.734	−226.119	8.431	−322.908	15.555	−480.960
−93.000	1.610	−118.893	2.654	−156.369	4.493	−212.283	7.830	−298.529	14.100	−436.518
−90.000	1.585	−114.101	2.568	−148.537	4.264	−199.151	7.275	−275.882	12.793	−396.213
−87.000	1.560	−109.384	2.485	−140.958	4.049	−186.685	6.764	−254.833	11.619	−359.626
−84.000	1.536	−104.740	2.405	−133.625	3.845	−174.847	6.292	−235.259	10.562	−328.384
−81.000	1.512	−100.168	2.327	−126.528	3.652	−163.604	5.855	−217.047	9.609	−296.152
−78.000	1.489	−95.668	2.253	−119.659	3.471	−152.922	5.452	−200.093	8.751	−268.834
−75.000	1.466	−91.237	2.181	−113.009	3.298	−142.771	5.079	−184.303	7.976	−243.564
−72.000	1.443	−86.874	2.111	−106.572	3.136	−133.122	4.734	−169.589	7.275	−220.706
−69.000	1.421	−82.579	2.044	−100.340	2.982	−123.948	4.415	−155.872	6.642	−199.865
−66.000	1.399	−78.350	1.979	−94.305	2.836	−115.222	4.119	−143.076	6.069	−180.792
−63.000	1.377	−74.186	1.917	−88.481	2.698	−106.922	3.845	−131.135	5.550	−163.375
−60.000	1.356	−70.087	1.857	−82.802	2.568	−99.024	3.591	−119.987	5.079	−147.443
−57.000	1.335	−66.049	1.798	−77.320	2.444	−91.508	3.355	−109.573	4.652	−132.856
−54.000	1.315	−62.074	1.742	−72.009	2.327	−84.352	3.136	−99.842	4.264	−119.490
−51.000	1.295	−58.160	1.688	−66.865	2.216	−77.538	2.932	−90.743	3.912	−107.235
−48.000	1.275	−54.305	1.635	−61.880	2.111	−71.048	2.744	−82.232	3.591	−95.990
−45.000	1.256	−50.510	1.585	−57.051	2.011	−64.866	2.568	−74.268	3.298	−85.663

Table B.13 (*Continued*)
HYPERBOLIC DECLINE CURVE FUNCTIONS FOR $n = 0.1$ (continued)

Decline rates listed, (a), are at time = 0 and in months^{-1}.

Time, months	a = 0.005		a = 0.010		a = 0.015		a = 0.020		a = 0.025	
	q/q_i	N_p	q/q_i	N_p	q/q_i	N_p	q/q_i	N_p	q/q_i	N_p
−42.000	1.236	−46.772	1.536	−52.370	1.917	−58.974	2.405	−66.812	3.032	−76.173
−39.000	1.218	−43.091	1.489	−47.834	1.827	−53.359	2.253	−59.829	2.790	−67.446
−36.000	1.199	−39.488	1.443	−43.437	1.742	−48.006	2.111	−53.266	2.568	−59.415
−33.000	1.181	−35.895	1.399	−39.175	1.661	−42.902	1.979	−47.153	2.366	−52.019
−30.000	1.163	−32.379	1.356	−35.043	1.585	−38.034	1.857	−41.401	2.181	−45.204
−27.000	1.146	−28.916	1.315	−31.037	1.512	−33.389	1.742	−36.005	2.011	−38.919
−24.000	1.128	−25.506	1.275	−27.153	1.443	−28.958	1.635	−30.940	1.857	−33.121
−21.000	1.111	−22.146	1.236	−23.386	1.377	−24.729	1.536	−26.185	1.715	−27.767
−18.000	1.095	−18.837	1.199	−19.733	1.315	−20.691	1.443	−21.719	1.585	−22.820
−15.000	1.078	−15.578	1.163	−16.190	1.256	−16.837	1.356	−17.522	1.466	−18.247
−12.000	1.062	−12.368	1.128	−12.753	1.199	−13.155	1.275	−13.576	1.356	−14.017
−9.000	1.046	−9.206	1.095	−9.419	1.146	−9.639	1.199	−9.866	1.256	−10.102
−5.000	1.031	−6.091	1.062	−8.184	1.095	−6.279	1.128	−6.376	1.163	−6.476
−3.000	1.015	−3.022	1.031	−3.045	1.046	−3.069	1.062	−3.092	1.078	−3.116
0.0	1.000	−0.0	1.000	−0.0	1.000	−0.0	1.000	−0.0	1.000	−0.0
3.000	0.985	2.976	0.970	2.955	0.956	2.933	0.942	2.912	0.928	2.890
15.000	0.928	14.452	0.862	13.934	0.801	13.442	0.744	12.977	0.692	12.534
27.000	0.875	25.263	0.766	23.688	0.672	22.255	0.591	20.949	0.520	19.755
39.000	0.824	35.454	0.682	32.367	0.566	29.687	0.472	27.296	0.394	25.206
51.000	0.777	45.061	0.608	40.099	0.478	35.920	0.379	32.377	0.301	29.352
63.000	0.733	54.123	0.543	46.996	0.405	41.210	0.305	36.462	0.232	32.528
75.000	0.692	62.672	0.485	53.157	0.344	45.697	0.247	39.763	0.179	34.979
87.000	0.653	70.742	0.434	58.668	0.293	49.514	0.201	42.442	0.140	36.883
99.000	0.617	78.380	0.389	63.602	0.250	52.769	0.164	44.626	0.110	38.371
111.000	0.583	85.554	0.349	68.025	0.214	55.551	0.135	46.413	0.086	39.540

Table B.13 (Continued)

HYPERBOLIC DECLINE CURVE FUNCTIONS FOR n = 0.2

Decline rates listed, (a), are at time = 0 and in months^{-1}.

Time, months	a = 0.005		a = 0.010		a = 0.015		a = 0.020		a = 0.025	
	q/q	N_p	q/q_i	N_p	q/q_i	N_p	q/q_i	N_p	q/q_i	N_p
−120.000	1.895	−166.878	3.944	−249.675	9.313	−413.371	26.302	−792.302	97.656	−1903.116
−117.000	1.863	−161.242	3.792	−238.073	8.685	−386.386	23.466	−717.748	81.238	−1635.682
−114.000	1.832	−155.700	3.647	−226.916	8.107	−361.209	20.990	−651.146	68.023	−1412.494
−111.000	1.801	−150.251	3.508	−216.186	7.575	−337.698	18.820	−591.502	57.306	−1225.053
−108.000	1.771	−144.894	3.376	−205.861	7.084	−315.720	16.914	−537.962	48.552	−1066.702
−105.000	1.741	−139.626	3.250	−195.923	6.631	−295.158	15.236	−489.790	41.355	−932.186
−102.000	1.712	−134.445	3.129	−186.356	6.212	−275.903	13.753	−446.353	35.401	−817.330
−99.000	1.684	−129.351	3.014	−177.143	5.824	−257.856	12.440	−407.103	30.447	−718.781
−96.000	1.656	−124.340	2.904	−168.267	5.465	−240.930	11.274	−371.566	26.302	−633.840
−93.000	1.629	−119.412	2.798	−159.716	5.132	−225.040	10.238	−339.328	22.815	−560.314
−90.000	1.602	−114.564	2.697	−151.474	4.824	−210.112	9.313	−310.028	19.869	−496.409
−87.000	1.576	−109.796	2.601	−143.528	4.537	−196.076	8.487	−283.351	17.368	−440.654
−84.000	1.551	−105.106	2.508	−135.865	4.271	−182.869	7.747	−259.019	15.236	−391.832
−81.000	1.526	−100.492	2.420	−128.474	4.023	−170.433	7.084	−236.790	13.410	−348.933
−78.000	1.501	−95.952	2.335	−121.343	3.792	−158.715	6.487	−216.449	11.840	−311.118
−75.000	1.477	−91.486	2.254	−114.460	3.577	−147.666	5.950	−197.807	10.486	−277.679
−72.000	1.453	−87.092	2.176	−107.817	3.376	−137.240	5.465	−180.698	9.313	−248.022
−69.000	1.430	−82.768	2.101	−101.402	3.189	−127.396	5.027	−164.971	8.295	−221.646
−66.000	1.407	−78.513	2.030	−95.207	3.014	−118.095	4.630	−150.494	7.407	−198.125
−63.000	1.385	−74.326	1.961	−89.222	2.850	−109.301	4.271	−137.151	6.631	−177.095
−60.000	1.363	−70.205	1.895	−83.439	2.697	−100.983	3.944	−124.837	5.950	−158.246
−57.000	1.341	−66.150	1.832	−77.850	2.554	−93.108	3.647	−113.458	5.351	−141.313
−54.000	1.320	−62.159	1.771	−72.447	2.420	−85.649	3.376	−102.930	4.824	−126.067
−51.000	1.299	−58.230	1.712	−67.222	2.294	−78.581	3.129	−93.178	4.357	−112.310
−48.000	1.279	−54.363	1.656	−62.170	2.176	−71.878	2.904	−84.134	3.944	−99.870
−45.000	1.259	−50.557	1.602	−57.282	2.065	−65.518	2.697	−75.737	3.577	−88.600

Table B.13 (Continued)

HYPERBOLIC DECLINE CURVE FUNCTIONS FOR $n = 0.2$ (continued)

Decline rates listed, (a), are at time $= 0$ and in months^{-1}.

Time, months	a = 0.005		a = 0.010		a = 0.015		a = 0.020		a = 0.025	
	q/q_i	N_p	q/q_i	N_p	q/q_i	N_p	q/q_i	N_p	q/q_i	N_p
−42.000	1.239	−46.809	1.551	−52.553	1.961	−59.481	2.508	−67.933	3.250	−78.369
−39.000	1.220	−43.120	1.501	−47.976	1.863	−53.747	2.335	−60.671	2.958	−69.065
−36.000	1.201	−39.489	1.453	−43.546	1.771	−48.298	2.176	−53.908	2.697	−60.590
−33.000	1.183	−35.913	1.407	−39.256	1.684	−43.117	2.030	−47.603	2.464	−52.855
−30.000	1.165	−32.392	1.363	−35.103	1.602	−38.188	1.895	−41.719	2.254	−45.784
−27.000	1.147	−28.925	1.320	−31.079	1.526	−33.497	1.771	−36.223	2.065	−39.311
−24.000	1.129	−25.512	1.279	−27.182	1.453	−29.030	1.656	−31.085	1.895	−33.376
−21.000	1.112	−22.151	1.239	−23.405	1.385	−24.775	1.551	−26.276	1.741	−27.925
−18.000	1.095	−18.840	1.201	−19.744	1.320	−20.720	1.453	−21.773	1.602	−22.913
−15.000	1.078	−15.580	1.165	−16.196	1.259	−16.852	1.363	−17.551	1.477	−18.297
−12.000	1.062	−12.369	1.129	−12.756	1.201	−13.163	1.279	−13.591	1.363	−14.041
−9.000	1.046	−9.206	1.095	−9.420	1.147	−9.642	1.201	−9.872	1.259	−10.111
−6.000	1.031	−6.091	1.062	−6.184	1.095	−6.280	1.129	−6.378	1.165	−6.478
−3.000	1.015	−3.023	1.031	−3.045	1.046	−3.069	1.062	−3.092	1.078	−3.116
0.0	1.000	−0.0	1.000	−0.0	1.000	−0.0	1.000	−0.0	1.000	−0.0
3.000	0.985	2.977	0.971	2.955	0.956	2.934	0.942	2.912	0.928	2.891
15.000	0.928	14.453	0.863	13.939	0.802	13.453	0.747	12.994	0.697	12.560
27.000	0.875	25.271	0.769	23.714	0.677	22.307	0.599	21.031	0.531	19.871
39.000	0.826	35.475	0.687	32.437	0.575	29.802	0.484	27.501	0.410	25.481
51.000	0.780	45.106	0.615	40.241	0.491	36.181	0.395	32.758	0.321	29.844
63.000	0.737	54.202	0.552	47.240	0.421	41.638	0.325	37.063	0.254	33.279
75.000	0.697	62.800	0.497	53.531	0.363	46.327	0.269	40.617	0.203	36.012
87.000	0.659	70.930	0.448	59.198	0.314	50.375	0.225	43.571	0.164	38.209
99.000	0.624	78.624	0.405	64.315	0.272	53.885	0.189	46.043	0.134	39.991
111.000	0.591	85.909	0.367	68.943	0.238	56.940	0.159	48.125	0.110	41.448

Table B.13 (*Continued*)
HYPERBOLIC DECLINE CURVE FUNCTIONS FOR n = 0.3

Decline rates listed, (a), are at time = 0 and in months^{-1}.

Time, months	a = 0.005		a = 0.010		a = 0.015		a = 0.020		a = 0.025	
	q/q_i	N_p	q/q_i	N_p	q/q_i	N_p	q/q_i	N_p	q/q_i	N_p
−120.000	1.938	−168.262	4.427	−261.857	13.309	−487.814	69.631	−1321.194	2154.372	−12253.668
−117.000	1.903	−162.501	4.225	−248.882	12.086	−449.761	56.573	−1132.769	1095.308	−7610.066
−114.000	1.868	−156.845	4.036	−236.494	11.004	−415.159	46.527	−978.756	624.354	−5116.105
−111.000	1.835	−151.290	3.857	−224.658	10.046	−383.612	38.682	−851.415	386.021	−3637.644
−108.000	1.802	−145.834	3.688	−213.342	9.193	−354.779	32.473	−745.035	253.599	−2696.236
−105.000	1.771	−140.474	3.529	−202.517	8.432	−328.363	27.500	−655.345	174.635	−2063.433
−102.000	1.739	−135.210	3.379	−192.157	7.751	−304.108	23.472	−579.095	124.863	−1619.600
−99.000	1.709	−130.038	3.237	−182.234	7.139	−281.789	20.180	−513.778	92.060	−1297.460
−96.000	1.679	−124.956	3.103	−172.727	6.589	−261.210	17.463	−457.442	69.631	−1056.955
−93.000	1.650	−119.962	2.975	−163.611	6.093	−242.199	15.203	−408.546	53.816	−873.108
−90.000	1.622	−115.055	2.855	−154.868	5.644	−224.604	13.309	−365.861	42.369	−729.720
−87.000	1.594	−110.232	2.741	−146.476	5.238	−208.292	11.710	−328.398	33.896	−615.936
−84.000	1.567	−105.491	2.632	−138.418	4.868	−193.142	10.353	−295.358	27.500	−524.276
−81.000	1.540	−100.831	2.529	−130.677	4.532	−179.050	9.193	−266.084	22.587	−449.460
−78.000	1.514	−96.251	2.432	−123.237	4.225	−165.921	8.196	−240.037	18.757	−387.674
−75.000	1.489	−91.747	2.339	−116.082	3.945	−153.672	7.336	−216.770	15.730	−336.116
−72.000	1.464	−87.318	2.251	−109.199	3.688	−142.228	6.589	−195.908	13.309	−292.689
−69.000	1.439	−82.964	2.166	−102.575	3.453	−131.520	5.938	−177.138	11.350	−255.802
−66.000	1.416	−78.682	2.086	−96.196	3.237	−121.489	5.369	−160.196	9.751	−224.229
−63.000	1.392	−74.471	2.010	−90.052	3.038	−112.081	4.868	−144.857	8.432	−197.018
−60.000	1.369	−70.328	1.938	−84.131	2.855	−103.245	4.427	−130.928	7.336	−173.416
−57.000	1.347	−66.254	1.868	−78.422	2.686	−94.937	4.036	−118.247	6.418	−152.824
−54.000	1.325	−62.245	1.802	−72.917	2.529	−87.118	3.688	−106.671	5.644	−134.763
−51.000	1.304	−58.302	1.739	−67.605	2.385	−79.750	3.379	−96.078	4.987	−118.842
−48.000	1.283	−54.422	1.679	−62.478	2.251	−72.799	3.103	−86.363	4.427	−104.743
−45.000	1.262	−50.604	1.622	−57.527	2.126	−66.237	2.855	−77.434	3.945	−92.203

Table B.13 (*Continued*)
HYPERBOLIC DECLINE CURVE FUNCTIONS FOR n = 0.3 (continued)

Decline rates listed, (a), are at time = 0 and in months^{-1}.

Time, months	a = 0.005		a = 0.010		a = 0.015		a = 0.020		a = 0.025	
	q/q_i	N_p	q/q_i	N_p	q/q_i	N_p	q/q_i	N_p	q/q_i	N_p
−42.000	1.242	−46.848	1.567	−52.746	2.010	−60.035	2.632	−69.209	3.529	−81.007
−39.000	1.223	−43.151	1.514	−48.125	1.903	−54.167	2.432	−61.618	3.169	−70.972
−36.000	1.203	−39.512	1.464	−43.659	1.802	−48.611	2.251	−54.600	2.855	−61.947
−33.000	1.184	−35.931	1.416	−39.341	1.709	−43.346	2.086	−48.098	2.580	−53.804
−30.000	1.166	−32.406	1.369	−35.164	1.622	−38.352	1.938	−42.065	2.339	−46.433
−27.000	1.148	−28.935	1.325	−31.123	1.540	−33.610	1.802	−36.458	2.126	−39.742
−24.000	1.130	−25.518	1.283	−27.211	1.464	−29.106	1.679	−31.239	1.938	−33.652
−21.000	1.113	−22.155	1.242	−23.424	1.392	−24.824	1.567	−26.373	1.771	−28.095
−18.000	1.096	−18.843	1.203	−19.756	1.325	−20.748	1.464	−21.830	1.622	−23.011
−15.000	1.079	−15.581	1.166	−16.203	1.262	−16.868	1.369	−17.582	1.489	−18.349
−12.000	1.062	−12.369	1.130	−12.759	1.203	−13.171	1.283	−13.606	1.369	−14.066
−9.000	1.046	−9.206	1.096	−9.421	1.148	−9.645	1.203	−9.878	1.262	−10.121
−6.000	1.031	−6.091	1.062	−6.185	1.096	−6.281	1.130	−6.380	1.166	−6.481
−3.000	1.015	−3.023	1.031	−3.045	1.046	−3.069	1.062	−3.092	1.079	−3.116
0.0	1.000	−0.0	1.000	−0.0	1.000	−0.0	1.000	−0.0	1.000	−0.0
3.000	0.985	2.977	0.971	2.956	0.956	2.934	0.942	2.912	0.929	2.891
15.000	0.929	14.455	0.864	13.944	0.804	13.463	0.750	13.011	0.701	12.584
27.000	0.876	25.278	0.771	23.740	0.682	22.357	0.606	21.110	0.541	19.981
39.000	0.827	35.495	0.692	32.506	0.583	29.931	0.496	27.696	0.425	25.741
51.000	0.782	45.149	0.622	40.378	0.502	36.429	0.411	33.116	0.340	30.305
63.000	0.740	54.279	0.562	47.473	0.435	42.041	0.343	37.625	0.275	33.978
75.000	0.701	62.922	0.508	53.886	0.379	46.918	0.290	41.413	0.226	36.972
87.000	0.664	71.112	0.462	59.700	0.332	51.181	0.247	44.622	0.187	39.443
99.000	0.630	78.878	0.420	64.986	0.293	54.926	0.211	47.363	0.157	41.503
111.000	0.598	86.248	0.384	69.805	0.259	58.232	0.182	49.720	0.133	43.257

Table B.13 (Continued)

HYPERBOLIC DECLINE CURVE FUNCTIONS FOR $n = 0.4$

Decline rates listed, (a), are at time = 0 and in months^{-1}.

Time, months	$a = 0.005$		$a = 0.010$		$a = 0.015$		$a = 0.020$		$a = 0.025$	
	q/q_i	N_p	q/q_i	N_p	q/q_i	N_p	q/q_i	N_p	q/q_i	N_p
−120.000	1.986	−169.771	5.129	−277.804	24.105	−638.814	3124.851	−10333.047		
−117.000	1.947	−163.871	4.844	−262.851	20.628	−571.906	965.021	−5063.512		
−114.000	1.910	−158.086	4.581	−248.718	17.815	−514.385	435.296	−3108.863		
−111.000	1.873	−152.412	4.338	−235.343	15.511	−464.509	238.203	−2139.914		
−108.000	1.837	−146.847	4.113	−222.670	13.603	−420.926	146.604	−1578.191		
−105.000	1.803	−141.387	3.903	−210.651	12.009	−382.578	97.655	−1218.741		
−102.000	1.769	−136.030	3.708	−199.236	10.664	−348.625	68.858	−972.486		
−99.000	1.736	−130.772	3.527	−188.386	9.521	−318.392	50.680	−795.125	99985.187	−66594.187
−96.000	1.704	−125.613	3.358	−178.062	8.542	−291.335	38.572	−662.402	3124.887	−8266.492
−93.000	1.673	−120.548	3.200	−168.228	7.699	−267.003	30.158	−560.031	771.339	−3532.949
−90.000	1.642	−115.575	3.052	−158.854	6.968	−245.028	24.105	−479.112	316.223	−2041.500
−87.000	1.613	−110.693	2.913	−149.909	6.330	−225.102	19.625	−413.838	164.111	−1355.633
−84.000	1.584	−105.898	2.783	−141.366	5.772	−206.967	16.230	−360.282	97.655	−974.994
−81.000	1.556	−101.190	2.662	−133.200	5.279	−190.406	13.603	−315.695	63.550	−738.298
−78.000	1.528	−96.564	2.547	−125.389	4.844	−175.233	11.535	−278.106	44.049	−579.393
−75.000	1.501	−92.020	2.439	−117.911	4.458	−161.292	9.882	−246.070	32.000	−466.665
−72.000	1.475	−87.556	2.338	−110.748	4.113	−148.447	8.542	−218.502	24.105	−383.289
−69.000	1.450	−83.169	2.242	−103.879	3.804	−136.580	7.444	−194.575	18.689	−319.580
−66.000	1.425	−78.858	2.152	−97.289	3.527	−125.591	6.533	−173.650	14.835	−269.604
−63.000	1.400	−74.621	2.067	−90.963	3.277	−115.390	5.772	−155.226	12.009	−229.547
−60.000	1.377	−70.456	1.986	−84.885	3.052	−105.902	5.129	−138.902	9.882	−196.856
−57.000	1.353	−66.361	1.910	−79.043	2.847	−97.059	4.581	−124.359	8.248	−169.765
−54.000	1.331	−62.335	1.837	−73.424	2.662	−88.800	4.113	−111.335	6.968	−147.017
−51.000	1.309	−58.376	1.769	−68.015	2.492	−81.073	3.708	−99.618	5.950	−127.696
−48.000	1.287	−54.483	1.704	−62.806	2.338	−73.832	3.358	−89.031	5.129	−111.122
−45.000	1.266	−50.654	1.642	−57.788	2.196	−67.034	3.052	−79.427	4.458	−96.776

Table B.13 (*Continued*)
HYPERBOLIC DECLINE CURVE FUNCTIONS FOR n = 0.4 (continued)

Decline rates listed, (a), are at time = 0 and in months^{-1}.

Time, months	a = 0.005		a = 0.010		a = 0.015		a = 0.020		a = 0.025	
	q/q_i	N_p	q/q_i	N_p	q/q_i	N_p	q/q_i	N_p	q/q_i	N_p
−42.000	1.245	−46.887	1.584	−52.949	2.067	−60.642	2.783	−70.683	3.903	−84.260
−39.000	1.225	−43.181	1.528	−48.282	1.947	−54.624	2.547	−62.695	3.441	−73.264
−36.000	1.205	−39.536	1.475	−43.778	1.837	−48.949	2.338	−55.374	3.052	−63.542
−33.000	1.186	−35.949	1.425	−39.429	1.736	−43.591	2.152	−48.645	2.722	−54.895
−30.000	1.167	−32.419	1.377	−35.228	1.642	−38.525	1.986	−42.443	2.439	−47.165
−27.000	1.149	−28.945	1.331	−31.168	1.556	−33.730	1.837	−36.712	2.196	−40.220
−24.000	1.131	−25.525	1.287	−27.241	1.475	−29.185	1.704	−31.403	1.986	−33.954
−21.000	1.113	−22.159	1.245	−23.444	1.400	−24.874	1.584	−26.475	1.803	−28.277
−18.000	1.096	−18.845	1.205	−19.768	1.331	−20.778	1.475	−21.889	1.642	−23.115
−15.000	1.079	−15.583	1.167	−16.209	1.266	−16.885	1.377	−17.614	1.501	−18.404
−12.000	1.063	−12.370	1.131	−12.763	1.205	−13.179	1.287	−13.621	1.377	−14.091
−9.000	1.046	−9.207	1.096	−9.423	1.149	−9.648	1.205	−9.884	1.266	−10.131
−6.000	1.031	−6.091	1.063	−6.185	1.096	−6.282	1.131	−6.381	1.167	−6.484
−3.000	1.015	−3.023	1.031	−3.046	1.046	−3.069	1.063	−3.093	1.079	−3.117
0.0	1.000	−0.0	1.000	−0.0	1.000	−0.0	1.000	−0.0	1.000	−0.0
3.000	0.985	2.977	0.971	2.955	0.956	2.934	0.942	2.912	0.929	2.891
15.000	0.929	14.456	0.864	13.949	0.806	13.473	0.753	13.027	0.705	12.608
27.000	0.877	25.286	0.774	23.764	0.687	22.406	0.613	21.187	0.550	20.086
39.000	0.829	35.515	0.696	32.572	0.591	30.055	0.507	27.881	0.439	25.986
51.000	0.784	45.191	0.629	40.510	0.513	36.665	0.425	33.455	0.357	30.738
63.000	0.743	54.354	0.570	47.695	0.449	42.423	0.360	38.153	0.295	34.631
75.000	0.705	63.042	0.519	54.223	0.395	47.475	0.309	42.158	0.247	37.869
87.000	0.670	71.288	0.474	60.175	0.350	51.936	0.267	45.604	0.209	40.596
99.000	0.637	79.122	0.434	65.620	0.312	55.900	0.233	48.595	0.179	42.919
111.000	0.606	86.575	0.399	70.617	0.279	59.440	0.204	51.210	0.155	44.915

Table B.13 (*Continued*)
HYPERBOLIC DECLINE CURVE FUNCTIONS FOR n = 0.5

Decline rates listed, (a), are at time = 0 and in months^{-1}.

Time, months	a = 0.005		a = 0.010		a = 0.015		a = 0.020		a = 0.025	
	q/q_i	N_p	q/q_i	N_p	q/q_i	N_p	q/q_i	N_p	q/q_i	N_p
−120.000	2.041	−171.428	6.250	−299.999	99.998	−1199.988				
−117.000	1.998	−165.371	5.806	−281.927	66.638	−955.094				
−114.000	1.956	−159.440	5.408	−265.115	47.562	−786.201				
−111.000	1.916	−153.633	5.050	−249.438	35.642	−662.682				
−108.000	1.877	−147.945	4.726	−234.782	27.701	−568.417				
−105.000	1.839	−142.372	4.432	−221.052	22.145	−494.115				
−102.000	1.802	−136.913	4.165	−208.163	18.108	−434.040				
−99.000	1.766	−131.561	3.921	−196.039	15.081	−384.464	9998.816	−9899.418		
−96.000	1.731	−126.316	3.698	−184.615	12.755	−342.855	624.982	−2399.967		
−93.000	1.698	−121.172	3.494	−173.831	10.928	−307.437	204.078	−1328.559		
−90.000	1.665	−116.129	3.306	−163.636	9.467	−276.922	99.999	−899.994		
−87.000	1.633	−111.182	3.133	−153.982	8.281	−250.359	59.171	−669.228		
−84.000	1.602	−106.329	2.973	−144.827	7.305	−227.026	39.062	−524.998		
−81.000	1.572	−101.567	2.825	−136.134	6.491	−206.369	27.701	−426.314		
−78.000	1.543	−96.894	2.687	−127.869	5.806	−187.951	20.661	−354.544	1599.910	−3119.913
−75.000	1.515	−92.308	2.560	−120.000	5.224	−171.428	16.000	−299.999	255.995	−1199.988
−72.000	1.487	−87.805	2.441	−112.500	4.726	−156.521	12.755	−257.142	99.999	−719.995
−69.000	1.460	−83.384	2.331	−105.343	4.295	−143.005	10.406	−222.580	52.892	−501.815
−66.000	1.434	−79.042	2.228	−98.507	3.921	−130.693	8.650	−194.117	32.653	−377.141
−63.000	1.409	−74.777	2.131	−91.971	3.594	−119.431	7.305	−170.270	22.145	−296.469
−60.000	1.384	−70.588	2.041	−85.714	3.306	−109.091	6.250	−150.000	16.000	−239.999
−57.000	1.360	−66.472	1.956	−79.720	3.051	−99.563	5.408	−132.558	12.098	−198.260
−54.000	1.336	−62.428	1.877	−73.973	2.825	−90.756	4.726	−117.391	9.467	−166.153
−51.000	1.314	−58.453	1.802	−68.456	2.623	−82.591	4.165	−104.082	7.610	−140.689
−48.000	1.291	−54.545	1.731	−63.158	2.441	−75.000	3.698	−92.308	6.250	−120.000
−45.000	1.270	−50.704	1.665	−58.064	2.278	−67.924	3.306	−81.818	5.224	−102.857
−42.000	1.248	−46.927	1.602	−53.164	2.131	−61.314	2.973	−72.414	4.432	−88.421

Table B.13 (Continued)

HYPERBOLIC DECLINE CURVE FUNCTIONS FOR $n = 0.5$ (continued)

Decline rates listed, (a), are at time = 0 and in months^{-1}.

Time, months	a = 0.005		a = 0.010		a = 0.015		a = 0.020		a = 0.025	
	q/q_i	N_p	q/q_i	N_p	q/q_i	N_p	q/q_i	N_p	q/q_i	N_p
−39.000	1.228	−43.213	1.543	−48.447	1.998	−55.124	2.687	−63.934	3.807	−76.097
−36.000	1.208	−39.560	1.487	−43.902	1.877	−49.315	2.441	−56.250	3.306	−65.454
−33.000	1.188	−35.967	1.434	−39.521	1.766	−43.854	2.228	−49.254	2.897	−56.170
−30.000	1.169	−32.432	1.384	−35.294	1.665	−38.710	2.041	−42.857	2.560	−48.000
−27.000	1.150	−28.954	1.336	−31.214	1.572	−33.856	1.877	−36.986	2.278	−40.755
−24.000	1.132	−25.532	1.291	−27.273	1.487	−29.268	1.731	−31.579	2.041	−34.286
−21.000	1.114	−22.163	1.248	−23.464	1.409	−24.926	1.602	−26.582	1.839	−28.475
−18.000	1.096	−16.848	1.208	−19.780	1.336	−20.809	1.487	−21.951	1.665	−23.226
−15.000	1.079	−15.584	1.169	−16.216	1.270	−16.901	1.384	−17.647	1.515	−18.462
−12.000	1.063	−12.371	1.132	−12.766	1.208	−13.187	1.291	−13.636	1.384	−14.118
−9.000	1.047	−9.207	1.096	−9.424	1.150	−9.651	1.208	−9.890	1.270	−10.141
−6.000	1.031	−6.091	1.063	−6.186	1.096	−6.283	1.132	−6.363	1.169	−6.486
−3.000	1.015	−3.023	1.031	−3.046	1.047	−3.069	1.063	−3.093	1.079	−3.117
0.0	1.000	−0.0	1.000	−0.0	1.000	−0.0	1.000	−0.0	1.000	−0.0
3.000	0.985	2.978	0.971	2.956	0.956	2.934	0.943	2.913	0.929	2.892
15.000	0.929	14.458	0.865	13.953	0.808	13.483	0.756	13.043	0.709	12.632
27.000	0.878	25.292	0.776	23.788	0.692	22.453	0.620	21.260	0.559	20.167
39.000	0.830	35.535	0.700	32.636	0.599	30.174	0.518	28.058	0.452	26.218
51.000	0.787	45.233	0.635	40.637	0.523	36.890	0.439	33.775	0.373	31.145
63.000	0.746	54.427	0.578	47.909	0.461	42.784	0.376	38.650	0.313	35.245
75.000	0.709	63.158	0.529	54.545	0.410	48.000	0.327	42.857	0.266	38.710
87.000	0.675	71.458	0.486	60.627	0.366	52.647	0.286	46.524	0.229	41.677
99.000	0.643	79.359	0.447	66.221	0.329	56.815	0.253	49.749	0.200	44.246
111.000	0.613	86.888	0.414	71.383	0.298	60.573	0.225	52.607	0.175	46.492

Table B.13 (Continued)

HYPERBOLIC DECLINE CURVE FUNCTIONS FOR n = 0.6

Decline rates listed, (a), are at time = 0 and in months^{-1}.

Time, months	a = 0.005		a = 0.010		a = 0.015		a = 0.020		a = 0.025	
	q/q_i	N_p	q/q_i	N_p	q/q_i	N_p	q/q_i	N_p	q/q_i	N_p
−120.000	2.104	−173.261	8.345	−334.117						
−117.000	2.056	−167.022	7.521	−310.353						
−114.000	2.009	−160.926	6.821	−288.866						
−111.000	1.964	−154.967	6.219	−269.326	99853.437	−16490.254				
−108.000	1.921	−149.141	5.698	−251.469	387.299	−1640.790				
−105.000	1.879	−143.442	5.244	−235.071	125.714	−985.732				
−102.000	1.838	−137.867	4.845	−219.950	64.610	−716.352				
−99.000	1.799	−132.411	4.492	−205.955	40.205	−563.736				
−96.000	1.761	−127.071	4.179	−192.958	27.803	−463.547				
−93.000	1.725	−121.841	3.899	−180.849	20.560	−391.875				
−90.000	1.690	−116.720	3.648	−169.534	15.925	−337.621				
−87.000	1.655	−111.702	3.422	−158.935	12.761	−294.871				
−84.000	1.622	−106.786	3.218	−148.981	10.496	−260.163				
−81.000	1.590	−101.967	3.032	−139.611	8.811	−231.320	387.302	−1230.596		
−78.000	1.559	−97.242	2.863	−130.772	7.521	−206.902	97.654	−656.244		
−75.000	1.529	−92.610	2.706	−122.419	6.509	−185.915	46.415	−455.196		
−72.000	1.500	−88.066	2.567	−114.509	5.698	−167.646	27.803	−347.660		
−69.000	1.472	−83.608	2.437	−107.005	5.038	−151.574	18.798	−279.163		
−66.000	1.444	−79.233	2.317	−99.877	4.492	−137.303	13.695	−231.068	2154.183	−2054.338
−63.000	1.418	−74.940	2.206	−93.094	4.035	−124.532	10.496	−195.122	125.715	−591.440
−60.000	1.392	−70.725	2.104	−86.630	3.648	−113.023	8.345	−167.059	46.415	−364.157
−57.000	1.367	−66.587	2.009	−80.463	3.317	−102.588	6.821	−144.433	24.987	−262.315
−54.000	1.343	−62.523	1.921	−74.570	3.032	−93.074	5.699	−125.735	15.925	−202.573
−51.000	1.319	−58.531	1.838	−68.933	2.784	−84.358	4.845	−109.975	11.174	−162.596
−48.000	1.296	−54.610	1.761	−63.535	2.567	−76.339	4.179	−96.479	8.345	−133.647
−45.000	1.273	−50.756	1.690	−58.360	2.376	−68.931	3.648	−84.767	6.509	−111.549
−42.000	1.252	−46.968	1.622	−53.393	2.206	−62.063	3.218	−74.490	5.244	−94.028

Table B.13 (Continued)

HYPERBOLIC DECLINE CURVE FUNCTIONS FOR $n = 0.6$ (continued)

Decline rates listed, (a), are at time $= 0$ and in months^{-1}

Time, months	a = 0.005		a = 0.010		a = 0.015		a = 0.020		a = 0.025	
	q/q_i	N_p	q/q_i	N_p	q/q_i	N_p	q/q_i	N_p	q/q_i	N_p
−39.000	1.230	−43.245	1.559	−48.621	2.056	−55.674	2.863	−65.386	4.331	−79.736
−36.000	1.210	−39.585	1.500	−44.033	1.921	−49.713	2.567	−57.254	3.648	−67.814
−33.000	1.190	−35.986	1.444	−39.617	1.799	−44.137	2.317	−49.938	3.123	−57.691
−30.000	1.170	−32.446	1.392	−35.363	1.690	−38.907	2.104	−43.315	2.708	−48.967
−27.000	1.151	−28.964	1.343	−31.262	1.590	−33.989	1.921	−37.285	2.376	−41.358
−24.000	1.133	−25.538	1.296	−27.305	1.500	−29.355	1.761	−31.768	2.104	−34.652
−21.000	1.115	−22.168	1.252	−23.484	1.418	−24.980	1.622	−26.696	1.879	−28.688
−18.000	1.097	−18.851	1.210	−19.793	1.343	−20.841	1.500	−22.016	1.690	−23.344
−15.000	1.080	−15.586	1.170	−16.223	1.273	−16.919	1.392	−17.681	1.529	−18.522
−12.000	1.063	−12.372	1.133	−12.769	1.210	−13.195	1.296	−13.652	1.392	−14.145
−9.000	1.047	−9.207	1.097	−9.425	1.151	−9.655	1.210	−9.896	1.273	−10.151
−6.000	1.031	−6.092	1.063	−6.186	1.097	−6.284	1.133	−6.385	1.170	−6.489
−3.000	1.015	−3.023	1.031	−3.046	1.047	−3.069	1.063	−3.093	1.080	−3.117
0.0	1.000	−0.0	1.000	−0.0	1.000	−0.0	1.000	−0.0	1.000	−0.0
3.000	0.985	2.978	0.971	2.956	0.957	2.934	0.943	2.913	0.929	2.892
15.000	0.929	14.459	0.866	13.958	0.810	13.493	0.759	13.059	0.713	12.654
27.000	0.878	25.299	0.779	23.812	0.696	22.499	0.626	21.330	0.567	20.283
39.000	0.832	35.555	0.704	32.698	0.606	30.288	0.527	28.226	0.464	26.439
51.000	0.789	45.273	0.641	40.760	0.533	37.104	0.451	34.078	0.388	31.529
63.000	0.749	54.499	0.586	48.113	0.473	43.128	0.391	39.120	0.330	35.822
75.000	0.713	63.270	0.538	54.853	0.423	48.497	0.343	43.516	0.285	39.499
87.000	0.679	71.622	0.497	61.057	0.381	53.318	0.304	47.389	0.249	42.691
99.000	0.648	79.586	0.460	66.791	0.346	57.676	0.271	50.833	0.219	45.493
111.000	0.619	87.190	0.427	72.108	0.315	61.638	0.244	53.918	0.195	47.976

Table B.13 (*Continued*)
HYPERBOLIC DECLINE CURVE FUNCTIONS FOR n = 0.7

Decline rates listed, (a), are at time = 0 and in months^{-1}.

Time, months	a = 0.005		a = 0.010		a = 0.015		a = 0.020		a = 0.025	
	q/q_i	N_p	q/q_i	N_p	q/q_i	N_p	q/q_i	N_p	q/q_i	N_p
−120.000	2.178	−175.304	13.708	−397.755						
−117.000	2.122	−168.854	11.494	−360.118						
−114.000	2.070	−162.566	9.825	−328.250						
−111.000	2.019	−156.434	8.531	−300.793						
−108.000	1.971	−150.450	7.502	−276.801						
−105.000	1.924	−144.609	6.667	−255.590						
−102.000	1.879	−138.904	5.979	−236.653						
−99.000	1.836	−133.332	5.403	−219.604						
−96.000	1.795	−127.885	4.916	−204.145						
−93.000	1.755	−122.560	4.499	−190.039	212.337	−886.686				
−90.000	1.717	−117.353	4.139	−177.095	63.019	−548.024				
−87.000	1.680	−112.258	3.825	−165.161	33.002	−412.159				
−84.000	1.644	−107.272	3.549	−154.108	21.178	−333.108				
−81.000	1.610	−102.391	3.306	−143.633	15.104	−279.556				
−78.000	1.577	−97.611	3.090	−134.245	11.494	−240.079				
−75.000	1.545	−92.929	2.896	−125.271	9.139	−209.358				
−72.000	1.514	−88.340	2.723	−116.847	7.502	−184.534				
−69.000	1.484	−83.843	2.566	−108.917	6.307	−163.909	125.278	−543.257		
−66.000	1.455	−79.434	2.424	−101.435	5.403	−146.403	39.704	−336.253		
−63.000	1.427	−75.110	2.295	−94.358	4.700	−131.290	21.178	−249.831		
−60.000	1.400	−70.869	2.178	−87.652	4.139	−118.064	13.708	−198.877		
−57.000	1.374	−66.707	2.070	−81.283	3.683	−106.354	9.825	−164.125	5212.395	−1604.670
−54.000	1.349	−62.623	1.971	−75.225	3.306	−95.889	7.502	−138.401	63.019	−328.814
−51.000	1.324	−58.613	1.879	−69.452	2.990	−86.457	5.979	−118.326	24.193	−213.442
−48.000	1.300	−54.676	1.795	−63.943	2.723	−77.898	4.916	−102.073	13.708	−159.102

Table B.13 (Continued)

HYPERBOLIC DECLINE CURVE FUNCTIONS FOR $n = 0.7$ (continued)

Decline rates listed, (a), are at time = 0 and in months^{-1}.

Time, months	a = 0.005		a = 0.010		a = 0.015		a = 0.020		a = 0.025	
	q/q_i	N_p	q/q_i	N_p	q/q_i	N_p	q/q_i	N_p	q/q_i	N_p
−45.000	1.277	−50.809	1.717	−58.677	2.494	−70.082	4.139	−88.548	9.139	−125.615
−42.000	1.255	−47.011	1.644	−53.636	2.295	−62.906	3.549	−77.054	6.667	−102.236
−39.000	1.233	−43.278	1.577	−48.805	2.122	−56.285	3.090	−67.123	5.150	−84.677
−36.000	1.212	−39.611	1.514	−44.170	1.971	−50.150	2.723	−58.423	4.139	−70.838
−33.000	1.192	−36.005	1.455	−39.717	1.836	−44.444	2.424	−50.717	3.424	−59.552
−30.000	1.172	−32.460	1.400	−35.434	1.717	−39.118	2.178	−43.826	2.896	−50.108
−27.000	1.152	−28.974	1.349	−31.311	1.610	−34.130	1.971	−37.613	2.494	−42.049
−24.000	1.134	−25.545	1.300	−27.338	1.514	−29.447	1.795	−31.971	2.178	−35.061
−21.000	1.115	−22.172	1.255	−23.506	1.427	−25.037	1.644	−26.818	1.924	−28.922
−18.000	1.097	−18.853	1.212	−19.805	1.349	−20.874	1.514	−22.085	1.717	−23.471
−15.000	1.080	−15.587	1.172	−16.230	1.277	−16.936	1.400	−17.717	1.545	−18.586
−12.000	1.063	−12.372	1.134	−12.773	1.212	−13.204	1.300	−13.669	1.400	−14.174
−9.000	1.047	−9.208	1.097	−9.427	1.152	−9.658	1.212	−9.903	1.277	−10.162
−6.000	1.031	−6.091	1.063	−6.186	1.097	−6.284	1.134	−6.386	1.172	−6.492
−3.000	1.015	−3.023	1.031	−3.046	1.047	−3.069	1.063	−3.093	1.080	−3.117
0.0	1.000	−0.0	1.000	−0.0	1.000	−0.0	1.000	−0.0	1.000	−0.0
3.000	0.985	2.978	0.971	2.956	0.957	2.934	0.943	2.913	0.930	2.892
15.000	0.930	14.460	0.867	13.963	0.811	13.502	0.762	13.074	0.717	12.676
27.000	0.879	25.306	0.781	23.835	0.700	22.543	0.633	21.398	0.575	20.376
39.000	0.833	35.574	0.708	32.759	0.612	30.398	0.537	28.387	0.476	26.649
51.000	0.791	45.313	0.647	40.878	0.542	37.310	0.463	34.367	0.402	31.893
63.000	0.752	54.568	0.593	48.310	0.484	43.455	0.405	39.564	0.346	36.366
75.000	0.717	63.380	0.547	55.148	0.436	48.969	0.359	44.137	0.302	40.243
87.000	0.684	71.782	0.507	61.467	0.396	53.953	0.320	48.205	0.267	43.647
99.000	0.654	79.806	0.471	67.333	0.361	58.490	0.289	51.854	0.238	46.669
111.000	0.626	87.481	0.440	72.796						

Table B.13 (Continued)
HYPERBOLIC DECLINE CURVE FUNCTIONS FOR n = 0.8

Decline rates listed, (a), are at time = 0 and in months^{-1}.

Time, months	a = 0.005		a = 0.010		a = 0.015		a = 0.020		a = 0.025	
	q/q_i	N_p	q/q_i	N_p	q/q_i	N_p	q/q_i	N_p	q/q_i	N_p
−120.000	2.265	−177.604	55.900	−618.029						
−117.000	2.201	−170.906	31.065	−494.085						
−114.000	2.140	−164.395	20.864	−418.012						
−111.000	2.083	−158.061	15.434	−364.300						
−108.000	2.028	−151.895	12.108	−323.350						
−105.000	1.976	−145.890	9.882	−290.568						
−102.000	1.926	−140.033	8.298	−263.423						
−99.000	1.878	−134.333	7.119	−240.379						
−96.000	1.832	−128.768	6.211	−220.440						
−93.000	1.789	−123.337	5.492	−202.926						
−90.000	1.747	−118.034	4.910	−187.353						
−87.000	1.707	−112.854	4.430	−173.366						
−84.000	1.668	−107.791	4.029	−160.695						
−81.000	1.631	−102.842	3.688	−149.133	87.305	−481.533				
−78.000	1.596	−98.001	3.396	−138.518	31.065	−329.390				
−75.000	1.562	−93.264	3.144	−128.717	17.783	−259.425				
−72.000	1.529	−88.629	2.923	−119.624	12.108	−215.567				
−69.000	1.497	−84.090	2.728	−111.153	9.028	−184.269				
−66.000	1.467	−79.644	2.556	−103.232	7.119	−160.252				
−63.000	1.438	−75.287	2.402	−95.799	5.831	−140.942				
−60.000	1.409	−71.017	2.265	−88.802	4.910	−124.902	55.900	−309.015		
−57.000	1.382	−66.831	2.140	−82.198	4.221	−111.254	20.864	−209.006		
−54.000	1.356	−62.725	2.028	−75.948	3.688	−99.422	12.108	−161.675		
−51.000	1.330	−58.697	1.926	−70.019	3.266	−89.015	8.298	−131.711		
−48.000	1.305	−54.744	1.832	−64.384	2.923	−79.749	6.211	−110.220	55.901	−247.212
−45.000	1.282	−50.864	1.747	−59.017	2.640	−71.419	4.910	−93.677	17.783	−155.655
−42.000	1.258	−47.054	1.668	−53.895	2.402	−63.866	4.029	−80.348	9.882	−116.228

Table B.13 (Continued)
HYPERBOLIC DECLINE CURVE FUNCTIONS FOR $n = 0.8$ (continued)

Decline rates listed, (a), are at time = 0 and in months^{-1}.

Time, months	a = 0.005		a = 0.010		a = 0.015		a = 0.020		a = 0.025	
	q/q_i	N_p	q/q_i	N_p	q/q_i	N_p	q/q_i	N_p	q/q_i	N_p
−39.000	1.236	−43.312	1.596	−49.001	2.201	−56.969	3.396	−69.259	6.637	−92.028
−36.000	1.215	−39.637	1.529	−44.314	2.028	−50.632	2.923	−59.812	4.910	−74.941
−33.000	1.194	−36.024	1.467	−39.822	1.878	−44.778	2.556	−51.616	3.852	−61.915
−30.000	1.173	−32.474	1.409	−35.509	1.747	−39.345	2.265	−44.401	3.144	−51.467
−27.000	1.154	−28.984	1.356	−31.362	1.631	−34.281	2.028	−37.974	2.640	−42.851
−24.000	1.134	−25.553	1.305	−27.372	1.529	−29.543	1.832	−32.192	2.265	−35.521
−21.000	1.116	−22.177	1.258	−23.527	1.438	−25.096	1.668	−26.948	1.976	−29.178
−18.000	1.098	−18.856	1.215	−19.818	1.356	−20.908	1.529	−22.157	1.747	−23.607
−15.000	1.080	−15.589	1.173	−16.237	1.282	−16.955	1.409	−17.754	1.562	−18.653
−12.000	1.063	−12.373	1.134	−12.776	1.215	−13.212	1.305	−13.686	1.409	−14.203
−9.000	1.047	−9.208	1.098	−9.428	1.154	−9.661	1.215	−9.909	1.282	−10.173
−6.000	1.031	−6.092	1.063	−6.186	1.098	−6.285	1.134	−6.388	1.173	−6.495
−3.000	1.015	−3.022	1.051	−3.046	1.047	−3.069	1.063	−3.093	1.080	−3.118
0.0	1.000	−0.0	1.000	−0.0	1.000	−0.0	1.000	−0.0	1.000	−0.0
3.000	0.985	2.977	0.971	2.956	0.957	2.934	0.943	2.913	0.930	2.892
15.000	0.930	14.462	0.868	13.967	0.813	13.511	0.764	13.089	0.720	12.697
27.000	0.880	25.313	0.783	23.858	0.704	22.586	0.638	21.464	0.583	20.465
39.000	0.834	35.592	0.712	32.817	0.619	30.504	0.545	28.541	0.486	26.849
51.000	0.793	45.352	0.652	40.993	0.551	37.507	0.474	34.642	0.415	32.239
63.000	0.755	54.636	0.600	48.500	0.495	43.767	0.418	39.986	0.361	36.882
75.000	0.720	63.486	0.556	55.430	0.448	49.417	0.373	44.726	0.318	40.946
87.000	0.688	71.937	0.517	61.859	0.409	54.555	0.336	48.975	0.284	44.549
99.000	0.659	80.019	0.482	67.849	0.376	59.260	0.305	52.818	0.255	47.779
111.000	0.632	87.762	0.452	73.450	0.347	63.592	0.279	00.000		

Table B.13 (*Continued*)
HYPERBOLIC DECLINE CURVE FUNCTIONS FOR n = 0.9

Decline rates listed, (a), are at time = 0 and in months^{-1}.

Time, months	a = 0.005		a = 0.010		a = 0.015		a = 0.020		a = 0.025	
	q/q_i	N_p	q/q_i	N_p	q/q_i	N_p	q/q_i	N_p	q/q_i	N_p
−120.000	2.370	−180.225								
−117.000	2.295	−173.229								
−114.000	2.224	−166.452								
−111.000	2.158	−159.879	2152.321	−1154.225						
−108.000	2.095	−153.501	53.133	−487.772						
−105.000	2.035	−147.306	25.095	−380.253						
−102.000	1.979	−141.286	16.102	−320.343						
−99.000	1.926	−135.429	11.736	−279.241						
−96.000	1.875	−129.730	9.178	−248.168						
−93.000	1.826	−124.179	7.505	−223.304						
−90.000	1.780	−118.768	6.330	−202.648						
−87.000	1.737	−113.495	5.461	−185.022						
−84.000	1.695	−108.347	4.794	−169.681						
−81.000	1.655	−103.323	4.266	−156.121						
−78.000	1.617	−98.417	3.839	−143.984						
−75.000	1.580	−93.622	3.486	−133.013						
−72.000	1.545	−88.934	3.190	−123.011	53.133	−325.182				
−69.000	1.511	−84.350	2.939	−113.627	19.664	−231.333				
−66.000	1.479	−79.864	2.723	−105.343	11.736	−186.160				
−63.000	1.448	−75.474	2.535	−97.463	8.262	−156.741				
−60.000	1.419	−71.173	2.370	−90.112	6.330	−135.098				
−57.000	1.390	−66.959	2.224	−83.226	5.107	−118.068	53.133	−243.887		
−54.000	1.363	−62.832	2.095	−76.751	4.266	−104.081	16.102	−160.172		
−51.000	1.336	−58.784	1.979	−70.643	3.654	−92.245				
−43.000	1.310	−54.815	1.875	−64.865	3.190	−82.007	9.178	−124.084		
−45.000	1.286	−50.920	1.780	−59.384	2.827	−73.002	6.330	−101.324		
−42.000	1.262	−47.098	1.695	−54.173	2.535	−64.976	4.794	−84.841	25.095	−152.102

Table B.13 (Continued)
HYPERBOLIC DECLINE CURVE FUNCTIONS FOR $n = 0.9$

Decline rates listed, (a), are at time = 0 and in months^{-1}.

Time, months	$a = 0.005$		$a = 0.010$		$a = 0.015$		$a = 0.020$		$a = 0.015$	
	q/q_i	N_p	q/q_i	N_p	q/q_i	N_p	q/q_i	N_p	q/q_i	N_p
−39.000	1.239	−43.346	1.617	−49.208	2.295	−57.743	3.839	−71.993	10.308	−105.101
−36.000	1.217	−39.663	1.545	−44.467	2.095	−51.167	3.190	−61.506	6.330	−81.059
−33.000	1.196	−36.045	1.479	−39.932	1.926	−45.143	2.723	−52.671	4.515	−65.082
−30.000	1.175	−32.490	1.419	−35.586	1.760	−39.589	2.370	−45.056	3.486	−53.205
−27.000	1.155	−28.995	1.363	−31.416	1.655	−34.441	2.095	−38.375	2.827	−43.801
−24.000	1.135	−25.560	1.310	−27.408	1.545	−29.645	1.875	−32.432	2.370	−36.045
−21.000	1.117	−22.182	1.262	−23.549	1.448	−25.158	1.695	−27.087	2.035	−29.481
−18.000	1.098	−18.860	1.217	−19.832	1.363	−20.944	1.545	−22.233	1.760	−23.754
−15.000	1.081	−15.591	1.175	−16.245	1.286	−16.973	1.419	−17.793	1.580	−18.724
−12.000	1.064	−12.375	1.135	−12.780	1.217	−13.221	1.310	−13.704	1.419	−14.234
−9.000	1.047	−9.209	1.098	−9.430	1.155	−9.665	1.217	−9.916	1.286	−10.184
−6.000	1.031	−6.092	1.064	−6.187	1.098	−6.287	1.135	−6.390	1.175	−6.498
−3.000	1.015	−3.023	1.031	−3.046	1.047	−3.070	1.064	−3.094	1.081	−3.118
0.0	1.000	−0.0	1.000	−0.0	1.000	−0.0	1.000	−0.0	1.000	−0.0
3.000	0.985	2.978	0.971	2.956	0.957	2.934	0.943	2.913	0.930	2.893
15.000	0.930	14.463	0.869	13.972	0.815	13.520	0.767	13.104	0.724	12.718
27.000	0.880	25.320	0.785	23.880	0.708	22.628	0.644	21.528	0.590	20.550
39.000	0.836	35.611	0.716	32.875	0.625	30.607	0.554	28.689	0.497	27.040
51.000	0.795	45.390	0.657	41.104	0.559	37.695	0.485	34.904	0.428	32.567
63.000	0.758	54.702	0.607	48.682	0.505	44.065	0.431	40.386	0.375	37.370
75.000	0.724	63.590	0.564	55.701	0.460	49.844	0.387	45.284	0.333	41.611
87.000	0.693	72.087	0.526	62.234	0.422	55.127	0.351	49.705	0.300	45.403
99.000	0.664	80.225	0.493	68.342	0.389	59.990	0.321	53.731	0.272	48.828
111.000	0.638	88.033	0.463	74.074	0.362	64.492	0.295	57.423	0.249	51.949

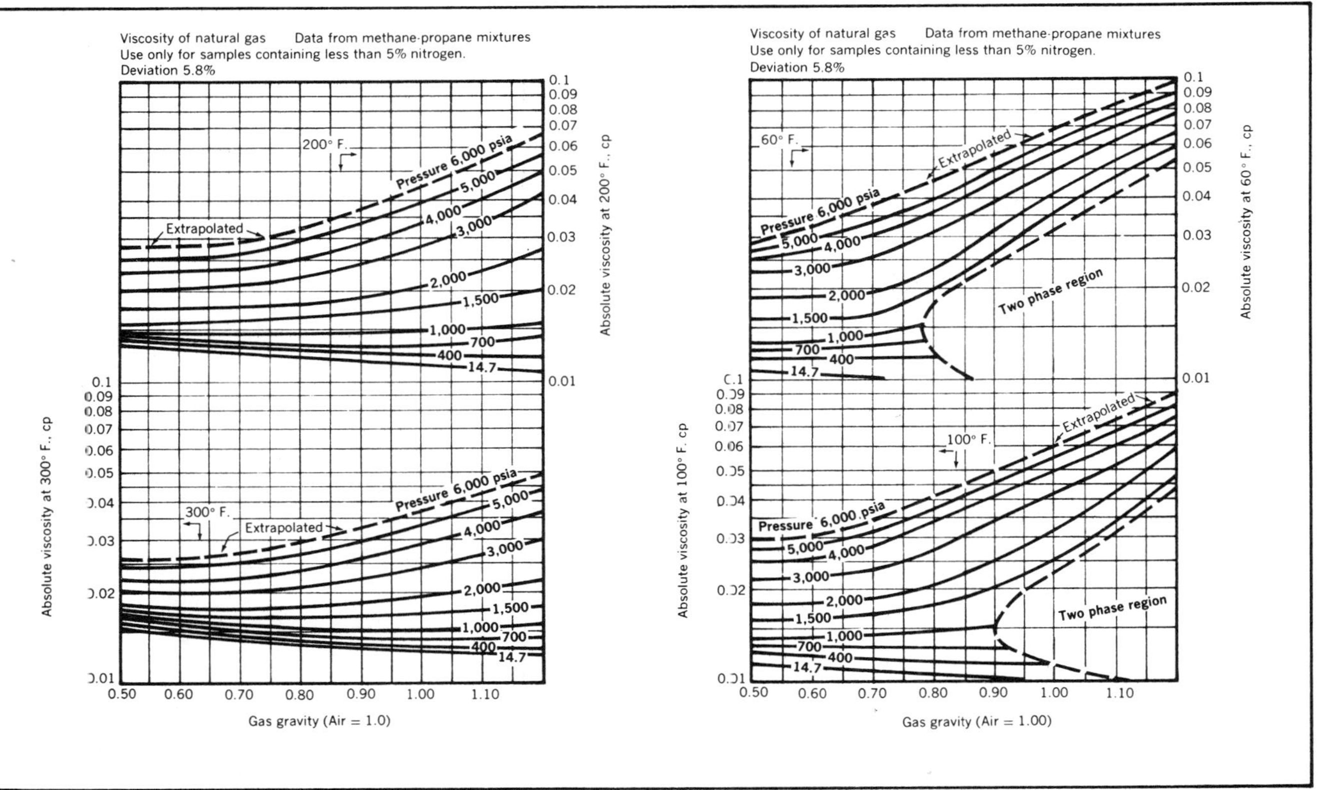

FIG. B.14 Viscosity of natural gas (From Bicher & Katz, Viscosity of Natural Gases, *AIME Transactions,* 155, 246, 1944)

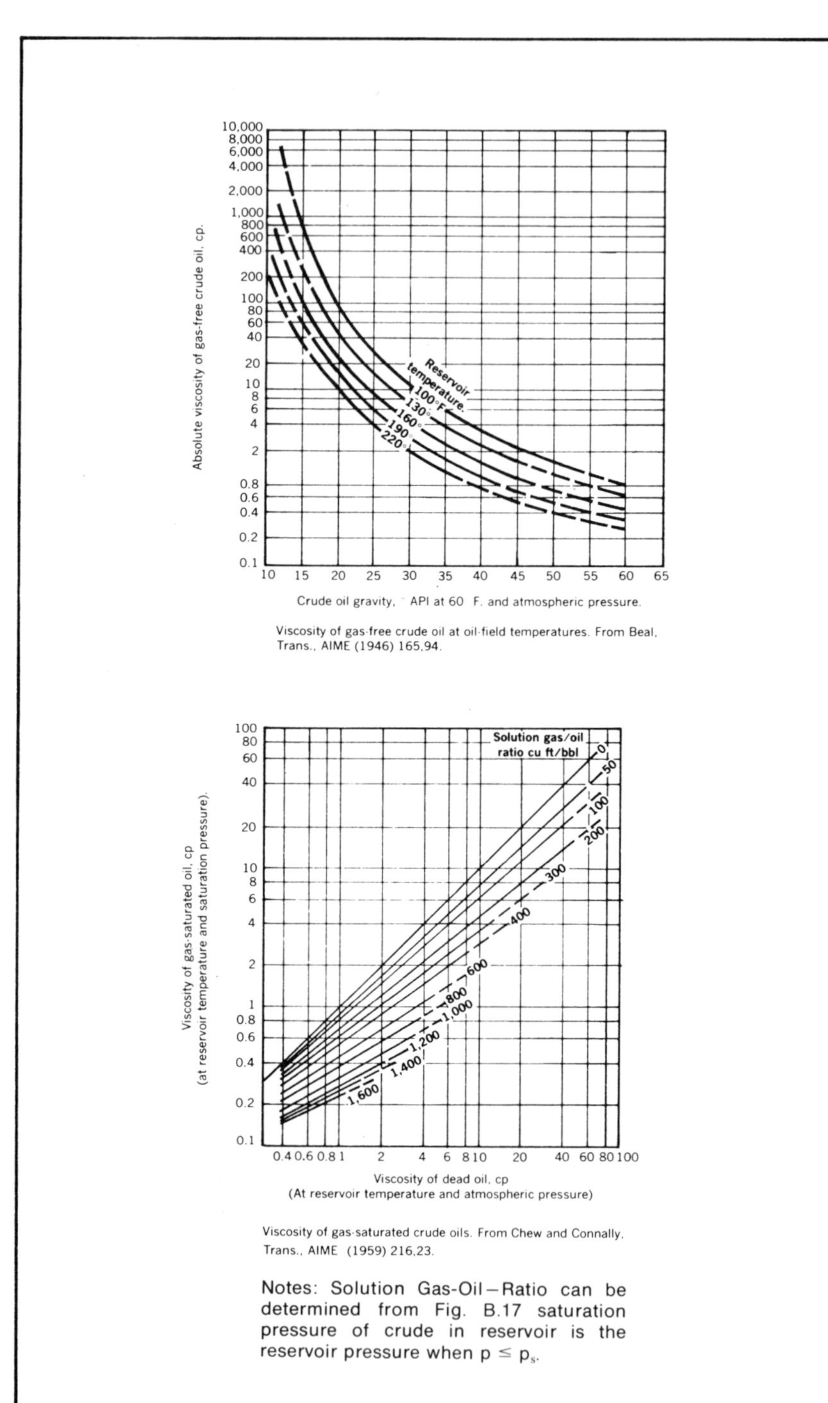

Viscosity of gas-free crude oil at oil-field temperatures. From Beal, Trans., AIME (1946) 165,94.

Viscosity of gas-saturated crude oils. From Chew and Connally, Trans., AIME (1959) 216,23.

Notes: Solution Gas-Oil—Ratio can be determined from Fig. B.17 saturation pressure of crude in reservoir is the reservoir pressure when $p \leq p_s$.

FIG. B.15 Reservoir oil viscosities

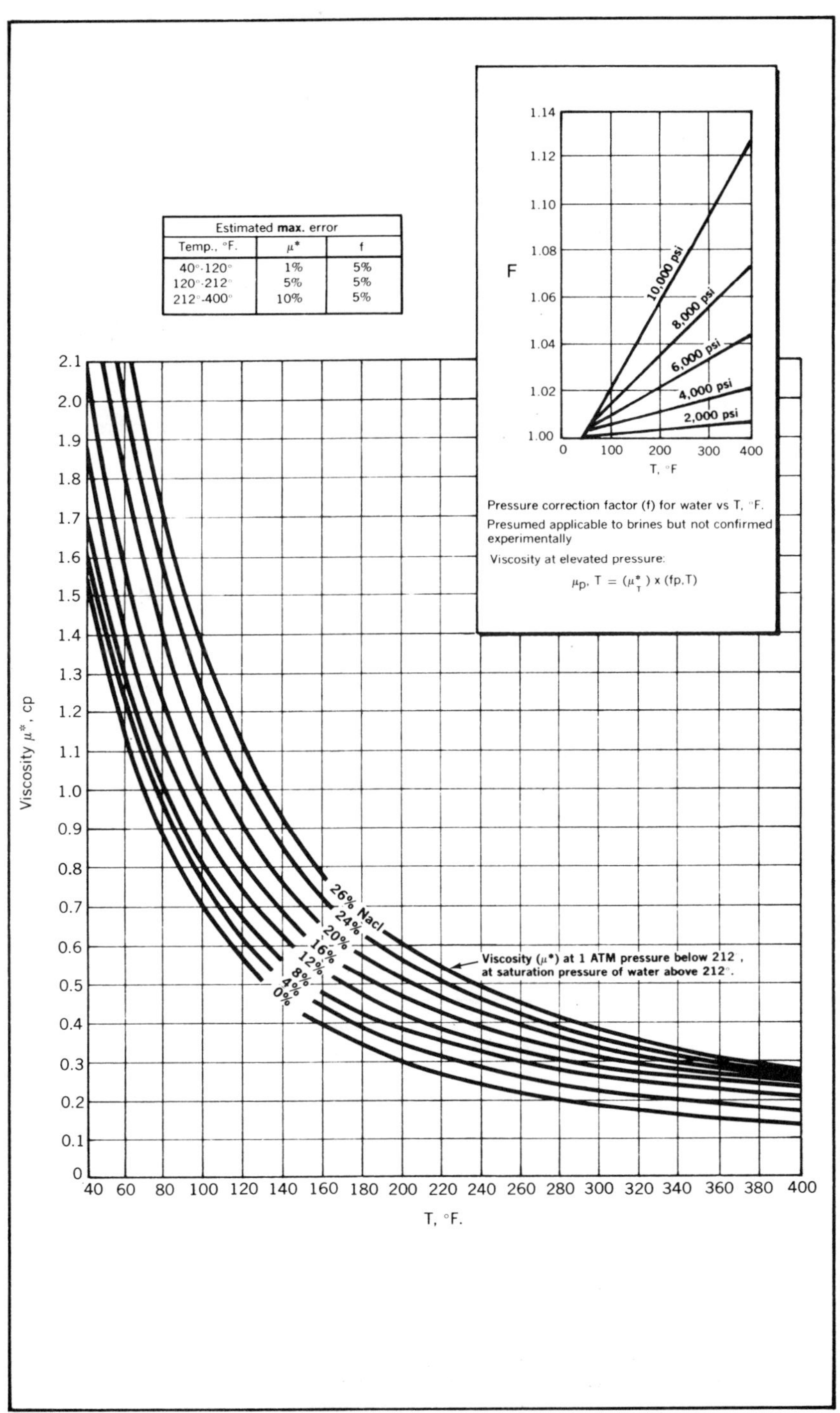

FIG. B.16 Reservoir water viscosities (From SPE Monograph No. 1, Chesnut, unpublished Shell Development Co. data, Courtesy SPE of AIME.)

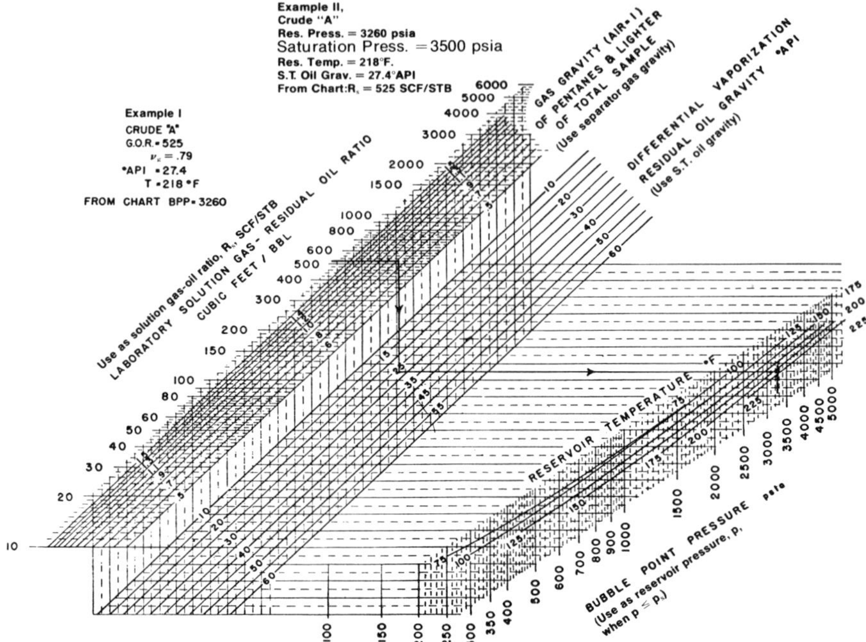

FIG. B.17 Gas in solution or bubble-point pressure (From Borden and Rzasa, "Correlation of Bottom Hole Sample Data," *Trans. AIME 192,* 19, 1951)

Table B.18

Gas equivalent (GE) of stock-tank condensate			
(1)	(2)	(3)	(4)
			GE of ST
Oil gravity,	Oil specific	Molecular	condensate
°API	gravity	weight	scf/STB
45	0.802	156	684
50	0.780	138	752
55	0.759	124	814
60	0.739	113	870
65	0.720	103	930

$$(2)\ °API = \frac{141.5}{\gamma_L} - 131.5$$

$$(3)\ MW = \frac{44.29\,\gamma_L}{(1.03 - \gamma_L)}$$

$$(4)\ GE = \frac{133{,}000\,\gamma_L}{MW}$$

Equation (3) was developed by C. S. Cragoe, "Thermo dynamic Properties of Petroleum Products" Bureau of Standards (1929) Miscellaneous Publication No. 97.

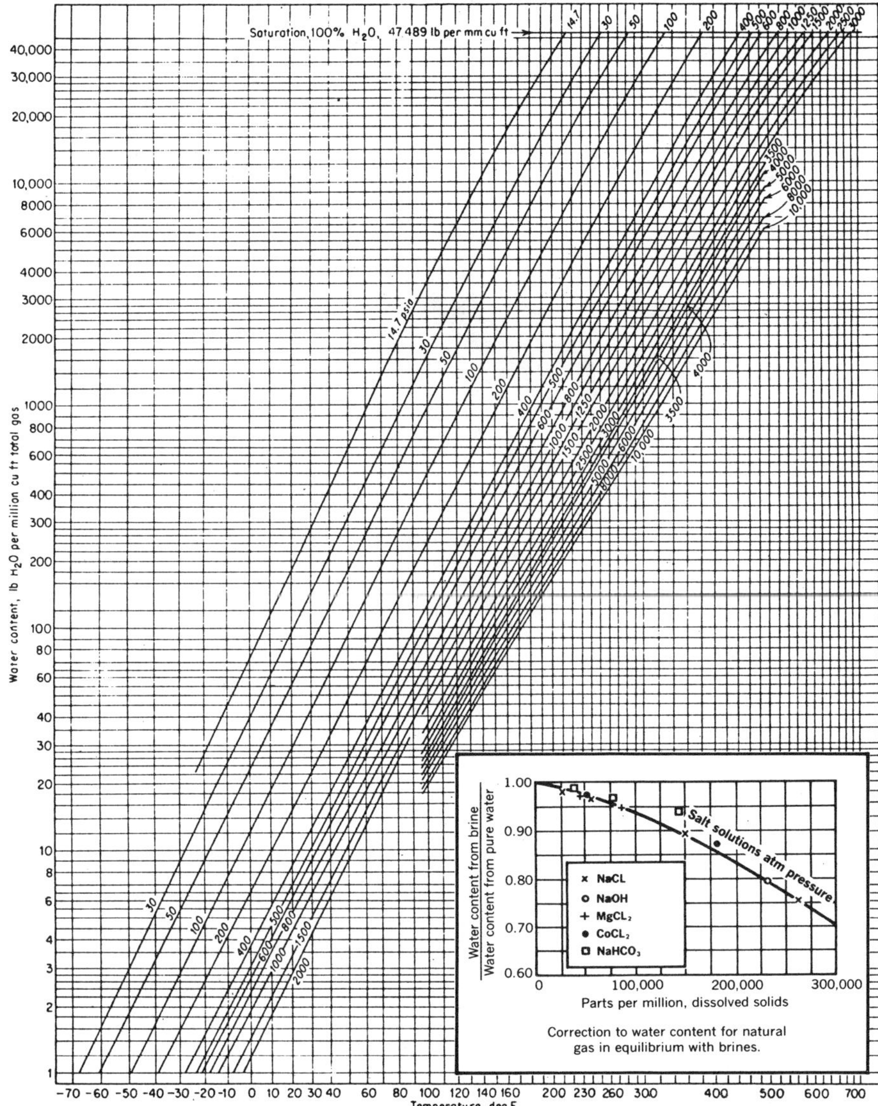

FIG. B.19 Water content of natural gas in the reservoir (*From Handbook of Natural Gas Engineering,* McGraw-Hill Book Co., 1959)

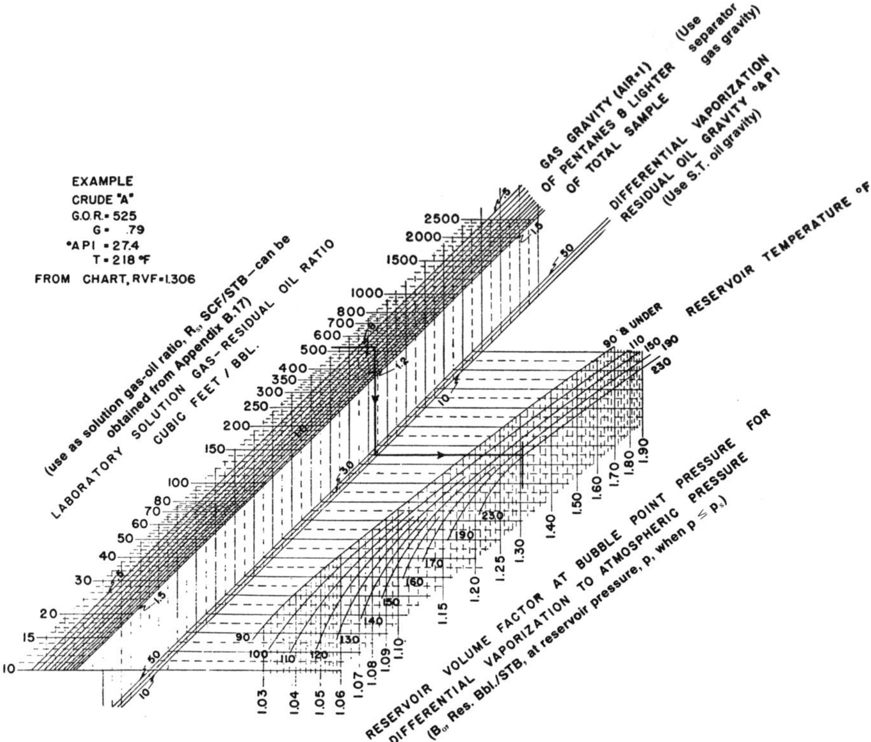

FIG. B.20 Oil formation volume factors (From Borden & Rzasa, "Correlation of Bottom Hole Sample Data," *Trans. AIME 192,* 19, 1951)

Appendix C

PROBLEM 1.1 SOLUTION—CALCULATION OF RELATIVE PERMEABILITY DATA FROM LAB TESTS

The absolute permeability can be calculated only when the core is 100% saturated with one fluid so use data when $S_w = 100\%$.

$$q_w' = \frac{k_w A'}{\mu_w} \frac{\Delta p'}{\Delta x'} \qquad \text{(1.2 applied to water flow)}$$

$$k_w = \frac{q_w' \mu_w}{A'} \frac{\Delta x'}{\Delta p'}$$

$$k_w = \frac{(0.5)(1.0)}{(5)} \frac{(3)}{(2-1)}$$

$= 0.3$ darcies or 300 md is the absolute permeability

at $S_w = 30\%$

$$k_o = \frac{q_o' \mu_o}{A} \frac{\Delta x'}{\Delta p} \qquad \text{(1.2 applied to oil flow)}$$

$$= \frac{(0.38)(1.25)(3)}{(5)(2-1)}$$

$= 0.285$ darcies or 285 md

Calculation of relative permeabilities

(1) S_w	(2) k_w	(3) k_o	(4) k_{rw}	(5) k_{ro}
100	0.300	0.0000	1.00	0.000
90	0.180	0.0000(critical)	0.60	0.000(critical)
80	0.090	0.0075	0.30	0.025
60	0.018	0.0750	0.06	0.250
40	0.006	0.1875	0.02	0.625
30	0.000(critical)	0.285	0.00(critical)	0.950

(1) Given in problem
(2) From k_w equation above
(3) From k_o equation above
(4) $k_w/0.300$
(5) $k_o/0.300$
"Critical" zeros refer to point where data first begins to deviate from zero.

PROBLEM 1.2 SOLUTION – RADIAL STEADY STATE FLOW FROM A DAMAGED WELL

A. To calculate Δp_{skin} use Equation 1.14.

$$q = \frac{7.08 k_{und.} h (p_e - p_w - \Delta p_{skin})}{\mu \ln(r_e/r_w)} \qquad (1.14)$$

$$(80)(1.25) = \frac{(7.08)(0.01)(6.9)(2,000 - 900 - \Delta p_{skin})}{0.708 \ln\left(\dfrac{700}{0.7}\right)}$$

$$\Delta p_{skin} = 100 \text{ psia}$$

B. If the Δp skin is reduced to zero, the production rate would be:

$$q = \frac{7.08 kh (p_e - p_w)}{\mu \ln(r_e/r_w)} \qquad (1.13)$$

$$= \frac{(7.08)(0.01)(6.9)(2,000 - p_w)}{(0.708) \ln \dfrac{(700)}{(7)}}$$

$$= 110 \text{ b/d}$$

$$\text{or} \quad \frac{110}{1.25} = 88 \text{ STB/d.}$$

C. Letting $r_w = 7.0$ and substituting into Equation 1.13

$$(80)(1.25) = \frac{(7.08)(0.010)(2,000 - p_w)}{(7.08) \ln(700/7)}$$

$$p_w = 1,333 \text{ psia}$$

$$\Delta p \text{ from 7 ft to 0.7 ft} = 1,333 - 900 = 433 \text{ psia}$$

PROBLEM 1.3 SOLUTION – DETERMINING AVERAGE PERMEABILITIES

Part A,
$$k_{avg} = \frac{\Sigma k_j h_j}{h} = \frac{(2)(0.2) + (3)(0.1) + (5)(0.01)}{2 + 3 + 5} \qquad (1.43)$$

$$= 0.075 \text{ darcies or 75 md}$$

$$q = \frac{7.08 kh}{\mu} \frac{(p_e - p_w)}{\ln \dfrac{r_e}{r_w}} \qquad (1.13)$$

$$= \frac{(7.08)(0.075)(10)(600 - 500)}{1.0 \ln \dfrac{100}{1/3}}$$

$$= 93 \text{ b/d}$$

Part B,
$$k_{ave} = \frac{\ln\left(\dfrac{r_e}{r_w}\right)_{total}}{\Sigma\left(\dfrac{\ln\left(\dfrac{r_o}{r_i}\right)_j}{k_j}\right)} \qquad (1.47)$$

$$= \frac{2.303 \log \frac{1,000}{1.0}}{\left(2.303 \times \frac{\log \frac{1,000}{10}}{0.075}\right) + \left(2.303 \times \frac{\log \frac{10}{1}}{0.0075}\right)}$$

$$= 0.019 \text{ darcies or } 19 \text{ md}$$

PROBLEM 1.4 SOLUTION — CORRECTING FOR STATIC PRESSURE DIFFERENCES

A. $\Delta p_{Flow} = \Delta p_{Total} - 0.433 \, \gamma \, \Delta D$ (1.49)

$$= (3,000 - 2,250) - (0.433)(1.0)(1,000)$$

$$= 750 - 433 = 317 \text{ psi}$$

$$q = \frac{7.08 kh(p_e - p_w)}{\mu \ln \frac{r_e}{r_w}} \tag{1.13}$$

$$= \frac{(7.08)(0.092)(10)(317)}{(0.708)\ln \frac{(4,000)}{0.4}}$$

$$= 317 \text{ res b/d}$$

B. All parameters are same except $\Delta p = 750$; thus, by ratio

$$q_{res\,bbl} = 317 \frac{(750)}{(317)} = 750 \text{ res b/d}$$

PROBLEM 2.1A SOLUTION — APPLICATION OF THE P_{td} FUNCTION

Calculate the diffusivity constant:

$$\eta = \frac{6.33k}{\phi \mu c} = \frac{(6.33)(10^{-4})}{(6.95)(10^{-2})(0.4)(6.33)(10^{-6})} \tag{2.8}$$

Calculate the reduced time:

$$t_D = \frac{\eta t \text{ days}}{r_w^2} = \frac{\eta t \text{ min}}{r_w^2 \, 1,440} = \frac{3.6t \text{ min}}{(0.25)(1.44)} = 3,600 \tag{2.10}$$

$$= 10 t_{min}$$

$$t_D = (10)(6) = 60$$

Read p_{tD} from Fig. 2.6 using the $(r_e/r_w) = \infty$ curve since it is obvious that after 6 min of production the well is still infinite acting.

$$\Delta p = \frac{0.141 q \mu}{kh} (p_{tD}) = \frac{(0.141)(18.0)(1.39)(0.4)}{(10^{-4})(141)} p_{tD}$$

$$= (100)p_{tD} = (100) \, 2.48 = 248 \text{ psia}$$

$$\text{(Equation 2.15 solved for } p_i - p_{w,t})$$

$p_{w,t} = p_i - \Delta p = 3,000 - 248 = 2,752$ psia which is the flowing pressure after 6 min of production if there is no well damage.

PROBLEM 2.1B SOLUTION—APPLICATION OF THE Ei FUNCTION

To find the pressure at any radius where the rate is not known we must use the Ei function. Since we only know the rate at the well radius in this example we must use the Ei function.

Pressure after 6 min at 5-ft radius

We must first test to see if we can use the Ei solution. From Problem 2.1A we know $\frac{\eta t}{r_w^2} = 60 < 100$, therefore we cannot use the Ei function and there is no solution.

Pressure after 1 hr at 5-ft radius

$$\frac{\eta t}{r_w^2} = 10\ (60) = 600 > 100$$

$$\frac{r_e^2}{4\eta} = \frac{(3,000)^2}{(4)(3,600)} = 625 > t = 1/24 \text{ days}$$

Therefore we can use the Ei solution

$$t_D = \frac{\eta t}{r^2} = \frac{(3,600)(1/24)}{(5)^2} = 6.0$$

$$\frac{1}{2}\left[-\text{Ei}\frac{1}{-4t_D}\right] = 1.33 \text{ (from function graph Fig. 2.8.)}$$

From Problem 2.1A:

$$(0.141q\mu/kh) = 100$$

$$p_{r,t} = p_i - \frac{0.141q\mu}{kh}\left(\frac{1}{2}\right)\left(-\text{Ei}\frac{-1}{4t_D}\right) \qquad (2.18)$$

$$p_{r,t} = 3,000 - (100)(1.33)$$

$$= 3,000 - 133 = 2,867 \text{ psia}$$

Pressure after 1 hr at 50-ft radius

From previous part $(\eta t/r_w^2) > 100$ and $(r_e^2/4\eta) > t$. Therefore we can use the Ei solution:

$$t_D = \frac{\eta t}{r^2} = \frac{(3,600)(1/24)}{(50)^2} = 0.06 \text{ and}$$

$$\frac{1}{2}\left[-\text{Ei}\frac{-1}{4t_D}\right] \approx 0.0 \text{ (from function graph)}$$

$$p_{r,t} = 3,000 - (100)(0). \text{ p at 50 ft and 1 hr} = 3,000 \text{ psia}$$

The pressure influence has not yet significantly affected the pressure at 50 ft.

PROBLEM 2.1C SOLUTION—CHOICE OF A PRESSURE FUNCTION SOLUTION

We will follow Table 2.1 to determine which pressure function to use.

Is $t_D > 100$?

In Problem 2.1A we found that t_D at the well was 60 when t was 6 min. Thus, when t is 100 min, t_D at the well will be 1,000.

Yes.

Is q_r known?
r is the well radius so q_r is known

Yes.

Is $t > r_e^2/4\eta$?

$$t = \left(\frac{100}{1,440}\right)_{day} \ngtr r_e^2/4\eta = \frac{(3,000)^2}{(4)(3,600)} = 625$$

No.

Therefore use the natural log equation for the pressure function

$$p_{tD} = \frac{1}{2}(\ln t_D + 0.809) = \frac{1}{2}(\ln 1,000 + 0.809) = 3.85 \tag{2.22}$$

$$p_{r,t} = p_i - \frac{0.141q_r\mu}{kh} p_{tD} \tag{2.16}$$

From Problem 2.1A $(0.141q_r\mu/kh) = 100$ p_w at 100 min, $= 3,000 - (100)(3.85) = 2,615$ psia.

Following Table 2.1 to find the best pressure function for the well pressure after 1,000 days of production:

Is $t_D > 100$?
In Problem 2.1A we found that t_D at the well was 60 when t was 6 min so when t is 1,000 days × 1,440. min/day or 1.44×10^6 min

$$t_D = 1.44 \times 10^7 > 100$$

Yes.

Is q_r known?
Since r is the well radius q_r is known.

Yes.

Is $t > r_e^2/4\eta$?
From above $r_e^2/4\eta = 625$; $t = 1,000 > 625$

Yes.

So must use p_{tD} by Equation 2.20

$$p_{tD} = \frac{2t_D}{r_D^2} + \ln r_D - \frac{3}{4} \tag{2.20}$$

$$r_D = \frac{3,000}{0.5} = 6,000$$

$$p_{tD} = \frac{(2)(1.44 \times 10^7)}{(6,000)^2} + \ln 6,000 - \frac{3}{4}$$

$$= 8.74$$

$$p_{r,t} = p_i - \frac{0.141q_r\mu}{kh} p_{tD} \tag{2.16}$$

From Problem 2.1A $(0.141q_r\mu/kh) = 100$ p_w after 1,000 days $= 3,000 - (100)(8.74) = 2,126$ psia.

PROBLEM 2.2 SOLUTION — CUMULATIVE WATER DISPOSAL IN AN AQUIFER

A. For reservoir to be infinite acting $(r_e^2/4\eta) >$ injection time of 100 days

$$\eta = \frac{6.33k}{\phi\mu c} = \frac{(6.33)(0.01)}{(0.2)(0.5)(6.33 \times 10^{-6})} = 10^5 \tag{2.8}$$

$$\frac{r_e^2}{4\eta} = \frac{(10,000)^2}{(4)(10^5)} = 250$$

$$250 > 100$$

Reservoir is infinite acting.

B. To find the total water injected.

$$Q = 1.12\phi cr^2h \; \Delta p \; Q_{tD} \tag{2.29}$$

$$t_D = \frac{(10^5)(100)}{(0.5^2)} = 4 \times 10^7$$

$$Q_{tD} = 4.610 \times 10^6 \text{ (from tables 2.2)}$$

Substituting in Equation 2.29.

$$Q = (1.12)(0.2)(6.33)(10^{-6})(14.1)(0.5)^2(2,200 - 1,000 - 200)(4.610)(10^6)$$
$$= 23,050 \text{ bbl}$$

PROBLEM 2.3A: SOLUTION — PRESSURE RESULTING FROM MORE THAN ONE WELL IN A RESERVOIR

$$\eta = \frac{6.33k}{\phi\mu c} = \frac{(6.33)(0.1)}{(0.2)(0.5)(6.33 \times 10^{-5})} = 10^5 \tag{2.8}$$

to find Δp caused by a rate of 540 STB/d. for 62.5 days at a radius of 2,500 ft. Note that 62.5 days is less than $r_e^2/4\eta = 10,000^2/(4 \times 10^5)$. So,

$$\Delta p_{2500.62.5} = \frac{0.141q_w\mu}{kh}\left[\left(\frac{1}{2}\right)\left(-\text{Ei}\,\frac{-1}{4t_{D1}}\right)\right] \qquad \text{(Equation 2.18 Solved for } p_i - p_{r,t})$$

$$t_{D1} = \frac{\eta t}{r^2} = \frac{(10^5)(62.5)}{(2,500)^2} = 1.0$$

from Fig. 2.8

$$\frac{1}{2}\left(-\text{Ei}\,\frac{1}{4t_{D1}}\right) = 0.5$$

$$\Delta p_{2500.62.5} = \frac{0.141(540)(1.39)(0.5)}{(0.1)(14.1)}(0.5) = (37.5)(0.5) = 18.7 \text{ psi}$$

To find Δp caused by a rate of 1,080 STB/d for 10 days at a radius of 0.5 ft,

$$\Delta p_{.5,10} = \frac{0.141q_r\mu}{kh}\left(\frac{1}{2}\right)(\ln t_{D2} + 0.809) \qquad \text{(Equations 2.21 and 2.16 combined)}$$

$$t_{D2} = \frac{\eta t}{r^2} = \frac{(10^5)(10)}{(0.5)^2} = 4 \times 10^6$$

$$\frac{1}{2}\,[\ln(4 \times 10^6) + 0.809] = 8.0$$

$$\Delta p_{.5,10} = \frac{(0.141)(1,080)(1.39)(0.5)}{(0.1)(14.1)}\,(8) = 600 \text{ psia}$$

p_w at well 2 = 3,000 − 19 − 600 = 2,381 psi.

PROBLEM 2.3B: SOLUTION – CALCULATION OF THE TIME NECESSARY TO OBTAIN A PARTICULAR PRESSURE DROP IN THE RESERVOIR

$$p_{r,t} = p_i - \frac{0.141q\mu}{kh}\left[\left(\frac{1}{2}\right)\left(-\text{Ei}\,\frac{-1}{4t_D}\right)\right] \tag{2.18}$$

$$7.5 = \frac{(0.141)(540)(1.39)(0.5)}{(0.1)(14.1)}\left[1/2\left(-\text{Ei}\,\frac{-1}{4t_D}\right)\right]$$

$$\left[1/2\left(-\text{Ei}\,\frac{-1}{4t_D}\right)\right] = 0.2$$

From Fig. 2.8 the t_D corresponding to this value is 0.38

$$t_D = \frac{\eta t}{r^2}$$

$$0.38 = \frac{(10^5)(t)}{(2,500)^2}$$

$$t = 24 \text{ days}$$

PROBLEM 2.3C SOLUTION – ACCOUNTING FOR VARIATIONS IN PRODUCING RATES

Δp at well 2 = (Δp for r = 2,500, Δq = 750, and t = 72.5)
   − (Δp for r = 2,500, Δq = (540 − 180) 1.39, and t = 10)
   + (Δp for r = 0.5, q = 1,500, and t = 20)

Δp for r = 2,500, Δq = 750, and t = 72.5

$$t_D = \frac{\eta t}{r^2} = \frac{(10^5)(72.5)}{(2,500)^2} = 1.16$$

from Fig. 2.8 $\quad 1/2\left(-\text{Ei}\,\frac{-1}{(4)(1.16)}\right) = 0.57$

$$\Delta p = \frac{0.141\mu q}{kh}\left(\frac{1}{2}\right)\left(-\text{Ei}\,\frac{-1}{4t_D}\right)$$

$$= \frac{(0.141)(0.5)}{(0.1)(14.1)}\,(750)(0.57)$$

$$= (0.05)(750)(0.57) = 21.4 \text{ psi}$$

Δp for $r = 2{,}500$, $\Delta q = 500$, and $t = 10$

$$t_D = \frac{(10^5)(10)}{(2{,}500)^2} = 0.16$$

from Fig. 2.8 $\; 1/2\left(-\text{Ei}\,\frac{-1}{(4)(0.16)}\right) = 0.04$

$$\Delta p = \frac{0.141\mu}{kh}q\left[1/2\left(-\text{Ei}\,\frac{-1}{(4)(0.16)}\right)\right]$$

$$= (0.05)(500)(0.04) = 1.0 \text{ psi}$$

Δp for $r = 0.5$, $q = 1{,}500$, and $t = 20$

$$t_D = \frac{(10^5)(20)}{(0.5)^2} = 8 \times 10^6$$

$$\frac{1}{2}\left[\ln t_D + 0.809\right] = \frac{1}{2}\left[\ln (8 \times 10^6) + 0.809\right] = 8.35$$

$$\Delta p = (0.05)(1500)(8.35) = 626.5 \text{ psi}$$

Substituting into the first equation, Δp at well $2 = 21.4 - 1.0 + 626.5 = 647$ psi then p_w at well $2 = 3{,}000 - 647 = 2{,}353$ psia.

PROBLEM 2.3D SOLUTION – SIMULATING BOUNDARY EFFECTS

$$\Delta p \text{ at well } 1 = (\Delta p \text{ for } q = 750, \ t = 52.5, \text{ and } r = 0.5)$$

$$+ (\Delta p \text{ for } q = 750, \ t = 52.5, \text{ and } r = (2)(1{,}250)$$

Δp for $q = 750$, $t = 52.5$, and $r = 0.5$

$$t_D = \frac{\eta t}{r^2} = \frac{(10^5)(52.5)}{(0.5)^2} = 21 \times 10^6$$

$$\frac{1}{2}\left[\ln t_D + 0.809\right] = \frac{1}{2}\left[\ln (21 \times 10^6) + 0.809\right]$$

$$= 8.82$$

$$\Delta p = \frac{0.141\mu q}{kh}\left(\frac{1}{2}\right)\left[\ln t_D + 0.809\right]$$

$$= \frac{(0.141)(0.5)}{(0.1)(14.1)}(750)(8.82)$$

$$= (0.05)(750)(8.82)$$

$$= 331 \text{ psi}$$

Δp for $q = 750$, $t = 52.5$, and $r = 2500$

$$t_D = \frac{(10^5)(52.5)}{(2500)^2} = 0.84$$

from Fig. 2.8 $\;\left(\frac{1}{2}\right)\left[-\text{Ei}\,\frac{-1}{(4)(0.84)}\right] = 0.45$

$$\Delta p = (0.05)(750)(0.45)$$

$$= 17 \text{ psi}$$

Substituting in the first equation, Δp_w at well $1 = 331 + 17 = 348$ psi and $p_w = 3000 - 348 = 2652$ psia.

PROBLEM 2.4 SOLUTION – ANALYSIS OF A PSEUDOSTEADY-STATE FLOW TEST

A Note constant producing rate and constant rate of pressure decline. Therefore, use pseudosteady-state equations

$$\left(\frac{\Delta p}{\Delta t}\right)_{\text{pseudo}} = \frac{1.79q}{r_e^2 \phi ch} \tag{2.41}$$

$$\therefore \phi h = \frac{1.8q}{r^2 c}\frac{\Delta t}{\Delta p_{\text{avg}}}$$

$$= \frac{(1.8)(850)(1.1)}{(4 \times 10^6)(10^{-5})}\left(\frac{1}{2}\right)$$

$$= 21 \ \phi \ \text{ft}$$

B
$$kh = \frac{q\mu \ln[(r_e/r_w) - 1/2]}{7.08(p_e - p_w)} \tag{2.51}$$

$$\ln r_e/r_w = \ln \frac{2000}{0.5}$$

$$= 2.303(\log 4{,}000) = (2.303)(3.603)$$

$$kh = \frac{(850)(1.1)(0.5)[(2.303)(3.603) - 1/2]}{(7.08)(3)(10^2)}$$

$$kh = 1.720 \text{ darcy-ft or } 1720 \text{ md-ft}$$

C
$$q = 7.08 \ \frac{kh}{\mu}\frac{(p_e - p_w)}{\ln (r_e/r_w)} \tag{1.13}$$

$$kh = \frac{q\mu(\ln r_e/r_w)}{7.08(p_e - p_w)}$$

$$= \frac{(850)(1.1)(0.5)(2.303 \times 3.603)}{7.08(3 \times 10^2)}$$

$$= 1.830 \text{ darcy-ft or } 1830 \text{ md-ft}$$

PROBLEM 3.1 SOLUTION – PRODUCTIVITY INDEX EVALUATION
PARTS A AND B

(1)	(2)	(3)	(4)	(5)
		"Initial"	When $N_p = 35{,}000$ STB	
p_w psig	q_o STB/d	PI STB/d/psig	PI STB/d/psig	q_o STB/d
2660.	90	0.265	0.157	
2380.	165	0.266	0.158	
1980.	270	0.265	0.157	

PROBLEM 3.1 SOLUTION (*Continued*)

(1)	(2)	(3) "Initial"	(4)	(5) When $N_p = 35{,}000$ STB
1600.	330	0.236	0.140	
1195.	373	0.207	0.123	197 Max. flow rate
0	**455	0.152	0.109	252 Max. pump rate

(1) From Fig. 3.1A.

(2) From Fig. 3.1A; **Maximum pumping rate obtained by extrapolation.

(3) $q/(p_e - p_w) = q/(3{,}000 - p_w)$ where q is in STB. Based on Equation 3.1. Where 3000 is obtained from Fig. 3.1A extrapolation.

(4) (PI when N_p is 35,000) = Initial PI $(k_o/B_o\mu_o)_{35{,}000}/(k_o/B_o\mu_o)_{Initial}$,

$$= \text{Initial PI}[0.5/(1.15 \times 1.1)]/[1.0/(1.25 \times 1.2)]$$
$$= \text{Initial PI} \times 0.5925$$

(5) $(q_o)_{35{,}000} = PI \times (p_e - p_w) = (PI)_{35{,}000} \times (2800 - p_w)$ From Equation 3.1

If only rates less than 270 b/d were used in the test the indicated maximum rate would be based on the PI of 0.265 STB/d/psi or $(3{,}000) \times (0.265) = 795$ STB/d.

$$DR = k_{undamaged}/k_{actual}$$

$$J = \frac{7.08 k_{act.}h}{B_o\mu_o \ [\ln (r_e/r_w) - 0.5]} \qquad \text{(Equation 3.2 where}$$
$$k_{act.} = k_o)$$

$$0.265 = \frac{7.08 k_{act.} \ 14.}{1.25 \times 1.2 \times [\ln (745/.333) - 0.5]}$$

$$k_{act.} = 0.0288 \text{ darcies or } 28.8 \text{ md}$$

$$DR = \frac{20.5}{28.8} = 0.712 \text{ —permeability around well bore has been improved.} \quad (3.5)$$

PART C

$$\left(\frac{\Delta p}{\Delta t}\right)_{pseudo} = \frac{1.8q}{\phi chr_e^2} \qquad (2.32)$$

$$3.0 = \frac{1.8 \times 75 \times 1.15}{(\phi cr_e^2) \ 14}$$

$$\phi cr_e^2 = \frac{3.7 \text{ ft}^2}{\text{psi}}$$

$$t_s = \frac{0.04(\phi cr_e^2)\mu}{k} \qquad (2.53)$$

$$= \frac{0.04 \times 3.7 \times 1.1}{(0.5 \times 0.0205)}$$

$$= 15.9 \text{ days}$$

PROBLEM 3.2 SOLUTION—PRESSURE BUILDUP FROM AN "UNCHANGING" WELL PRESSURE

Determine the undamaged k from the slope of the buildup plot.

$$\text{m from the plot} = \frac{1,484 - 1,424}{1.0 - 0.3} = 86$$

$$m = \frac{0.1625q\mu}{kh} \tag{3.12}$$

$$k = \frac{0.1625q\mu}{mh} = \frac{(0.1625)(199)(1.15)(2.0)}{(86)(22)}$$

$$= 0.04 \text{ darcy or 40 md}$$

ϕc can be determined from the time travel equation and the observation that the nearest boundary is 300 ft away.

$$t_s = \frac{0.04\phi\mu cr_e^2}{k} \tag{2.61}$$

$$\phi c = \frac{t_s k}{0.04\mu r_e^2} = \frac{(7.5/24)(0.04)}{(0.04)(2.0)(300)^2}$$

$$= 1.737 \times 10^{-6}$$

For convenience the Δp_{skin} calculation will be based on the shut-in pressure on the straight line extrapolation at 10 hr (log $= 1.0$).

$$\Delta t_D \text{ for 10 hr} = \frac{6.33k\Delta t}{\phi c \mu r_w^2} \tag{3.8}$$

$$= \frac{(6.33)(0.04)(10/24)}{(1.737)(10^{-6})(2.0)(1/3)^2} = 2.73 \times 10^5$$

$$\ln \Delta t_D = (5)(2.3) + 1.0 = 12.5$$

$$p_w - p_{wf} = 0.867m\left(\frac{1}{2}\right)(\ln \Delta t_D + 0.809) + \Delta p_{skin} \tag{3.13}$$

$$\Delta p_{skin} = p_w - p_{wf} - 0.867m\left(\frac{1}{2}\right)(\ln \Delta t_D + 0.809)$$

$$= 1,484 - 702 - (0.867)(86)\left(\frac{1}{2}\right)(12.5 + 0.8)$$

$$= 1,484 - 702 - 495$$

$$= 287 \text{ psi}$$

$$\Delta p_{skin} = 0.867ms \tag{3.15}$$

$$s = \frac{\Delta p_{skin}}{0.867m} = \frac{287}{(0.867)(86)} = 3.85$$

The rate after skin removal can be determined by ratio since with all other parameters equal the rate is proportional to the effective pressure drop.

$$q_{\text{zero skin}} = q_{\text{STB}} \frac{(p_{\text{unchanging}} - p_{\text{wf}})}{(p_{\text{unchanging}} - p_{\text{wf}} - \Delta p_{\text{skin}})}$$

$$= \frac{(199)(1{,}482 - 702)}{(1{,}482 - 702 - 287)}$$

$$= 315 \text{ STB/d}$$

PROBLEM 3.3A SOLUTION – PRESSURE BUILDUP ANALYSIS OF AN "OLD" WELL

2 Points for Δp_q versus log Δt plot will check Fig. 3.6
at t = 24 hr

$$\left(\frac{\Delta p}{\Delta t}\right)_{\text{pseudo}} = \frac{1 \text{ psi}}{\text{hr}} \text{ or } \frac{24 \text{ psi}}{\text{day}}$$

$$\Delta p_q = p_w - p_{\text{wf}} + \Delta t \left(\frac{\Delta p}{\Delta t}\right)_{\text{pseudo}} \tag{3.22}$$

$$\Delta p_q = 2{,}675 - 1{,}123 + 24 = 1{,}576 \text{ psia at 24 hr}$$

at t = 30 hr

$$\Delta t \left(\frac{\Delta p_w}{\Delta t}\right) = \frac{1 \text{ psi}}{\text{hr}} (30 \text{ hr}) = 30 \text{ psia}$$

$$\Delta p_q = 2{,}687 - 1{,}123 + 30 = 1{,}594 \text{ psia}$$

Calculation of (ϕc) product

$$\frac{\Delta p}{\Delta t} = \frac{1.8q \, B_o}{r_e^2 \, \phi h c} \quad \text{Where q is in STB} \tag{2.41}$$

$$r_e = [(43{,}560)(40)/\pi]^{0.5} = 745 \text{ ft}$$

$$(\phi c) = \frac{1.8q \, B_o}{r_e^2 h} \frac{dt}{dp}$$

$$= \frac{(1.8)(280)(1.31)}{(7.45)^2(10^4)(40)} \frac{1}{24}$$

$$(\phi c) = 1.24 \times 10^{-6}$$

Calculation of permeability, k

$$\left.\begin{array}{l} m = \dfrac{0.1625qB_o\mu}{kh} \\[2em] k = \dfrac{0.1625q \, B_o\mu}{mh} \end{array}\right\} \quad \text{where q is in STB} \quad \text{(modified Eq. 3.12)}$$

$$m = \text{slope} = \frac{1{,}594 - 1{,}304}{\log 31 - \log 1} = \frac{290}{1.491} = 197$$

$$k = \frac{0.1625(280)(1.31)(2.0)}{(197)(40)} = 0.0151 \text{ darcies or 15.1 md}$$

Beginning of pseudosteady state:

$$t_s = \frac{0.04\phi c\mu r_e^2}{k} \tag{2.63}$$

$$r_e^2 = 40 \times 43{,}560/\pi = 5.55 \times 10^5$$

$$t_s = \frac{(4)(10^{-2})(1.24)(10^{-6})(2)(5.55)(10^5)}{0.0151}$$

$$t_s = 3.65 \text{ days}$$

Calculation of Δp_{skin} and s will be based on Δp_q of 1,576 psia at a shut-in time of 24 hr

$$\Delta p_q = 0.867m(0.5)(\ln \Delta t_D + 0.809) + \Delta p_{skin} \tag{3.21}$$

$$\ln \Delta t_D + 0.809 = \ln (6.33k\Delta t/\phi\mu cr_w^2) + 0.809$$

$$= \ln(6.33 \times 0.0151 \times 1.0/1.24 \times 10^{-6} \times 2 \times 0.333^2) + 0.809$$

$$= 13.6$$

$$1{,}576 = 0.867 \times 197 \times 0.5 \times 13.6 + \Delta p_{skin} \tag{3.21}$$

$$\Delta p_{skin} = 415 \text{ psi}$$

$$S = \frac{\Delta p_{skin}}{0.867m} \tag{3.15}$$

$$= \frac{415}{0.867 \times 197} = 2.43$$

PROBLEM 3.3B SOLUTION—DETERMINING THE AVERAGE PRESSURE IN THE DRAINAGE AREA OF A PSEUDOSTEADY-STATE WELL

To find p_s we must first evaluate p^*. Equation 3.40 will be used in preference to Equation 3.39 because of the short producing time.

$$p^* = p_{wf} + \Delta p_{skin} + (0.867m)(0.5)(\ln t_D + 0.809) \tag{3.40}$$

$$\ln t_D = \ln \left(\frac{6.33kt}{\phi\mu cr_w^2}\right)$$

$$= \ln\left[\frac{6.33 \times 0.0151 \times 10}{(1.24 \times 10^{-6} \times 2 \times 0.333^2)}\right]$$

$$= \ln (3.47 \times 10^6) = 15.064$$

$$p^* = 1{,}123 + 415 + 0.867 \times 197 \times 0.5 \times (15.064 + 0.809)$$

$$= 2{,}894 \text{ psia.}$$

$$\frac{(p_i - p^*)}{m} = \frac{(2{,}960 - 2{,}894)}{197} = 0.335$$

From Fig. 3.9B

$$\frac{(p_i - p_s)}{m} = 1.175$$

$$\frac{(2{,}960 - p_s)}{197} = 1.175 \qquad p_s = 2{,}729 \text{ psia}$$

If we assume a radial drainage system we can calculate p_s as

$$p_s = p_{wf} - 0.439m + (\Delta p_q)_{ts} \tag{3.43}$$

$$= 1,123 - (0.439)(197) + 1,685$$

$$= 2,722 \text{ psia compared with the } p_s \text{ of } 2,729 \text{ psia calculated from the Odeh data.}$$

To find the damage ratio we must first evaluate p_e.

$$p_e = 0.217m + p_s = 0.217 \times 197 + 2,729$$

$$= 2,772 \text{ psia}$$

$$DR = \frac{(p_e - p_{wf})}{(p_e - p_{wf} - \Delta p_{skin})}$$

$$= \frac{(2,772 - 1,123)}{(2,772 - 1,123 - 415)}$$

$$= 1.34$$

PROBLEM 3.4 SOLUTION – ANALYSIS OF A TWO-RATE PRESSURE BUILDUP

$$m = \frac{0.1625\Delta q\mu}{kh} \tag{3.49}$$

$$k_o = \frac{0.1625\Delta q\mu}{mh} \qquad\qquad m \text{ is 8.75 at "shut-in times" greater than 4 hr.}$$

$$= \frac{(0.1625)(14.)(1.322)(0.39)}{(8.75)(10)}$$

$$k_o = 0.0134 \text{ darcies or } 13.4 \text{ md}$$

Base the skin calculations on the $\Delta p_{\Delta q}$ value after 1 day of shut-in.

$$\Delta t_D \text{ for 1 day} = \frac{6.33k\Delta t}{\phi\mu cr_w^2} \tag{3.8}$$

$$= \frac{(6.33)(0.0134)(1)}{(0.2)(0.39)(1.379)(10^{-4})(0.265)^2}$$

$$= 1.12 \ (10^5)$$

$$\ln \Delta t_D = 11.6$$

$$\Delta p_{\Delta q} = 0.867m \left(\frac{1}{2}\right)(\ln \Delta t_D + 0.809) + \Delta p_{skin} \tag{3.50}$$

$$\Delta p_{skin} = (2,211 - 1,963) - 0.867(8.75)\left(\frac{1}{2}\right)(11.6 + 0.8)$$

$$\Delta p_{skin} = 201 \text{ psi.}$$

$$S = \frac{\Delta p_{skin}}{0.867m} = \frac{201}{(0.867)(8.75)} = 26.5 \tag{3.15}$$

To adjust m an Δp_{skin} from a Δq to a q base,

$$m \text{ for } q = (78)(1.322) = \frac{78}{14}(8.75) = 48.75$$

$$\Delta p_{skin} \ @ \ q = (78)(1.322) = 0.867 \ (48.75)(26.5)$$

$$= 1{,}123 \ \text{psi.}$$

$$p_s = p_{wf} + \Delta p_{skin} + 0.867m \left(\ln \frac{r_e}{r_w} - \frac{3}{4} \right) \tag{3.46}$$

$$= 1{,}963 + 1{,}123 + (0.867)(48.75) \left(\ln \frac{1{,}490}{0.265} - 0.75 \right)$$

$$= 3{,}419 \ \text{psia.}$$

PROBLEM 3.5 SOLUTION—PRESSURE FALLOFF ANALYSIS

PLOT p_w vs Δt

To find the undamaged permeability

$$\text{Slope } m = \frac{0.1625\mu i}{kh} \quad \text{where } k = \text{undamaged permeability} \tag{3.54}$$

$$\text{from plot } m = 82.0$$

$$82.0 = \frac{(0.1625)(250)(1.0)}{k(50)}$$

$$K = 0.00995 \ \text{darcies or } 10 \ \text{md}$$

To find Δp skin

$$p_w = p_{wf} - \frac{0.141 \ \mu i}{kh} \left(\frac{1}{2} \right) \ (\ln \Delta t_D + 0.809) - \Delta p_{skin} \tag{3.52}$$

$$@t = 15 \ \text{min}$$

$$t_d = \frac{6.33kt \ \text{days}}{\phi\mu c r_w^2} = \frac{6.33 \ (0.01) \ t_{min}}{\left(0.2)(1.0)(5 \times 10^{-6})(1/16)(1{,}440 \ \frac{min}{day} \right)}$$

$$= 703 \times t_{min}$$

$$4 = 500 - \frac{(0.141)(1.0)(250)}{(0.01)(50)} \frac{\ln(703 \times 15) + 0.809}{2} - \Delta p_{skin}$$

$$\Delta p_{skin} = 141 \ \text{psi}$$

$$\Delta p_{skin} = \frac{0.141 \ \mu i}{kh} s$$

$$141 = \frac{(.141)(250)(1.0)}{(.01)(50)} s$$

$$s = 2.0$$

To find r_e at $t = 15$ minutes use Equation 2.53

$$t_s = \frac{0.04\phi\mu c r_e^2}{k} \tag{2.63}$$

$$r_e^2 = \frac{kt_{days}}{0.04\phi\mu c} = \frac{0.01 \ t_{days}}{(0.04)(0.2)(1.0)(5 \times 10^{-6})}$$

$$r_e = \frac{(t_{days})^{1/2}}{2(10^{-3})} = 13.2\,(t_{min})^{1/2}$$

$$r_e = 13.2\,(15)^{1/2} = 51.0 \text{ ft}$$

i when $\Delta p_{skin} = 0.0$

BHP in injection well $= 500 + (0.433)(3,000) = 1,300$ psig
New rate $= 250(1,300 - 0)/(1,300 - 0 - 141) = 271$ b/d

PROBLEM 3.6 SOLUTION — DETERMINING THE DISTANCE TO A RESERVOIR BOUNDARY FROM A DRAWNDOWN TEST

A. From graph, $\Delta p_{(10\ days)} = 67$ psi

$$m = 80 \text{ psi/cycle}$$

$$m = \frac{0.1625 q \mu}{kh} \quad \text{i.e.} \quad \frac{kh}{\mu} = \frac{(0.1625)(80)(1.25)}{80} = 0.20$$

$$\text{therefore } \frac{k}{\mu} = 0.02 \text{ when k is in darcies}$$

B. Also, $\dfrac{1}{2}\left[-Ei\left(\dfrac{-1}{4t_D}\right)\right] = \dfrac{\Delta p'}{0.867m} = \dfrac{(1.151)(67)}{(80)} = 0.97$ (Equation 3.64 where $p_w{}' - p_w = \Delta p'$)

Then from Fig. 2.8

$$t_D = 3.0$$

But

$$t_D = \frac{\eta t}{(2d)^2} \quad \text{and} \quad t_D = \frac{6.33 k t_{days}}{\phi \mu c (2d)^2} \tag{3.63}$$

$$(2d)^2 = \frac{\eta t}{t_D} = \frac{(6.33)(0.02)(10)}{(0.2)(10 \times 10^{-6})(3.0)} = 2.11(10^5)$$

$$d^2 = 52750.$$
$$d = 229 \text{ ft}$$

C. Using the pressure at 10 days again.
$$p_w = p_i - 0.867m0.5\,(\ln t_{Dreal} + 0.809) - \Delta p_{skin} - (p_w{}' - p_w)$$

(Equation 3.61 with $p_w{}' - p_w$ and 0.867m substituted according to Equations 3.62 and 3.12)

$$0.5(\ln t_{Dreal} + 0.809) = 0.5\left(\ln \frac{(6.33)(0.02)(10)}{(0.2)(10^{-5})(1/16)} + 0.809\right) = 8.47$$

$$2337 = 2800 - (0.867)(80)(8.47) - \Delta p_{skin} - 67.$$

$$\Delta p_{skin} = -191$$

$$s = \frac{\Delta p_{skin}}{0.867m} = \frac{-191}{(0.867)(80)} = -2.76 \tag{3.15}$$

PROBLEM 3.7 SOLUTION—ANALYZING DST DATA

A.

$$\text{For: } t = 60 \text{ min}; \quad \frac{t + \Delta t}{\Delta t} = \frac{80 + 60}{60} = 2.33$$

$$t = 100 \text{ min}; \quad \frac{t + \Delta t}{\Delta t} = \frac{80 + 100}{100} = 1.8$$

$$t = 80 \text{ min}; \quad \frac{t + \Delta t}{\Delta t} = \frac{80 + 80}{80} = 2.0$$

p_i from plot at $(\Delta t + t)/\Delta t = 1.0$ (Fig. 3.8)

$$p_i = 4,500 \text{ psia}$$

$$m \text{ from plot} = 300 \text{ psi/cycle}$$

B.

$$\text{Hydrostatic mud pressure} = (0.433)\left(\frac{\text{mud wt.}}{8.33}\right) \times \text{bomb depth}$$

$$= (0.433)\left(\frac{9.8}{8.33}\right)(10,000) = 5,100 \text{ psig}$$

$$\text{Drill pipe capacity} = \frac{(3.826)^2}{1,000} = 0.0147 \text{ bbl/ft}$$

$$\text{Fluid recovery} = (0.0147)(7,500) = 110 \text{ bbl}$$

$$q_{STB} = \frac{110}{80/1440} = 1,980 \text{ b/d}$$

$$(k_o/\mu_o) = \frac{0.1625qB}{hm} = \frac{(0.1625)(1,980)(1.75)}{(20)(300)}$$

$$(k_o/\mu_o) = 0.0938 \text{ darcies/cp or } 93.8 \text{ md/cp}$$

$$k_{omd} = (93.8)(0.4) = 37.5 \text{ md}$$

C.

$$p_w = (0.825)(0.433)10,000 = 3,572 \text{ psig}$$

$$q_{undamaged} = \frac{7.08kh(p_e - p_w)}{B\mu\left(\ln \dfrac{r_e}{r_w} - \dfrac{1}{2}\right)} \tag{3.68}$$

$$= \frac{(7.08)(0.0938)(20)(4,500 - 3,572)}{1.75 \, (\ln 2,980 - 1/2)}$$

$$= \frac{939 \text{ STB}}{d}$$

D.

$$p_w - p_{wf} = 0.867m(0.5)(\ln t_D + 0.809) + \Delta p_{skin} - m \log \left(\frac{t + \Delta t}{\Delta t}\right) \tag{3.36}$$

$$\text{when } \Delta t = 100 \text{ min}; \ p_w = 4,422$$

$$\ln t_D = \ln \left[\frac{(6.33)(0.0938)(80/1,440)}{(0.2)(5 \times 10^{-5})(0.25)^2}\right] = 10.87$$

$$4,422 - 2,700 = (0.867)(300)(0.5)(10.87 + 0.809) + \Delta p_{skin} - 300 \log 1.8$$

$$\Delta p_{skin} = 1{,}270 \text{ psia}$$

$$S = \frac{\Delta p_{skin}}{0.867m} = \frac{1{,}270}{(0.867)(300)} = 4.88 \tag{3.15}$$

E.
$$t_s = \frac{0.040\phi\mu c r_i^2}{k} \tag{3.72}$$

$$r_e = \sqrt{\frac{(80)(0.0938)}{(1{,}440)(40)(0.2)(5 \times 10^{-5})}} = \sqrt{1.30 \times 10^4}$$

$$= 114 \text{ ft}$$

PROBLEM 4.1 SOLUTION—DETERMINING GAS CHARACTERISTICS

A.
$$_pP_c = \sum_{j=1}^{j=n} (p_{cj}\, MF_j)$$

$$MF = \text{Vol. fraction for gas}$$

$$_pP_c = (673)(0.8) + (708)(0.15) + (550)(0.05)$$

$$= 672 \text{ psia}$$

$$_pT_c = \sum_{j=1}^{j=n} (T_{cj})(MF_j)$$

$$= (343)(0.8) + (549)(0.15) + (766)(0.05)$$

$$= 395° R.$$

$$_pP_r = \frac{3{,}000}{672} = 4.46$$

$$_pT_r = \frac{170 + 460}{395} = 1.59$$

B. From Appendix B.7

$$z = 0.82$$

C.
$$MW = \sum_{j=1}^{j=n} (MF)_j\, (MW)_j$$

$$= (0.8)(16) + (0.15)(30) + (0.05)(58)$$

$$= 20.2$$

$$\gamma = \frac{20.2}{29} = 0.70$$

D. From Appendix B.8

$$_pP_c = 668; \quad _pT_c = 390$$

$$_pP_r = \frac{3{,}000}{668} = 4.49$$

$$_pT_r = \frac{170 + 460}{390} = 1.62$$

$$z = 0.83$$

E.
$$B_g = \frac{0.00504zT}{p}$$

$$= \frac{(5.04)(10^{-3})(0.82)(6.3)(10^2)}{(3.0)(10^3)}$$

$$= (8.68)(10^{-4}) \text{ res. bbl/scf}$$

F.
$$\rho = \frac{2.7\gamma p}{zT}$$

$$= \frac{(2.7)(0.7)(3.0)(10^3)}{(0.82)(6.3)(10^2)}$$

$$= 10.98 \text{ lb/ft}^3$$

G.
$$c \approx \frac{1}{p} = \frac{1}{3,000} = \frac{(10)(10^{-1})}{(3)(10^3)} = (3.3)(10^{-4})$$

from Appendix B.5, $c_r = 0.2$

$$c_g = 0.2/672 = (3.0)(10^{-4})$$

H. At $100°$ F., when $\gamma = 0.7$ and $p = 3,000$, $\mu_g = 0.023$

$$@ \ 200° \text{ F.}, \ \mu_g = 0.021$$

$$@ \ 170° \text{ F.}, \ \mu_g = 0.021 + \frac{30}{100}(0.002) = 0.0216 \text{ cp.}$$

I. From Appendix B.12

$$\frac{\text{Well fluid grav.}}{\text{Trap gas grav.}} = 1.16$$

Well fluid grav. = Reservoir fluid grav.
$$= (1.16)(0.7) = 0.81$$

From Appendix B.8, condensate data

$$_pP_c = 657 \text{ psia}$$

$$p_r = \frac{3,000}{657} = 4.57$$

$$_pT_c = 408° \text{ R.}$$

$$T_r = \frac{170 + 460}{408} = 1.54$$

From Appendix B.7

$$z = 0.81$$

PROBLEM 4.2: SOLUTION – DETERMINING GAS PRODUCTION FOR USE IN MATERIAL BALANCE

G_p = scf dry gas + scf original reservoir water vapor + scf of condensate vapor + scf vent gas.   (Equivalent of Eq. 4.25)

From Appendix B.18

$$GE = 814 \text{ scf/STB}$$

Condensate gas $= (814)(150) = 122{,}000$ scf

Water Vapor in original gas (Appendix B.19) $= (350 \text{ lb/MMcf} (0.89) = 312 \text{ lb/MMcf}.$

$$\frac{15 \text{ bbl.} \times 350 \text{ lb/bbl.}}{10.122 \text{ MMcf}} = 519 \text{ lb/MMcf} = \text{water production}$$

$$SCF \text{ water vapor} = \frac{(312)}{(350)}(7390)(10.122)$$
$$= 66{,}680 \text{ SCF}$$

$$G_p/\text{Day} = 10. + .067 + .122 + .030 \text{ (vent gas)}$$
$$= 10.219 \text{ MMcfd}$$

$$W_p/\text{day} = \frac{(519 - 312)(10.122)}{350} = 5.98 \text{ bw/d}$$

PROBLEM 4.3A SOLUTION—CALCULATION OF BOTTOM HOLE PRESSURE FROM STATIC PRESSURE MEASUREMENTS USING NOMOGRAPHS

Thermal Gradient $= (110 - 90)/3{,}676 = 5.45°\,\text{F}/1{,}000$ ft. Interpolate between the specific gravity charts and thermal gradient charts. From FIG. C4.3: Moving down vertically from the surface temperature of 90° F to the well head pressure of 2,715 psia on the exactly horizontal lines, read on the hyperbolic curves a pressure of 223 psia. Using the bottom limits, notice 90° F corresponds to 3,250 ft equivalent well depth. At the total equivalent well depth, $3{,}676 + 3{,}250 = 6{,}926$ ft, proceed vertically upward to the intersection of the slightly inclined line which

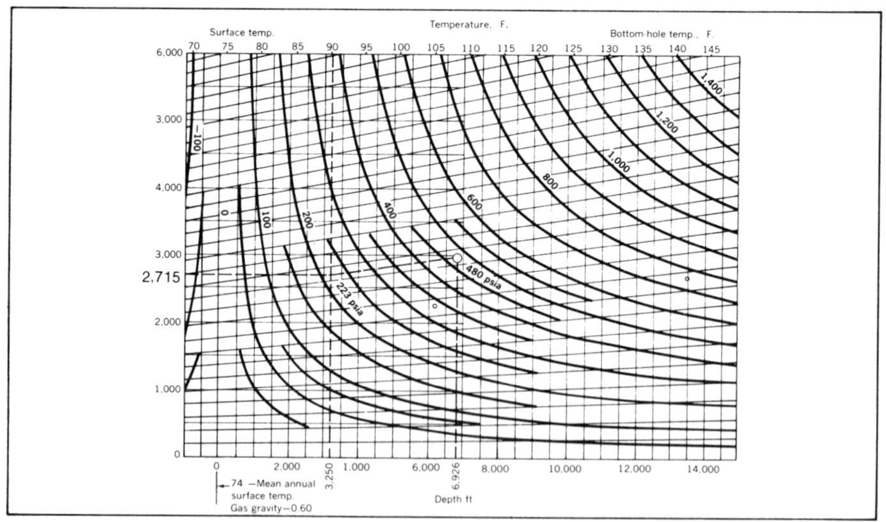

FIG. C.4.3 Static bottom-hole-pressure chart

passes through the surface-condition point established above. At this point we read 480 psia. Pressure of 3,676 ft column of 0.6 sp gr gas with a gradient of 5° F/1,000 ft, p = 480 − 223 = 257 psia.

Using the same procedure we find other pressures from Fig. 4.3B, C, and D.

$$0.6 \text{ sp gr gas at } 10° F/1,000 \text{ ft} = 370 − 110 = 260 \text{ psia}$$
$$0.7 \text{ sp gr gas at } 10° F/1,000 \text{ ft} = 450 − 137 = 313 \text{ psia}$$
$$0.7 \text{ sp gr gas at } 5° F/1,000 \text{ ft} = 600 − 263 = 337 \text{ psia}$$

By interpolation:

Pressure of 0.68 sp gr gas at 5° F/1,000 ft column $= 257 + (337 − 257)(0.80)$
$$= 321 \text{ psia}$$

Pressure of 0.68 sp gr gas at 10° F/1,000 ft column $= 260 + (313 − 260)(0.80)$
$$= 302 \text{ psia}$$

Pressure of 0.68 sp gr gas at 5.45° F/1,000 ft column $= 321 − (321 − 302)(0.45/5)$
$$= 319.29 \text{ psia}$$

Therefore the BHP $= 2,715 + 319 = 3,034$ psia which compares with a computer-calculated value of 3,034 psia.

PROBLEM 4.3B SOLUTION — CALCULATION OF STATIC BHP FROM SURFACE PRESSURE MEASUREMENTS

$$p_2 = p_1 \, e^{(0.01875 \, \gamma_g D/Z_{avg}T_{avg})} \tag{4.31}$$

$$= p_1 \, e^{(0.01875 \times .68 \times 3676/Z_{avg} \times 560)} = p_1 \, e^{(0.0838/z)}$$

z can be determined by trial and error using Appendix B.7 and B.8.

$$\text{Assume } p_2 = 3035; \quad p_{avg} = \frac{3035 + 2715}{2} = 2875$$

$$p_r = 2875/668 = 4.303; \quad T_r = \frac{560}{380} = 1.474$$

$$z = 0.77$$

Checking the p_2 assumption

$$p_2 = 2,715 \, e^{(0.0838/0.77)} = 3,027$$

Now assume $p_2 = 3,027; \; p_{avg} = (3,027 + 2,715)/2; \; p_r = \dfrac{2,871}{668} = 4.3$

$$z = 0.77$$

$$p_2 = 2,715 \, e^{(0.0838/0.77)} = 3,027 \text{ psia}$$

PROBLEM 4.4 SOLUTION — APPLICATION OF GRAPHICAL GAS MATERIAL-BALANCE TECHNIQUES

To construct the graphical material-balance plot we must first determine the p/z values. Using Appendices B.7 and B.8 for a gas gravity of 0.68,

$$_\text{p}P_c = 667.5 \text{ psia}; \quad _\text{p}T_c = 385; \quad T_R = \frac{560}{385} = 1.45$$

G_p,MMcf	p,psia	p_r	z	p/z
1,800	3,461	5.18	0.796	4,348
3,900	3,370	5.05	0.790	4,266
5,850	3,209	4.81	0.778	4,125
9,450	3,029	4.54	0.765	3,959

The p/z vs. G_p plot is shown in Fig. C4.4.

To determine the initial pressure read

$$\frac{p_i}{z_i} = 4{,}458 \text{ at } G_p = 0.0; \text{ So } \frac{(p_i/z_i)}{p_c} = \frac{p_{ri}}{z_i} = \frac{4{,}458}{667.5} = 6.678$$

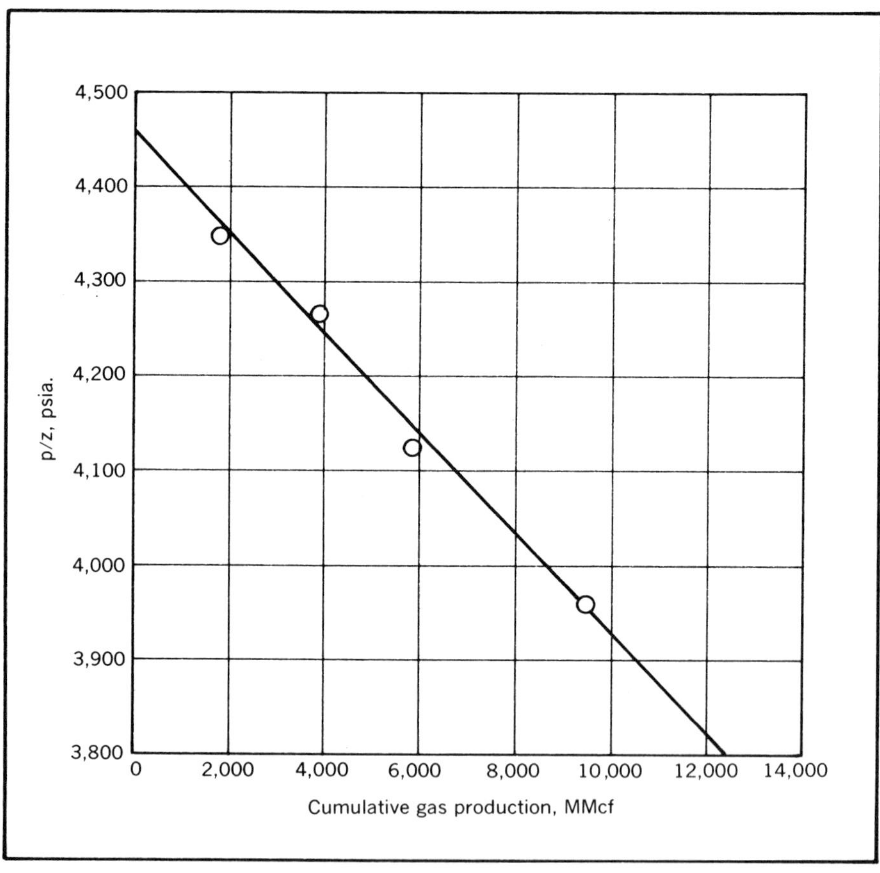

FIG. C.4.4 Plot of p/z vs. cumulative gas production, Problem 4.4

From Appendix B.9

$$z = 0.81$$
$$p_i = 4,458 \times 0.81$$
$$p_i = 3611 \text{ psia}$$

The best straight-line fit of the (p/z) vs. G_p plot gives

$$\frac{p}{z} = 4,458 - 0.053\ G_p, \text{ where } 0.053 = (4,458 - 3,800)/12,400$$

When $p/z = 0.0$, $G = \dfrac{4,458}{0.053} = 84.113$ MMscf (Original gas in place)

Cumulative production at the end of the contract = (20 MMcfd)(365 days/year) (5 years) + 9,450 MMcf = 45,950 MMcf.

$$\text{when } G_p = 45,950$$
$$(p/z) = 4,458 - 0.053\ G_p$$
$$= 4,458 - (0.053)(45,950)$$
$$= 2,023 \text{ psia}$$
$$(p_r/z) = \frac{2,023}{.667.5} = 3.031$$

From Appendix B.9

$$z = 0.775$$
$$p = (2,023)(0.775) = 1,568 \text{ psia}$$

at the completion of the contract.

PROBLEM 4.4A SOLUTION – CALCULATION OF WATER ENCROACHMENT IN A GAS RESERVOIR

Time months	t_D	*Q_{tD}	Average reservoir pressure psia	Pressure at contact	Δp
0	–	–	2,500	2,500	0
6	0	0	2,450	2,500	47
12	5	4.539	2,356	2,406	87
18	10	7.411	2,276	2,326	70
24	15	9.949	2,216	2,266	

$$Q \text{ at 12 months} = 192,000 = B(47)(4.539)$$
$$B = 192,000/(47)(4.539) = 900 \text{ Bbl./psi}$$

$$Q \text{ at 18 months} = 1.12\phi chr^2\ \Sigma\Delta p\ Q_{tD}$$
$$= 900\ [(47)(7.411) + (87)(4.539)$$
$$= 668,889 \text{ bbl}$$

$$Q \text{ at 24 months} = 900\ [(47)(9.949) + (87)(7.411) + (70)(4.539)]$$
$$= 1,287,081 \text{ bbl}$$

* Q_{tD} is read from Table 2.2 for each t_D.

PROBLEM 4.5 SOLUTION — TURBULENCE EFFECTS IN GAS WELLS

$$B = 3.5 \times 10^{10} \text{ (from Appendix B.10)}$$

$$\Delta(p^2)_{\text{turbulence}} = \frac{3.161(10^{-12})\, B\gamma q_g^2 zt \left(\dfrac{1}{r_2} - \dfrac{1}{r_1}\right)}{h^2}$$

$$= \frac{(3.161)(10^{-12})(3.5)(10^{10})(0.76)(3.9)^2(10^6)(0.97)(7.12)(10^2)\left(\dfrac{1}{1/3} - \dfrac{1}{550}\right)}{(30)^2}$$

$$= \frac{(3.161)(3.5)(0.76)(10^6)(3.9)^2(0.97)(7.12)(3 - 0)}{(9)(10)^2}$$

$$= (293)(10^4) = (2.93)(10^6)$$

$$\Delta p^2_{\text{viscous}} = \frac{1.424\mu z_R\, t_f\, q_g\, \ln(r_1/r_2)}{kh}$$

$$= \frac{(1.424)(2.7(10^{-2})(0.97)(7.12)(10^2)(3.9)(10^3)\ln\left(\dfrac{550}{1/3}\right)}{(1.5)(10^{-3})(3.0)(10)}$$

$$= 17.05 \times 10^6$$

A.
$$(p_w)^2_{\text{no turbulence}} = (4{,}583)^2 - 17.05 \times 10^6$$

$$p_w = 1988 \text{ psia}$$

B.
$$(p_w)_{\text{turbulence}} = \sqrt{(4{,}583)^2 - (17.05)(10^6) - (2.93)(10^6)}$$

$$= 1{,}012 \text{ psia}$$

$$\Delta p_{w\ \text{turbulence}} = 1{,}988 - 1{,}012 = 976 \text{ psia}$$

C.
$$\Delta p_{\text{skin}} = 1{,}400 - 976 = 424 \text{ psia}$$

PROBLEM 4.6 SOLUTION — CONVENTIONAL GAS WELL BACK-PRESSURE TEST

A.

Stabilized BHP, psia	$(p_w)^2$	$\Delta(p^2)$	$\log\Delta(p^2)$	q_g Mscfd	$\log q_g$
2,800	7.84×10^6				
2,670	7.13×10^6	0.71×10^6	5.851	1800.	3.255
2,590	6.71×10^6	1.13×10^6	6.053	2700.	3.431
2,500	6.25×10^6	1.59×10^6	6.201	3600	3.556
2,425	5.88×10^6	1.96×10^6	6.292	4500	3.653

See Fig. 4.5 for the data plot.

$$\log 5{,}000 \text{ Mscfd} = 3.699$$
$$\log\Delta(p^2) = 6.349$$
$$\Delta p^2 = 2.23 \times 10^6$$
$$p_e^2 - p_w^2 = 2.23 \times 10^6$$
$$(4.0 \times 10^6) - (2.23 \times 10^6) = p_w^2$$
$$p_w^2 = 1.77 \times 10^6$$
$$\underline{p_w = 1{,}340 \text{ psia}}$$

B. $\qquad t_s = \dfrac{0.04\phi\,\mu c r_e^2}{k}$ $\qquad\qquad r_e^2 = (2250)(43560)/3.14$

$$t_s = \frac{(0.04)(0.15)(0.021)(0.000357)(2{,}250)(43{,}560)}{(0.074)(3.14)} = \underline{\underline{18.9 \text{ days}}}$$

PROBLEM 4.7 SOLUTION – ADJUSTMENT OF DELIVERABILITY FOR WELL SPACING CHANGE

$$\text{For 2 wells } r_{e2} = \left[\frac{\dfrac{4{,}500}{2}(43{,}560)}{3.14}\right]^{1/2} = 5{,}580 \text{ ft}$$

$$\text{For 5 wells } r_{e5} = \left[\frac{\dfrac{4{,}500}{5}(43{,}560)}{3.14}\right]^{1/2} = 3{,}530 \text{ ft}$$

$$\log q_g = \log C + n \log (p_e^2 - p_w^2) \tag{4.43}$$

$$\text{where } C = \frac{0.703kh}{uz_r\, t_f \ln(0.606\, r_e/r_w)} \tag{4.44}$$

$$\frac{C_5}{C_2} = \frac{\ln(0.606\, r_{e2}/r_w)}{\ln(0.606\, r_{e5}/r_w)} \qquad\qquad C_5 = C_2 \frac{\ln(0.606\, r_{e2}/r_w)}{\ln(0.606\, r_{e5}/r_w)}$$

Calculate C_2: Fig. C4.7 is the plot of data from Problem 4.6

$$@ \ q_g = 1{,}800 \text{ Mscfd}, \ \log \Delta(p^2) = 5.851$$
$$\log q_g = \log C + n \log (\Delta(p)^2)$$
$$3.255 = \log C_2 + (0.900)(5.851)$$
$$\underline{\underline{C_2 = 0.00975}}$$

$$C_5 = (0.00975)\,\frac{\ln\!\left(0.606\,\dfrac{5{,}580}{0.25}\right)}{\ln\!\left(0.606\,\dfrac{3{,}530}{0.25}\right)}$$

$$\underline{\underline{C_5 = 0.01025}}$$

At contract completion $p_e = 1{,}568$ psia and the five wells are producing at a total rate of 20 MMscfd; so one well is producing 4 MMscfd

$$\log q_g = \log C + 0.900 \log\Delta(p^2)$$

$$\log \Delta(p^2) = \frac{\log 4{,}000 - \log 0.01025}{0.900} = \frac{3.602 - (-1.989)}{0.900} = \frac{5.591}{0.900} = 6.21$$

$$\Delta(p^2) = 1.621 \times 10^6$$
$$(1{,}568)^2 - (p_w)^2 = 1.621 \times 10^6$$
$$(p_w)^2 = (1{,}568)^2 - 1.621 \times 10^6 = 2.459 \times 10^6 - 1.621 \times 10^6 = 0.838 \times 10^6$$
$$\underline{\underline{p_w = 915 \text{ psia}}}$$

Curve for 5 wells is parallel to the one for 2 wells and passes through $q_g = 4{,}000$ Mscfd, $\Delta(p^2) = 1.621 \times 10^6$

$$\log 4{,}000 = 3.602; \ \log \Delta(p^2) = 6.21$$

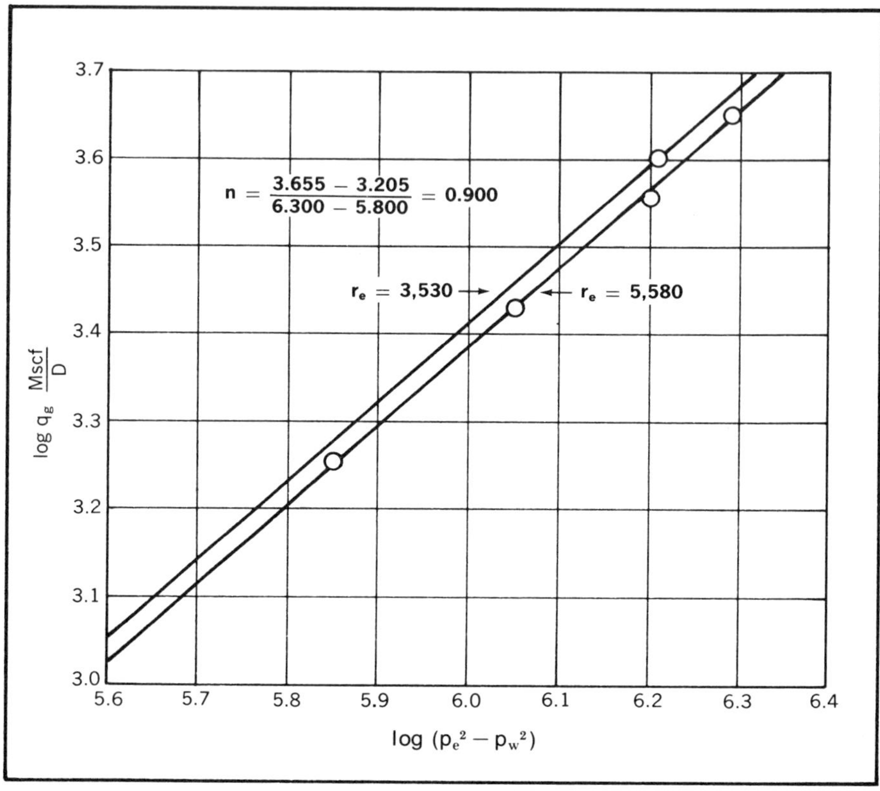

FIG. C.4.7 Data from Problem 4.6

PROBLEM 4.8 SOLUTION — USE OF ISOCHRONAL DATA

A.
$$r_e = \left(\frac{38.5kt^*}{\phi\mu c}\right)^{1/2}$$

$$r_e = \left(\frac{38.5 \times 0.074 \times \frac{1}{24}}{0.15 \times 0.021 \times 0.000357}\right)^{1/2}$$

$$r_e = (10.55 \times 10^4)^{1/2}$$

$$\underline{\underline{r_e = 325 \text{ ft}}}$$

B.

q_g,mscfd	$\log q_g$	BHP at q,psia	BHP^2	$\Delta(p^2)$	$\log \Delta(p^2)$
		2,800	7.84×10^6		
1,800	3.255	2,710	7.35×10^6	0.49×10^6	5.690
2,700	3.431	2,659	7.06×10^6	0.78×10^6	5.892
3,600	3.556	2,605	6.79×10^6	1.05×10^6	6.021
4,500	3.653	2,545	6.48×10^6	1.36×10^6	6.134

$$\log q_g = \log c + n \log \Delta(p^2) \qquad (4.43)$$

At 1,800 msfd

$$\log 1{,}800 = \log C + 0.900 \log (.49 \times 10^6)$$

$$3.25 = \log C + (0.900)(5.690)$$

$$\log C = 3.25 - 5.12$$

$$\log C = -1.87$$

$$C = 0.0135$$

$$\underline{C_{60\ min} = 0.0135}$$

$$r_e \text{ for 2 wells} = \sqrt{\frac{(2{,}250)(43{,}560)}{\pi}} = 5{,}580 \text{ ft}$$

$$\frac{C_{5.580}}{C_{60\ min}} = \frac{\ln(0.606 r_{60\ min}/r_w)}{\ln(0.606 r_{5,580}/r_w)}$$

$$C_{5.580} = C_{60\ min} \frac{\ln(0.606 r_{60\ min}/r_w)}{\ln(0.606 r_{5,580}/r_w)}$$

$$C_{5.580} = 0.0135 \frac{\ln(0.606 \times 325/0.25)}{\ln(0.606 \times 5{,}580/0.25)} = 0.0135 \frac{\ln(788)}{\ln(13{,}510)} = 0.0135 \frac{6.66}{9.51}$$

$$C_{5.580} = 0.00945$$

At $q_g = 1{,}800$

$$\log 1{,}800 = \log 0.00945 + n \log \Delta(p^2)$$

$$3.255 = -2.024 + 0.900 \log \Delta(p^2)$$

$$\log \Delta(p^2) = \frac{5.279}{0.900}$$

$$\log \Delta(p^2) = 5.865$$

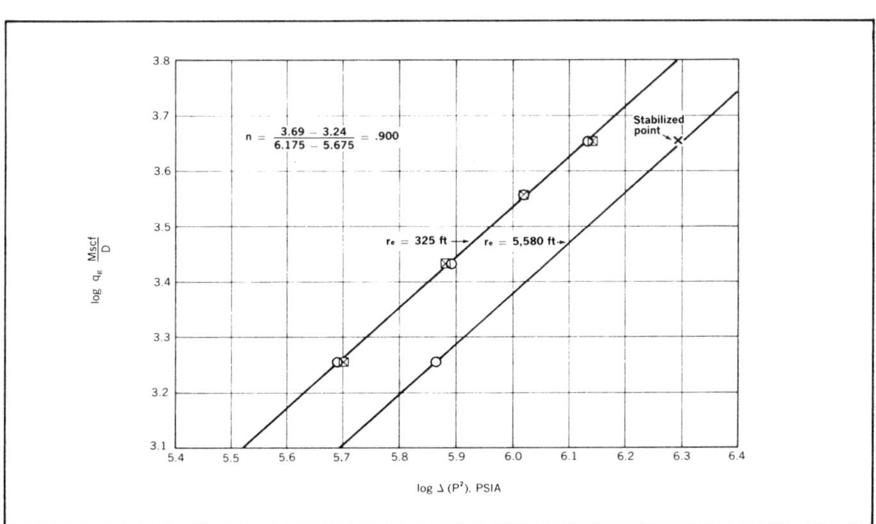

FIG. C.4.8 Solution for Problem 4.8

Therefore: The stabilized back-pressure curve for $r_e = 5,580$ ft passes through the point log $1,800 = 3.255$, log $\Delta(p^2) = 5.865$ and is parallel to the curve for $r_e = 325$ ft.

C.
$$q_g = 4,500 \text{ Mscfd}$$
$$\text{BHP} = 2,425 \text{ psia}$$
$$\log q_g = \log 4,500 = 3.653$$
$$\log \Delta(p^2) = \log (p_e^2 - p_w^2) = \log (2,800^2 - 2,425^2)$$
$$= \log (7.84 \times 10^6 - 5.89 \times 10^6)$$
$$= \log (1.95 \times 10^6)$$
$$= 6.290$$
$$\therefore \log q_g = 3.653$$
$$\log \Delta(p^2) = 6.290$$

The point is plotted on the graph (Fig. C4.8) and does fall near the $r_e = 5,580$ curve.

PROBLEM 4.9 SOLUTION—DETERMINING APPROXIMATE ISOCHRONAL DATA FROM CONVENTIONAL DRAWDOWN DATA

$$t_D = \frac{6.33kt}{\phi\mu cr_w^2} = \frac{6.33(0.074)(t_{min})}{(0.15)(0.021)(0.000357)(0.25)^2(1,440)}$$

$$= 4.62(10^3)t_{min}$$

$$t_{60} = 27.7(10^4) \qquad \ln t_{60} = 12.5$$

$$t_{120} = 55.5(10^4) \qquad \ln t_{120} = 13.22$$

$$t_{180} = 83.2(10^4) \qquad \ln t_{180} = 13.64$$

$$t_{240} = 111.0(10^4) \qquad \ln t_{240} = 14.00$$

for 60 min

$$\frac{\Delta(p^2)_{actual}}{\Delta(p^2)_{desired}} = \frac{1,800(\ln t_{60} + 0.809)}{1,800(\ln t_{60} + 0.809)} \qquad \text{(Similar to Eq. 4.65)}$$

from test data $\Delta(p^2)_{actual} = (2,800)^2 - (2,710)^2$

$\Delta(p^2)_{desired} = 0.5(10^6)$; log $0.5(10^6) = 5.699$ which is plotted vs. log $1,800 = 3.255$ in Fig. C4.8.

for 120 min

$$\frac{\Delta(p^2)_{actual}}{\Delta(p^2)_{desired}} = \frac{1,800(\ln t_{120} + 0.809) + 900(\ln t_{60} + 0.809)}{2,700(\ln t_{60} + 0.809)}$$

from test data $\Delta(p^2)_{actual} = 2,800)^2 - (2,653)^2$

$\Delta(p^2)_{desired} = 0.77(10^6)$; log $0.77(10^6) = 5.886$ is plotted in Fig. C4.8 opposite log $2,700 = 3.431$

for t = 180 min

$$\frac{\Delta(p^2)_{actual}}{\Delta(p^2)_{desired}} = \frac{1,800(\ln t_{180} + 0.809) + 900(\ln t_{120} + 0.809) + 900(\ln t_{60} + 0.809)}{3,600(\ln t_{60} + 0.809)}$$

from data $\Delta(p^2)_{actual} = (2,800)^2 - (2,596)^2$

$\therefore \Delta(p^2)_{\text{desired}} = 1.05(10^6)$; $\log 1.05(10^6) = 6.021$ is plotted in Fig. C4.8 opposite $\log 3{,}600 = 3.556$

for $t = 240$ min

$$\frac{\Delta\ (p^2)_{\text{actual}}}{\Delta\ (p^2)_{\text{desired}}} =$$

$$\frac{1{,}800(\ln t_{240} + 0.809) + 900(\ln t_{180} + 0.809) + 900(\ln t_{120} + 0.809) + 900(\ln t_{60} + 0.809)}{4{,}500(\ln t_{60} + 0.809)}$$

from data $(p^2)_{\text{actual}} = (2{,}800)^2 - (2{,}532)^2$

$\therefore \Delta(p^2)_{\text{desired}} = 1.34(10^6)$; $\log 1.34(10^6) = 6.127$ is plotted in Fig. C4.8 opposite $\log 4{,}500 = 3.653$

PROBLEM 4.10 SOLUTION – DETERMINING ISOCHRONAL DATA FROM CONTINUOUS FLOW TEST DATA BY NEGATIVE SUPERPOSITION

We must first evaluate the turbulence constant, B, before the isochronal data can be evaluated. This is accomplished by using Equation 4.74 and the data resulting from the initial rate change from 0.0 to 1,000 Mcfd and the data resulting from the rate change from 5,000 Mcfd to 4,000 Mcfd that occurred at the end of 2 hr.

$$B = \frac{[\Delta(p_w^2)/\Delta q_g]_{\text{high}} - [\Delta(p_w^2)/\Delta q_g]_{\text{low}}}{(q_{g1} + q_{g2})_{\text{high}} - (q_{g1} + q_{g2})_{\text{low}}} \tag{4.74}$$

In applying this equation, $\Delta(p_w^2)$ is the pressure drop caused by the subject rate change alone, which is the difference between the extrapolated pressure, without the last rate change, and the observed pressure.

$$\Delta(p_w^2) \text{ in Equation (4.74)} = (p_w')^2 - (p_w)^2$$

$$B = \frac{[(1{,}354)^2 - (1{,}090)^2]/(5{,}000 - 4{,}000) - [(2{,}000)^2 - (1{,}888)^2]/(1{,}000 - 0.0)}{(5{,}000 + 4{,}000) - (0 + 1{,}000)}$$

$$= 0.0264 \text{ psi}^2/\text{Mcfd}$$

To determine the isochronal data we will first recognize that the $\Delta(p^2)$ caused by producing at 1,000 Mcfd for 1 hr is obtained directly from the data.

$$\Delta(p^2)_{1{,}000,1 \text{ hr}} = (2{,}000)^2 - (1{,}888)^2$$

$$= 435{,}000 \text{ psi}^2$$

To determine the isochronal $\Delta(p_w^2)$ caused by the other rate changes we must first determine the $\Delta(p_w^2)$ due only to viscous flow by using Equation 4.70 to correct for turbulence and then add back the effect of turbulence if the rate change was from 0.0 to a rate of Δq_{gn} as it would be in an isochronal test. For example, we will first calculate the isochronal effect of a rate of 4,000 Mcfd for 1 hr as indicated by the rate change of 4,000 Mcfd that occurs at a time of 1 hr. First we will calculate the total pressure change caused by the rate changing from 1,000 Mcfd to 5,000 Mcfd acting for 1 hr, by using the extrapolated well pressure at 2 hr. See Fig. C4.10A.

$$\Delta(p_w^2) = (p_w')^2 - (p_w)^2$$

$$= (1{,}883)^2 - (1{,}130)^2$$

$$= 2{,}269{,}000 \text{ psi}^2$$

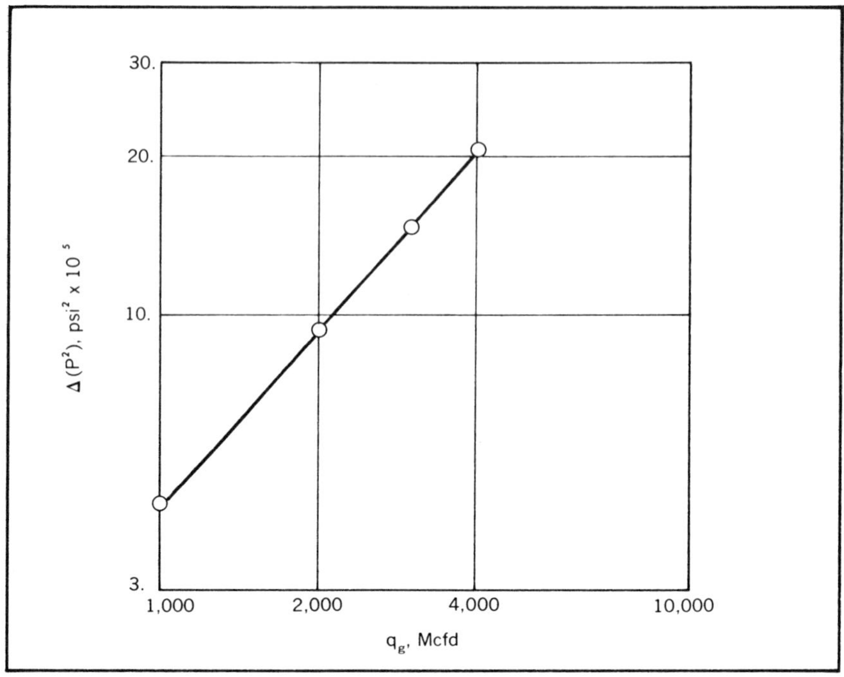

FIG. C.4.10A   One-hour isochronal data calculated from Fig. C.4.10B

But this pressure drop includes turbulence due to the rate of 5,000 Mcfd and can be corrected to the viscous pressure drop by applying Equation 4.70 and recognizing that the first term is the viscous pressure drop term.

$$\Delta(p^2)_{visc} = \Delta(p^2) - B\Delta(q_{gn}^2)$$

$$= 2,269,000 - 0.0264\left[(5,000)^2 - (1,000)^2\right] \qquad (4.70A)$$

$$\Delta(p^2)_{visc} = 2,269,000 - 634,000$$

$$= 1,635,000 \text{ psi}^2$$

Now using Equation 4.70A above, we can add the effect of turbulence to the effect of the viscous flow to obtain the pressure drop caused by an isochronal rate change from 0.0 to 4,000 Mcfd acting for 1 hr.

$$\Delta(p^2)_{4,000,1\ hr} = \Delta(p^2)_{visc} + B\,\Delta(q_{gn}^2)$$

$$= 1,635,000 + 0.0264\left[(4,000)^2 - 0.0^2\right]$$

$$= 2,057,000 \text{ psi}^2$$

When the rate changes of 3,000 Mcfd at 3 hr and 2,000 Mcfd at 4 hr are analyzed similarly we find that

$$\Delta(p^2)_{2,000,1\ hr} = 924,000 \text{ psi}^2$$

$$\Delta(p^2)_{3,000,1\ hr} = 1,464,000 \text{ psi}^2$$

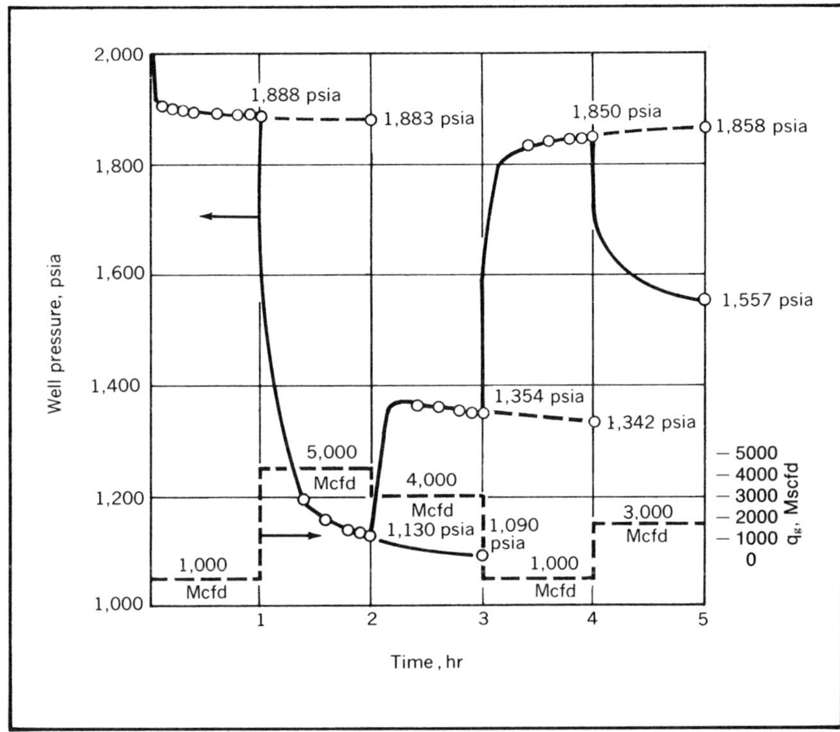

FIG. C.4.10B Data from a continuous flow test of a gas well (computer generated)

When these isochronal data are plotted as in Fig. C4.10B a reasonably accurate straight line is obtained. As noted in the problem statement the turbulence constant, B, can be used with the initial drawdown data to determine the skin factor S, and the group of "constants" indicated by Equation 4.68. This would permit the prediction of the deliverability at the state of depletion existing at the time of the test.

PROBLEM 4.11 SOLUTION—PRESSURE DROPS IN THE PRODUCING SYSTEM TO CALCULATE THE WELLHEAD PRESSURE

$$Q_{scf/hr} = 18.062 \frac{T_{sc}}{P_{sc}} \left[\frac{(p_1^2 - p_2^2)d_{in}^{16/3}}{GT_{ave}L_{mi}Z_{ave}} \right]^{0.5} \tag{4.76}$$

$$T_{ave} = \frac{90 + 70}{2} = 80° \text{ F. or } 540° \text{ R.}$$

for $q = 4$ MMscfd $= 166,667$ scf/hr

$$166,667 = 18.062 \frac{520}{14.65} \left[\frac{(p_1^2 - 800^2)(3.068)^{16/3}}{(0.7)(540)(z)(1)} \right]^{0.5}$$

$$p_1^2 = \left[\frac{(14.65)(166,667)}{(18.062)(520)}\right]^2 \left[\frac{(0.7(540)z}{(3.068)^{16/3}}\right] + 640,000$$

$$p_1^2 = 6.42(10^4)z + 640,000$$

By trial and error assume $p_1 = 800$, $p_{ave} = \dfrac{800 + 800}{2} = 800$

therefore $p_r = \dfrac{800}{666} = 1.2$ $T_r = \dfrac{540}{390} = 1.38$ so $z = 0.84$

Calculating, we find $p_1 = 833$ psia.

so assume $p_1 = 833$ psia; $p_{ave} = \dfrac{833 + 800}{2} = 817$, so $p_r = \dfrac{817}{666} = 1.225$

$z = 0.84$ so $p_1 = 833$ psia which matches assumption

Using a wellhead pressure of 833 psia we can find the BHP from the graphical method used in Problem 4.3. Note that the attached figures are for $2\frac{1}{2}$-in. tubing so the conversion chart (Fig. 4.16) must be used to find that 4 MMscfd in 3-in tubing is equivalent to 2.35 MMscfd in 2.5-in. tubing.

An interpolation is necessary between charts for static BHP and 4 MMscfd flowing BHP.

From Fig. 4.3C: $180 - 80 = 100$ psia gas column pressure
From Fig. 4.12: $260 - 120 = 140$ psia gas column pressure

Interpolating for the 2.35 MMscfd rate gives

$$100 + \frac{2.35}{4.0}(140 - 100) = 123.5 \text{ psia weight pressure of gas column.}$$

$\text{BHP} = 833 + 123.5 = 956.5$ psia

SIMILARLY

for 5 MMscfd in 3-in pipe (same as 2.95 MMscfd in 2.5-in. pipe)

Wellhead pressure $= 851$ psia

$$\frac{\text{Gas column pressure} = 133 \text{ psia}}{\text{BHP} \qquad = 984 \text{ psia}}$$

for 6.67 MMscfd in 3-in pipe (same as 4.0 MMscfd in 2.5-in. pipe)

Wellhead pressure $=\quad 887$ psia

$$\frac{\text{Gas column pressure} =\quad 147 \text{ psia}}{\text{BHP} \qquad = 1,036 \text{ psia}}$$

for 10.0 MMscfd in 3-in. pipe (same as 5.9 MMscfd in 2.5-in. pipe)

Wellhead pressure $=\quad 986$ psia

$$\frac{\text{Gas column pressure} =\quad 196 \text{ psia}}{\text{BHP} \qquad = 1,182 \text{ psia}}$$

PROBLEM 4.12 SOLUTION – DETERMINING GAS WELL SPACING

At contract completion, the following information is known:

$$q = 20 \text{ MMscfd}; \ p_e = 1,568 \text{ psia}; \ p_e^2 = 2.459(10^6)$$

If there were five wells at completion: $q = 4000$ Mscfd/well

$$\log 4000 = 3.602$$

From Fig. 4.18B,

$$\text{For five wells} \log \Delta(p^2) = 6.210$$
$$\Delta(p^2) = 1.622(10^6)$$
$$(p_e^2 - p_w^2) = 1.622(10^6)$$
$$p_w^2 = 2.459(10^6) - 1.622(10^6)$$
$$p_w^2 = 0.837(10^6)$$
$$p_w = 915 \text{ psia}$$

This well pressure is less than the 956 psia needed for a flow rate of 4.0 MMscfd through the tubing and flow line as shown in Fig. 4.18A. Thus, we cannot fulfill the contract with five wells using this equipment. If there were six wells at completion: $q = 3,333$ Mscfd/well. From $\log 3,667 = 3.565$

$$\log \Delta(p^2) \text{ for six wells} = 6.165$$
$$\Delta(p^2) = 1.461 \times 10^6$$
$$p_e^2 - p_w^2 = 1.461(10^6)$$
$$p_w^2 = 2.459(10^6) - 1.461(10^6)$$
$$p_w^2 = 0.998(10^6)$$
$$p_w = 1.000(10^3)$$
$$p_w = 1,000 \text{ psia}$$

$p_w = 1,000$ psia which is more than the 948 psia needed for a flow rate of 3.667 MMscfd as indicated in Fig. 4.18A. Therefore: Six equally spaced wells are needed to meet this contract, if the equipment investigated must be used. Use of larger tubing and/or flow lines would undoubtedly prove advisable.

PROBLEM 4.13 SOLUTION – PRESSURE BUILDUP ANALYSIS AND DETERMINING THE DISTANCE TO A RESERVOIR BARRIER

A. $t = \dfrac{248}{3.9} = 63.6$ days or 1526 hr

$$@ \text{ 5 hr } p_w = 4,011 \text{ psia and } \frac{\Delta t}{t + \Delta t} = \frac{5}{1,531} = 3.27 \times 10^{-3}$$

$$@ \text{ 10 hr } p_w = 4,094 \text{ psia and } \frac{\Delta t}{t + \Delta t} = \frac{10}{1,536} = 6.51 \times 10^{-3}$$

from the plotted points.

$$m = 275 \text{ per/cycle}$$

$$\text{and } p_w \text{ at } \frac{\Delta t}{t + \Delta t} = 10^{-2} \text{ is } 4,145$$

Then at $\dfrac{t}{t + \Delta t} = 1.0$,

$$p_w = 4,145 + 2(275)$$
$$= 4,695 < 4,990$$
$$\therefore \text{ reservoir is finite acting}$$

For calculation of k/μ from m see part B below.

$$k = (0.0836)(.027)$$
$$k = 22.6 \times 10^{-4} \text{ darcys}$$
$$\text{or } 2.26 \text{ md}$$

$$p_w = p_{wf} + 0.867m\left[\frac{1}{2}(\ln \Delta t_D + 0.809) + s\right] \qquad \text{(equations 3.13}$$
$$\text{and 3.15)}$$

Using p_w when $\Delta t = 0.5$ days

$$\Delta t_D = \frac{6.33 \ k\Delta t}{\phi\mu cr_w^2} \tag{3.8}$$

$$= \frac{(6.33)(8.36)(10^{-2})(0.5)}{(5)(10^{-2})(3.3)(10^{-4})(1/9)}$$

$$= 14.4 \times 10^4$$

$$4{,}118 = 998 + (0.867)(275)\left[\frac{1}{2}(\ln 14.4 \times 10^4 + 0.809) + s\right]$$

$$s = 6.74$$

B. At $\Delta t = 68$ hr

p_w extrapolated $= 4{,}320$

$p_w = 4{,}378$; Note that Fig. 4.20 does not extend to this time.

$$\Delta p' = 58 \text{ psi}$$

$$\Delta p' = \frac{0.141q\mu}{kh}\left[\frac{1}{2}\left(Ei\frac{-1}{4t_D}\right)\right]$$

$$58 = (0.867)(275)\left[\frac{1}{2} Ei\frac{-1}{4t_D}\right]$$

$$\left[\frac{1}{2} Ei\frac{-1}{4t_D}\right] = 0.243$$

from graph of t_D vs. $\frac{1}{2} Ei\frac{-1}{4t_D}$

$$t_D = 0.43$$

$$t_D = \frac{6.33(k/\mu)t}{\phi c(2d)^2} \tag{3.63}$$

Must evaluate k/μ from m.

$$m = \frac{0.1625q\mu}{kh} \tag{3.12}$$

Must evaluate q at well in reservoir b/d

$$T_c = 410°\text{F.}$$

$$p_c = 665 \text{ psia}$$

$$T_R = \frac{253 + 460}{410} = 1.74$$

assume $p_{avg} = (4{,}695 + 998)/2 = 2{,}847$

$$p_R = \frac{2{,}847}{665} = 4.28$$

from Fig. B.7, $z = 0.867$

$$B_g = \frac{(5.04) \times (10^{-3})(0.867)(7.13 \times 10^2)}{2{,}847} = 0.00109 \qquad (4.12)$$

$$q = (10.9 \times 10^{-4})(3.9 \times 10^6) = 4{,}251 \text{ b/d}$$

$$275 = \frac{(1.625 \times 10^{-1})(4.251 \times 10^3)}{(30)(k/\mu)}$$

$$\frac{k}{\mu} = \frac{0.0836}{} \text{ darcy/cp}$$

from $p_{avg} = 2{,}847$ $c_r = 0.22$ from Fig. B.5

$$c_g = \frac{0.22}{665} = 3.3 \times 10^{-4}$$

substituting into equation 3.63

$$.43 = \frac{(6.33) \qquad (8.36)(10^{-2})\dfrac{68}{24}}{(5.0) \times 10^{-2})(4)(d^2)(3.3 \times 10^{-4})}$$

$$d^2 = 5.27 \times 10^4$$

$$d = 230 \text{ ft}$$

C. $$(\Delta p/\Delta t)_{pseudo} = \frac{5.615q}{c\phi V_p} = \frac{q}{cV_{pbbls}} \qquad (2.42)$$

for $p_s = 4{,}600$ $p_r = \dfrac{4{,}600}{665} = 6.92.$

from Appendix B.5

$$c_r = 0.1 \ \& \ c_g = \frac{0.1}{665} = 1.5 \times 10^{-4}$$

$$V_{p(bbl)} = \frac{3025}{(6)(1.5 \times 10^{-4})} = 3.36 \times 10^6 \text{ bbl}$$

D. $$GB_{gi} = 3.36 \times 10^6$$

for $p = 4990$ psia

$$p_r = \frac{4990}{665} = 7.5$$

from Appendix B.7

$$z = 1.0$$

$$B_{gi} = \frac{(5.04)(10^{-3})(1.0)(7.12)(10^2)}{(4.990)(10^3)} = 7.19 \times 10^{-4} \qquad (4.12)$$

$$G = \frac{3.36 \times 10^6}{7.19 \times 10^{-4}} = 4.67 \times 10^9$$

$$GB_{gi} = (G - G_p) B_g$$

$$\frac{p}{z} = \frac{(4.670 - 0.248)10^9}{(4.67)(10^9)(1.0)/4,990}$$

$$= \frac{4.422}{4.67} (4,990) = 4,725$$

$$\frac{p_r}{z} = \frac{p/z}{p_c} = \frac{4725}{665} = 7.11$$

from Appendix B.9   z = .97

$$p = 4725 \times .97 = 4583 \text{ psia}$$

PROBLEM 5.1 SOLUTION – DETERMINING THE STATIC SATURATION DISTRIBUTION

To find the free water level, read the threshold pressure for the bottom zone (#V) as 0.45 psi. This corresponds to the 100% saturation point in the reservoir.

$$P_c = 0.433 \, \Delta\gamma h$$

$$0.45 = 0.433 \left(\frac{65.3 - 56.2}{62.4}\right) h \qquad (5.2)$$

$$h = \frac{(0.45)(62.4)}{(0.433)(9.1)} = 7.13 \text{ ft}$$

Free water level = 4,053 + 7 = 4,060 ft
Find $\Delta P_{c \, res}$ equivalent of 1 ft

$$\frac{0.45}{7.13} = (\Delta P_{c \, res})_{1 \, ft}$$

$$(\Delta P_{c \, res})_{1 \, ft} = \frac{0.45}{7.13} = 0.063 \text{ psi/ft}$$

(1)	(2)	(3)	(4)	(5)
Depth	Height above free water level, h, ft	Equivalent P_c	Zone	S_w, %
4,053	7.0	.45	V	100 (critical)
4,049	11.0	.69	V	55
4,047	13	.82	V	47
4,045	15	.94	IV	80
4,042	18	1.07	IV	67
4,039	21	1.33	IV	53
4,036	—	Shale		100
4,033	—	Shale		100
4,030	30	1.89	III	18
4,021	39	2.46	III	16
4,018	42	2.65	II	23

(1) Depth	(2) Height above free water level, h, ft	(3) Equivalent P_c	(4) Zone	(5) S_w, %
4,009	51	3.21	II	22
4,006	54	3.40	I	16
4,000	60	3.78	I	16

(2) h = 4060 − Depth
(3) Equivalent P_c = h (.063)
(5) S_w is read from P_c curve

PROBLEM 5.2 SOLUTION — CONVERSION OF LAB P_c TO RESERVOIR P_c DATA

ZONE I   3,998 − 4,007
Avg. perm. 564 md; use Curve 4, perm. 569 md.

ZONE II   4,007 − 4,019
Avg. perm. 166 md; use Curve 3 with avg. perm. 157 md.

ZONE III   4,019 − 4,031
Avg. perm. 591 md; use Curve 4 with perm. 569 md.

ZONE IV   4,037 − 4,046
Avg. perm. 10.2 md; use Curve 1 with perm. 11.2 md.

ZONE V   4,046 − 4,055
Avg. perm. 72 md; No close curve available. Interpolate between curves 2 and 3; $\frac{2}{3}$ of distance

$$72 - 34 = 38$$
$$157 - 72 = 85$$

$$\frac{38}{38 + 85} \approx \frac{2}{3}$$

$$(P_c)_{res} = (P_c)_{lab} \frac{(\sigma \, \mathrm{Cos}\, \theta)_{res}}{(\sigma \, \mathrm{Cos}\, \theta)_{lab}} \tag{5.6}$$

$$= (P_c)_{lab} \frac{(28)(1)}{(70)(1)} = 0.4 \, (P_c)_{lab}$$

	(1) $(P_c)_{lab}$ at 50% S_w	(2) $(P_c)_{res}$ at 50% S_w
Curve 1	3.65	1.46
2	2.53	1.01
3	1.35	0.54
4	0.80	0.32
Interpolated curve	1.75	0.70

(1) From Fig. 5.7.
(2) = 0.4 × Col. (1)

PROBLEM 5.3 SOLUTION—USING THE J FUNCTION TO AVERAGE CAPILLARY PRESSURE DATA

(1) Core no.	(2) P_c for $S_w = 50\%$	(3) k, md	(4) θ	(5) J for $S_w = 50\%$
1	3.66	11.2	0.147	0.46
2	2.50	34.0	0.174	0.50
3	1.38	157.	0.208	0.54
4	0.85	569.	0.275	0.55

(1), (2), (3), (4) from Fig. 5.7

(5)
$$J = \frac{P_c \, (k_{md}/\theta)^{0.5}}{\sigma \, \text{Cos} \, \theta} \tag{5.14}$$

$$= \frac{P_c \, (k_{md}/\theta)^{0.5}}{(70)(1.0)}$$

Note that the average J value for $S_w = 50\%$ from Fig. 5.8 is about 0.5. To calculate the P_c curve for Zone II of Problem 5.1 from the J function—

$$P_c = \frac{J \, \sigma \, \text{Cos} \, \theta}{(k_{md}/\theta)^{0.5}} \tag{5.14 modified}$$

$$= \frac{J \, (28)(1.0)}{(166/0.208)^{0.5}}$$

$$= (0.99) \, J$$

(1) S_w, %	(2) J	(3) $P_c = 0.99$ J
100	0.35	0.35
70	0.40	0.40
54	0.45	0.45
44	0.60	0.59
30	1.45	1.43
20	3.15	3.12

(1) Assumed
(2) From Fig. 5.8

PROBLEM 5.4A SOLUTION—CALCULATION OF A FRACTIONAL FLOW CURVE

$$f_d = \frac{1 - (0.488k \, k_{ro}A \, |\Delta\gamma| \, \sin \, \alpha/\mu_o q_t)}{1 + (k_o/k_w)(\mu_w/\mu_o)} \tag{5.19}$$

$$= \frac{1 - (0.488 \times 0.108 \times k_{ro} \times 240{,}000 \times 0.04 \times \sin \, 15.5)/(1.51 \times 2{,}830)}{1 + (k_o/k_w)(0.83/1.51)}$$

$$= \frac{1 - 0.0316k_{ro}}{1 + 0.5497(k_{ro}/k_{rw})}$$

S_w	k_{ro}	k_{rw}	(k_{ro}/k_{rw})	$(1 - 0.0316 k_{ro})$	$1 + 0.5497(k_{ro}/k_{rw})$	f_w
79	0.00	0.63	0.000	1.000	1.000	1.000
75	0.02	0.54	0.037	0.9994	1.0203	0.980
65	0.09	0.37	0.243	0.9972	1.1336	0.880
55	0.23	0.23	1.000	0.9927	1.5497	0.641
45	0.44	0.13	3.385	0.9861	2.8607	0.345
35	0.73	0.06	12.167	0.9769	7.6882	0.127
25	0.94	0.02	47.000	0.9703	26.8359	0.036
16	0.98	0.00		0.9690		0.000

See Fig. 5.14 for a plot of f_w vs. S_w

PROBLEM 5.4B SOLUTION — APPLICATION OF THE BUCKLEY-LEVERETT EQUATION

$$\Delta X_j = \left(\frac{5.615 q_t}{\phi A}\right) \Delta t \left(\frac{df_w}{dS_w}\right)_j \tag{5.15}$$

$$= \frac{(5.615 \times 2830)}{(0.215 \times 240{,}000)} \Delta t \left(\frac{df_w}{dS_w}\right)_j$$

$$= 0.308 \, \Delta t \left(\frac{df_w}{dS_w}\right)_j$$

Table of Δx values, ft

S_w	$\left(\dfrac{df_w}{dS_w}\right)^*$	Δx at ½ year (182.5 days)	Δx at 1 year (365 days)	Δx at 2 years (730 days)	$x_{init.}$ **	x at 2 years	x at ½ year
0.79	0.36	20	40	81	10	91	30
0.75	0.56	31.	63.	126.	12	138	43
0.70	0.99	56.	111.	222.	15	237	71
0.65	1.50	84.	169.	338.	18	356	102
0.60	2.27	128.	255.	510.	22	532	150
0.55	3.00	168.	336	672	26	698	194

* Evaluated graphically on Fig. C-5.4B. ** From initial saturation curve of Fig. 5.16. Figure 5.16 is a plot of the calculated saturation profiles.

PROBLEM 5.4C SOLUTION — EVALUATION OF THE FRONTAL POSITION BY MATERIAL BALANCE

$$\frac{5.615 \, q_t \, t}{\phi A} = \Sigma[(S_d - S_{di})\Delta X] \tag{5.23}$$

when $t = 182.5$ days (½ year)

$$\frac{5.615 \, q_t \, t}{\phi A} = 5.615 \times 2{,}830 \times 182.5 / 0.215 \times 240{,}000$$

$$= 56.2$$

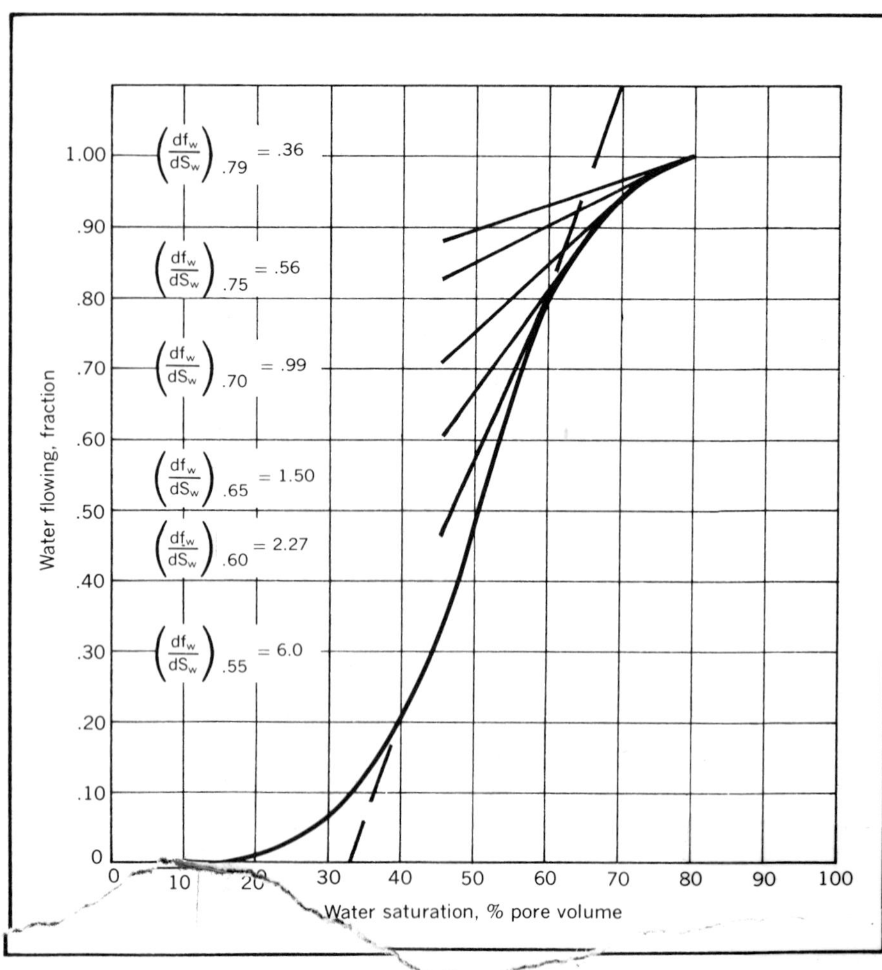

FIG. C.5.4B Solution, Problem 5.4B

Graphically integrating as in Fig. 5.17

ΔX	$\Sigma\Delta X$	S_d	S_{di}	$\Delta X(S_d - S_{di})$	$\Sigma\Delta X(S_d - S_{di})$
10	10	0.895	0.895	0	0
20	30	0.790	0.620	3.4	3.4
20	50	0.765	0.430	6.7	10.1
20	70	0.720	0.320	8.0	18.1
20	90	0.686	0.265	8.4	26.5
20	110	0.654	0.235	8.4	34.9
20	130	0.628	0.216	8.2	43.1
20	150	0.605	0.206	8.0	51.1
*20	170	0.583	0.200	*7.7	*58.8

* Too large

To calculate size of last increment to satisfy Equation 5.23

$$56.2 = 51.1 + (0.583 - 0.200) \Delta X$$
$$\Delta X = 13.$$
$$X_f = 150 + 13 = 163 \text{ ft}$$
$$\text{and } S_{wf} = 58\%$$

PROBLEM 5.5 SOLUTION – USING THE WELGE GRAPHICAL METHOD

A. Refer to Fig. C 5.5. The tangent drawn through $S_{di} = 35$ touches the fractional flow curve at $S_w = 0.56$ pore volumes. Thus, the water saturation at the front prior to breakthrough is 56% pore volume.

B. Prior to breakthrough of water at the producing face, oil recovery is equal to water injected less initial gas saturation (all expressed in pore volumes). After breakthrough the following method may be used.

The cumulative injection (pore volume) is related to the saturation at the producing face as:

$$\frac{W_i}{V_p} = \left(\frac{dS_w}{df_w}\right)_{p.f.} = \frac{1}{\left(\frac{df_w}{dS_w}\right)_{p.f.}} \tag{5.30}$$

Also, by the graphical method shown in Fig. C 5.5 we can determine the average saturation of the invaded zone. The average change in the water saturation less the initial gas saturation is equal to the oil recovery expressed in pore volumes.

(1) S_w at prod. face	(2) $\left(\dfrac{df_w}{dS_w}\right)_{p.f.}$	(3) S_w	(4) ΔS_w	(5) $B_o N_{pf}/V_p$	(6) W_i/V_p
0.560	3.42	0.642	0.292	0.242	0.292
0.600	2.36	0.670	0.320	0.270	0.424
0.650	1.45	0.702	0.352	0.302	0.690
0.700	0.73	0.728	0.378	0.328	1.370

$B_o N_{pf}/V_p$ is plotted versus W_i/V_p in Fig. C 5.5A.
(1) Assumed
(2) Calculated slope of tangent in Fig. C 5.5 e.g., when S_w at prod. face is 0.56
   $(df_w/dS_w)_{p.f.} = (1 - 0)/(0.642 - 0.35) = 3.42$
(3) Read from Fig. C 5.5.
(4) $\Delta S_w = S_w - 0.35$
(5) $B_o N_{pf}/V_p = \Delta S_w - 0.05$
(6) Calculated from Equation 5.30 above
Col. 5 plotted vs. Col. 6 in Fig. C-5.5A

PROBLEM 6.1 SOLUTION – MATERIAL BALANCE EXERCISE

A. Gas material balance

$$G\, B_{gi} = (G - G_p)\, B_g \tag{6.3}$$
$$(4 \times 10^8)(0.001) = [(4 \times 10^8) - G_p]\,(0.0011)$$
$$G_p = 36{,}400{,}000 \text{ scf}$$

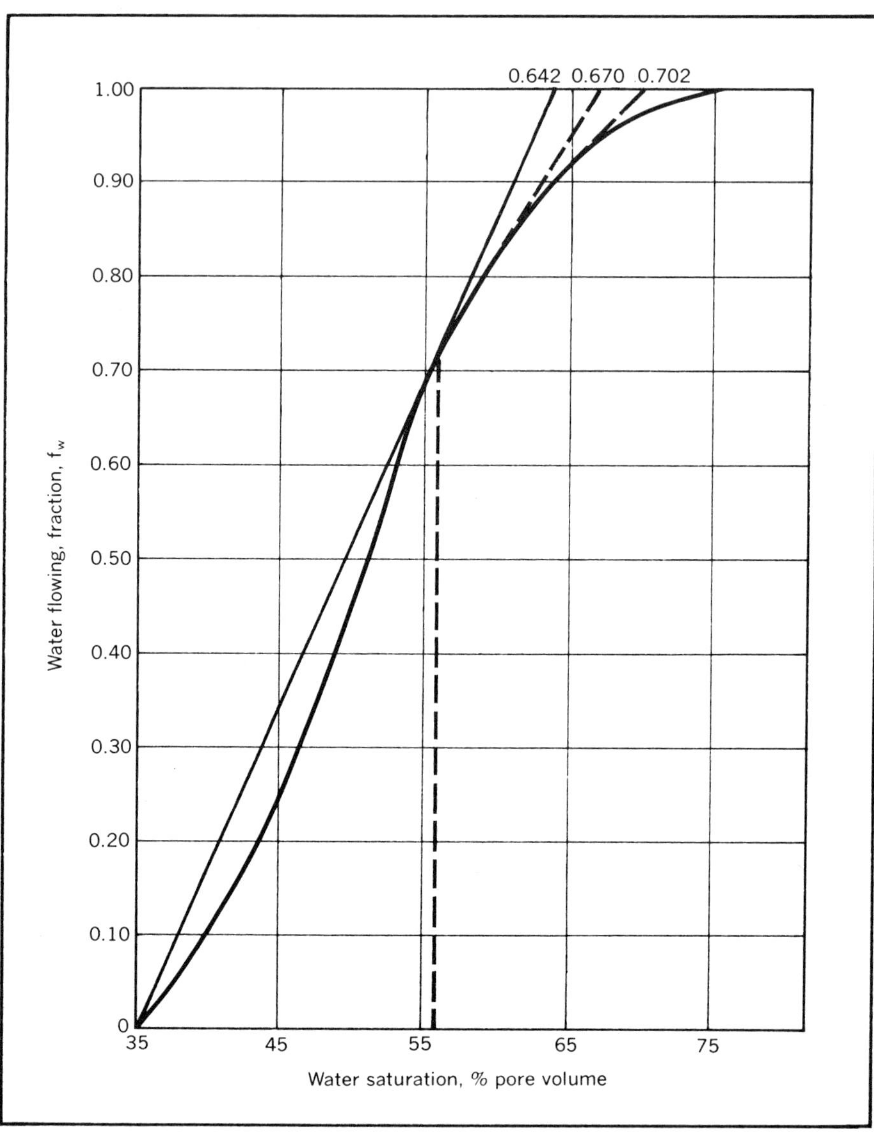

FIG. C.5.5 Fractional flow curve for Problem 5.5 solution

B. Gas-cap expansion

$$\text{Gas expansion} = (G - G_{pc})\, B_g - G\, B_{gi} \qquad (6.4A)$$
$$= [(4 \times 10^8) - 0]\,(0.0011) - (4 \times 10^8)(0.001)$$
$$= 40,000 \text{ reservoir bbl or } 40,000/0.0011 = 36,400,000 \text{ scf}$$

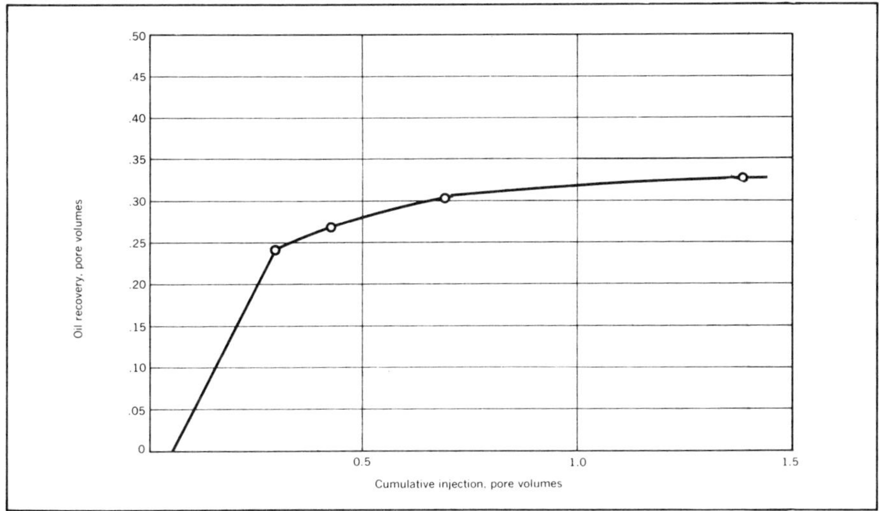

FIG. C-5.5A   Solution to Problem 5.5 — Oil recovery vs. cumulative injection

C. Solution-gas-drive material balance

$$N = \frac{N_p B_o + B_g(G_{ps} - N_p R_s)}{B_o - B_{oi} + (R_{si} - R_s)B_g} \qquad (6.8)$$

$$G_{ps} = R_p N_p = 600\ N_p;\ \text{then}$$

$$4 \times 10^6 = \frac{N_p\ 1.32 + 0.0011\ (600\ N_p - 550\ N_p)}{1.32 - 1.34 + (600 - 550)\ 0.0011}$$

$$N_p = 101{,}800\ \text{STB}$$

D. Material balance with a gas cap or water drive

Oil zone shrinkage $= (W_e - W_p) + $ Change in Gas Cap Volume

$$= N\ B_{oi} - [(N - N_p)B_o + [NR_{si} - (N - N_p)R_s - G_{ps}]B_g] \qquad (6.9)$$

$$= (4 \times 10^6)\ 1.34 - [(4 \times 10^6) - 130{,}800]\ 1.32$$
$$- [(4 \times 10^6)\ 600 - (4 \times 10^6 - 130{,}800)\ 550$$
$$- (600)(130{,}800)] \times 0.0011$$

$$= 39{,}850\ \text{res. bbl} \approx 40{,}000\ \text{res. bbl}$$

E. Determining N by material balance

$$N = \frac{N_p B_o + B_g(G_p - N_p R_s) - G(B_g - B_{gi}) - (W_e - W_p)}{B_o - B_{oi} + (R_{si} - R_s)\ B_g} \qquad (6.11)$$

$$N = \frac{(130{,}800)(1.32) + 0.0011\ [(600 \times 130{,}800) - (130{,}800 \times 550)] - 4 \times 10^8(0.0011 - 0.001)}{1.32 - 1.34 + (600 - 550)\ 0.0011}$$

$$N = 4{,}000{,}000\ \text{STB}$$

PROBLEM 6.2A SOLUTION—PVT LAB DATA EXERCISE

$$(V_o)_{ST} = 500 \text{ cc}$$

$$R_{si} = \frac{44,500}{500} = \text{cc/cc or bbl/bbl}$$

$$= \left(\frac{44,500}{500}\right) 5.615 \text{ cu ft/bbl} = 500 \text{ scf/bbl}$$

At 2,000 psia

$$B_o = \frac{V_o}{(V_o)_{ST}} = \frac{650}{500} = 1.3$$

$$B_t = \frac{V_t}{(V_o)_{ST}} = \frac{650}{500} = 1.3$$

$$R_s = R_{si} = 500 \text{ scf/STB}$$

At 1,500 psia

$$B_o = \frac{669}{500} = 1.34$$

$$B_t = \frac{669}{500} = 1.34$$

$$R_s = R_{si} = 500 \text{ scf/STB}$$

At 1,000 psia

$$B_o = \frac{650}{500} = 1.3$$

$$B_t = \frac{650 + 150}{500} = 1.6$$

$$\text{Liberated gas} = (150)\left(\frac{1,000}{14.7}\right)\left(\frac{520}{655}\right)\left(\frac{1.0}{.91}\right) = 8,902 \text{ cc}$$

$$R_s = (44,500 - 8,902)\, 5.615/500 = 400 \text{ scf/STB}$$

At 500 psia

$$B_o = \frac{615}{500} = 1.23$$

$$B_t = \frac{615 + 700}{500} = 2.63$$

$$\text{Liberated gas} = (700)\left(\frac{500}{14.7}\right)\left(\frac{1.0}{0.95}\right) = 19,897 \text{ cc}$$

$$R_s = (44,500 - 19,897)\, 5.615/500 = 276 \text{ scf/STB}$$

Oil compressibility

$$c = \frac{\Delta V/V}{\Delta p} = \frac{(699 - 650)/669}{2,000 - 1,500}$$
$$= 5.7 \times 10^{-5}/\text{psi}$$

PROBLEM 6.2B SOLUTION—DETERMINING PVT DATA EMPIRICALLY

The gas in solution at 500, 1000, and 1,500 psia can be determined from Appendix B.17. Using R_s and Appendix B.20 the B_o values can be determined for $p \leqq p_s$.

p, Psia	R_s, scf/STB	B_o
14.7	0	1.000
500	90	1.058
1,000	220	1.108
1,500	370	1.155
2,000	*370	**1.149

* Since 2,000 psia is above the saturation pressure, R_s will be the same as R_s when $p = p_s$.

** To obtain B_o for $p > p_s$ it must be calculated from the oil compressibility, c_o. The oil compressibility is determined from Appendix B.4. by determining the oil specific gravity at p_s; finding an equivalent specific gravity at 60° F. from Fig. B.4C; using this to determine the critical pressure and temperature from Fig. B.4B; calculating the corresponding reduced values; determining c_r from Fig. B.4A; and calculating c_o from c_r.

$$\text{Res. oil sp gr} = (\text{ST oil wt.} + \text{dissolved gas wt.})/(B_{os})(\text{water density})$$
$$\text{ST oil wt.} = (\text{sp gr oil})(350 \text{ lb/bbl})$$
$$\text{sp gr oil} = 141.5(131.5 + 40)$$
$$\text{ST oil wt.} = [141.5/(131.5 + 40)] \ 350$$
$$= 289 \text{ lb}$$
$$\text{Dissolved gas wt.} = [\text{MW air}/(\text{SCF/Mole})]R_s(\text{Gas sp. gr.})$$

$$\text{Dissolved gas wt.} = \left(\frac{29}{379}\right)(370)(0.7) = 20.$$
$$\text{Res. oil sp gr} = (289 + 20)/(1.155 \times 350) = 0.764$$

Entering Fig. B.4C with this specific gravity and the reservoir temperature, establish a density-temperature correlation line near cyclohexane. Read the established correlation line at 60° F. as 0.795 sp gr.

From this specific gravity read from Fig. B.4B

$$T_c = 920° \text{ R.}$$
$$P_c = 440 \text{ psia}$$
Then
$$T_r = (125 + 460)/920 = 0.635$$
$$P_r = 1,500/440 = 3.41$$

From Fig. B.4A

$$c_r = 0.0059.$$
$$c_o = 0.0059/440 = 1.3 \times 10^{-5}/\text{psia}$$
$$B_o \text{ at } 2,000 \text{ psia} = B_{os} - B_{os}c_o \ (2,000 - p_s)$$
$$= 1.155 - 1.155 \times 1.3 \times 10^{-5} \ (2,000 - 1,500)$$
$$= 1.147$$

PROBLEM 6.3 SOLUTION—PREDICTING THE BEHAVIOR OF A GAS-DRIVE RESERVOIR

A.
$$S_o = \frac{(1 - S_{wc})(N - N_p)\, B_o}{N\, B_{oi}} \tag{6.26}$$

$$S_o = \left(1 - \frac{N_p}{N}\right)\frac{B_o\,(1 - S_{wc})}{B_{oi}}$$

$$S_o = \left(1 - \frac{1{,}179{,}000}{10{,}025{,}000}\right)\left(\frac{1.233}{1.315}\right)(1 - 0.22)$$

$$\underline{\underline{S_o = 0.646}} \;\therefore\; S_L = 0.646 + 0.22 = 0.866$$

$$R = R_s \left(\frac{k_g}{k_o}\right)\left(\frac{\mu_o}{\mu_g}\right)\left(\frac{B_o}{B_g}\right) \tag{6.24}$$

$$\frac{k_g}{k_o} = (R - R_s)\Big/\left[\left(\frac{\mu_o}{\mu_g}\right)\left(\frac{B_o}{B_g}\right)\right]$$

$$\frac{k_g}{k_o} = (2{,}080 - 450)\Big/\left[(102.61)\left(\frac{1.233}{1.616 \times 10^{-3}}\right)\right]$$

$$\frac{k_g}{k_o} = 0.0207$$

B.
$$N_{pn} = \frac{N[B_o - B_{oi} + (R_{si} - R_s)B_g] + G(B_g - B_{gi}) - B_g[G_{p(n-1)} - (R_n + R_{n-1})N_{p(n-1)}/2]}{B_o - B_g R_s + (R_n + R_{n-1})\, B_g/2} \tag{6.32}$$

For p = 1200

The numerator = 10025000.[1.224 − 1.315 + (650 − 431).001807] − .001807
$$\times\; [1.127 \times 10^9(2080 + R_{1200})1179000./2$$
$$= 324\,1299 + 1065.23\, R_{1200}$$

The denominator = 1.224 − .001807 × 431 + (2080 + R_{1200}).001807/2
$$= 2.324 + .0009035\, R_{1200}$$

$$N_{p1200} = \frac{3241299. + 1065.23\, R_{1200}}{2.324 + .0009035\, R_{1200}}$$

$R_{1,200\ est}$ = 2,520 from GOR versus pressure extrapolation

$$N_{p1200} = \frac{3241299. + 1065.23(2520)}{2.324 + .0009035(2520)}$$
$$= 1{,}288{,}000 \text{ STB}$$

$$\Delta G_p = \left(\frac{R_{1,302} + R_{1,200\ est}}{2}\right)(\Delta N_p)$$

$$\Delta G_p = 1.097 \times 10^5 \left(\frac{2{,}080 + 2{,}520}{2}\right)$$

$$\Delta G_p = 2.52 \times 10^8$$

$$\Delta G_p = 252 \times 10^6 = 252 \text{ MMscf}$$

$$G_p = 252 + 1123 = 1375 \text{ MMscf}$$

$$S_o = (1 - S_{wc})\left(\frac{B_o}{B_{oi}}\right)\left(1 - \frac{N_p}{N}\right) \tag{6.26}$$

$$S_o = (0.78)\left(\frac{1.224}{1.315}\right)\left(1 - \frac{N_p}{10,025,000}\right) = 0.726\left(1 - \frac{N_p}{10,025,000}\right)$$

$$= 0.726\,[1 - (1,288.000/10,025,000)]$$

$$S_o = 0.635 \therefore S_L = 0.634 + 0.22 = 0.854$$

From S_L vs. k_g/k_o curve for field data: $\dfrac{k_g}{k_o} = 0.0274$

$$R_{1,200} = R_s + \left(\frac{k_g}{k_o}\right)\left(\frac{\mu_o}{\mu_g}\right)\left(\frac{B_o}{B_g}\right)$$

$$R_{1,200} = 431 + (k_g/k_o)(108.96)\left(\frac{1.224}{1.807 \times 10^{-3}}\right) = 431 + 73,806(k_g/k_o)$$

$$R_{1,200} = 431 + (73,806)(0.0274)$$

$$R_{1,200} = 2453 \text{ compared with an estimated value of } 2,520$$

$$R_{1,200} \text{ calculated} < R_{1,200} \text{ estimated}$$

Repeat calculations using $R_{1,200}$ estimate $= R_{1,200}$ calculated until $R_{1,200}$ calculated $= R_{1,200}$ assumed.

PROBLEM 7.1A SOLUTION — USING CONSTANT-PERCENTAGE DECLINE TO CALCULATE THE FUTURE LIFE AND RATES OF A WELL

The constant decline rate can be calculated from the straight-line slope of Fig. 7.2. From Dec. '66 thru Dec. '73 (85 months) the rate changes from 35 to 10 STB/day.

$$a = (2.3)\frac{\log 35 - \log 10}{85} = 0.0147/\text{month}$$

In June '72 the rate is 13.2 b/d or

$$q_i = 13.2 \text{ b/d} \times 30.4 \text{ days/month}$$
$$= 401.3 \text{ bbl/month}$$
$$q = q_i\,e^{-at} \qquad (7.11)$$
$$q_{5\text{ years}} = 401.3e^{-0.0147(5\times12)}$$
$$= 166 \text{ bbl/month or } 6.5 \text{ b/d}$$

To find the remaining life use Equation 7.11 to find the time of decline to the economic limit of 1.0 b/d or 30.4 bbl/month from 401.3 bbl/month

$$30.4 = 401.3e^{-0.0147\,t}$$
$$t = 176.0 \text{ months}$$

PROBLEM 7.1B SOLUTION — USE OF A RATE VERSUS CUMULATIVE PLOT DURING CONSTANT-PERCENTAGE DECLINE

Quadrupling the rate (53/13.2) will also quadruple the decline rate if the mobile oil remains the same. After frac, $q_i = 53$ and $a = 0.0147 \times 4 = 0.0588/\text{month}$. Then to find the remaining life

$$q = q_i \, e^{-at}$$
$$30.4 = (53 + 30.4) \, e^{-0.0588(t)}$$
$$t = 68 \text{ month}$$

By fracturing, the life is decreased $176 - 68 = 108$ months. To see the effect of the fracture treatment on the reserves apply Equation 7.12 before and after the fracture treatment. Without the fracture treatment

$$\Delta N_p = (q_1 - q_2)/a \qquad (7.12)$$

Reserves $= (13.2 - 1.0) \, 30.4/0.0147 = 25,100$ STB. After the fracture treatment reserves $= (53 - 1) \, 30.4/0.0588 = 26,700$ STB

The reserves are increased only $26,700 - 25,100$ or $1,600$ STB but the reduction in the life gives a much quicker return of the profits plus saving much of the operating expense.

The remaining primary mobile oil $= q_1/a \qquad (7.16)$
$$= (13.2)(30.4)/0.0147 \text{ or}$$
$$= (4)(13.2)(30.4)/(4)(0.0147) = 27,300 \text{ STB.}$$

SOLUTION PROBLEM 7.2—APPLICATION OF THE HYPERBOLIC-TYPE DECLINE CURVES

The data are plotted on an overlay of Fig. 7.5 with the scales chosen as indicated in the solution illustration Fig. C7.2. When this data plot was fit to curves of $n = 0.3$, 0.5, and 0.7 (Figs. 7.4, 7.5, and 7.6) it was found that a fit could be obtained for all three n's. Thus, the average, $n = 0.5$, was employed as shown in the illustration. In practice the plot would be compared with curves representing several additional n's to determine the best fit.

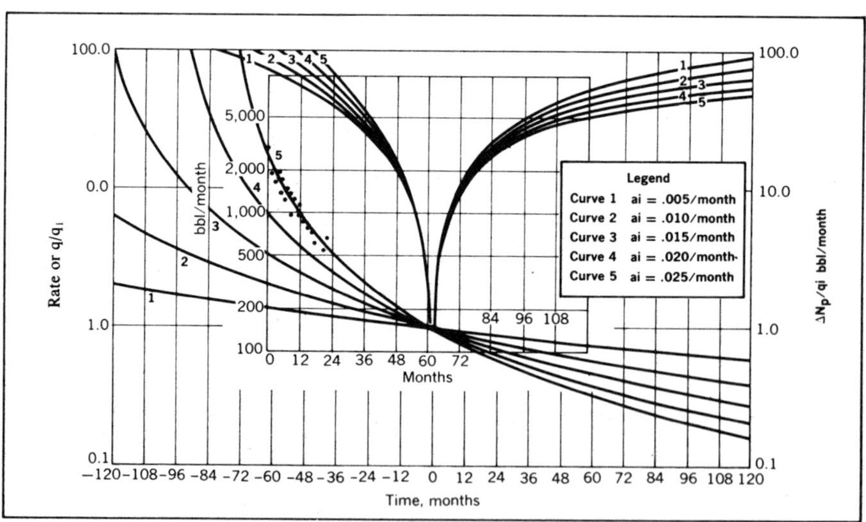

FIG. C.7.2 Hyperbolic decline

From the solution illustration fit we note:

t (chart) = 0 for a
real time of about 62 months
when the rate is
q_i = 150 bbl/month and
a_i = 0.025/month (characteristic of curve 5)

If the economic limit is 100 bbl/month: the remaining life (after 19 months) = (62 − 19) + 18 = 61 months, where 18 months is the chart time where the fit rate curve (curve 5) intersects the economic limit rate of 100 bbl/month. To find the remaining reserves (after 19 months) we will find the reserves from 19 months to 62 months (chart time of 0.0) and from 62 months to the economic limit (a chart time of 18 months). These can be determined by employing the cumulative curves (curve 5) of Fig. 7.5. read $\Delta N_p/q_i$ bbl/month at a chart time of −43 months corresponding to a real time of 19 months and at a chart time of +18. Now calculate the reserves.

$$\Delta N_{p\ 19} \rightarrow {}_{62} = (\Delta N_p/q_{i\ bbl/month})_{-43} {}^* q_i$$
$$= 90 {}^* 150 = 13{,}500\ bbl$$
$$\Delta N_{p\ 62} \rightarrow {}_{EL} = (\Delta N_p/q_{i\ bbl/month})_{+18} {}^* q_i$$
$$= 14.8 {}^* 150 = 2{,}220\ bbl$$
$$Reserves = 13{,}500 + 2{,}220 = 15{,}720\ bbl$$

If the economic limit is 10 bbl/month: Now the economic limit falls off the prepared graph and it is necessary to use equations with q_i and a_i to obtain the life and reserves. The remaining life from a chart time of 0 (real time of 62 months) can be found by

$$q = q_i/(1 + na_i t)^{1/n} \tag{7.22}$$
$$10 = 150/(1 + (0.5 {}^* 0.025\ t)^{1/0.5}$$
$$t = 230\ months.$$
$$remaining\ life = (62 − 19) + 230$$
$$= 273\ months.$$

The reserves from a chart time of 0.0 can also be found by

$$\Delta N_p = [q_i{}^n/a_i\ (1 − n)][q_i{}^{1-n} − q^{1-n}]$$
$$(\Delta N_p)_0 \rightarrow {}_{EL} = [150^{0.5}/0.025\ (1 − 0.5)][150^{1-0.5} − 10^{1-0.5}]$$
$$= 8{,}902.$$
$$Reserves = 13{,}500 + 8{,}902 = 22{,}402\ bbl$$

PROBLEM 8.1 SOLUTION–SATURATION DISTRIBUTION IN A WATERFLOOD

The total liquid saturation may be found at any location at any time by applying Equation 5.15 to the displacement of the oil by water,

$$\Delta X_{Swj} = \frac{5.615 \Delta t\ q_t}{\phi A}(df_w/dS_w)_{Swj} \tag{1}$$

and then applying Equation 5.15 to the displacement of free gas by the oil bank.

$$\Delta X_{SLj} = \frac{5.615 \Delta t\ q_t}{\phi A}(df_o/dS_L)_{SLj} \tag{2}$$

$(df_w/dS_w)_{Swj}$ = slope of f_w vs. S_w curve at a particular saturation
$(df_o/dS_L)_{SLj}$ = slope of f_o vs. S_L curve at a particular saturation
X = distance from injectors in feet

Now find the time for the first production increase to occur. This will be when the oil bank reaches the producing well. The initial liquid saturation is

$$S_L = S_o + S_w = 0.40 + 0.40 = 0.80$$

then from Fig. C8.1A $(df_o/ds_L) = 5.7$ at the producing well when the oil bank reaches the well.

from Equation (2) above:

$$t = \frac{x_{SLj}\phi A}{5.615q_t}\left(\frac{dS_L}{df_o}\right) = \frac{(1,000)(0.2)(150,000)}{(5.615)(3,000)}\frac{1}{5.7}$$
$$t = 313 \text{ days}$$

The liquid saturations must be between the frontal saturation (0.915, the intersection of the curve and the tangent line that passes through the 0.80 initial saturation) and the 0.97 which is the maximum saturation. (See Fig. C8.1A)

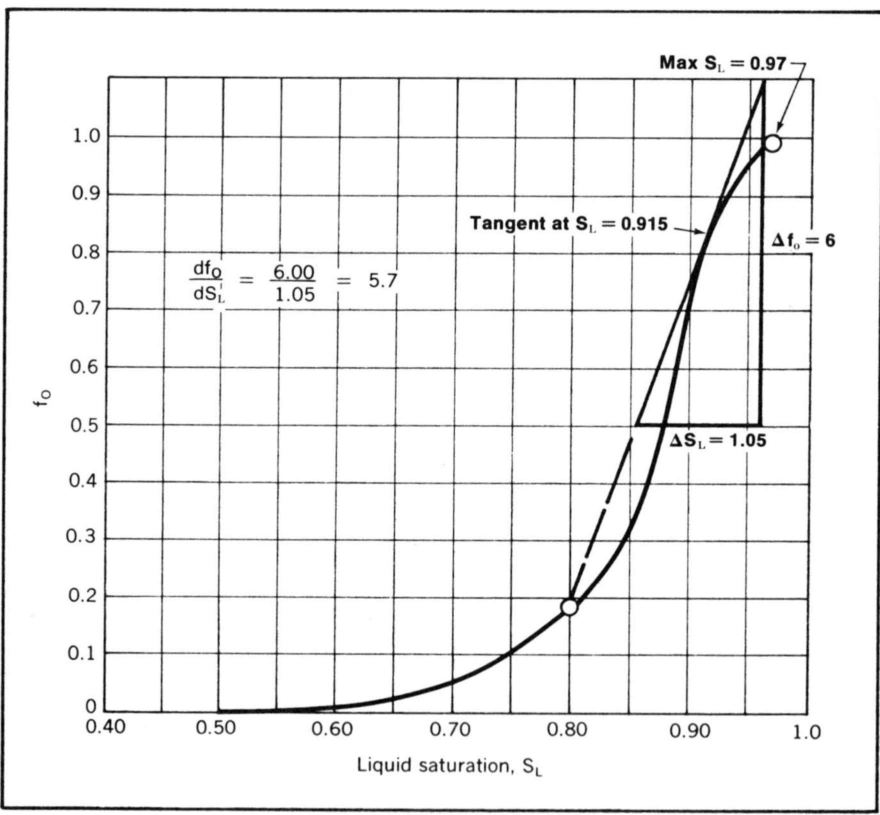

FIG. C.8.1A　Solution, Problem 8.1

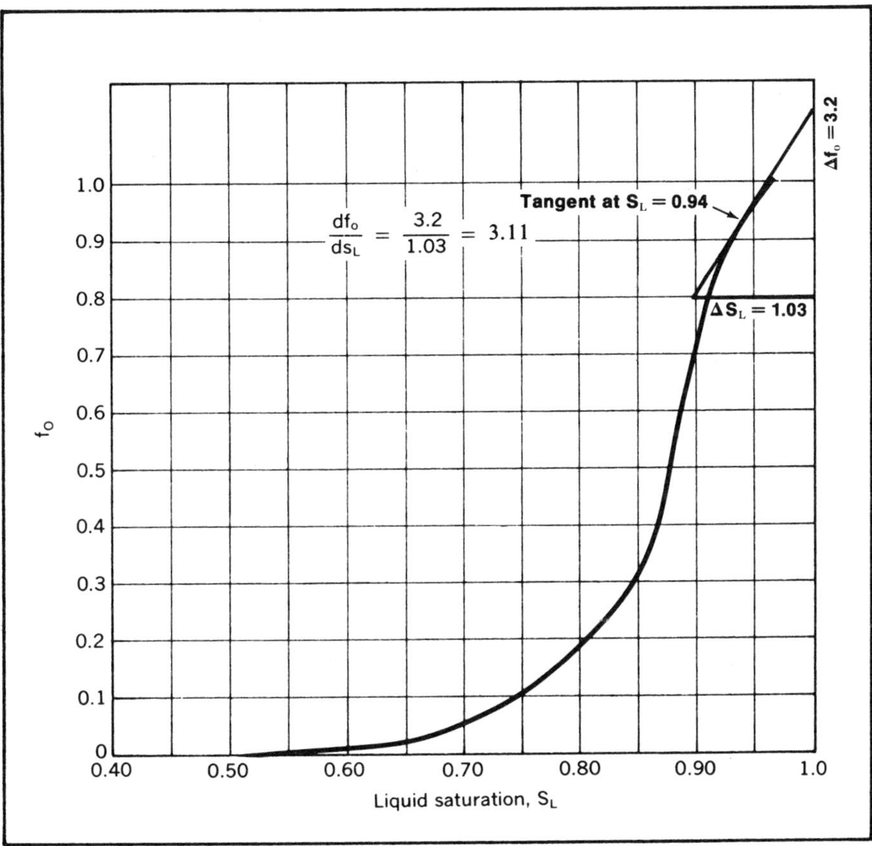

FIG. C.8.1B Solution, Problem 8.1

To find the position where $S_L = 0.94$

$$X_{SLj} = \frac{5.615 t q_t}{\phi A}(df_o/dS_L)_{SLj} \qquad \text{(From Eq. (2))}$$

where $(df_o/dS_L)_{SLj} = 3.11$. (See Fig. C8.1B)

$$X_{SLj} = \frac{(5.615)(313)(3,000)(3.11)}{(0.2)(150,000)} = 548 \text{ ft}$$

Now determine several points of water saturation vs. distance between the frontal saturation of 0.70 and a maximum saturation of 0.90 (Fig. C8.1C).

$$X_{SWj} = \frac{5.615 t q_t}{\phi A}(df_w/dS_w)_{SWj} \qquad \text{(From Eq. (2))}$$

At the frontal saturation of $S_w = 0.7$, $(df_w/dS_w)_{SWj} = 2.64$. (See Fig. C8.1D)

$$X_{0.70} = \frac{(5.615)(313)(3,000)(2.64)}{(0.2)(150,000)} = 464 \text{ ft}$$

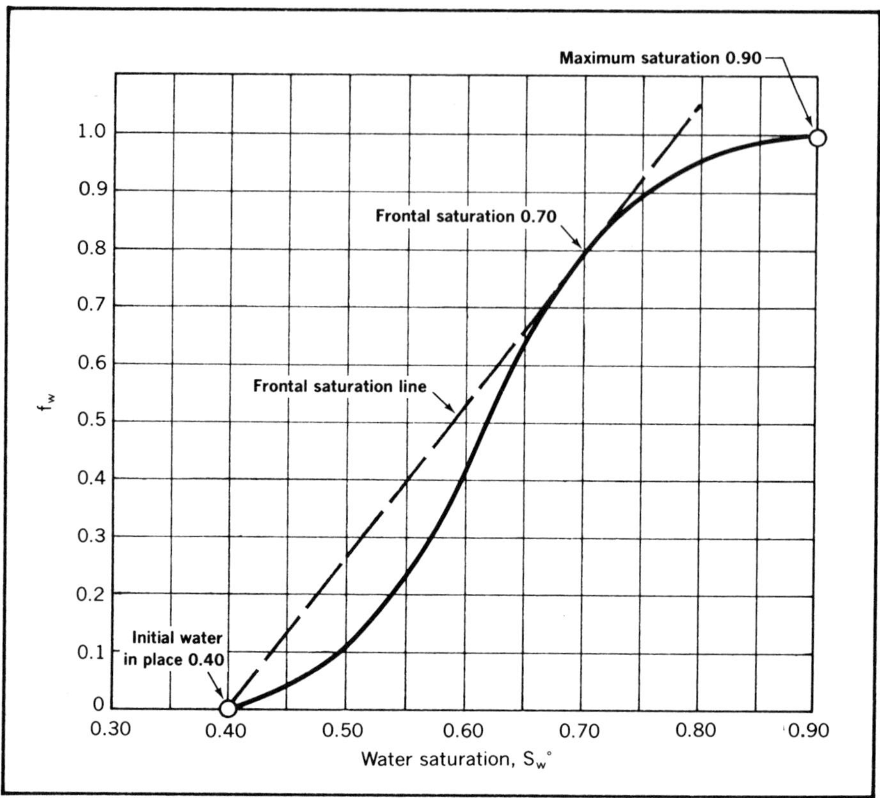

FIG. C.8.1C   Solution, Problem 8.1

Similarly we can determine the position of S_w for values between 0.70 and 0.90 as

$$X_{0.75} = 288 \text{ ft}$$
$$X_{0.80} = 151 \text{ ft}$$
$$X_{0.85} = 72 \text{ ft}$$

The water saturation for distances greater than $X_f = 464$ will be the initial S_w of 0.40. Saturations can then be plotted as in Fig. C8.1E.

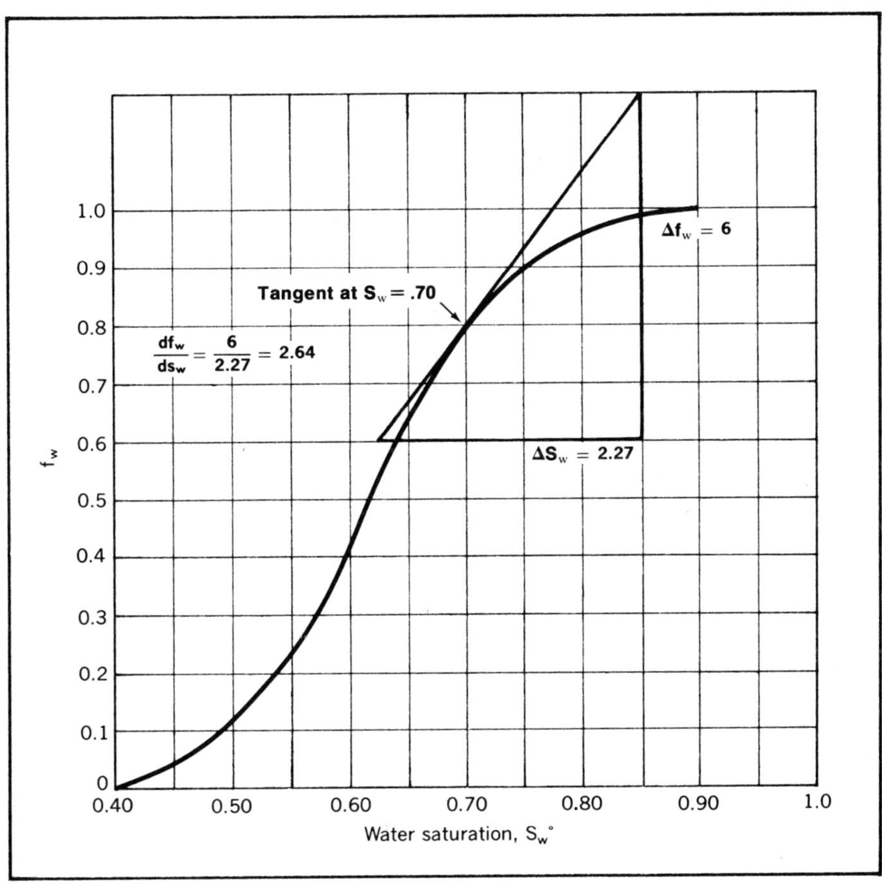

FIG. C.8.1D   Solution, Problem 8.1

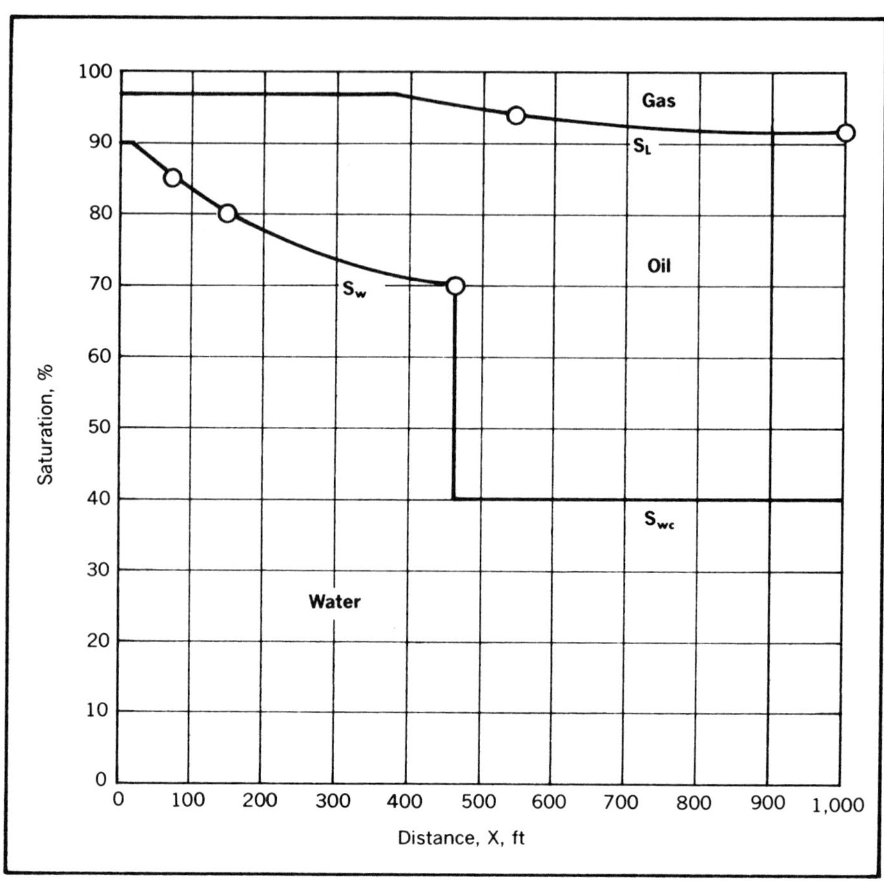

FIG. C.8.1E   Solution, Problem 8.1

PROBLEM 8.2A SOLUTION—PRIMARY DRAINAGE VOLUME REFER TO FIG. C8.2

Plot	Est. avg. thickness, ft	Estimated area, acres	Acre-ft
NE of 1	3	3	9
NE of 2	3	3	9
NE of 3	2	2	4
1	5	9	45
2	10	9	90
3	9	10	90
NE of 6	2	3	6
SW of 1	5	5	25
4	16	10	160
5	18	10	180

Plot	Est. avg. thickness, ft	Estimated area, acres	Acre-ft
6	10	10	100
NE of 10	2	3	6
7	10	10	100
8	19	10	190
9	10	10	100
10	8	10	80
NE of 13	1	2	2
SW of 7	2	2	4
11	10	10	100
12	10	10	100
SE of 9	2	5	10
13	4	10	40
SW of 11	3	4	12
SE of 11	5	6	30
SE of 12	2	3	6
SE of 13	1	3	3

$$\text{Total volume} = 1{,}501 \text{ acre-ft}$$
$$V_{BP} \approx 1{,}500 \text{ acre-ft}$$

$$N = 7{,}758 \; \phi \, (1 - S_{wc}) \, V_{BP}/B_{oi} \tag{8.3}$$
$$N = (7{,}758)(0.2)(1 - 0.2)1{,}500/1.2$$
$$N = 15.516 \times 10^5 \text{ STB}$$
$$S_{oP} = (N - N_{pP})B_{os}(1 - S_{wc})/NB_{oi} \tag{8.2}$$
$$= (15.516 - 3.645) \times 10^5 \times 1.1 \times (1 - 0.2)/(15.516 \times 10^5 \times 1.2)$$
$$= 0.561$$

The best five-spot conversion pattern is as shown in Fig. C8.2. It totals $62\frac{1}{2}$ acres. The alternative pattern totals $52\frac{1}{2}$ acres.

The gross swept volume is estimated as

$$= (20 \times 15) + (20 \times 17) + (5 \times 15) + (5 \times 14) + (10 \times 11) + (2.5 \times 8)$$
$$= 915 \text{ acre-ft} \approx 900 \text{ acre-ft}$$

PROBLEM 8.2B SOLUTION – CALCULATION OF TOTAL ULTIMATE FLOOD RECOVERY

Assume the total displacement efficiency at the economic limit is Stiles R at the economic water cut.

$$f_w = \frac{A\,c_j}{[A\,c_j + (c_t - c_j)]} \tag{8.5}$$

$c_t = \Sigma(kh) = 5059$ md-ft from the given data

$$A_{eff} = \frac{B_o \left[1 + \left(\left(\frac{k_r}{\mu}\right)_w \Big/ \left(\frac{k_r}{\mu}\right)_o\right)\right]}{2} \tag{8.7A}$$

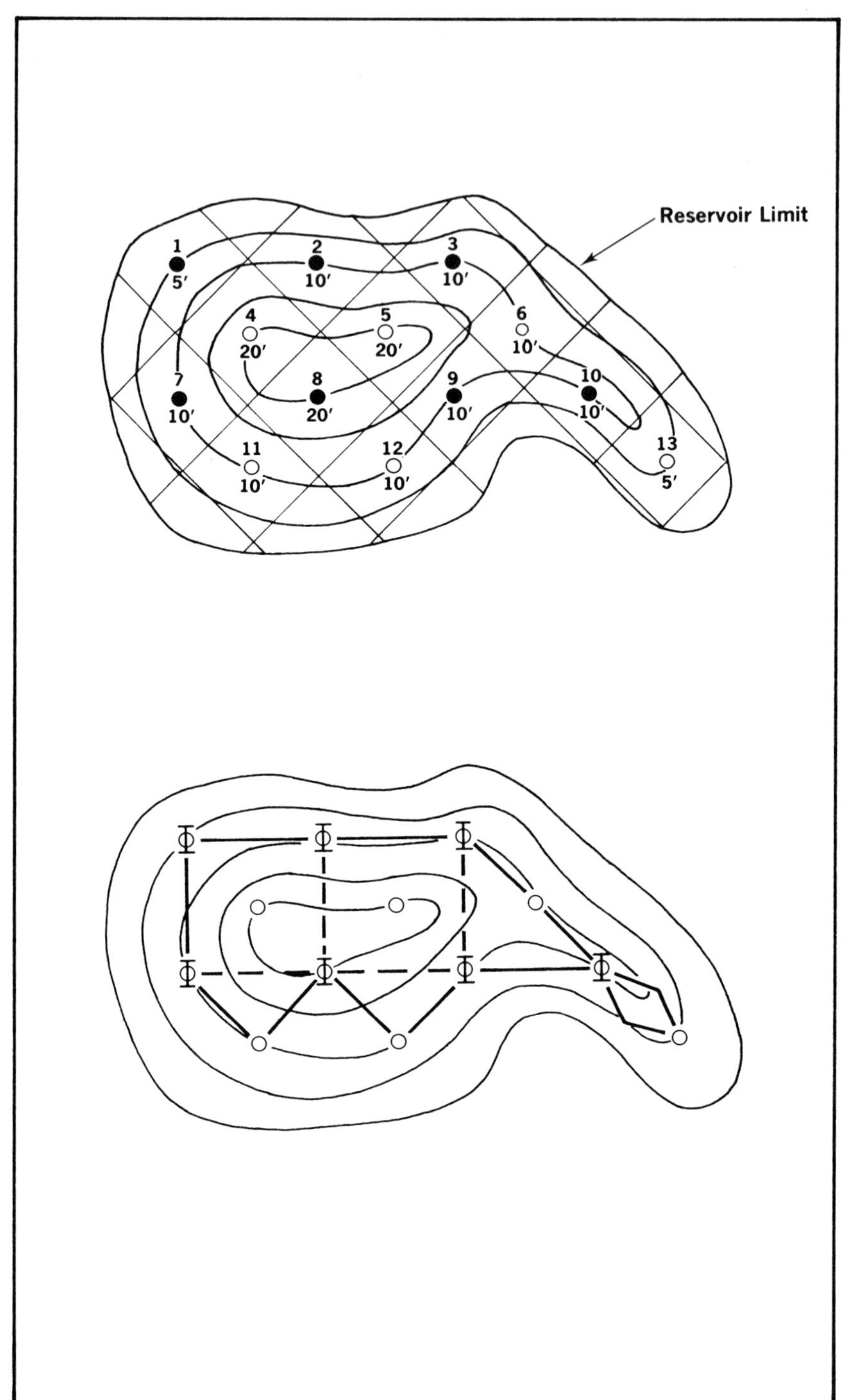

FIG. C.8.2 Solution, Problem 8.2

Reading k_{rw} and k_{ro} from Fig. 8.11A

$$A_{eff} = \frac{1.1\left[1 + \left(\dfrac{.67}{.7}\middle/\dfrac{.98}{6.38}\right)\right]}{2} = 3.98$$

$$0.95 = \frac{3.98\ c_j}{[3.98\ c_j + 5{,}059 - c_j]}$$

$$c_j = 4{,}183 \text{ md-ft}$$

$$c_t - c_j = 5{,}059 - 4{,}183 = 876 \text{ md-ft}$$

To find h_j corresponding to c_j sum the kh values from the least permeable in an increasing permeability direction until 876 md-ft is reached. The corresponding thickness will be h_j.

$$15 + 35 + (46 \times 2) + 50 + 54 + 57 + 61 + 70 + 76 + (85 \times 2) + 87 + 109$$
$$= 876 \text{ md ft}$$

This corresponds to a thickness of 14 ft.

$$R = \frac{h_t k_j + (c_t - c_j)}{h_j k_j} \text{ and } h_j = 29 - 14 = 15 \text{ ft}$$

$$= \frac{(15 \times 127) + 876}{(29 \times 127)}$$

$$= 0.76$$

This efficiency is high due to the lower permeability limit being so high. If the formation down to a permeability of 1.0 md is considered, the efficiency would be more normal.

$$N_{pf} = 7{,}758\ \phi\ [(S_{op}/B_{op}) - (S_{or} - B_{or})]\ E_t V_{sw} \tag{8.1}$$

From problem solution 8.2A we know $S_{op} = 0.561$; $V_{sw} = 900$ acre-ft and $N = 155.2 \times 10^4$ STB

$$N_{pf} = 7.758 \times 0.2 \times \left[\left(\frac{0.561}{1.1}\right) - \left(\frac{0.19}{1.1}\right)\right] \times 0.76 \times 900$$

$$N_{pf} = 357{,}900 \text{ STB}$$

$$N_{pf} + N_{pP} = 357{,}900 + 364{,}500$$

$$= 722{,}400 \text{ STB.}$$

$$\left(\frac{N_{pf}}{N}\right) = \frac{357{,}900}{1552000.} = 0.23$$

$$\frac{(N_{pf} + N_{pP})}{N} = \frac{722{,}400}{1552000.} = 0.47$$

Oil originally in swept area $= (900/1{,}500)\ 1552000.$
$$= 931200 \text{ STB.}$$

$$\frac{N_{pf}}{931200} = \frac{357{,}900}{931200}$$

$$= 0.38$$

N_{pP} from gross swept flood volume $= (900/1{,}500) \times 364{,}500 = 218{,}700.$

$$\frac{(N_{pf} + N_{pp} \text{ from swept volume})}{931200}$$

$$= \frac{(357,900 + 218,700)}{931200}$$

$$= 0.62$$

PROBLEM 8.3A: CALCULATION OF REDUCED TIME CURVES FOR USE IN A WATER-FLOOD PREDICTION

(1)	(2)	(3)	(4)	(5)	(6) t_{r1} $\frac{.25}{.025}\Sigma$	(7) t_{r2} $\frac{.2}{.01}\Sigma$	(8) t_{r3} $\frac{.15}{.002}\Sigma$
(W_i/V_p)	$\Delta(W_i/V_p)$	i_r	$(i_r)_{avg.}$	$\Sigma i_r \Delta(W_i/V_p)$			
0	0	1.0	—	0	0	0	0
0.2	0.2	1.0	1.0	0.2	2.	4.0	15.0
0.33	0.13	4.33	4.33	0.763	7.63	15.26	57.23
0.45	0.12	3.165	3.75	1.213	12.13	24.26	90.98
0.60	0.15	2.63	2.90	1.648	16.48	32.96	123.60
0.84	0.24	2.33	2.48	2.243	22.43	44.86	168.23
1.20	0.36	2.18	2.26	3.057	30.57	61.14	229.28
2.68	1.48	2.00	2.09	6.150	61.50	123.00	461.25
3.68	1.00	2.00	2.0	8.150	81.50	163.00	611.25

(1) Assumed.
(2) Change in Col. (1).
(3) Read from Fig. 8.14 for Col. (1) value.
(4) Average of Col. (3) over interval of Col. (2).
(5) Sum of products of Col. (2) × Col. (4).
(6), (7), (8) Application of Equation 8.21 to the individual zones. The summation is Col. (5).

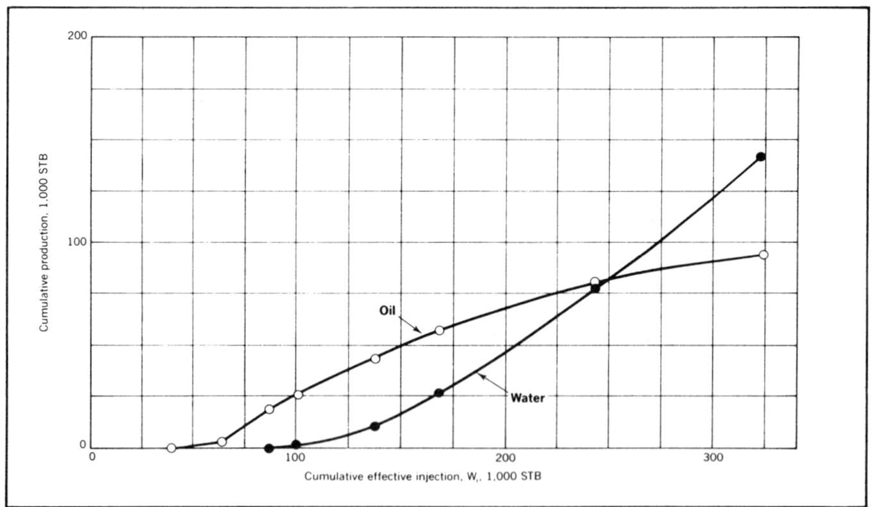

FIG. C.8.3B   Solution, Problem 8.3B

PROBLEM 8.3B SOLUTION—CALCULATION OF OIL AND WATER PRODUCTION RATES FROM REDUCED-TIME CURVES

(1)* t_r: Zone	(2)* W_i/V_p	(3)* $N_{pf}B_o/V_p$	(4)* V_p, MSTB	(5)* W_i, MSTB	(6)* N_{pf}, MSTB	(7)* W_p, MSTB
2:1	.20	0.0 (Crit)	97			0.0
:2	.10	0.0	155			0.0
:3	.027	0.0	175			0.0
2: Total				39.6	0 (Crit)	0.0
4:1	.24	0.04	97			0.0
:2	.20	0.0 (Crit)	155			0.0
:3	.05	0.00	175			0.0
4: Total				63.0	3.9	0.0
7.63:1	.33	0.13	97			0.0 (Crit)
:2	.24	0.04	155			0.0
:3	.10	0.00	175			0.0
7.63: Total				86.7	18.8	0.0 (Crit)
10:1	.39	0.173	97			1.6
:2	.26	0.060	155			0.0
:3	.13	0.000	175			0.0
10: Total				100.9	26.1	1.6
15:1	.55	0.249	97			9.8
:2	.32	0.120	155			0.0
:3	.20	0.000	175			0.0
15: Total				138	43	9.8
20:1	.73	.290	97			23.3
:2	.39	.173	155			2.6
:3	.21	.010	175			0.0
20: Total				168	57	25.9
30:1	1.17	.338	97			61.3
:2	.55	.249	155			15.7
:3	.25	.050	175			0.0
30: Total				243	80	77.0
40:1	1.65	.360	97			105.7
:2	.74	.291	155			38.6
:3	.28	.080	175			0.0
40: Total				324	94	144.3

(1)* t_r is assumed
(2)* from Fig. 8.16
(3)* from Fig. 8.15
(4)* Given data
(5)* $= \Sigma[(W_i/V_p)V_p]$----------------(8.22)
(6)* $= \Sigma(N_{pf}B_o/V_p)(V_p/B_o)$----------(8.23)
(7)* $= \Sigma[(W_i/V_p) - (N_{pf}B_o/N_p) - S_{gi}]\,V_p$: Equation 8.24. Set negative terms to 0.0.

PROBLEM 8.3C SOLUTION – CALCULATION OF INDIVIDUAL STRATA RECOVERY CURVES

$$M = (k_{rw}/k_{ro})(\mu_o/\mu_w) = (0.55/0.98)(4.15/0.7) = 3.33 \text{ and } 1/M = 1/3.33 = 0.3$$

at Gas fillup (W_i/V_p) $= 0.2 = S_{gi}$

at Water B.T. DVI $= 0.55$ at $1/M$ of 0.3 from Fig. 8.10B

Displacement volume $= (1.0 - S_{wc} - S_{or}) V_p = (1.0 - 0.15 - 0.25) V_p = 0.6 V_p$

(1) Displacement volumes injected	(2) (W_i/V_p)	(3) E_p	(4) (B_oN_{pf}/V_p)
0	0	0	0.0
–	0.2	–	0.0(critical)
0.55	0.33	0.55	0.130
0.75	0.45	0.68	0.208
0.90	0.55	0.74	0.244
1.00	0.60	0.77	0.262
1.10	0.66	0.80	0.280
1.20	0.72	0.82	0.292
1.30	0.78	0.83	0.298
1.40	0.84	0.85	0.310
1.50	0.90	0.86	0.316
1.75	1.05	0.88	0.328
2.00	1.20	0.90	0.340
2.25	1.35	0.91	0.346
2.50	1.50	0.92	0.352
	2.68*	1.00	0.400

(1) : Assumed

(2) $= (0.6)$ (Col. 1)

(3) : Read from Fig. 8.10B at DVI of Col. 1 and $1/M$ of 0.3.

(4) $= (0.6)(\text{Col. 3}) - (0.2)$

* Determined by extrapolation of Fig. 8.15 plotted points to a recovery value of 0.4.

PROBLEM 8.3D SOLUTION – CALCULATION OF INDIVIDUAL STRATA INJECTIVITY CURVES

$$M = (k_{rw}/k_{ro})(\mu_o/\mu_w) = (0.55/0.98)(4.15/0.7) = 3.33$$

at gas fillup $(W_i/V_p) = 0.2 = S_{gi}$

at water B.T., DVI $= 0.55$ at $1/M$ of 0.3 from Fig. 8.10B

1.0 displacement volume $= (1.0 - S_{wc} - S_{or}) V_p = (1.0 - 0.15 - 0.25) V_p = 0.6 V_p$

From Problem 8.3C solution

(1) DVI	(2) (W_i/V_p)	(3) E_p	(4) f_w	(5) i_r
0	0	0	0.0	1.0
–	0.2	–	0.0	4.330
0.55(B.T.)	0.33	0.55	0.0(critical)	4.330
0.75	0.45	0.68	0.50	3.165
0.90	0.55	0.74	0.67	2.770
1.00	0.60	0.77	0.73	2.630
1.10	0.66	0.80	0.78	2.510
1.20	0.72	0.82	0.82	2.420
1.30	0.78	0.83	0.83	2.400
1.40	0.84	0.85	0.86	2.330
1.50	0.90	0.86	0.88	2.280
1.75	1.05	0.88	0.90	2.230
2.00	1.20	0.90	0.925	2.180
2.25	1.35	0.91	0.940	2.140
2.50	1.50	0.92	0.950	2.110
	2.68	1.00	1.000	2.000

(4) Read from Fig. 8.10A at sweep efficiency of Col. (3) and $1/M = 0.3$.

(5) Until gas fillup $i_r = 1.0$; after gas fillup $i_r = 4.33 - 2.33$ (Col. 4) according to Equation 8.40.

PROBLEM 8.4A SOLUTION – PREDICTING TOTAL FLOOD RECOVERY BY ANALOGY

For the old flood:

$$\text{Total recovery, STB/acre-ft} = \frac{\text{STB Primary prod.}}{\text{acre-ft Primary drainage}} + \frac{\text{STB Waterflood prod.}}{\text{acre-ft gross swept vol.}}$$

$$= \frac{2,410,000}{10,000} + \frac{1,570,000}{8,000} \tag{8.42}$$

$$= 441 \frac{\text{reservoir bbl}}{\text{ACRE-FT}} = 437 \times 1.01$$

where $437 = \text{STB}; \; Bo = 1.01$

$$\text{Fractional recovery} = \frac{441}{0.18(7758)} = 0.316$$

Now:

To determine the total recovery of the new flood assume fractional

$$\text{Total recovery}_{old} = \text{fractional total recovery}_{new} = 0.316$$

For the new flood:

$$0.316 = \frac{(\text{Tot. rec./acre-ft})B_{os}}{\text{pore vol/acre-ft}} = \frac{(\text{Tot. rec./acre-ft}) \, 1.0}{7,758 \, (0.20)}$$

$$\text{Total recovery/acre-ft} = 490 \frac{\text{STB}}{\text{acre-ft}}$$

$$490 \frac{\text{STB}}{\text{acre-ft}} = \frac{\text{Primary tot. prod.}}{\text{Primary acre-ft}} + \frac{\text{Secondary tot. prod.}}{\text{Secondary acre-ft}} \tag{8.42}$$

$$\frac{\text{Secondary Tot. Prod.}}{14,000} = 490 \frac{\text{STB}}{\text{acre-ft}} - \frac{3,645,000}{15,000} \frac{\text{STB}}{\text{acre-ft}}$$

Secondary total production = 3,458,000 STB

PROBLEM 8.4B SOLUTION—PREDICTION OF RATE VS. TIME BY ANALOGY

We will first determine the effective injection rate for the new flood.

$$p_{wi} = 0.433 \, \gamma D + p_{wh} \tag{8.44}$$

$$p_{wi(old)} = (0.433)(1.0)(3,000) + 600 = 1,899 \text{ psia}$$

$$p_{wi(pro)} = (0.433)(1.0)(2,500) + 800 = 1,883 \text{ psia}$$

$$i_{pro} = i_{old} \frac{(kh/\mu)_{pro}}{(kh/\mu)_{old}} \frac{(p_{wi} - p_{wp})_{pro}}{(p_{wi} - p_{wp})_{old}} \frac{\left(\ln \dfrac{d}{r_w} - 0.619\right)_{old}}{\left(\ln \dfrac{d}{r_w} - 0.619\right)_{proposed}} \tag{8.43}$$

The geometry terms cancel and

$$i_{pro} = 1,700 \frac{(0.075 \times 780/1.0)}{(0.1 \times 401/1.0)} \frac{(1,883 - 15)}{(1,899 - 15)} = 2,458 \text{ b/d}$$

Effective injection rate = $0.95 \times 2,458 = 2,335$ b/d

The oil producing rate versus time can then be calculated in tabular form as shown in accompanying table.

Read from data graphs of old flood

(1) Time, years	(2) Rate, b/d	(3) Current inj. rate, b/d	(4) Cumulative injection, bbl	(5) $\dfrac{\text{b/d}}{\text{cur. eff. inj.}}$ $\dfrac{\text{col.(2)}}{0.90 \text{ col.(3)}}$	(6) New rate, b/d Col.(5) × (2,335)	(7) $\dfrac{\text{cum. eff. inj.}}{\text{sec. rec}}$ $\dfrac{0.9 \text{ col.(4)}}{1,570,000}$	(8) New time, years $\dfrac{\text{Col.(7) }3,458,000}{365} \dfrac{}{0.95} \times \dfrac{1}{2,458}$
.50	94	1,500	191,625	0.0696	163	0.1078	0.44
1.50	1,370	3,000	1,286,625	0.507	1184	0.718	2.91
2.25	1,170	2,200	1,888,875	0.592	1382	1.061	4.30
3.50	580	2,000	2,801,375	0.322	752	1.572	6.38
5.25	250	1,600	3,823,375	0.174	406	2.146	8.70
5.75	195	1,800	4,151,875	0.120	280	2.340	9.49
7.50	101	1,500	5,110,000	0.075	175	2.870	11.64

Subject Index

Absolute permeability, 5
Absolute temperature, 180
Angle of contact, 268
Apparent velocity, 4
Aquifer:
  constant, 355
  dimensionless size, 61
Average permeability, 36–40
Average water saturation, 302–304

Back-pressure testing, 217–222, 226–229
Bottom-hole pressure:
  flowing, 174, 247–252
  gradients, 331
  static, 196–205, 331
Bottom-hole sampling, 335
Boundary effect:
  gas reservoir, 260–263
  oil reservoir, 94–97, 160–169
  pressure build-up, 156–157, 262
  pressure drawdown, 163
Bubble point, 324, 326, 488
Buckley Leverett effect, 290, 387–390
Build-up tests:
  after a rate change, 144–149
  after shut in, 116–135
  during drill stem test, 171–175
Bulk volume, 320, 394

Capacity, 105
Capacity distribution, 405–406
Capillary pressure:
  averaging, 277–283
  correlation with permeability, 274–280
  definition, 268–270
  forces, 268–270
  gas saturation from, 272–274
  J function, 278–280
  methods of measuring, 271–272
  pore size distribution from, 271
  reservoir, 275–277
  water saturation from, 272–274
Combination drive, 329–330
Compressibility:
  effective, 83–85
  of natural gases, 189–191, 460
  of oil, 24, 84, 458

Compressibility (*Continued*)
  of rock, 84, 330–333, 460
  of water, 84, 457
  reduced gas, 190–191, 460
Compressibility factor, 25, 181–188, 461
Condensate, 185–186, 466
Connate water, 10, 194, 195
Contact angle, 268
Core analysis data interpretation, 266–268
Cricondentherm, 184
Critical point, 183, 184
Critical pressure, 181–183, 462
Critical temperature, 181–184, 462

Damage ratio, 113–114
Darcy equation, 3–5
Darcy units, 4
Datum level, 41
Decline curves:
  Constant rate, 368–373
  Constant percentage, 368–373
  Harmonic, 384
  Hyperbolic, 373–384, 467–484
  Rate vs. cumulative, 371–373
Deliverability curve, 219
Deviation factors, 181–188, 461
Dew point curve, 184
Differential liberation of a gas, 336
Diffusivity constant, 54
Diffusivity equation:
  gas, 221–223
  liquid, 54
Dimensionless flow rate, 83
Dimensionless pressure drop, 59–60
Dimensionless time, 59–60
Dipping reservoirs, 40–43
Displaceable volume, 404, 422
Displacement:
  efficiency, 294
  immiscible, 283
  miscible, 363, 436, 443
Dissolved gas drive, 338–352
Drainage boundaries, 160–163, 260–263
Drainage radius:
  equivalent, 29
  pseudo steady state equivalent
    of isochronal time, 227

Drainage volume:
  during primary life, 429
  during secondary life, 394–398
Drawdown tests, 162–163, 217–218, 226–232
Drives:
  combination, 328–330
  gas cap, 328–330, 338–352
  solution gas, 324, 338–352
  water, 328–360
Drill stem test analysis, 169–177

Economic limit, 30
Effective permeability, 7, 43–45
Effective radius, 29, 227
Efficiency:
  conformance, 402
  displacement, 294, 409–410
  horizontal sweep, 400–402
  pattern sweep, 400–403
  stratigraphic sweep, 403–410
  vertical sweep, 403–410
Ei function solution, 64–67
Electric analyzer, 359
Equilibrium saturation, 388–389
Exponential integral, (see Ei function solution)

Falloff test, 149–152
Faults, distance to, 160–163
Field units, 4
Fingering, 311
Five spot pattern, 25, 395, 400, 401, 403, 404
Flash liberation of gas, 336
Flood front, 284, 388
Flood patterns, 395, 400, 401
Flow capacity:
  from drill stem test, 173
  from flow after flow test, 234
  from four point test, 217–218
  from isochronal test, 226–229
  from PI test, 108–112
  from pressure build-up, 116–135
  from pressure drawdown, 163–164
  from pressure fall off, 149–152
  from two rate test, 144–149
Flow test:
  drill stem, 169–177
  flow after flow, 234
  isochronal, 226–229
  pressure drawdown, 163
  productivity index, 108–112
Fluid displacement mechanisms, 283
Fluid distribution in reservoir, 265–266, 283–288
Fluid flow equations:
  gas, 25–29, 211–262
  pseudo steady state, 97–106, 124–129, 214–221
  unsteady state, 54–97, 223–243
Fluid injection, 82–83, 386–444
Fluid properties:
  gas, 180–191, 457, 460, 461, 462, 463, 464, 465, 466, 485, 488, 489

Fluid properties (*Continued*)
  oil, 45, 335–337, 458, 465, 486, 488, 490
  water, 45, 457, 487, 489
Fluid recovery rate from DST, 173–174
Fluid saturations, 265–266, 283–288
Fluid viscosities:
  gas, 191, 485
  oil, 45, 486
  water, 45, 487
Formation damage, 22, 113–114, 115
Formation dip, effects of, 40–43, 292
Formation volume factor:
  gas, 188
  oil, 24, 45, 490
  total, 327–328
  water, 45, 457
Four spot pattern, 401
Fractional flow curve, 290–294
Fractional flow equations, 290, 292
Free gas volume, 328–329
Frontal advance, 284–288, 400–402
Frontal drive mechanisms, 284–288

Gas balance, 191–192, 328
Gas cap drive, 322, 328–330
Gas cap fingering, 311
Gas condensate, 193–194, 466, 488
Gas coning, 312–316
Gas Deviation factor, 181–188, 461, 463
Gas Density, 188–189, 465
Gas Drive Reservoirs, 338–349
Gas equivalent:
  condensate, 193
  water, 196
Gas formation volume factor, 188
Gas injection, 362
Gas law, 180
Gas material balance, 191–211
Gas-oil contact, 266–268
Gas-oil ratio:
  cumulative produced, 335
  flowing, 339–340
  producing, 339–340
  solution in oil, 327, 488
  solution in water, 457
Gas properties:
  compressibility, 189–190, 5
  compressibility factor, 181–186, 461
  critical pressure, 182, 465
  critical temperature, 182, 465
  gravity, 188
  pseudo critical pressure, 184–185, 462
  pseudo critical temperature, 184–185, 462
  reduced pressure, 182
  reduced temperature, 182
  solubility, 327, 457, 488
  viscosity, 191, 485
  water content, 194, 489
Gas reservoirs:
  determining size, 206, 260–263
  material balance, 191–211
  with a water drive, 207–211

Gas saturations:
  before flooding, 388
  equilibrium, 388–389
  initial, 265–266, 272–275
Gas-water contact, 266–268
Gas well:
  back pressure test, 217–218
  deliverability, 218–221
  ischronal test, 226–232
  pressure build-up, 241–244
  pressure drawdown, 217–232
  two rate flow test, 243
Geological zonation, 276–277, 410
Geothermal gradient, 200
Grain volume, 330–331
Gravity:
  gas, 188, 465
  oil, 41, 292
Gravity drainage, 384
Gravity forces, 41, 292

Harmonic decline, 384
Humping pressure build-up data, 154–156
Hurst and Van Everdingen solutions, 61–64,
  72–83
Hydrocarbons, table of properties, 465
Hydrodynamic interface tilt, 307–309

Ideal gas law, 180
Image well, 95
Immiscible displacement, 284–288
Infinite acting reservoir:
  constant pressure solutions, 77–79
  constant rate solutions, 60–67
Initial reservoir pressure, 16, 331
Injection rates, 82, 430–432
Injection wells:
  for water disposal, 82
  predicting performance, 430–432
  pressure analysis in, 149–152
In-situ combustion, 386
Interfacial tension, 268–277
Interference effects, 86–87
Interference tests, 86–87, 160–169
Internal gas drive (see solution gas drive)
Interstitial water, 194, 195
Inverted well patterns, 401
Irreducible water saturation, 8, 265
Isochronal tests, 226–232
Isopach maps, 399, 546
Isopotential distribution, 26

J function, 278–280

Klinkenberg effect, 43

Laminar flow (nonturbulent flow), 5
Layer selection, 276–277, 410–413
Liberation process, 324, 326, 336
Line drive, 401
Linear Buckley Leverett equation, 290, 294–
  296
Linear flow, 21, 71

Liquid compressibility, 57, 84, 457, 458
Liquid viscosity, 45–46, 486, 487
Logging initial saturations, 265

Map, isopach, 399, 546
Mass balance, 318
Mass flow rate, 15, 25
Material balance:
  above saturation pressure, 322–324
  difficulties, 335–338
  gas cap drive, 328–330
  gas reservoir, 191–211
  general equation, 333
  Muskat, 348–349
  PVT data, 335–337
  solution gas drive, 338–352
  Tarner, 344–345
  water drive, 328–330, 337, 353–358
Maximum efficient rate, 363
Mercury injection, 271
Micellar flood, 443
Micro emulsions, (see micellar flood)
Minimum (irreducible) water saturation, 8,
  265
Miscible displacement, 363
Miscible floods, 363, 443
Miscible slug flood, 363
Mobility:
  connate water, 388
  displacing fluid, 400, 408
  gas, 5, 389
  oil, 388, 389, 408
  water, 400, 408
Mobility ratio:
  definition, 163, 400, 409
  effect on displacement efficiency, 438, 439
  effect on sweep efficiency, 400, 403, 404,
    437–438
  effect on vertical (stratification) sweep,
    409, 438
  effect on water cut, 403, 404, 409, 438
Models:
  coning, 316
  digital computer, 359–360
  electric analog, 359
  geological (zonation), 276–277, 410
  Hawkin's hydraulic, 53
  reservoir, 34–36, 359
Mole, 180, 189
Molecular weight, 189, 465
Mud filtrate, in cores, 267
Multicomponent gases, 183–185
Multiphase flow, 7–8
Multiple boundaries, 165–167
Multiple rate flow tests, 144–149
Muskat's material balance, 348–349

Natural gas:
  composition, 183
  deviation factors, 181–188
  pseudo critical properties, 183–184, 462
  super compressibility (see deviation fac-
    tors)

Natural gas (*Continued*)
  viscosity, 191, 485
  water content, 194, 489
Nine spot pattern, 401
Nomenclature list, 445–449
Non Darcy (turbulent) flow, 210–217, 225–226
Non-ideal gas (actual gas), 181–186
Non-wetting phase:
  saturation distribution, 264–275, 283–290
  relative permeability, 7–13

Oil bank, 389, 390
Oil cut (1.0 − watercut), 403–406, 413
Oil displacement:
  by frontal mechanisms, 283–288
  by gas, 283–288, 338–352
  by immiscible fluids, 283–288
  by internal gas drive, 338–349
  by miscible fluids, 363
  by water, 283–288, 386–443
  efficiency, 294, 438
Oil formation volume factor, 24, 45, 490
Oil in place, 323, 328, 329, 330, 333, 335, 342, 393
Oil recovery:
  at water breakthrough, 402–404
  by gas drive, 329, 338–352
  by water drive, 329, 337, 353–358
  effect of initial gas saturation, 363
  effect of mobility ratio, 294, 400–401, 403, 404
  effect of permeability variation, 403–419
  effect of producing rate, 363
  effect of trapped gas, 363–364
  effect of viscosity, 294, 400–401, 403, 404
  efficiency, 294, 360–364
  in five spot pattern, 402–404
  performance prediction of, Chapter 5, 6, 7, 8
Oil saturation:
  after primary production, 393
  before flooding, 393
  equations, 340, 393
  from capillary pressures, 272–275
  initial, 265–275
  reduction by trapped gas, 363–364
  residual after flood, 399–400
Oil shrinkage, 24, 45, 490
Oil viscosity, 24, 486
Oil-water interface, 266–268, 306–316
Oil-wet reservoirs, 8
Open flow potential (from drawdown tests), 215–221
Original oil in place, (see Oil in place)
Overburden pressure, 84, 331

Partially penetrating wells, 24
Patterns, waterflooding, 401
Pattern flow, 25, 26
Perfect ideal gas law, 180
Peripheral flood pattern, 401

Permeability:
  absolute, 5–7
  average, 36–40
  damaged, 22, 23, 113
  dimensionless distribution, 411, 412
  distribution, 5–7, 403–410
  directional variation, 314
  effective, 7–8, 43–45
  effect on displacement efficiency, 294
  fracture, 6
  from pressure build-up, 116–135
  from pressure draw down, 163
  from pressure falloff, 149–152
  from productivity index, 113
  histogram, 6–7
  klinkenberg effect, 43
  normal distribution curve, 6–7
  porosity relationship, 278–280
  relative (see Relative permeability)
  undamaged, 118–152, 163
  units, 4
  variations, 6
Phase diagram, 183, 184
Physical properties of hydrocarbons, 465
Pilot waterflood evaluation, 428–436
Point source solution. 64–67
Poisuelles equation, 279
Pore compressibility, 84, 330–333, 460
Pore volume by material balance (see Original Oil in place)
Porosity, 57, 98, 279, 331, 332
Pore size distribution, 268, 270, 271
Pore volume, 57, 98
Pore volume change, 330–333
Pressure:
  average for radial pseudo steady state flow, 141–142
  average for radial steady state flow, 143
  average drainage area, 135–140
  bottom hole from surface, 174, 197–205, 247–252, 432
  bubble point (see saturation pressure)
  capillary (see Capillary pressure)
  critical, 181–182
  datum correction, 42
  initial reservoir, 16, 17, 331
  pseudo critical, 184–186, 462
  reduced, 182
  saturation, 324, 326, 328, 488
  static corrections, 40–41, 196–205
  static gas well, 196–205
  static gradient, 40–41, 196–197
Pressure build up:
  gas well, 241–244
  Horner method, 130–135
  infinite acting well, 129–136
  pseudo steady state wellm 124–129, 137–143
  two rate, 144–149
  steady state well at shut in, 117–124, 143
  unchanging well pressure at shut in, 117–124

Pressure draw down tests:
  constant rate, 163–164, 217, 226–229
  gas, 226–229
Pressure interference:
  between wells, 86
  from boundaries, 94–97, 160–167
Pressure maintenance by gas injection, 362–363
Pressure maintenance by water injection, 362–363
Pressure-temperature diagram, 183, 184
Pressure-volume-temperature (PVT) relationships:
  gas formation volume factor, 188
  gas in solution in oil, 327, 337, 488
  gas in solution in water, 457
  oil formation volume factor, 24, 337, 490
  water formation volume factor, 457
Productivity index, 108–116
Pseudo critical pressure, 184–186, 462
Pseudo critical temperature, 184–186, 462
Pseudo steady state:
  characteristics, 18–20, 97–99, 101, 103
  pressure build up analysis, 124–129, 137–143
  radial liquid flow equation, 100
  radial gas flow equation, 216

Radial flow:
  average pressure, 141–143
  Buckley-Leverett equation, 290, 387–390
  diffusivity equations, 54, 221–223
  gases, 22, 27–28, 211–241
  incompressible fluids, 21, 22, 25
  pseudo steady state, 100, 216
  steady state, 20–24
  unsteady state, 16–18, 54–83, 207–210
Radius of investigation, 174
Recovery:
  gas reservoir, 205–207
  gas cap drive reservoir, 360–363
  gas injection effect, 362–363
  gravity drainage, 384
  mobility ratio effect, 436–440
  permeability distribution effect, 403–410
  produced gas-oil ratio effect, 339–342, 362
  solution gas drive, vii, 361
  water drive reservoir, viii, 353–358
  water flooding, Chapter 8
Reduced pressure, 182
Reduced temperature, 182
Relative permeability:
  averaging, 7
  characteristics, 6
  definition, 6
  empirical calculation of, 450, 451
  field data calculation of, 345
  gas, 6, 10, 13, 41, 450
  oil, 6, 10, 13, 289, 450, 451
  ratio, 292, 340, 404, 409, 451
  three phase, 7, 12
  water, 6, 9, 289
  wettability effect, 6

Reserves:
  decline curve calculations, 370, 375, 380
  gas reservoir, 205
  gas cap drive reservoir, 352
  solution gas drive reservoir, 338–352
  water drive reservoir, 353–359
  water flooding, Chapter 8
Reservoir fluid sampling, 335
Reservoir heterogeneity, 5–7, 276–277, 403–410
Reservoir limit tests, 160–168
Reservoir pressure, 16, 17, 331
Reservoir simulation, 34, 53, 276, 359, 410
Residual gas saturation, 388
Residual oil saturation:
  definition, 399
  from conventional core saturations, 399
  from lab analysis, 399
  from restored state analysis, 399
Restored state lab analysis, 399
Rock properties:
  porosity, 57, 98, 279, 331, 332
  fluid saturation, 7–14
  permeability, 3–14
  wettability, 8–9

Sampling of reservoir fluids, 335
Saturation:
  connate water, 10, 194, 195
  distribution during displacement, 283–288
  equilibrium gas, 388–389
  gas, 265–266, 272–275, 388
  gradients, 265–266
  irreducible water, (see connate water)
  nonwetting phase, 8–9
  oil, 265–275, 340, 363–364, 393, 399
  residual oil, 399
  vertical distribution, 265–266
  wetting phase, 8–9
Saturation pressure, 324, 326, 328
Schilthius method, 342
Secondary recovery, Chapter 8
Semi steady state flow (see pseudo steady state)
Seven spot pattern, 401
Skin effect:
  definition, 22, 23
  from drill stem test, 169–179
  from pressure falloff, 149–152
  from pressure buildup, 116–135
Skin factor (see skin effect)
Solution gas drive, 338–352
Specific gravity of gas, 189
Stabilization time, 102
Stabilized flow (see pseudo steady state flow
Standard conditions, 181
Statistical reservoir zonation, 411
Steady state flow:
  five spot, 22, 34
  gas, 28
  linear, 21
  liquid, 21
  radial, 21

Steady state flow (*Continued*)
seven spot, 22
spherical, 25
Steam flooding, 386
Stratification of reservoirs
geologic, 276–277, 410
statistical, 411
Streamline distribution, 26
Superposition, 85–96
Surface tension:
definition, 268
laboratory, 276
reservoir, 276
Sweep efficiency
areal (horizontal), 400–403
conformance, 402
stratification, 403–410
total, 400
vertical, 403–410
Symbols, 445–449

Tarner method, 344–345
Thermal secondary recovery, 386
Tilted water-oil contact
coning, 312
during displacement, 309
hydrodynamic, 307
initial, 307
Tortuosity, 279
Total volume factor, 327, 335
Two phase volume factor (see total volume factor)
Two rate flow test, 144
Transient pressure behavior (see Infinite acting reservoir)
Transition Zone, 265, 270

Undersaturated reservoirs, 322
Unsteady state flow:
constant pressure case, 72–82
constant rate case, 60–71
diffusivity equation, 54–58
gases, 221–225
general solutions, 59

Vaporization, 324, 326, 336
Viscosity:
gas, 191, 485
reservoir oil, 24, 486
water, 487

Water:
compressibility, 84, 457
connate, 10, 194, 195
formation volume factor, 45, 457
interstitial (see connate)
vapor in gas, 194, 489
viscosity, 487
Water breakthrough, 390, 400, 402, 403, 404
Water coning, 312–317
Water cut equation, 290, 292, 404
Water drive reservoirs, 353–359
Water disposal, 82
Water flooding:
conformance efficiencies, 402
injection rates, 430–432
pattern sweep efficiencies, 400–403
performance prediction, 413–427, 433–435
pilot results, 428–435
recovery, 392–402, 428–429
variations, 436–443
vertical sweep efficiencies, 403–412
Water fingering, 311
Water influx:
by material balance, 355
constant (aquifer constant), 355
into gas reservoirs, 207–210
into oil reservoirs, 353–358
unsteady state calculations, 354–355
Water injection, 82, 149–152, 430–432
Water injectivity ratio, 416, 423–427
Water-oil contact, 266, 306–312
Water saturation, 10, 194, 195, 265–266, 283–288
Welge method, 299–305
Wellbore damage (see damage ratio and skin effect)
Well spacing, 219–220
Well tests:
gas, 217–218, 226–241, 260–263
drill stem, 169–178
injection, 149
interference, 86–87, 160–169
pressure buildup, 116–159, 241–243
pressure drawdown, 162–163, 217–218, 226, 232
productivity index, 108–115
Wettability, 8–9

z factor (see gas deviation factor)